PROGRESS IN

Molecular Biology and Translational Science

Volume 94

PROGRESS IN

Molecular Biology and Translational Science

Genes and Obesity

edited by

Claude Bouchard

Human Genomics Laboratory
Pennington Biomedical Research Centre
Baton Rouge, Louisiana, USA

Volume 94

ELSEVIER

AMSTERDAM • BOSTON • HEIDELBERG • LONDON
NEW YORK • OXFORD • PARIS • SAN DIEGO
SAN FRANCISCO • SINGAPORE • SYDNEY • TOKYO
Academic Press is an imprint of Elsevier

Academic Press is an imprint of Elsevier
32 Jamestown Road, London, NW1 7BY, UK
Radarweg 29, PO Box 211, 1000 AE Amsterdam, The Netherlands
30 Corporate Drive, Suite 400, Burlington, MA 01803, USA
525 B Street, Suite 1900, San Diego, CA 92101-4495, USA

This book is printed on acid-free paper. ∞

Library of Congress Cataloging-in-Publication Data
A catalog record for this book is available from the Library of Congress

British Library Cataloguing in Publication Data
A catalogue record for this book is available from the British Library

ISBN: 978-0-12-375003-7
ISSN: 1877-1173

For information on all Academic Press publications
visit our website at elsevierdirect.com

Printed and Bound in the USA

10 11 12 13 10 9 8 7 6 5 4 3 2 1

Contents

Contributors

Numbers in parentheses indicate the pages on which the authors' contributions begin.

Peter Arner, Department of Medicine, Karolinska Institutet, Karolinska University Hospital, Huddinge, Stockholm, Sweden (39)

Philip L. Beales, Molecular Medicine Unit, UCL Institute of Child Health and Great Ormond Street Hospital for Children, London, United Kingdom (125)

Claude Bouchard, Human Genomics Laboratory, Pennington Biomedical Research Center, Baton Rouge, Louisiana, USA (1)

Javier Campión, Department of Food Sciences, Physiology and Toxicology, University of Navarra, Pamplona, Spain (291)

Shelley Cole, Department of Genetics, Southwest Foundation for Biomedical Research, San Antonio, Texas, USA (197)

Anthony G. Comuzzie, Department of Genetics, Southwest Foundation for Biomedical Research, and Group Leader, Chronic Disease and Genomics, Southwest National Primate Research Center, San Antonio, Texas, USA (197)

Ingrid Dahlman, Department of Medicine, Karolinska Institutet, Karolinska University Hospital, Huddinge, Stockholm, Sweden (39)

Thomas A. Drake, Department of Pathology and Laboratory Medicine, David Geffen School of Medicine at UCLA, Los Angeles, California, USA (9)

Janis S. Fisler, Department of Nutrition, University of California, Davis, California, USA (373)

Johannes Hebebrand, Department of Child and Adolescent Psychiatry and Psychotherapy, University of Duisburg-Essen, Essen, Germany (241)

Robert A. Hegele, Departments of Medicine and Biochemistry, Schulich School of Medicine & Dentistry, and Blackburn Cardiovascular Genetics Laboratory, Robarts Research Institute, University of Western Ontario, London, Ontario, Canada (159)

Paul B. Higgins, Department of Genetics, Southwest Foundation for Biomedical Research, San Antonio, Texas, USA (197)

Anke Hinney, Department of Child and Adolescent Psychiatry and Psychotherapy, University of Duisburg-Essen, Essen, Germany (241)

Christopher T. Johansen, Departments of Medicine and Biochemistry, Schulich School of Medicine & Dentistry, and Blackburn Cardiovascular Genetics Laboratory, Robarts Research Institute, University of Western Ontario, London, Ontario, Canada (159)

Tisha R. Joy, Departments of Medicine and Biochemistry, Schulich School of Medicine & Dentistry, Robarts Research Institute, University of Western Ontario, London, Ontario, Canada (159)

Antti Knaapila, Monell Chemical Senses Center, Philadelphia, Pennsylvania, USA (213)

Robert A. Koza, Pennington Biomedical Research Center, Baton Rouge, Louisiana, USA (75)

Leslie P. Kozak, Pennington Biomedical Research Center, Baton Rouge, Louisiana, USA (75)

Matthew B. Lanktree, Departments of Medicine and Biochemistry, Schulich School of Medicine & Dentistry, and Blackburn Cardiovascular Genetics Laboratory, Robarts Research Institute, University of Western Ontario, London, Ontario, Canada (159)

Joshua P. Lewis, University of Maryland School of Medicine, and Geriatric Research and Education Clinical Center, Veterans Administration Medical Center, Baltimore, Maryland, USA (349)

J. Timothy Lightfoot, Department of Kinesiology, University of North Carolina at Charlotte, Charlotte, North Carolina, USA, and Huffines Institute, Department of Health and Kinesiology, Texas A&M University, College Station, TX, USA (271)

J. Alfredo Martínez, Department of Food Sciences, Physiology and Toxicology, University of Navarra, Pamplona, Spain (291)

Fermin Milagro, Department of Food Sciences, Physiology and Toxicology, University of Navarra, Pamplona, Spain (291)

Trudy Moore-Harrison, Department of Kinesiology, University of North Carolina at Charlotte, Charlotte, North Carolina, USA (271)

Danielle Renee Reed, Monell Chemical Senses Center, Philadelphia, Pennsylvania, USA (213)

Susann Scherag, Department of Child and Adolescent Psychiatry and Psychotherapy, University of Duisburg-Essen, Essen, Germany (241)

Alan R. Shuldiner, University of Maryland School of Medicine, and Geriatric Research and Education Clinical Center, Veterans Administration Medical Center, Baltimore, Maryland, USA (349)

Saroja Voruganti, Department of Genetics, Southwest Foundation for Biomedical Research, San Antonio, Texas, USA (197)

Craig H. Warden, Rowe Program in Genetics; Department of Neurobiology, Physiology and Behavior, and Department of Pediatrics, University of California, Davis, California, USA (373)

Preface

A volume entirely devoted to the recent advances on the genetics and molecular biology of obesity is highly relevant to the serial Progress in Molecular Biology and Translational Science (PMBTS). We have been able to assemble a panel of distinguished authors for this volume, and I express my gratitude to them for a timely delivery of their contributions.

The leadership of the PMBTS publication series has been a delight to work with. I express my thanks to Dr. Michael Conn, Editor of the PMBTS serial, who first came up with the suggestion of a volume devoted to obesity. I also benefited greatly from the support of Janice Hackenberg and subsequently Lisa Tickner, Acquisition Editors, Delsy Retchagar, Developmental Editor, and Vijayaraj Purush, Project Manager, all at Elsevier Inc. They were very supportive during the development of the book manuscript.

Finally, I would not have been able to undertake the task of serving as editor for this volume without the outstanding and competent support of Nina Laidlaw and later Allison Templet in my office at the Pennington Biomedical Research Center. They worked diligently with each author to ensure that the manuscript was complete and met all the requirements of the publisher.

Claude Bouchard

Genetics and Genomics of Obesity: Current Status

CLAUDE BOUCHARD

Human Genomics Laboratory, Pennington Biomedical Research Center, Baton Rouge, Louisiana, USA

It is commonly recognized that the prevalence of overweight and obesity is increasing around the world and that the obese are becoming more severely obese. Current estimates suggest that there are more than one billion adults who are overweight or obese worldwide. Recognizing the threat associated with excess weight, the World Health Organization has identified overweight as one of the main risk factors for the overall burden of disease in the world and one of the top five in developed nations.[1]

There are four major classes of factors contributing to the ongoing epidemic of obesity, which is ultimately the result of widespread energy imbalance favoring storage of the energy surplus not expended. These four broad classes of factors can be labeled as follows: built environment, social environment, behavior, and biology.[2] We have proposed before that the built environment (e.g., reliance on the automobile, building design, lack of safe sidewalks) and the social environment (e.g., advertising, pressure to consume) are such that the global environment has become "obesogenic" not only in developed countries but also in most developing areas of the world.[3] As the concept implies, an obesogenic environment favors the adoption of obesogenic behavior (e.g., consumption of large-portion-size meals, high-fat diets, high sugar intake, many hours spent watching TV, playing video games, or sitting at a computer). Common sense suggests that the obesogenic environment and behavior are fueling the acute rise in the prevalence of overweight and obesity that the world is currently experiencing. However, a complete definition of the circumstances fueling this rise needs to incorporate the concept of biological predisposition as well.

Progress in Molecular Biology
and Translational Science, Vol. 94
DOI: 10.1016/S1877-1173(10)94001-5

1877-1173/10 $35.00

A large body of research conducted in animal models and in humans over the past five decades has demonstrated beyond a shadow of a doubt that there are large individual differences in the propensity to gain weight in a variety of obesogenic conditions.[4] Strong evidence has been found for the contention that, even though it is not the sole causal source, genetic variation has much to do with the risk of becoming obese, particularly severe obesity. Hundreds of research reports have dealt with the critical issue of finding the exact site of the biological vulnerability to obesity. However, it is fair to summarize that (a) there is no consensus on this topic as of yet for the common forms of obesity, although results from the genome-wide association studies suggest that the regulation of appetite and satiety is playing a particularly critical role, (b) there are multiple paths to excessive energy storage involving several organs and systems, and (c) there is considerable heterogeneity in the biological determinants of excessive weight gain among individuals of a given species.

I. Familial Risk Level and Heritability Coefficients

It is commonly accepted that genetic differences among people play an important role in the risk of becoming obese, and the global evidence for the genetic contribution to obesity, as derived from genetic epidemiology models, has been reviewed elsewhere.[4] A first line of evidence is the observation that single gene syndromic and nonsyndromic disorders account for as much as 5% of obesity cases worldwide. Heritability estimates are generally derived from pairs of identical and fraternal twins raised together (but in some studies raised apart), nuclear family members from two generations, and adopted offspring raised by foster parents (but occasionally with information on their biological parents). The large body of data available to date reveals that the magnitude of the heritability estimates varies substantially among these various designs, with the twin models generating the highest values and the adoption studies the lowest coefficients.

Another line of evidence for the presence of a genetic component in adult BMI comes from the computation of the lambda coefficient or familial risk ratio.[5] Few studies have dealt with the familial risk ratio of human obesity. Lee and collaborators have computed such a ratio for various levels of BMI in first-degree relatives of probands in the World Health Organization classes of obesity.[6] The familial risk was found to increase with the severity of obesity, reaching 5 and higher with BMI levels of 40 and more. This suggests that an adult with a father, mother, brother, or sister with a BMI of about 40 is about five times more at risk of becoming obese with a BMI of 40 or more compared to individuals in the population who have only normal-weight first-degree relatives. Importantly, for a BMI level around 30, the familial risk is in the range of 2–3.

Although the consensus among obesity scientists is that the genetic component of obesity is quite high (in the range of 50–90%), one should keep an open mind regarding the possibility that the heritability values are highly inflated. For instance, some studies have found much lower heritability levels for BMI. Among them are studies performed using partial-adoption and full-adoption designs[7,8] and a report based on data aggregated from twins, nuclear families, and foster parents with adopted offspring.[9] Fueling this debate is the surprising observation that with about 20 loci identified through genome-wide association studies,[10] and with the predicted diminishing return in the effect size of the additional loci that may reach genome-wide significance in the presence of extraordinarily large sample sizes, only about 2% of the variance in BMI was accounted for.[11] Ultimately, this issue will be resolved only with a better understanding of the genetic architecture of the predisposition to obesity, that is, when all the loci have been properly identified and documented and when the role of epigenetic events in the genetic risk have been defined.

We are of the view that the current global obesity epidemic is fueled by a changing social and physical environment that encourages consumption and discourages expenditure of energy, behaviors that are poorly compatible with the genome that we have inherited.[2] Will it ever be possible to take advantage of the advances in our understanding of the genetic basis of obesity in order to identify the individuals at risk of becoming obese (not the single gene obesity cases in which such a screening is already widely recognized as a major advance) before they gain a large amount of body weight and adiposity? There is as of yet no consensus on this eventuality. In the meantime, the advances that have been made and will continue to accrue regarding our understanding of the genetic architecture of obesity will undoubtedly lead to new and exciting research on the biology and behavior of energy balance regulation.

II. Scope of the Volume

In this volume of *Progress in Molecular Biology and Translational Science*, we review the latest evidence for the contribution of genetic factors to the risk of obesity. Genes and pathways potentially involved and the behavior that they influence are explored. The role of genetic variation in white and brown adipose tissue biology, excess adipose tissue mass, syndromic and nonsyndromic obesity cases, and lipodystrophies is highlighted. The case is made that a systems biology approach is needed in order to arrive at a comprehensive and integrative solution to the enormously complex question of the biological predisposition to obesity. This volume also discusses our current understanding of the genetic basis of eating disorders, eating behavior, and physical activity level. The results of the findings from the recent genome-wide association studies of obesity and of epigenetic studies are reviewed. Further, the role of genetic differences in the modulation of risk factors for common chronic diseases and morbidities in the presence of an obese state is introduced. The final chapter comments upon the translational opportunities from what we have learned in animal and human studies on the genetics of obesity.

Chapter 2 by Thomas Drake from the University of California, Los Angeles, focuses on genes and pathways contributing to obesity from a systems biology perspective. Systems biology utilizes high-throughput data from multiple sources to develop models of biologic processes. The author argues that the increasing ability to generate high-throughput genetic, epigenetic, and transcriptomic data of various types will result in a greater reliance on systems approaches to better understand the processes by which genetic variation influences disease-related traits.

In Chapter 3, Ingrid Dahlman and Peter Arner from the Karolinska Institute in Stockholm deal with the genetics of adipose tissue biology, with an emphasis on morphology and function. Adipose tissue morphology and lipolytic properties are thought to be influenced by strong genetic determinants. Polymorphisms in numerous adipose-expressed genes have been evaluated for association with relevant adipose tissue traits, including morphological traits, adipocyte lipolysis, adipogenesis and lipid storage markers, and circulating levels of adipose tissue-secreted hormones and adipokines. These genes and allelic effects are reviewed in this chapter.

Leslie Kozak and Robert Koza from the Pennington Biomedical Research Center in Baton Rouge, Louisiana, discuss the genetics of brown adipose tissue in Chapter 4. Brown adipose tissue has evolved as a mechanism for heat production based upon uncoupling of mitochondrial oxidative phosphorylation. The development of brown adipocytes originates from two apparently independent pathways: one from muscle progenitor cells in the fetus leading to fully functional cells at birth, the other emerging in white fat depots at weaning. Although the latter regresses, it can be induced in adult mice upon adrenergic stimulation. Alleles at genetic loci in mice are associated with a 100-fold difference in brown adipocyte induction in white fat upon adrenergic stimulation. The recent surge of interest in human brown adipose tissue was brought about by technological advances allowing more accurate imaging of the tissue. Whether brown adipose tissue can be activated to enhance its energy expenditure potential is of great interest in obesity research and has spurred new studies of the biology of this tissue in humans.

Chapter 5 is devoted to single gene disorders. Philip Beales from the Institute of Child Health and Great Ormond Street Hospital for Children in London discusses the evidence for single defective genes that have been implicated in human obesity. These disorders are classified as nonsyndromic or syndromic obesity. Taking advantage of several monogenic disorders, researchers have identified and dissected key regulatory pathways of energy balance. Among them, defective alleles in *LEP*, *LEPR*, and *MC4R* have provided critical insights into the biology and behavior of human obesity.

Matthew Lanktree, Christopher Johansen, Tisha Joy, and Robert Hegele from the University of Western Ontario develop the topic of human lipodystrophies in Chapter 6. Lipodystrophy defines a group of heterogeneous disorders characterized by selective or generalized atrophy of anatomical adipose tissue stores. Lipodystrophies can be acquired or inherited. The loss of the ability to store lipids in common adipose tissue depots can result in significant metabolic disorders. This review focuses on clinical characterization of a suspected lipodystrophy case, clinical manifestations, molecular findings and the pathogenic basis of different forms of lipodystrophy, therapeutic options for lipodystrophy patients, and genetic advances that may be helpful in the identification of new pathogenic mechanisms.

Chapter 7 by Anthony Comuzzie, Paul Higgins, Saroja Voruganti, and Shelly Cole from the Southwest Foundation for Biomedical Research in San Antonio, Texas, discusses the various strategies for identifying genes contributing to human obesity. Recent findings from candidate gene, genome-wide linkage, and genome-wide association studies are reviewed, with an emphasis on strengths and weaknesses of the evidence. A large number of genes with small effect sizes is the most likely scenario to account for the genetic component of obesity. Uncovering all such genes requires very large cohorts and multiple replication studies. Such resources have been assembled, and the discovery process should deliver the panel of the most important genes and alleles in the not too distant future.

The genetics of taste and smell is discussed by Danielle Reed and Antti Knaapila from the Monell Chemical Senses Center in Philadelphia in Chapter 8. The senses of taste and smell have evolved so that humans recognize the bitter taste of poisons and the sour taste and bad smell of spoiled foods. But food selection is influenced by more than avoiding the bad; it is also motivated by seeking the good, such as fat and sugar. Yet what constitutes the best food and drink is often a matter of opinion. Genetic studies in humans and experimental animals strongly suggest that the liking of sugar and fat is influenced by genotype. Some progress has been made in defining the genes and their alleles associated with the positive and negative aspects of food and flavor. This chapter provides a genetic and evolutionary perspective on food perception and preference.

In Chapter 9, Anke Hinney, Susann Scherag, and Johannes Hebebrand from the University of Duisburg-Essen in Germany focus on the genetics of anorexia nervosa and bulimia nervosa. A complex etiology encompassing environmental and genetic factors is common to eating disorders, and the evidence from twin and family studies suggests that there is a strong genetic component. However, candidate gene studies derived mainly from pathways associated with appetite and satiety in the brain have yielded very few positive results, and findings substantiated in meta-analyses are scant. Genome-wide linkage approaches have been performed to identify genes with no *a priori* evidence for their relevance in eating disorders with limited success. Currently, there is high expectation in the field, as the results of the first genome-wide association study of eating disorders should be available soon.

Trudy Moore-Harrison and Timothy Lightfoot from the University of North Carolina at Charlotte report on the genetics of physical activity level in Chapter 10. Physical activity level is a complex, multifactorial behavior with multiple determinants. A sedentary mode of life is a risk factor for gaining weight and becoming obese, and human studies and animal models have indicated that genetic differences play a role in the inclination to be active or sedentary. Several genomic quantitative trait loci have been suggestively associated with physical activity level in both rodent and human studies. Some reports have revealed pleiotropic effects between the genetic determinants of physical activity level and body weight regulation. Early results from genome-wide association studies suggest that it may be possible to identify the genes and sequence variants associated with the propensity to be physically active.

Chapter 11 deals with epigenetic events and obesity and was written by Javier Campión, Fermin Milagro, and Alfredo Martínez from the University of Navarra in Pamplona, Spain. Epigenetics, defined as the study of heritable changes in gene expression that occur without a change in the DNA sequence, has emerged as a potentially important determinant of obesity. There is suggestive evidence to the effect that DNA methylation profile and histone epigenetic modifications could be helpful in the prediction of obesity risk, as well as in the ability to gain or lose weight. In this chapter, the concept of epiobesogenes is introduced and defined as a set of obesogenes that are potentially exerting their deleterious effects as a result of epigenetic events. The importance of maternal nutrition and early postnatal nutritional circumstances with respect to epigenetic events is highlighted.

In Chapter 12, Joshua Lewis and Alan Shuldiner from the University of Maryland School of Medicine address the critical issue of whether the metabolic complications of obesity are precipitated by genetic factors. The mechanisms linking obesity to its clinical outcomes are complex and not fully understood. Among potential causal pathways, attention has been devoted to insulin resistance and hyperinsulinemia, hypertriglyceridemia, dysfunctional adipose tissue with abnormal secretion of adipokines, elevated inflammatory markers, ectopic lipid deposition in skeletal muscle, liver, and other organs, and others. Since not all obese individuals develop the metabolic complications of obesity, attempts to identify the genetic predisposition to excess adiposity have been undertaken, with an emphasis on candidate genes. These genes are reviewed in detail in this chapter. The requirements for such studies to be successful in the future are also defined.

In the final contribution to this volume (Chapter 13), Craig Warden and Janis Fisler from the University of California, Davis, ask whether what we have learned from animal and human genetic studies of obesity can be translated into practical applications. Genetic testing is already a reality for several single gene disorders causing obesity. A central issue is determining the conditions under which genotype information would be useful for preventing or treating the common form of obesity. Currently, family history is more predictive than specific genes and sequence variants for defining the risk of obesity, but individual testing could, in the future, guide obesity therapy and perhaps even public health policy designed to prevent excessive weight gain. A potentially serious ethical problem relates to prenatal testing, assuming that obesity gene screening tests will eventually become available.

In summary, human biology has been shaped by the ability to survive famines, to gorge during periods of food abundance, to adapt to cold or hot climates, to hunt or fight predators, and to escape from or attack adversaries, as well as by the advantages conferred by a large or a small body size, by the immune response capacity, by random allele selection, by sexual selection, and by other forces and events.[2] Identifying the genes and alleles that were selected for during our evolutionary journey constitutes an enormously complex undertaking. However, the fundamental principle remains valid: if a predisposition to obesity exists, and it seems to be the case as suggested by a reasonable body of data, it is faithfully recorded in our genome, and it is potentially modulated by epigenetic events currently under intensive investigation. Nonetheless, if it is not in our genome, it is not part of the inherited vulnerability that we observe today in some human lineages and family lines.

References

1. World Health Organization. *The global burden of disease 2004 update.* Geneva: World Health Organization; 2008.
2. Bouchard C. The biological predisposition to obesity: beyond the thrifty gene scenario. *Int J Obes* 2007;**31**:1337–9.
3. Lakka HM, Bouchard C. Etiology of obesity. In: Buchwald H, Cowan GSM, Pories WJ, editors. *Surgical management of obesity.* Philadelphia: Saunders; 2006. p. 18–28 [chapter 3].
4. Loos RJF, Bouchard C. Obesity—is it a genetic disorder? *J Intern Med* 2003;**254**:401–25.

5. Risch N. Linkage strategies for genetically complex traits. I. Multilocus models. *Am J Hum Genet* 1990;**46**:222–8.
6. Lee JH, Reed DR, Price RA. Familial risk ratios for extreme obesity: implications for mapping human obesity genes. *Int J Obes Relat Metab Disord* 1997;**21**:935–40.
7. Sorensen TIA, Stunkard AJ. Overview of the adoption studies. In: Bouchard C, editor. *The genetics of obesity.* Boca Raton, FL: CRC Press; 1994. p. 49–63.
8. Stunkard AJ, Sørensen TI, Hanis C, et al. An adoption study of human obesity. *N Engl J Med* 1986;**314**:193–8.
9. Bouchard C, Pérusse L, Leblanc C, Tremblay A, Thériault G. Inheritance of the amount and distribution of human body fat. *Int J Obes* 1988;**12**:205–15.
10. Li S, Zhao JH, Luan J, Luben RN, Rodwell SA, Khaw KT, et al. Cumulative effects and predictive value of common obesity-susceptibility variants identified by genome-wide association studies. *Am J Clin Nutr* 2010;**91**:184–90.
11. Bouchard C. Defining the genetic architecture of the predisposition to obesity: a challenging but not insurmountable task. *Am J Clin Nutr* 2010;**91**:5–6.

Genes and Pathways Contributing to Obesity: A Systems Biology View

Thomas A. Drake

Department of Pathology and Laboratory Medicine, David Geffen School of Medicine at UCLA, Los Angeles, California, USA

Systems biology is an expanding discipline that utilizes high-throughput data from multiple sources to develop models of biologic processes. This chapter reviews basic systems biology concepts and how they inform the traditional search for genes and pathways involved in disease pathogenesis. Systems approaches yield networks representing interactions and relationships among elements, with subnetworks or modules being functional groupings of these. For genetics of common disease-related traits such as obesity, integrative genetics is representative of a top-down systems biology approach, bringing together high-throughput genotyping, global tissue mRNA expression data, and phenotypic data. Coexpression network analysis yields network models that allow identification of groups of coexpressed genes that can be related to particular traits. Combining this with genetics has shown that genetic variation at disease associated loci act though influencing such modules. Constraint-based reconstruction of metabolic networks represents a bottom-up systems biology approach that can be utilized to model the effects of genetic variation. The progressively

Progress in Molecular Biology
and Translational Science, Vol. 94
DOI: 10.1016/S1877-1173(10)94002-7

1877-1173/10 $35.00

increasing ability to generate high-throughput data of various types will promote continued application of systems approaches to better understand the processes by which genetic variation influences disease-related traits.

I. Introduction

The pathogenesis of obesity is complex at all levels of biology: genetics, cell and tissue biology, physiology, and behavior. A systems biology view attempts to identify and understand how the multiple components interact with the ultimate goal of developing working models that can predict the behavior of the system as a whole. As attested to by the presence of this volume itself, a lot is known about the genetic aspects of obesity, but much more remains to be discovered. This chapter will discuss how a systems biology perspective informs current avenues of research in the genetics of obesity and why it will become increasingly important in the future.

Medical genetics is fundamentally interested in understanding the relationship between genetic variation and human health and disease. The primary goals are identifying the specific genetic variations and the biologic consequences that they produce, or as commonly put, discovering the genes and pathways involved in producing phenotypic variation, and the factors that influence them. In the case of obesity, the phenotype is increased adipose tissue mass and a physiological state with altered metabolism, activity, endocrine function, and so forth, affecting multiple organ systems. At the biologic level, there are many steps and interactions among molecules, organelles, and cells that sit between the genetic variation and the phenotype we observe as obesity. These are traditionally organized conceptually as "pathways" that can be schematized to depict the individual elements, functions, and interactions—metabolic pathways, such as lipogenesis; signaling pathways, such as insulin signaling; etc.—where traditional cell and molecular biology approaches centered on specific genes and pathways, combined with biochemistry and physiology, have generated most of what we know.

This has yielded substantial information at the pathway level, but as yet limited information as to how the multitude of involved pathways interacts in various physiological settings. A major reason for this has been the inability or difficulty to assess multiple elements and pathways simultaneously in a high-throughput manner. This has changed over the past decade, first with the sequencing of whole genomes and development of global gene expression microarrays, and more recently, by advances in high-throughput methods for determining protein interactions and quantification, metabolite quantification, specific protein phosphorylation, and various DNA modifications and

DNA–protein interactions, as a few examples. Concomitant with technical advances have been computing and analytical advances. Together, these have enabled study of the behavior of multiple elements that make up a biologic process, which is the focus of systems biology. Just as the behavior of a pathway has properties that are different from those of individual genes, the behavior of a system has properties that are different from the genes and pathways that compose it, which is often referred to as the system having "emergent properties."[1] Thus, systems biology is a discipline providing a characterization of biologic processes that is unique, and arguably closer to the phenotype. Obesity research in the coming years will continue to expand on understanding specific genes and pathways, but in the context of a systems perspective. As discussed below, systems-level analyses yield networks and more complex models, which are not directly interpretable as an extension of the pathway view of biology. A challenge for the future is to develop analyses and models that bridge the gap between high-level system network modeling and the lower level pathway modeling.[2] Following a brief review of systems biology concepts, examples of several types of systems-based research approaches relevant to obesity genetics will be discussed from the perspective of how these are novel, and also complement current gene and pathway-based approaches.

II. Systems Biology: A Brief Overview

Systems biology is a now widely used term, with varying connotations. In the loosest sense, it is applied in settings where biological inferences have been drawn from the analysis of large data sets, such as expression microarrays. More appropriately, it is an approach to studying biological systems.[3] The common aspect of various definitions of systems biology is the goal of understanding how a biological system functions as a whole and integrating its multiple components, relationships, and interactions, in which the performance and analysis of large data sets is but one part. From the genetics perspective, this is directed toward understanding how genetic variation results in phenotypic variation (Fig. 1).

High-throughput technologies are utilized to obtain the largest possible measurement of individual components (often referred to as a "parts list"), and functional assessments of these components under a variety of conditions (system perturbations). The large data sets collected and appropriate analytical approaches necessitate use of sophisticated computational tools as another essential aspect.[4] The scope of a given "system" for study can be relatively discrete, such as a particular cellular or subcellular function, or more global such as a cell, tissue, or organism. Although in principle, all relevant molecular components would be included, in practice, technical or financial feasibility

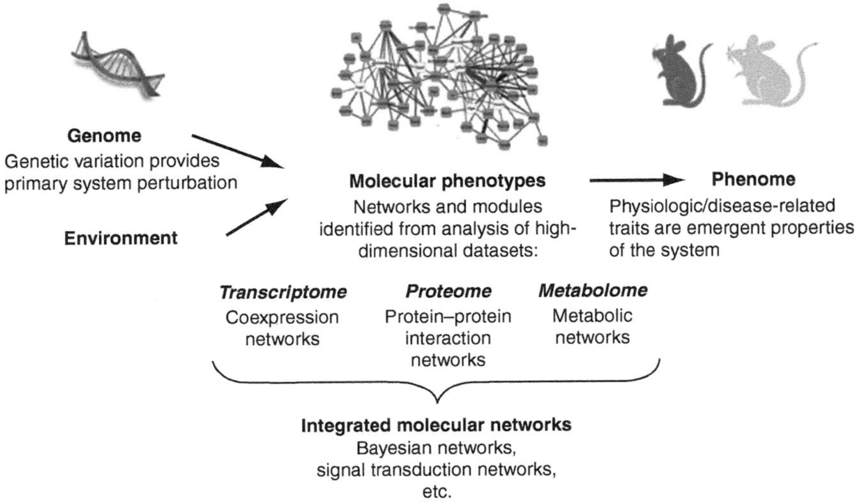

FIG. 1. A systems biology perspective of genome–phenome relationships, where the network structures of intermediate molecular phenotypes play a central role.

usually limits large-scale measurements to several categories, such as messenger RNA, protein interactions, or metabolites. Analysis of these large data sets allows formulation of biological networks from which models of the process can be developed.[2] Models allow hypothesis generation of behavior in novel conditions, and experiments devised to test and revise the model as needed.[3,5]

The mathematical determination of interrelationships among elements in such data sets generates networks that can help identify, or map, the elements of the system (e.g., sets of genes or proteins or metabolites) that act in concert in a common biologic process.[2] A network is a mathematical, and as visually presented, graphical depiction of these interrelationships (Fig. 2). Networks consist of elements (termed "nodes" in network nomenclature) that exhibit functional connections ("edges"). These occur in all areas of life. Everyday examples that may help to intuitively understand what they are include the internet, with servers as nodes and links between them as edges; social networks, with individuals as nodes, and instances of personal interaction, dating; e-mailing; coauthorship as edges; or airline transportation, with airports as hubs and flights between them as edges. In biology, the connections can be identified using coregulation of gene expression, physical interactions (as in protein complexes), or metabolic relationships (e.g., the intermediates in glycolysis). "Undirected" networks are simply connected nodes, with no causal direction (such as reversible enzymatic reaction), whereas "directed" networks have edges with defined causal relationships between nodes. An important

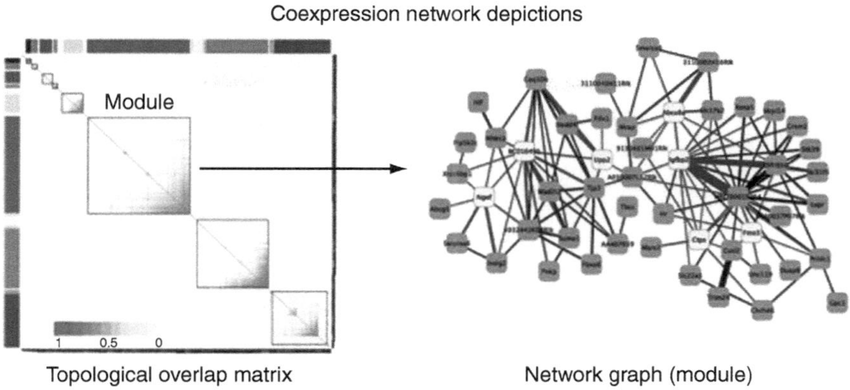

FIG. 2. Networks can be depicted schematically in different ways. A topographical overlap matrix (TOM) plot can be used to show the full set of modules identified in a coexpression network. This is a symmetric plot where both x and y axes represent transcripts, clustered in a bidirectional manner based on a measure of correlation. A network graph is often used to display the relationship among transcripts composing a module, where nodes are the genes/transcripts, and the edges (lines connecting nodes) represent a level of correlation between them. In this case, the width of the edges represents the relative degree of correlation. One can see that a smaller number of nodes have many connections, and are thus identified as "hubs," while the majority have fewer.

property observed for many biologic networks is scale-free topology, in which a smaller number of nodes have many edges (these nodes are termed "hubs") but most nodes have few edges, and these are arranged into sets of nodes referred to as modules, or subnetworks, where the nodes composing a module are more connected among one another than with nodes of other modules.[6,7] Studies in a variety of biological networks have shown that the more highly connected hub nodes of a module are more critical to network function.[8] Properties of networks and "location" of the specific elements, therefore, provide biologically relevant information about the system under study. These concepts will be further discussed and illustrated below.

An implication of a systems biology approach is that multiple component types are (or should ultimately be) studied, since functioning systems are obviously composed of more than one type of molecule. Thus far, only DNA and RNA can be surveyed at anything near a global level, and so expression array-based studies have predominated, in conjunction with DNA sequence information. We are hopefully near a turning point where protein, metabolite, and other molecular assays can be similarly performed and broadly available. Thus, the ability to measure and integrate data from the various "-omes"—genome, transcriptome, proteome, metabolome, lipidome, phenome, etc.—is a necessary long-term goal for systems biology (Fig. 1).

From a clinical genetics perspective, "integrative genetics"—the integration of high-throughput-obtained genotype determination and transcript quantification, in conjunction with phenotype data—has been the most widely applied systems biology-based approach, and is the major focus of discussion here. However, there are other important systems biology approaches relevant to genetics of obesity, which will be brought to the readers' attention.

III. Integrative Genetics as a "Top-Down" Systems Biology Approach

Traditional genetics studies in mice have sought to identify and characterize individual genes that influence specific phenotypic traits. A classic example is the *Ob* mutation that causes a Mendelian form of obesity in mice, and through mapping and positional cloning was found to be a mutation in leptin.[9] Quantitative trait locus (QTL) mapping using inter- and backcross populations has sought to identify genes harboring naturally occurring variation between inbred strains. Such variants typically have a much smaller effect on the phenotype, making positional cloning more difficult.[10] Hence, very few genes influencing body fat have been identified in this manner, despite many genomic loci being identified. Many more genes have been characterized through transgenic or knock-out methods to induce overexpression or loss of gene function. However, in the mouse, these have been limited to candidate genes initially brought to consideration by traditional biochemical studies. Likewise, genome-wide association studies (GWASs) in human populations have identified relatively few specific genes so far.[11] Nevertheless, it is likely that the common forms of most complex diseases, like obesity, result from the interactions of multiple genes having small to moderate effects, and, therefore, cannot be realistically modeled by single-gene perturbations. This provides another rationale for studying genetic variation in a population setting.

Advances in high-throughput methods for genotyping, measuring mRNA abundance, and computational analyses allowed systems biology approaches to be applied to these traditional genetics methods to gain deeper insights into the genetics of obesity in the animal models and, more recently, humans. Integrative genetics (sometimes referred to as genetical genomics[12]) combines analysis of natural genetic variation with global gene expression data, and other phenotype data[13] (Fig. 3). In a genetically segregating population, such as an F2 intercross in which each mouse has a unique combination of genetic variants from the original strains, or a human population, the endogenous genetic variation is sufficient to provide the necessary system perturbation. Complex traits, having multiple genetic determinants, will exhibit substantial variation

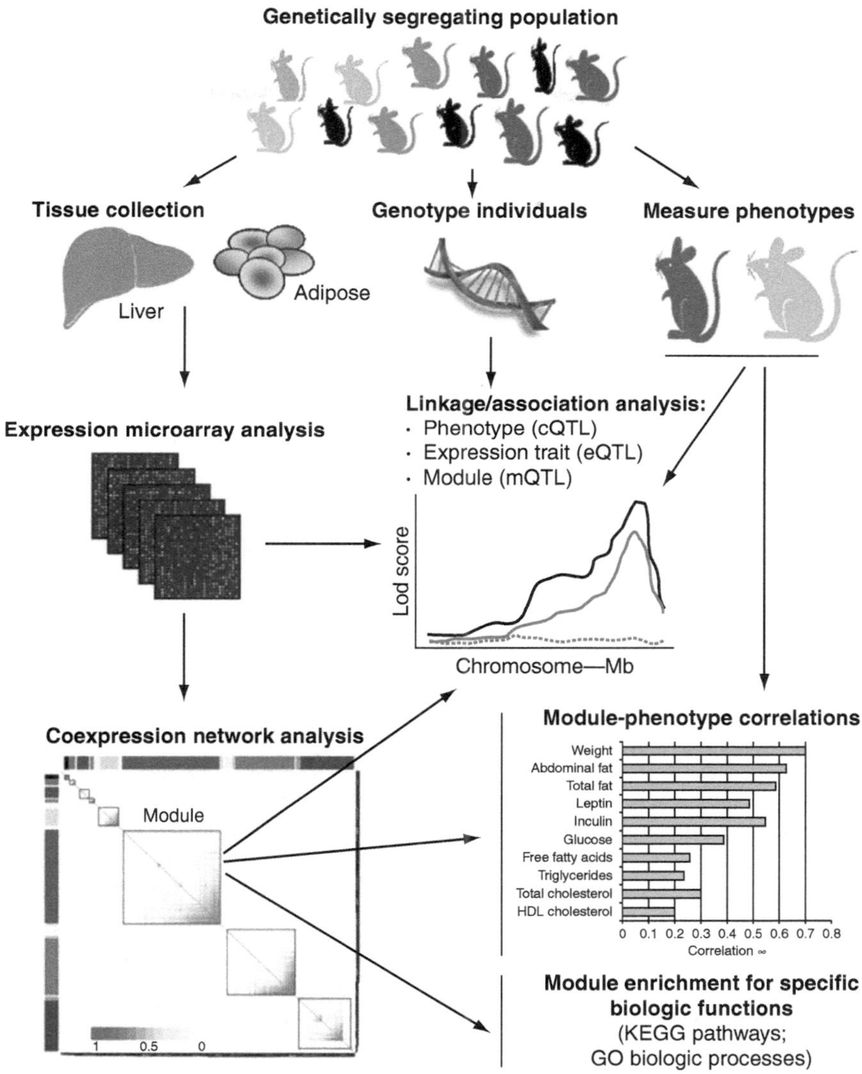

FIG. 3. Schematic of the elements of an integrative genetics study in mice.

across the population. Thus, measurement of transcript levels in such populations allows the identification of coregulated genes and of relationships between transcript levels and clinical traits. In addition, this integrative approach relates DNA variation to transcript abundance, allowing identification of likely primary (*cis*-acting) and secondary (*trans*-acting) genetic effects controlling transcript abundance.[14] The integration of genetic and transcript abundance data can be used to predict causal relationships between transcripts and traits such as obesity.[15,16] Combined with coexpression network analysis, genes, pathways, and networks causal for disease can be identified.[17–19]

A. Coexpression Networks

Coexpression networks provide a representation of the relationship among transcripts present in a population sample. In such a network, the transcripts are the nodes and the connections between them (the edges) are based on the correlation between transcripts (or a derivation thereof). Coexpression networks represent a "top-down" high-level approach where the data dictate the structure of the network independent of any preexisting pathway level-defined relationships. This has been referred to as "reverse engineering" the underlying biology, in contrast to a network constructed on the basis of existing knowledge, such as metabolic networks to be discussed later.[18] The mathematical details of constructing coexpression networks vary, but are all fundamentally based on correlation between transcripts. One analytical method most familiar to the author is weighted gene coexpression network analysis (WGCNA).[20] This generates an undirected network in which the degree of correlation between transcripts is used to weight the relationship between nodes in constructing the network, as opposed to using a specific threshold. Additionally, a power transformation is used to optimize the network to conform to a scale-free structure. Clustering analysis is used to establish which sets of transcripts are most closely correlated based on adjacency. Such a network can be visualized as a topological overlap matrix plot (Fig. 2). Closely clustered transcripts are defined as a "module." One can adjust the criteria for relatedness to obtain smaller and more numerous modules, or fewer larger modules. Each transcript has a calculated connectivity (based on the number and strength of relationships with other transcripts in the module). The most highly connected transcripts in a module are defined as "hubs."

A module, therefore, represents a set of transcripts that are highly intercorrelated, presumably not by chance and having a biologic basis for its formation. One can think of a module as a recognizable subunit of a larger network, as loosely being the systems biology equivalent of a pathway in biochemistry. Although the mechanistic basis for any given module may be unknown, it would be expected that at least some modules would include functionally important traditionally defined pathways, which is indeed the case. Cellular functions that a given module may be associated with are assessed by determining whether there is overrepresentation of transcripts belonging to defined gene sets representing specific pathways (e.g., KEGG pathway sets) of specific Gene Ontology (GO) categories, for example. Other functional associations could be sought by evaluating overrepresentation of gene expression signatures obtained from knockout or drug-treated animals or cells, or by genes differentially expressed in specific cells or tissues. One is also particularly interested in which modules are associated with the phenotypic traits of interest in the population being studied. This can be done by examining the correlation of the individual genes composing a module with the traits, or by examining the correlation of a parameter representative of the expression of all genes in a module, such as the first principle component. This is referred to as a module significance score in WCGNA.

These concepts and analyses are exemplified by a study of genetic determinants for obesity and metabolic traits that incorporated global gene expression data from multiple tissues of a mouse F2 intercross[21] (Fig. 3). A population of approximately 300 males and females F2 mice was phenotyped for various plasma lipid and body composition traits (among others), and expression microarrays for approximately 23,000 transcripts performed on samples of adipose tissue, liver, skeletal muscle, and whole brain. Coexpression networks of the 3000 most varying transcripts were constructed for males and females separately. A topological overlap plot for adipose tissue is shown in Fig. 2. Modules are indicated by the bars on the *x* and *y* axes, and the corresponding boxed regions of the TOM plot. One can observe a series of distinct modules of varying size. The relationship of one module to the various phenotypic traits as determined by module significance scores is shown. Multiple body composition and glucose/insulin-related traits are correlated with this module. Among the hub, genes for this module are peroxisome proliferator-activated receptor gamma (*Pparg*), upstream transcription factor 1 (*Usf1*), and leptin (*Lep*), and an analysis using DAVID software showed GO categories of lysosome, catalytic activity, phosphorylation, and GTPase regulator activity to be overrepresented. Thus, applying coexpression network analysis to microarray data allows for substantially more information and understanding of expression data than would be gleaned from simply determining which transcripts are most highly correlated.

Here, coexpression networks were constructed for individual tissues, which is clearly relevant and typically what is done. However, metabolic disorders such as obesity have well-established endocrine signaling and other cross-tissue influences at work in their pathogenesis, and so cross-tissue coexpression analyses are a logical extension to make. One study attempted to do this by developing the network in a manner similar to that for a single tissue, except that it computed correlations of each transcript in one tissue to all those of the other tissues.[22] Within tissue networks were constructed for comparison. Adipose, liver, and hypothalamus were so examined in a population of approximately 300 F2 mice. Tissue-to-tissue coexpression networks were significantly different from within tissue networks, both in terms of module composition and the sets of genes included in each. Approximately 40% of the genes in the tissue to tissue networks did not appear in the within tissue networks, indicating that such analyses are complementary. "Asymmetric connectivity" was noted, more commonly between hypothalamus and either of the peripheral tissues, where a given transcript in hypothalamus was connected to many transcripts in the corresponding tissues but not vice versa. Such findings suggest that hypothalamus exerts a greater cross-tissue controlling effect than the others.

In these examples, transcript data were obtained from a large number of tissue samples, where the perturbation was genetic variation, and coexpression networks constructed for each tissue and sex. However, the approach can be utilized in almost any setting where a sufficient number of samples and perturbations are present. For example, tissue samples have multiple cell types and it may be desirable to study particularly cell types depending on the disease or process of interest. An instructive example of this is provided by work of Gargalovic and colleagues[23] who were interested in the role of oxidized phospholipids in the pathogenesis of atherosclerosis and were studying the response of primary endothelial cells to these agonists. Utilizing a set of primary cell lines from 12 unrelated heart transplant donors, transcript profiling was performed on cells that were untreated and ones exposed to oxidized phospholipids for varying times. Thus, the perturbations included endogenous genetic variation, agonist treatment, and time. Coexpression networks were constructed with the set of 1043 genes (probe sets) that were differentially expressed between control and treatment groups. This analysis yielded 15 distinct modules of highly correlated genes. Several modules were found to be enriched for relevant functional genes sets, including one for sterol biosynthesis where insulin induced gene 1 (*INSIG1*), a key regulator of that process, had the highest connectivity and represented the main hub. Another module of interest was enriched for transcripts encoding molecular chaperones and heat shock proteins, many of which are known to be induced by endoplasmic reticulum stress as part of the unfolded protein response (UPR) pathway. The module also contained other UPR targets and was significantly enriched for *UPR* target genes. Further analysis and study of hub genes in this and a related module enriched for *UPR* genes led to the identification and characterization of a novel *UPR* target gene. To illustrate the advantage of the network approach, the authors used simple correlation and analyzed all network genes on the basis of their correlation with a key *UPR*-related transcription factor, *ATF4,* in the treated cells. The novel gene ranked as the 99th most correlated gene with ATF4. However, the novel gene and *ATF4* were both among the top six hub genes in one of the relevant modules. Therefore, using simple correlation, this gene would likely have been missed. The power of this network approach is further highlighted by the fact that a relatively small sample of individuals was used. An extension of this work to include cells from approximately 100 different donors has shown that this number is sufficient to perform association mapping and thus incorporate genetic data in the analyses.[24]

B. Integration of Genetic Analyses

The integration of global gene expression measurements with genetic analysis provides further information as to how genetic variation influences trait expression. However, before further discussion of networks, it is important to review basic aspects of the quantitative genetics of transcript levels. Integrative genetics, or the genetics of gene expression, has expanded substantially within the past decade, and is the subject of a number of informative reviews.[14,25–33] Integrative genetics simply treats expression data as quantitative traits along with the traditional physiologic or clinical trait data in linkage or association studies. This has been applied to a number of human and animal model settings where there are sufficient subject numbers and genetic segregation among them. These include tissue, peripheral blood cells, and derived cell lines from humans and a variety of tissues and cells in animals. Animal model systems include traditional crosses of inbred strains, sets of inbred and recombinant inbred strains, and outbred populations. In all systems studied, transcripts show substantial genetic influences with effect sizes that are generally greater than for complex traits.

The basic concepts are relatively straightforward. Since one can associate a specific transcript with the genomic location of its gene, one can categorize expression QTL (eQTL) as distal (*trans*) when the eQTL maps to a location distant from the gene, and local (*cis*) when it is sufficiently near the gene. The latter can be caused by genetic variation within the gene itself (a true *cis* effect) or locally acting *trans* effects, or feedback settings. Studies suggest that at least 60–80% of local eQTLs are true *cis* effects.[34] Colocalization of eQTLs with clinical trait QTLs or other QTLs (other transcripts, metabolites, proteins, etc.) allow the hypothesis that the colocalization may be a consequence of shared genetic determination, as opposed to simply close linkage. A local eQTL colocalizing with a trait QTL suggests that a candidate gene may harbor the genetic variation responsible for the trait QTL. Distal eQTLs colocalizing with clinical trait QTLs suggest that these transcripts may be related to the trait "downstream" of the genetic variation(s) responsible for the trait.

At a given locus linked with a trait, the underlying variation in DNA alters the abundance or function of the product through various possible mechanisms, including transcript abundance. These primary effects influence the expression of multiple other genes, which together effect phenotypic variation. In fact, it is not uncommon to find multiple transcripts showing colocalization of linkage with a clinical trait, some whose genes physically reside at the locus (i.e., are local eQTLs), but typically more are distal eQTLs. For example, an analysis of genes that were differentially expressed in liver between lean and obese mice from an F2 population showed that over a dozen metabolic pathways were disproportionately represented.[35] When eQTLs for genes

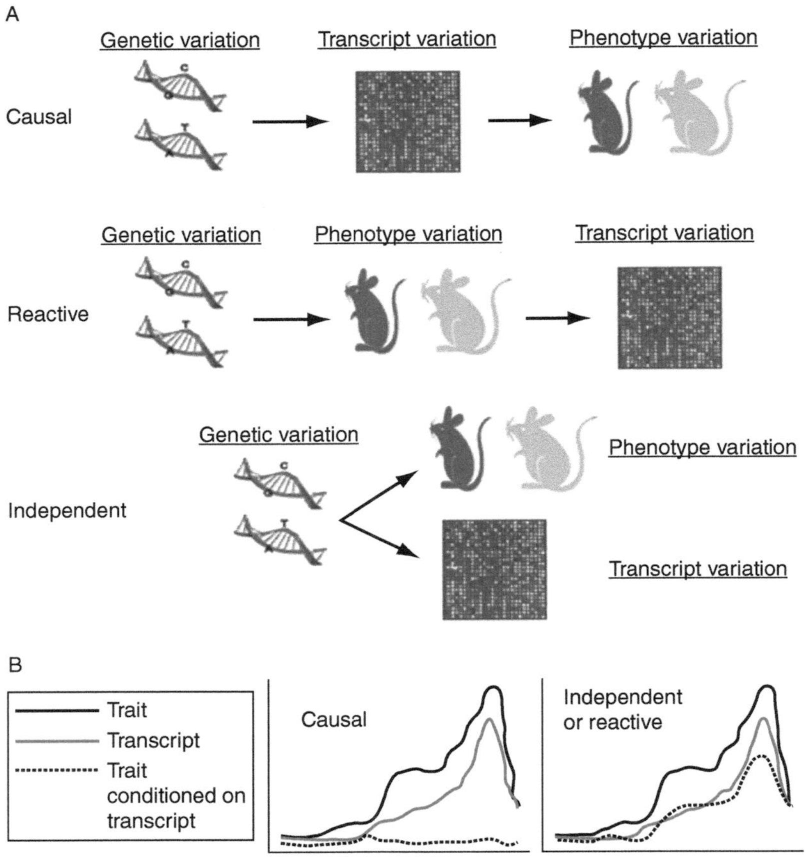

FIG. 4. "Causality" testing in integrative genetic studies: (A) the basic possible relationships among genotype, transcript level, and trait being tested for a given locus where one observes QTLs or associations for both transcript levels and trait and (B) schematic QTL plots from an F2 mouse cross of transcript and trait on one chromosome, and resulting trait QTLs after it has been conditioned on transcript levels. The QTL for the trait is significantly diminished or abolished when a causal relationship exists, but not otherwise.

composing these pathways were mapped, there was clear clustering at four loci, two of which overlapped QTLs for obesity traits. The large majority of these were distal eQTLs, implying that they were secondarily influenced by whatever the primary genetic variations were at these loci.

Such colocalized eQTLs could represent transcripts that are causal for the clinical trait, reactive to the trait, or unrelated (Fig. 4A). Mathematical assessments of the likelihood of these different possible relationships at a given locus between genetic variation, transcript levels, and clinical traits have been developed, and are commonly referred to as causality tests.[15,16] Conceptually, one is evaluating whether the correlation between genotype and clinical trait is conditional on a given transcript or vice versa (Fig. 4B). These are the simplest possible relationships, where it is assumed that DNA variation is always primary. Epigenetic, feedback, and interaction effects have not been incorporated into the models as yet. Nevertheless, an experimental study has provided validation for the causality test developed by Schadt and colleagues.[16] Utilizing transgenic or knockout models, 9 of 10 genes predicted to be causal for body fat were found to influence related body composition traits.[36] Over half of these had no prior evidence for a physiological role relating to obesity.

A report from Chen *et al.* provides an example of application of the causality test.[37] In an F2 cross, multiple transcripts show eQTLs that colocalize with a major locus influencing multiple metabolic and body composition traits. When applied to all the transcripts with eQTLs at this locus, which is broad and very likely encompasses multiple causal genetic variants, over 50 local eQTLs and hundreds of distal eQTLs tested as being causal for the body weight trait. Although linkage in this population leads to some false positives, there are obviously many causal genes, most of which are being acted upon by a smaller number of local variants.

C. Genetic Variation Acts by Influencing Networks

Data such as these support a network model of genetic control of traits as shown simplistically in Fig. 5. The integration of network analysis data with genetic and causality test results is further supportive and provide insight into how genetic variation influences trait expression.[37] As described above, coexpression network analysis from genetically segregating populations generates a series of modules, and the relationship of a given module to clinical traits can be determined by correlation-based methods. Applying genetic data and analyses provides evidence that network modules serve to mediate or "transmit" the impact of the underlying genetic variation on trait expression.[18,37] Genes that test as causal for a given trait would be expected to be overrepresented in such modules, which is indeed the case. Conversely, hub genes of clinically relevant modules would be more likely than less connected genes to have eQTLs that colocalize with trait QTLs, which is also observed. In a related manner, treating a module significance statistic as a quantitative trait shows that QTLs for clinically relevant modules colocalize with respective clinical trait QTLs.[38]

These observations are consistent with the concept that a given locus influencing a trait does so through the coordinated impact on the expression of multiple genes, producing a "molecular phenotype." The sum of these across multiple loci yields the extensive genetically determined variation in gene

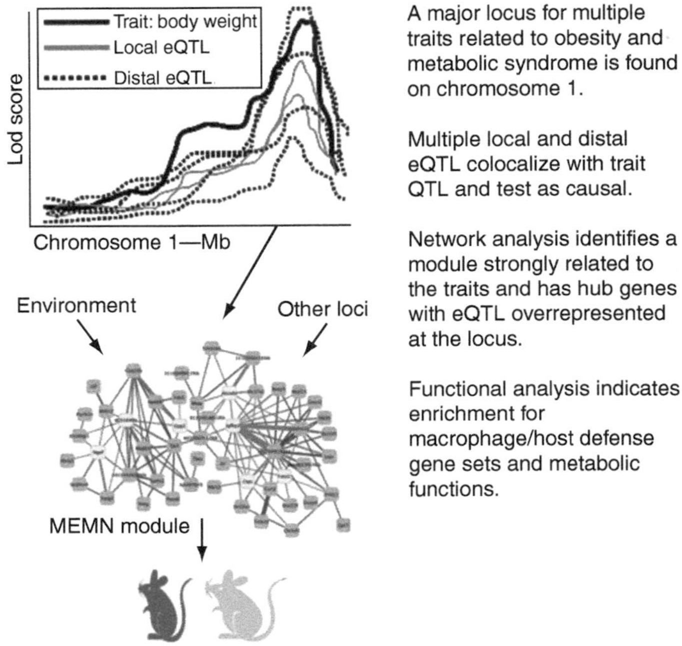

FIG. 5. Genetic variation acts by influencing networks. An integrative genetics analysis of an F2 mouse cross supports a model, whereby genetic loci influence trait phenotype through their effects on specific coexpression network modules, in this case, one strongly associated with macrophages and inflammation (designated the macrophage-enriched metabolic network or MEMN).[37] A smaller number of genetic variants at the locus influence a much larger number of other transcripts secondarily to effect phenotypic differences. (*Note:* the actual MEMN has many more genes than are depicted in the network icon used in this figure).

expression that one observes in human and animal populations. The integration of genetic, causal gene, and coexpression analyses, thus, allows the deconstruction ("reverse engineering") of these complex relationships and the identification of critical specific modules or subnetworks involved in mediating genetic influences on traits.[18,39] Characterization of the biologic associations of such modules yields insight into possible mechanisms. Continuing with the above example, the adipose tissue module with the strongest significance for multiple metabolic traits showed highly significant enrichment for a GO category set of genes involved in host defense mechanisms, and for genes showing strong expression in bone marrow-derived macrophage cells and in spleen. Liver coexpression analysis yielded an analogous module. This implicated macrophages and inflammation as important cells and processes affecting both adipose tissue and liver as being involved in obesity and multiple other metabolic traits, consistent with a body of related literature.[40–42] It also allowed the identification of novel genes that were likely to be important in obesity pathogenesis. A transcriptional network derived from human adipose tissue samples showed the presence of an analogous module, indicating that this is not unique to the mouse model system.[43]

Thus, in this example, a network of genes was shown to be influenced by a genetic locus, which in turn influenced trait expression. Experimental validation was provided by studying the effect of genetic manipulation of selected macrophage-related module genes (none of which were physically located within the chr 1 locus), via knockout or transgenic models on trait expression, and observing in each case significant variation in obesity-related traits.[36,37] The implications for both traditional genetic and cell biology-based approaches for understanding disease pathogenesis are profound. For traditional genetics, which is focused on identifying the causative genes at a given locus, the added information provided by the network analyses aids in identifying such causal genes and implicating biologic pathways influenced by these genes. For cell biology-based approaches, the identification of modules or subnetworks enriched for particular pathways or processes provides a large set of interacting partners to incorporate into the known biochemistry. From a therapeutic perspective, the implication is that therapeutic intervention should target elements that will favorably influence network properties as opposed to simply individual genes or pathways. Products of genes that compose a relevant module or subnetwork are likely to be appropriate targets, rather than ones selected solely on the basis of strength of correlation with a trait of interest.

D. Application to Human Genetics

The integrative genetics approaches described have largely been established through studies in mice and other model organisms. How do these concepts and approaches apply to human studies? The fundamental aspects have been established to be applicable. Firstly, the extent of genetic control of gene expression is as extensive in humans as it is in model organisms, with detection of both local and distal transcript associations, although in contrast to mouse F2 populations, local associations exceed distal associations. This has been studied in a variety of settings, including primary and derived cell lines from endothelium and lymphocytes,[44–48] brain cortex,[49] adipose tissue,[43] and liver.[50] Secondly, coexpression network analyses have proved to be informative for disease-related traits.[23,43,51]

The most definitive human study to date that combines global gene expression, high-density SNP determination, and coexpression network analysis in conjunction with a relevant human condition comes from the Icelandic population, where blood and adipose tissue biopsies were obtained in conjunction with measurement of obesity-related traits of BMI, % body fat, and waist–hip ratio.[43] While less than 10% of transcripts detected in blood showed correlation with obesity traits, over 50% of adipose transcripts were correlated, with approximately 2800 of these explaining more than 10% of the variation in BMI. Over 3000 local eSNPs were detected in adipose tissue, which are 650 more than seen in blood. Coexpression networks for men and women were constructed from a set of the most differentially expressed genes in adipose tissue. One module of just under 900 genes was highly conserved between males and females, and strongly associated with obesity, with over 90% of transcripts being significantly correlated with obesity traits. The module was highly enriched for macrophage and immune-related GO biologic processes, and for genes associated with bone marrow-derived macrophages, and lymphoid tissues, supporting inflammation in adipose tissue as being a key factor in the pathogenesis of obesity. Further support was obtained by showing that the human macrophage-related module highly overlapped the comparable module in mice that was described above. Additionally, local eSNPs for genes in this module were significantly associated, albeit modestly, with obesity traits of BMI and % body fat. Altogether, this study supports the network model of disease, where genetic perturbations (disease risk alleles) generally perturb a larger set of genes composing a biologically relevant subnetwork (module).

Human GWASs identify SNPs associated with disease or physiologic traits, but usually lack data other than proximity of a SNP to adjacent genes and any associated gene annotation regarding function to select which among them is more likely to be causal. For genes where altered transcript levels mediate the allelic differences, the integration of data from independent human eSNPs and mouse coexpression data sets can be informative. Another study of genetic and gene expression analysis of human tissue, in this case, liver, demonstrated how these can used in conjunction to inform human GWASs (Fig. 6). Schadt and colleagues applied high-density SNP typing with global gene expression determination to a set of over 400 human liver tissue samples.[50] In this case, clinical data were not available. Similar to the above study, over significant 3000 local eSNPs were identified; distal eSNPs were only about 15% of this number. It should be noted that in this and the above study, the relatively low number of distal eQTLs detected relative to the mouse studies should most likely not be interpreted as an absence of their occurrence, but rather an inability to detect them for statistical reasons related to lack of power given the large number of SNPs and expression traits being tested, with the added factor of a more complex genetic background where there are multiple loci that influence trait expression. In applying these data, analysis of the human liver eSNP data set for genes whose expression is significantly associated, the SNPs identified from a GWAS can suggest a more likely candidate. Secondly, to bring in functional information, the role of the candidate genes can be examined in the mouse hepatic coexpression network, where transcript levels can be correlated with

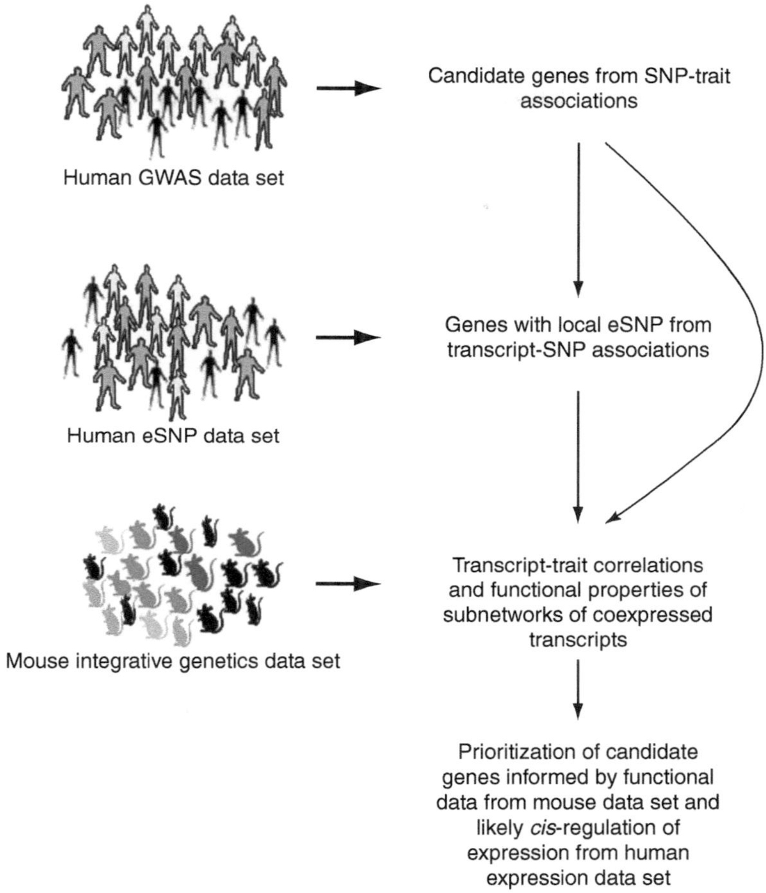

FIG. 6. Use of human and mouse integrative genetics data sets to inform identification of candidate genes from human GWASs. Most appropriately, the human eSNP and mouse data sets are from relevant tissues for the disease being evaluated by the GWAS.

relevant traits, and local network properties can be evaluated. An analogous approach was taken by a group studying bone density, where local eQTLs obtained from a set of human osteoblast-derived cell lines were utilized to inform an independent GWAS of bone density.[52]

An example provided was evaluation of a SNP strongly associated with Type 1 diabetes in a large GWAS, v-erb-b2 erythroblastic leukemia viral oncogene homolog 3 (avian) (*ERBB3*) was inferred as the likely causal gene based on known function related to immune signaling. However, in the human liver eSNP data set, *ERBB3* had no evidence for genetic determination of transcript levels, while a flanking gene, ribosomal protein S26 (*RPS26*), did. In the mouse coexpression network, an evaluation of the genes in the local networks containing each showed significant enrichment for Type 1 diabetes-related genes in the local network containing *RPS26*, but not *ERBB3*. Additional support for *RPS26* came from examination of tissue atlases, where its expression levels were significantly greater in pancreatic islets than those of *ERBB3*. Although these data are not definitive in identifying which is the causal gene, they certainly provide independent support for prioritizing further studies.

In a more recent extension of this type of integrative analyses to inform GWAS results, the authors combined multiple coexpression data sets from mice, rats, and human liver tissue samples.[53] Conserved and species-specific interactions (gene–gene correlations) were identified and coexpression networks developed. Approximately 10% of gene–gene coexpression relationships were conserved across the three species, and modules from networks constructed of conserved gene interactions tended to represent fundamental cell functions such as sterol biosynthesis, cell adhesion, and cell–cell signaling. Candidate genes for plasma lipid traits from two GWASs were examined using the logic described above. Using criteria of the presence of eSNPS in the human liver data set and membership in conserved coexpression modules enriched for lipid-related genes, 395 genes were identified as being the most likely causal genes, from a starting set of approximately 1400 genes. Again, the general concept is one of applying evidence from multiple gene expression data sets to inform GWAS results. This study is also an important reminder that differences in gene expression between species are significantly greater than differences in genome sequence. This underscores the need for the inclusion of intermediate expression phenotypes from relevant tissues in GWASs whenever possible, and where that is difficult to obtain, development of species- and tissue-specific eQTL/eSNP data sets, to expand on the existing tissue expression profiles.

E. Integrating Metabolite and Transcriptional Profiling

Individual metabolites such as glucose and free fatty acids have long been analyzed and used as intermediate phenotypic traits in the setting of genetic analysis of disease. Technical advances are now bringing concurrent measurement of up to hundreds of metabolites and proteins in the integrative genetics setting.

Attie and colleagues have applied the integrative genetics approach to include metabolites in the mouse, and show how it can be used to predict causal regulators of metabolism.[54] The levels of 67 metabolites were profiled in liver samples from 60 F2 mice, along with genotype, transcript, and phenotype data. These included amino acids, urea cycle and TCA cycle intermediates, and acyl-carnitines. Correlation-based clustering showed functionally related sets of metabolites, and approximately one-fifth of the metabolites demonstrated significant linkage to one or more loci. Several loci showed linkage for multiple-related metabolites, suggesting common regulation. Transcript levels of genes selected for being involved in metabolism showed correlation overall with respective metabolite groups. There were also subsets of tight clusters of correlated metabolites and transcripts with shared QTLs. From these data, a causal network was developed, centered on glutamate, composed of seven genes with solute carrier family 1 (glial high-affinity glutamate transporter), member 2 (*Slc1a2*), an amino acid transporter, identified as one of several genes likely regulating glutamate levels. This provides a proof of principle for further such studies.

Recently, a human GWAS reported associations from an analysis of 163 metabolic traits in serum, noting that a number of these mapped near genes are known to be involved in the cellular metabolism of the metabolite, such as enzymes or transporters.[55] Another study evaluated genetic determinants of 33 sphingolipid species in over 4000 subjects.[56] Although these studies did not incorporate network-based analyses, it is easy to extrapolate how data from studies such as these could be examined in the manner described above for integrating human tissue and mouse data sets to inform disease-related trait GWAS. The wealth of biochemical data concerning metabolism provides a significantly greater degree of "annotation" useful for interpreting results than is available for complex traits such as disease phenotypes. Metabolic traits may be found to be an important bridge between the top-down integrative genetics approach and the bottom-up metabolic network approach described in the subsequent section.

IV. Metabolic Networks as a "Bottom-Up" Systems Biology Approach

Integrative genetics represents a "top-down" systems biology approach. This is necessary when there are relatively little data available defining mechanistic relations between network elements. Where such data are available as is the case for a substantive portion of cell metabolism, a "bottom-up" approach is feasible for constructing networks. Palsson and colleagues among others have developed this to a sophisticated degree for many organisms, including humans, as recently reviewed.[57,58] In this approach, termed constraint-based reconstruction and analysis (COBRA), networks are constructed from known molecular components, in this case, metabolites, and the enzymes that act on them, structured such that metabolites are the nodes, and biochemical reactions with respective associated enzymes that transform metabolites representing the edges. Across a series of different organisms, the species genomic sequence with its identifiable genes defines the possible enzyme components. The combination of existing annotation and pathway databases, combined with manual literature-based annotation allows compilation of the genes, proteins, reactions, and metabolites in a structured database, ready for network reconstruction. A derived metabolic network from these data is represented mathematically in a matrix format of stoichiometric coefficients. In such a matrix, the rows are metabolite components and columns are the reaction links, with each corresponding cell containing the coefficients of substrates and products. Constraints can be placed to represent specific regulatory, environmental, or other factors that would influence these. This allows for studying a derived network from a topological perspective (i.e., what pathways exist and how are they structured in relation to one another) and for computational modeling of metabolic states. The latter is of particular relevance to understanding disease-related phenotype effects.

A global human metabolic network of over 3000 well-defined metabolic and transport reactions has been developed and studied in several contexts.[59–62] The increasing availability of high-throughput-derived data, such as transcript, protein, or metabolite profiling, allows for modifying this global network to reflect cell- and tissue-specific conditions. One example of this utilized expression microarray data from skeletal muscle obtained pre- and post-gastric bypass surgery to assess the potential effects on the procedure on energy metabolism.[62] The analysis suggested that there would a relative decrease in mitochondrial bioenergetics.

Genetic or epigenetic processes affecting gene expression or protein function can also be modeled.[59,63] The potential effects of a SNP or mutation affecting either of these could be studied *in silico*.[63] A recent study analyzed the potential effects of altered imprinting for nine known or predicted imprinted genes.[59] Four were found to have altered metabolic effects consistent with known phenotypes, including ATPase, class V, type 10A (*ATP10A*), which is associated with obesity in mice and humans.

This approach, therefore, is highly complementary to the top-down approaches discussed above, and well suited for obesity-related genetic research. Database and computational tools for metabolic systems biology studies are publically available, allowing for broader use, and the specifics described above are but one example of work in this field. Beyond metabolism, the general approach of modeling in this manner could be extended to cell signaling or other relevant networks as these become progressively better defined at the reaction level. Li *et al.* have utilized the COBRA approach to model human toll-like receptor-signaling pathways, and identified eight novel candidate targets for pathway inhibition.[64]

V. Other Areas of Relevance and Future Direction

A. Animal Models

Many of the integrative genetic studies performed to date in mice utilized F2 intercross populations. These simplify the genetic variation to the two parental strains, but have several drawbacks from a systems biology perspective. One is the very significant linkage effect, which makes mapping resolution relatively poor. Perhaps more important is the fact that each animal is genetically unique, which limits, for practical and ethical reasons, the number and timing of phenotypic measures that can be obtained, and requires genotyping of every animal. The use of inbred strains overcomes these issues, since as many animals as needed can be studied, and high-density genotype data are publicly available. However, it has been limiting in that the number of recombinant inbred strains available is insufficient for adequate power to map traits, as is the number of informative individual strains.

This has changed with several advances. One is the development of the collaborative cross, which will be nearing completion in the next few years.[65,66] This is intended to generate a set of 1000 strains with a heterogeneous genetic makeup from eight parental strains, and having a much improved mapping resolution. A second is the development of improved analytical tools and study design for association-based mapping in sets of inbred strains. Recently, a "hybrid mouse diversity panel" (HMDP) was developed and tested, composed of approximately 30 regular inbred strains and 70 recombinant inbred strains from several existing recombinant inbred strain sets, all of which are commercially available.[67] The analytical hurdle that was overcome was the effective correction for strain relatedness, which if not performed, yields significant false positives. An analysis of plasma lipid traits showed excellent mapping resolution and good power. With such inbred strains where the genetic makeup is constant, data of all sorts can be collected at varying time points and under varying conditions, and from multiple tissues or isolated cells. Significant phenotypic data for current strains already exist in the Mouse Phenome Database[68] (http://www.jax.org/phenome). As various studies are performed and data made public in such a manner, an ever increasing rich characterization will be available, avoiding the need for starting anew for each new study. Thus, such resources as the HMDP and collaborative cross are ideally suited for systems biology-related research.

Although mammals are closer to humans than other available animal models, research over the past decade has shown that diverse "lower" species often share some basic characteristics related to energy and metabolism that are relevant to obesity.[69–76] For example, the fruit fly (*Drosophila melanogaster*) has a fat body that stores triglycerides, and cells analogous to pancreatic beta cells and hepatocytes. An insulin-signaling network among these exists that regulates fat body lipid stores. Such model organisms also have characteristics and "tools" available so that for some purposes, not the least of which are rapid reproduction and short life span, they may be preferable to mice or other mammals. A comprehensive RNAi reagent set exists for *Caenorhabditis elegans* to knockdown genes *in vivo*, and *Drosophila* has numerous mutant and other strains available for genetic analyses. As a vertebrate, *Danio rerio* (Zebrafish) has some physiological features in common with mammals that the invertebrate models lack, and are amenable to knockdown studies with antisense morpholino reagents. These models have the requisite features of varying genetics, known genome sequence, and the ability to study phenotypes and molecular variation across populations, for systems biology-type studies.

B. Integration of Diverse Data Types

Genome and transcriptional data dominate genetics-oriented systems biology-based research. This will change as technologies develop and mature for other important facets of cell and molecular biology. Of direct relevance for genetic networks is and will be the ability to identify epigenetic and other gene regulatory features on a larger scale than has been possible. DNA sequence provides information on putative transcription factor-binding sites, but direct data from actual transcription factor-binding site occupancy will provide much needed supporting information. This has been done on a global scale in some instances using ChIP-on-chip or ChIP-seq methods, such as for USF1/2 binding.[77] Similarly, identification of histone-binding sites improves prediction of enhancer and promoter regions. Identification of DNA methylation sites on a global scale is now feasible with high-throughput sequencing.[78] Protein–protein interaction data can also be useful. Such data can be incorporated into the integrative genetics analyses discussed above to generate a more accurate and detailed genetic network using Bayesian network approaches.[79] A Bayesian network can incorporate diverse data types to inform and improve the accuracy of the probability of a given relationship between nodes, which is at the basis of Bayesian network construction.

Also advancing are methodologies for larger scale quantification of proteins and metabolites in cells and tissues.[80–82] Both mass spectrometry-based and protein chip approaches can be applied for protein quantification, and mass spec and NMR methodologies exist for metabolite analysis. Methodologies for assessing signaling pathway activation by measuring ratios of phosphorylated versus nonphosphorylated intermediates are also available on a limited scale.[83] For top-down type systems approaches, Bayesian networks allow inclusion of these data for creating integrated causal networks.[79]

C. Computation, Data Sharing, and Modeling

The generation, storage, and analysis of high-throughput data sets are nontrivial even before getting to the point of the kinds of network-based analyses discussed here. Expertise in data management, computation, and bioinformatic analysis within research groups will be increasingly necessary to utilize systems biology-related data sets. Likewise, there is a communal need for maintaining and improving publicly available databases, and for the research community to share data publicly. Large-scale experiments with high-throughput data sets will remain expensive to perform, and untenable to obligate repeating in the long run on account of data sharing concerns. The more important it becomes to integrate data to get the most information, the greater the importance of tools and databases to be available to facilitate identification and retrieval of relevant data. For example, while the integrative genetic studies discussed above typically deposit microarray data in GEO or Ensembl, these typically do not include the associated genotyping or trait data. For data analysis and display, the systems biology community has developed standards for data notation with SMBL so that various software programs such as Cytoscape can be used for modeling.[84] Ultimately, it is imagined that it will be feasible for researchers—whether involved in cell and molecular biology, physiology, or genetics—to have access to and ability to use systems biology data sets and software.

D. Approaches to Validation of Model Predictions

Although systems biology is sometimes viewed as a more hypothesis-generating mode of doing science, the testing or validation of hypotheses from the models developed is a fundamental tenet of the approach. As presented above, these various systems biology-based studies predict a role of modules, or subnetworks, in addition to specific genes and pathways. One approach for validating the relevance of a given module is replication in one or more independent experiments. These are as yet limited, as such studies are expensive and time consuming, but as more are performed, this will improve. An example from the studies discussed above is the replication of the macrophage-enriched network, which was observed in both mouse liver and adipose tissues, and human adipose tissue, and, in each of these, strongly correlated with obesity and metabolic traits.[37,43] There is also a large body of published work on the role of macrophages and inflammation in adipose tissue and obesity that is supportive.[40,41]

A variation on this is to examine GWAS association data to see if genes composing an apparently relevant module show an increased likelihood of stronger statistical association for a trait than a control set of genes, as was mentioned above.

In vivo genetic manipulation of individual genes predicted to be important has also been used, but these are obviously not high-throughput approaches, and such validation has been performed on only a small number of predicted candidate genes. One study assessed the impact of transgenic or knockout manipulation of nine genes predicted from an integrative genetic study to be causal for the obesity phenotype in mice.[36] Eight of these were found to have an effect on physiologic phenotype, thus validating the analytical approach used to predict them. Additionally, gene expression analyses of liver showed partially overlapping signatures among them, along with common patterns of GO biologic process and pathway enrichment. The latter included metabolic pathways related to energy and lipid metabolism. These genes were also members of a subnetwork of a Bayesian network of liver gene expression constructed from independent F2 crosses. Other individual genes have been validated in a comparable manner.[43,85,86]

Knockdown experiments using siRNA in cell culture systems are commonly used to screen or assess functional relevance of specific genes, and provide a better opportunity for screening the larger number of candidates generated by network approaches. An example of integrating pathway, protein–protein interaction, and expression data into a hybrid Bayesian network comes from a study that was attempting to validate predicted causal genes in a cell culture system. Off target effects are not uncommon and can confound interpretation. Tu *et al.* applied a novel network-based approach for identifying true positive novel genes involved in insulin signaling in adipocytes from such a screening set.[87] An anti-lipolysis assay was employed to identify insulin resistance in the adipocyte cell line 3T3-L1 associated with siRNA knockdown of a set of over 300 selected genes, over half of which were genes predicted to be involved in obesity from the integrative genetics studies discussed above. Approximately, a third showed impaired lipolysis compared with control siRNA treated cells. A step-wise process was employed to develop a network to help identify which of those were most likely to be real hits. Elements of KEGG pathways that intersected with any of the assay positive genes were used to develop an initial network. Protein–protein interaction data sets were then used to expand the network. The final network was obtained by "trimming" the expanded network based on the set of genes that showed no effect in the siRNA screen. Directed relationships were incorporated when known. Thus, a Bayesian network of 185 genes and 17 small molecules was obtained, derived from the integration of pathway and protein–protein interaction data sets that had been functionally associated with insulin signaling. This included 79 of the 126 siRNA screen positive genes. The authors went on to show that this network set was highly enriched among genes differentially expressed in adipose tissue in four knockout models that had insulin resistance. Additionally, a novel gene previously unassociated with insulin signaling was validated by observing insulin resistance *in vivo* in a knockout model. In this study, the network was related to insulin signaling by the anti-lipolysis functional screen. The approach could obviously be applied in an analogous manner to other cellular functions relevant to adipocyte biology and obesity. Demonstration of influencing a relevant phenotype *in vitro* is at least an intermediate step in prioritizing candidate genes for more detailed *in vivo* validation.

VI. Conclusions

This review has attempted to present a perspective on what differentiates systems biology approaches from more traditional types of investigation, and how these have been applied to the genetics of obesity. Integrative genetics has been the most relevant and active area of such investigation thus far. It should be noted that there are many more areas of systems biology-related research that are relevant to obesity, such as gene regulatory networks, but have yet to specifically incorporate genetic variation as an integral aspect, and therefore not discussed here. These studies represent early advances in characterizing the systems-level transcriptional network and relating it to genetic variation and trait expression. Conclusions that can be drawn from these relevant to the study of obesity include the following:

(1) Subnetworks (modules) can be identified that represent tightly correlated sets of transcripts that in turn may be correlated with relevant physiological traits. These are often associated with particular biologic processes and pathways, which provides insight into possible mechanisms through which a module may be acting. The more highly connected hub genes of a module can be identified and in general are thought to be more important in module function.

(2) By integrating genetics with coexpression analysis, genetic loci that influence traits can be seen to do so through subnetworks or modules. A limited number of genetic variants at a given locus (local eQTLs or eSNPs) influence a larger set of transcripts, which in turn can be seen to influence traits. Causal relationships can be defined based on the flow if information from DNA to transcript to trait, and the incorporation of these data allows for improved network construction, and separation of causal transcripts from ones simply correlated with a trait for other reasons.

(3) Human and mouse integrative genetics data sets can inform interpretation of human GWASs through the identification of eSNPS in human tissues and functional gene networks from mouse data sets. Most human GWASs cannot collect and analyze gene expression in relevant

tissues. For relevance to obesity, adipose tissue, liver, muscle, and hypothalamus are important among other tissues, and the collection of such tissue-specific human and mouse data sets that combine gene expression and genetic data, and where possible as in mouse, functional data, will be useful to develop.

(4) Identification of modules and causal genes associated with obesity-related traits has generated a large number of genes predicted to be relevant to the biology of obesity. These fall into two broad categories. One is represented by the macrophage-enriched metabolic network set of genes identified in both liver and adipose tissue in mice, and in adipose tissues in humans. The second encompasses a number of basic metabolic pathways involving carbohydrate, amino acid, lipid, and energy metabolism. However, many genes, including a number of those validated by transgenic approaches, are not obviously associated with these based on existing knowledge. Substantial effort will be necessary to establish specific functional cellular roles for these. In the interim, these data provide a novel type of gene annotation for otherwise uncharacterized genes as well as the suggestion of novel functions for known genes.

Systems biology is a nascent field that has only recently been supported by the necessary technological advances for large-scale high-throughput data generation. It is now an important area of investigation in many areas of biology, attracting significant research activity. Although certainly not capturing all relevant papers, a PubMed search shows a 100-fold increase in publications using the term over the past decade, from just 19 in 2001 to nearly 1800 in 2009. Institutes and departments of systems biology have been formed at academic institutions, and journals devoted to systems biology established. It is reasonable to expect that this "new" discipline will continue to expand and become an essential approach for all areas of biology, complementing the traditional disciplines.

References

1. Weiss JN, Yang L, Qu Z. Systems biology approaches to metabolic and cardiovascular disorders: network perspectives of cardiovascular metabolism. *J Lipid Res* 2006;**47**:2355–66.
2. Ideker T, Lauffenburger D. Building with a scaffold: emerging strategies for high- to low-level cellular modeling. *Trends Biotechnol* 2003;**21**:255–62.
3. Ideker T, Galitski T, Hood L. A new approach to decoding life: systems biology. *Annu Rev Genomics Hum Genet* 2001;**2**:343–72.
4. Kitano H. Computational systems biology. *Nature* 2002;**420**:206–10.
5. Kitano H. Systems biology: a brief overview. *Science* 2002;**295**:1662–4.

6. Albert R. Scale-free networks in cell biology. *J Cell Sci* 2005;**118**:4947–57.
7. Barabasi AL. Scale-free networks: a decade and beyond. *Science* 2009;**325**:412–3.
8. Jeong H, Mason SP, Barabasi AL, Oltvai ZN. Lethality and centrality in protein networks. *Nature* 2001;**411**:41–2.
9. Halaas JL, Gajiwala KS, Maffei M, Cohen SL, Chait BT, Rabinowitz D, et al. Weight-reducing effects of the plasma protein encoded by the obese gene. *Science* 1995;**269**:543–6.
10. Flint J, Valdar W, Shifman S, Mott R. Strategies for mapping and cloning quantitative trait genes in rodents. *Nat Rev Genet* 2005;**6**:271–86.
11. Lindgren CM, Heid IM, Randall JC, Lamina C, Steinthorsdottir V, Qi L, et al. Genome-wide association scan meta-analysis identifies three Loci influencing adiposity and fat distribution. *PLoS Genet* 2009;**5**:e1000508.
12. Jansen RC, Nap JP. Genetical genomics: the added value from segregation. *Trends Genet* 2001;**17**:388–91.
13. Drake TA, Schadt EE, Davis RC, Lusis AJ. Integrating genetic and gene expression data to study the metabolic syndrome and diabetes in mice. *Am J Ther* 2005;**12**:503–11.
14. Rockman MV, Kruglyak L. Genetics of global gene expression. *Nat Rev Genet* 2006;**7**:862–72.
15. Aten JE, Fuller TF, Lusis AJ, Horvath S. Using genetic markers to orient the edges in quantitative trait networks: the NEO software. *BMC Syst Biol* 2008;**2**:34.
16. Schadt EE, Lamb J, Yang X, Zhu J, Edwards S, Guhathakurta D, et al. An integrative genomics approach to infer causal associations between gene expression and disease. *Nat Genet* 2005;**37**:710–7.
17. Benson M, Breitling R. Network theory to understand microarray studies of complex diseases. *Curr Mol Med* 2006;**6**:695–701.
18. Schadt EE. Molecular networks as sensors and drivers of common human diseases. *Nature* 2009;**461**:218–23.
19. Wu S, Lusis AJ, Drake TA. A systems-based framework for understanding complex metabolic and cardiovascular disorders. *J Lipid Res* 2009;**50**(Suppl.):S358–63.
20. Langfelder P, Horvath S. WGCNA: an R package for weighted correlation network analysis. *BMC Bioinformatics* 2008;**9**:559.
21. van Nas A, Guhathakurta D, Wang SS, Yehya N, Horvath S, Zhang B, et al. Elucidating the role of gonadal hormones in sexually dimorphic gene coexpression networks. *Endocrinology* 2009;**150**:1235–49.
22. Dobrin R, Zhu J, Molony C, Argman C, Parrish ML, Carlson S, et al. Multi-tissue coexpression networks reveal unexpected subnetworks associated with disease. *Genome Biol* 2009;**10**:R55.
23. Gargalovic PS, Imura M, Zhang B, Gharavi NM, Clark MJ, Pagnon J, et al. Identification of inflammatory gene modules based on variations of human endothelial cell responses to oxidized lipids. *Proc Natl Acad Sci USA* 2006;**103**:12741–6.
24. Romanoski CE, Lee S, Kim MJ, Ingram-Drake L, Plaisier CL, Yordanova R, et al. Systems genetics analysis of gene-by-environment interactions in human cells. *Am J Hum Genet* 2010;**86**:399–410.
25. Chen M, Kendziorski C. A statistical framework for expression quantitative trait loci mapping. *Genetics* 2007;**177**:761–71.
26. Cookson W, Liang L, Abecasis G, Moffatt M, Lathrop M. Mapping complex disease traits with global gene expression. *Nat Rev Genet* 2009;**10**:184–94.
27. Drake TA, Schadt EE, Lusis AJ. Integrating genetic and gene expression data: application to cardiovascular and metabolic traits in mice. *Mamm Genome* 2006;**17**:466–79.
28. Farber CR, Lusis AJ. Integrating global gene expression analysis and genetics. *Adv Genet* 2008;**60**:571–601.
29. Franke L, Jansen RC. eQTL analysis in humans. *Methods Mol Biol* 2009;**573**:311–28.

30. Li Y, Breitling R, Jansen RC. Generalizing genetical genomics: getting added value from environmental perturbation. *Trends Genet* 2008;**24**:518–24.
31. Pomp D, Nehrenberg D, Estrada-Smith D. Complex genetics of obesity in mouse models. *Annu Rev Nutr* 2008;**28**:331–45.
32. Skelly DA, Ronald J, Akey JM. Inherited variation in gene expression. *Annu Rev Genomics Hum Genet* 2009;**10**:313–32.
33. Tesson BM, Jansen RC. eQTL analysis in mice and rats. *Methods Mol Biol* 2009;**573**:285–309.
34. Doss S, Schadt EE, Drake TA, Lusis AJ. Cis-acting expression quantitative trait loci in mice. *Genome Res* 2005;**15**:681–91.
35. Ghazalpour A, Doss S, Sheth SS, Ingram-Drake LA, Schadt EE, Lusis AJ, et al. Genomic analysis of metabolic pathway gene expression in mice. *Genome Biol* 2005;**6**:R59.
36. Yang X, Deignan JL, Qi H, Zhu J, Qian S, Zhong J, et al. Validation of candidate causal genes for obesity that affect shared metabolic pathways and networks. *Nat Genet* 2009;**41**:415–23.
37. Chen Y, Zhu J, Lum PY, Yang X, Pinto S, MacNeil DJ, et al. Variations in DNA elucidate molecular networks that cause disease. *Nature* 2008;**452**:429–35.
38. Ghazalpour A, Doss S, Zhang B, Wang S, Plaisier C, Castellanos R, et al. Integrating genetic and network analysis to characterize genes related to mouse weight. *PLoS Genet* 2006;**2**:e130.
39. Lusis AJ, Attie AD, Reue K. Metabolic syndrome: from epidemiology to systems biology. *Nat Rev Genet* 2008;**9**:819–30.
40. Iyer A, Fairlie DP, Prins JB, Hammock BD, Brown L. Inflammatory lipid mediators in adipocyte function and obesity. *Nat Rev Endocrinol* 2010;**6**:71–82.
41. Maury E, Brichard SM. Adipokine dysregulation, adipose tissue inflammation and metabolic syndrome. *Mol Cell Endocrinol* 2010;**314:**1–16.
42. Nishimura S, Manabe I, Nagai R. Adipose tissue inflammation in obesity and metabolic syndrome. *Discov Med* 2009;**8**:55–60.
43. Emilsson V, Thorleifsson G, Zhang B, Leonardson AS, Zink F, Zhu J, et al. Genetics of gene expression and its effect on disease. *Nature* 2008;**452**:423–8.
44. Dimas AS, Deutsch S, Stranger BE, Montgomery SB, Borel C, Attar-Cohen H, et al. Common regulatory variation impacts gene expression in a cell type-dependent manner. *Science* 2009;**325**:1246–50.
45. Cheung VG, Spielman RS, Ewens KG, Weber TM, Morley M, Burdick JT. Mapping determinants of human gene expression by regional and genome-wide association. *Nature* 2005;**437**:1365–9.
46. Dixon AL, Liang L, Moffatt MF, Chen W, Heath S, Wong KC, et al. A genome-wide association study of global gene expression. *Nat Genet* 2007;**39**:1202–7.
47. Goring HH, Curran JE, Johnson MP, Dyer TD, Charlesworth J, Cole SA, et al. Discovery of expression QTLs using large-scale transcriptional profiling in human lymphocytes. *Nat Genet* 2007;**39**:1208–16.
48. Stranger BE, Nica AC, Forrest MS, Dimas A, Bird CP, Beazley C, et al. Population genomics of human gene expression. *Nat Genet* 2007;**39**:1217–24.
49. Myers AJ, Gibbs JR, Webster JA, Rohrer K, Zhao A, Marlowe L, et al. A survey of genetic human cortical gene expression. *Nat Genet* 2007;**39**:1494–9.
50. Schadt EE, Molony C, Chudin E, Hao K, Yang X, Lum PY, et al. Mapping the genetic architecture of gene expression in human liver. *PLoS Biol* 2008;**6**:e107.
51. Hagg S, Skogsberg J, Lundstrom J, Noori P, Nilsson R, Zhong H, et al. Multi-organ expression profiling uncovers a gene module in coronary artery disease involving transendothelial migration of leukocytes and LIM domain binding 2: the Stockholm Atherosclerosis Gene Expression (STAGE) study. *PLoS Genet* 2009;**5**:e1000754.
52. Grundberg E, Kwan T, Ge B, Lam KC, Koka V, Kindmark A, et al. Population genomics in a disease targeted primary cell model. *Genome Res* 2009;**19**:1942–52.

53. Wang K, Narayanan M, Zhong H, Tompa M, Schadt EE, Zhu J. Meta-analysis of inter-species liver co-expression networks elucidates traits associated with common human diseases. *PLoS Comput Biol* 2009;**5**:e1000616.
54. Ferrara CT, Wang P, Neto EC, Stevens RD, Bain JR, Wenner BR, et al. Genetic networks of liver metabolism revealed by integration of metabolic and transcriptional profiling. *PLoS Genet* 2008;**4**:e1000034.
55. Illig T, Gieger C, Zhai G, Romisch-Margl W, Wang-Sattler R, Prehn C, et al. A genome-wide perspective of genetic variation in human metabolism. *Nat Genet* 2010;**42**:137–41.
56. Hicks AA, Pramstaller PP, Johansson A, Vitart V, Rudan I, Ugocsai P, et al. Genetic determinants of circulating sphingolipid concentrations in European populations. *PLoS Genet* 2009;**5**: e1000672.
57. Palsson B. Metabolic systems biology. *FEBS Lett* 2009;**583**:3900–4.
58. Mo ML, Palsson BO. Understanding human metabolic physiology: a genome-to-systems approach. *Trends Biotechnol* 2009;**27**:37–44.
59. Sigurdsson MI, Jamshidi N, Jonsson JJ, Palsson BO. Genome-scale network analysis of imprinted human metabolic genes. *Epigenetics* 2009;**4**:43–6.
60. Oberhardt MA, Palsson BO, Papin JA. Applications of genome-scale metabolic reconstructions. *Mol Syst Biol* 2009;**5**:320.
61. Shlomi T, Cabili MN, Herrgard MJ, Palsson BO, Ruppin E. Network-based prediction of human tissue-specific metabolism. *Nat Biotechnol* 2008;**26**:1003–10.
62. Duarte NC, Becker SA, Jamshidi N, Thiele I, Mo ML, Vo TD, et al. Global reconstruction of the human metabolic network based on genomic and bibliomic data. *Proc Natl Acad Sci USA* 2007;**104**:1777–82.
63. Jamshidi N, Palsson BO. Systems biology of SNPs. *Mol Syst Biol* 2006;**2**:38.
64. Li F, Thiele I, Jamshidi N, Palsson BO. Identification of potential pathway mediation targets in Toll-like receptor signaling. *PLoS Comput Biol* 2009;**5**:e1000292.
65. Chesler EJ, Miller DR, Branstetter LR, Galloway LD, Jackson BL, Philip VM, et al. The Collaborative Cross at Oak Ridge National Laboratory: developing a powerful resource for systems genetics. *Mamm Genome* 2008;**19**:382–9.
66. Churchill GA, Airey DC, Allayee H, Angel JM, Attie AD, Beatty J, et al. The Collaborative Cross, a community resource for the genetic analysis of complex traits. *Nat Genet* 2004;**36**:1133–7.
67. Bennett BJ, Farber CR, Orozco L, Kang HM, Ghazalpour A, Siemers N, et al. A high-resolution association mapping panel for the dissection of complex traits in mice. *Genome Res* 2010;**20**:281–90.
68. Grubb SC, Maddatu TP, Bult CJ, Bogue MA. Mouse phenome database. *Nucleic Acids Res* 2009;**37**:D720–30.
69. Bharucha KN. The epicurean fly: using *Drosophila melanogaster* to study metabolism. *Pediatr Res* 2009;**65**:132–7.
70. DiAngelo JR, Birnbaum MJ. Regulation of fat cell mass by insulin in *Drosophila melanogaster*. *Mol Cell Biol* 2009;**29**:6341–52.
71. Flynn III EJ, Trent CM, Rawls JF. Ontogeny and nutritional control of adipogenesis in zebrafish (*Danio rerio*). *J Lipid Res* 2009;**50**:1641–52.
72. Jones KT, Ashrafi K. *Caenorhabditis elegans* as an emerging model for studying the basic biology of obesity. *Dis Model Mech* 2009;**2**:224–9.
73. Mak HY, Nelson LS, Basson M, Johnson CD, Ruvkun G. Polygenic control of *Caenorhabditis elegans* fat storage. *Nat Genet* 2006;**38**:363–8.
74. Ruden DM, Lu X. Evolutionary conservation of metabolism explains how Drosophila nutrigenomics can help us understand human nutrigenomics. *Genes Nutr* 2006;**1**:75–83.

75. Schlegel A, Stainier DY. Lessons from "lower" organisms: what worms, flies, and zebrafish can teach us about human energy metabolism. *PLoS Genet* 2007;**3**:e199.
76. Watts JL. Fat synthesis and adiposity regulation in *Caenorhabditis elegans*. *Trends Endocrinol Metab* 2009;**20**:58–65.
77. Rada-Iglesias A, Ameur A, Kapranov P, Enroth S, Komorowski J, Gingeras TR, et al. Whole-genome maps of USF1 and USF2 binding and histone H3 acetylation reveal new aspects of promoter structure and candidate genes for common human disorders. *Genome Res* 2008;**18**:380–92.
78. Ansorge WJ. Next-generation DNA sequencing techniques. *N Biotechnol* 2009;**25**:195–203.
79. Zhu J, Zhang B, Smith EN, Drees B, Brem RB, Kruglyak L, et al. Integrating large-scale functional genomic data to dissect the complexity of yeast regulatory networks. *Nat Genet* 2008;**40**:854–61.
80. Wingren C, Borrebaeck CA. Antibody-based microarrays. *Methods Mol Biol* 2009;**509**:57–84.
81. Oresic M. Metabolomics, a novel tool for studies of nutrition, metabolism and lipid dysfunction. *Nutr Metab Cardiovasc Dis* 2009;**19**:816–24.
82. Gstaiger M, Aebersold R. Applying mass spectrometry-based proteomics to genetics, genomics and network biology. *Nat Rev Genet* 2009;**10**:617–27.
83. Voshol H, Ehrat M, Traenkle J, Bertrand E, van Oostrum J. Antibody-based proteomics: analysis of signaling networks using reverse protein arrays. *FEBS J* 2009;**276**:6871–9.
84. Shannon P, Markiel A, Ozier O, Baliga NS, Wang JT, Ramage D, et al. Cytoscape: a software environment for integrated models of biomolecular interaction networks. *Genome Res* 2003;**13**:2498–504.
85. Mehrabian M, Allayee H, Stockton J, Lum PY, Drake TA, Castellani LW, et al. Integrating genotypic and expression data in a segregating mouse population to identify 5-lipoxygenase as a susceptibility gene for obesity and bone traits. *Nat Genet* 2005;**37**:1224–33.
86. Wu S, Mar-Heyming R, Dugum EZ, Kolaitis NA, Qi H, Pajukanta P, et al. Upstream Please provide volume and page numbers for this reference. transcription factor 1 influences plasma lipid and metabolic traits in mice. *Hum Mol Genet* 2009;**19**:597–608.
87. Tu Z, Argmann C, Wong KK, Mitnaul LJ, Edwards S, Sach IC, et al. Integrating siRNA and protein–protein interaction data to identify an expanded insulin signaling network. *Genome Res* 2009;**19**:1057–67.

Genetics of Adipose Tissue Biology

INGRID DAHLMAN AND
PETER ARNER

Department of Medicine, Karolinska Institutet, Karolinska University Hospital, Huddinge, Stockholm, Sweden

Adipose tissue morphology and release of free fatty acids, as well as peptide hormones, are believed to contribute to obesity and related metabolic disorders. These adipose tissue phenotypes are influenced by adiposity, but there is also a strong hereditary impact. Polymorphisms in numerous adipose-expressed genes have been evaluated for association with adipocyte and clinical phenotypes. In our opinion, some results are convincing. Thus *ADRB2* and *GPR74* genes are associated with adipocyte lipolysis, *GPR74* also with BMI; *PPARG* and *SREBP1*, which promote adipogenesis and lipid storage, are associated with T2D and possible adiposity; *ADIPOQ* and *ARL15* are associated with circulating levels of adiponectin, *ARL15* also with coronary heart disease. We anticipate that the use of complementary approaches such as expression profiling and RNAi screening, and studies of additional levels of gene regulation, that is, miRNA and epigenetics, will be important to unravel the genetics of adipose tissue function.

Progress in Molecular Biology
and Translational Science, Vol. 94
DOI: 10.1016/S1877-1173(10)94003-9

1877-1173/10 $35.00

I. Human White Adipose Tissue in Health and Disease

Research in recent years has established that white adipose tissue (WAT) through secretion of free fatty acids (FFA), prostaglandins, and peptides (adipokines) with signaling functions is an important regulator of whole body metabolic homeostasis. WAT constitutes the major store of energy in the body. The amount of stored energy and WAT can vary enormously between subjects, that is, from subjects with cachexia to subjects with obesity, but also over time within a subject. This is associated with changes in WAT phenotypes that, besides possibly contributing to development of obesity, are believed to be of importance for development of insulin resistance and type 2 diabetes (T2D).[1,2] Clear-cut but rare examples that primary defects in the function of WAT can cause disease are mutations in the gene encoding the WAT hormone leptin, which result in extreme morbid obesity (described elsewhere in this book), and in the gene encoding peroxisome proliferator receptor gamma (PPARG), which cause disturbances in fat cell differentiation, partial lipodystrophy, as well as a strong predisposition to T2D.[3,4] WAT is composed of fat cells, that is, adipocytes, and the stroma vascular fraction that comprises endothelial cells, fat cell precursors, fibroblasts, and inflammatory cells, primarily macrophages; changes in each of these cellular fractions can be affected by obesity and play a primary role in the development of disease. The differences in metabolism between different adipose tissue depots are described in detail elsewhere and not further discussed here.[5]

Which adipose tissue phenotypes are implicated in the development of obesity and metabolic complications such as insulin resistance, fatty liver disease, dyslipidemia, and atherosclerosis? First, disturbances in lipid turnover, FFA mobilization, and adipose peptide secretion in obesity are believed to, directly or indirectly, cause insulin resistance in liver and skeletal muscle.[2,6] Obesity is associated with increased spontaneous breakdown of the triglycerides (TAGs) (lipolysis) within adipocytes in all fat depots. However, excess WAT is also associated with decreased catecholamine-stimulated lipolysis in subcutaneous adipose tissue and reduced antilipolytic effect of insulin; these are the main hormonal regulators of lipolysis in man. Second, adipose tissue hypertrophy is a risk factor for cardiovascular and metabolic disorders.[7,8] Among subjects with similar amounts of fat, the number of fat cells can vary substantially; hypertrophic adipose tissue is characterized by few and large fat cells as opposed to hypercellular tissue.[7] The risk associated with hypertrophic WAT may be mediated by altered adipokine secretion and enhanced lipid mobilization.[9,10] Third, it has recently been proven that adult humans have small but active brown adipose tissue (BAT) depots that can release energy as heat at amounts sufficient to affect whole body energy expenditure.[11,12] However, not all obese individuals develop obesity-related metabolic or cardiovascular disorders potentially because they maintain normal adipose tissue architecture and function.

Of what relevance are polymorphisms in genes expressed in adipose tissue for WAT and clinical phenotypes? Human genetic epidemiological studies have demonstrated a hereditary component for some WAT phenotypes, for example, lipolysis, adipokine secretion, and fat cell size.[13–15] Thus, there is clearly a genetic predisposition to fat cell phenotype. Furthermore, weight-reduced individuals display adipose hypercellularity and altered lipolysis, as well as lower leptin secretion compared to subjects who have never been obese, which indicate that genes regulating these adipose phenotypes may have a primary role in obesity development.[1,16] Gene polymorphisms that regulate WAT function can highlight critical regulatory steps, or bottlenecks, for WAT phenotypes implicated in development of obesity and metabolic complications. Of note, genes known to play important roles in WAT are not among susceptibility genes for obesity and T2D identified recently in genome-wide association studies (GWAS). However, genes regulating WAT function may be important for genetic susceptibility to obesity and metabolic complications because (a) GWAS studies have incomplete coverage of the human genome and have not been published for all disorders, (b) gene–environment interactions have hardly been investigated, and (c) WAT genes could modify the clinical phenotypes.

Human WAT can be investigated in different ways. Usually, tissue pieces or pure fat cells isolated from the whole tissue are incubated *in vitro* and metabolites, adipokine secretion, gene expression, signaling properties, or other aspects of function are analyzed. It is also possible to do *in vivo* studies. Lipolysis and adipokines can be investigated *in situ* with microdialysis. By cannulation of the vein-draining abdominal subcutaneous WAT, this depot can be examined directly by measuring arterio-venous differences in concentration of adipokine or other compounds. The latter techniques can also be combined with isotopes to study local lipid and glucose metabolism.

This chapter focuses on genetic regulation of adipose tissue phenotypes that in our opinion are the most clinically relevant, that is, lipid turnover and mobilization, secretion of peptides, morphology, and energy expenditure. We review the common variation in these traits in the population but not the causes of extreme monogenic forms of these traits. For each phenotype, we will describe key genes, their function, and evidence for association between polymorphisms in these genes and adipose tissue and/or clinical phenotypes. As regards clinical associations, we have focused on obesity and T2D. Importantly, the underlying genetic causes of adipose tissue dysfunction are not well defined. One reason is that cohorts for genetic studies of adipose tissue phenotypes are small, which results in low power and inconsistent results between genetic studies. Another factor to consider is that it is difficult to make valid independent confirmation since the methods to study human WAT differs between research laboratories. However, numerous papers report on analysis of association between polymorphisms in adipose-expressed genes and various traits. We have, therefore, been forced to set a limit for this chapter and have chosen to focus on genes that have been extensively studied, that is, there are either large or several genetic association reports for the gene. Thus, we by no way present an exhaustive list of genes investigated for impact on human adipose tissue function. We leave out genes for which all genetic association studies have been negative if these studies have had low power. The latter results are difficult to evaluate. We will focus on human studies since important species-specific differences exist in adipose tissue phenotypes. Furthermore, we will describe emerging evidence that specific established susceptibility genes for common obesity act primarily in WAT. Finally, we will summarize available results on gene regulation of WAT function and briefly mention potential future directions of research.

II. Genetics of Fat Cell Lipid Storage and Mobilization

A. Lipid Storage

Adipose tissue is an active lipid-storing organ; around 50–200 g of lipids are renewed daily. A small genetic impact on lipid storage can, therefore, have a large impact on TAG stores. TAGs arise in the adipocyte from three major routes: *de novo* lipogenesis from nonlipid precursors, uptake of FFA from the plasma, and reesterification of FFA derived from adipocyte lipolysis.[2] There is little evidence that common genetic variation controls the buildup of TAG in adipose tissue. Lipoprotein lipase (LPL) originating from adipocytes is responsible for hydrolyzing TAG in lipoprotein particles and thus releasing FFA that can be taken up from the adipose microcirculation by fat cells. The *LPL* gene is located on chromosome 8p22. Monogenic *LPL* deficiency causes hereditary hyperchylomicronemia syndrome.[17] Also common variation of adipose tissue LPL activity in the population is genetically regulated; there is a coherent within twin pair response in LPL activity following short-term overfeeding.[18] Polymorphisms in the *LPL* gene have been reported to be associated with plasma enzyme activity and insulin resistance; however, these results are preliminary and need to be replicated.[19,20] FFAs are translocated into fat cells and subsequently, within the cells, transported bound to fatty acid-binding protein 4 (FABP4, AP2). The *FABP2* gene is located on chromosome 8q21 and has a functional genetic variation, which results in reduced transcriptional activity and decreased adipose tissue gene expression.[21] A genetic study in a large population revealed that this variant is associated with lower serum TAG levels, as well as a significantly reduced risk for coronary heart disease and T2D compared with subjects homozygous for the wild-type allele.[21] As far as we know, there is no evidence of common gene polymorphisms that regulate specific steps in subsequent esterification of FFAs to TAGs.

However, susceptibility genes for obesity and T2D with unknown molecular function have been investigated for genetic association with lipogenesis. Calpain 10 (*CAPN10*) is a susceptibility gene for T2D that is encoded on chromosome 2q37 and has unknown function.[22] The *CAPN10* mRNA expression in subcutaneous fat has been reported to be significantly reduced in subjects with the UCSNP-43 G/G genotype.[23] The UCSNP-43 polymorphism has also been found to be linked to basal and insulin-stimulated lipogenesis, but not lipolysis, in human fat cells.[24] Under these experimental conditions, lipogenesis above all reflects insulin action on glucose transport. The results, therefore, suggest that *CAPN10* might be in control of insulin action on glucose metabolism.

B. Lipolysis

During lipolysis, intracellular TAGs undergo hydrolysis through the action of three major lipases: hormone-sensitive lipase (LIPE), adipose triglyceride lipase (ATGL), and monoacylglycerol lipase (MGL).[2] *In vivo* and *in vitro* studies have shown that the development or maintenance of increased adipose tissue stores might be promoted by a blunted lipolytic response after beta-adrenergic stimulation.[2] In this context, it is noteworthy that there is strong evidence for a genetic impact on human fat cell lipolysis. Thus, there is a significant within pair resemblance in basal and epinephrine-induced lipolysis in monozygotic twins.[13,14] Furthermore, the significant within twin pair resemblance in changes of basal and epinephrine-induced lipolysis in response to overfeeding and very low calorie diet is compatible with genotype–environment interaction[14,18,25] and indicates that some individuals might be particularly prone to develop metabolic disturbances in response to overfeeding.

Polymorphisms in what genes may explain the hereditary impact on adipocyte lipolysis? Considerable progress has been made in elucidating the signals and steps that regulate lipolysis in humans (Fig. 1). Hormones exert a tight control on lipolysis; catecholamines and heart-derived natriuretic peptides stimulate, and insulin inhibits, lipolysis via receptors on the fat cell surface. Four catecholamine receptors are expressed on adipocytes; stimulatory beta 1 (ADRB1), beta 2 (ADRB2), and beta 3 (ADRB3) adrenoreceptors; and inhibitory alpha 2 (ADRA2) adrenoreceptors. Upon ligand-binding, beta-adrenoreceptors associate with stimulatory G proteins (Gs), which activate adenylate cyclase (AC) which converts adenosine triphosphate (ATP) to cyclic adenosine monophosphate (cAMP). cAMP in turn stimulates protein kinase A (PKA),

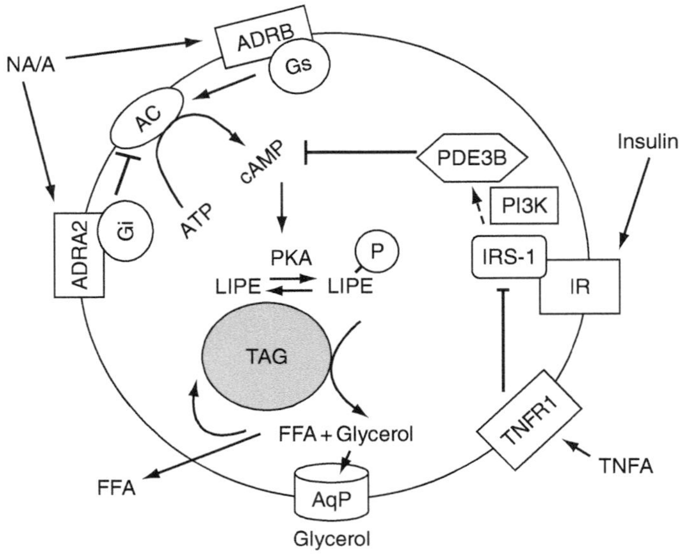

FIG. 1. Regulation of adipocytes lipolysis. A, adrenaline; AqP, aquaporin; IR, insulin receptor; IRS-1, insulin receptor substrate 1; NA, noradrenaline; P, phosphate; P13K, phosphoinositide 3-kinase; PDE3B, phosphodiesterase 3B. Remaining molecules are presented in the text.

which phosphorylates and activates hormone-sensitive lipase (LIPE) and translocates perilipin (PLIN) from the lipid droplet. LIPE hydrolyzes TAGs and diacylglycerides. Inhibitory ADRA2s are coupled to inhibitory G proteins Gi. Natriuretic peptides stimulate fat cell lipolysis via natriuretic peptide receptors that signal via a protein kinase G pathway that phosphorylates and activates LIPE. ATGL is a novel lipase that has been shown to participate in regulation of fat cell lipolysis. Little is known about the signaling pathways that regulate ATGL. However, similar to LIPE, ATGL is activated by catecholamines. ATGL hydrolyses TAGs to diglycerides more efficiently than LIPE. Additional local factors of importance for regulating lipolysis include proteins on the surface of the lipid droplet, for example PLIN, adipose differentiation-related protein (ADRP), fat-specific protein 27 (FSP27, CIDEC), cell death-inducing DFFA-like effector A (CIDEA), and CGI-58, which interacts with ATGL.[2]

Considering the physiological importance of catecholamines in regulation of fat cell lipolysis, it is not surprising that polymorphisms in the adrenoreceptor genes have been extensively studied regarding *in vitro* function and impact on lipolysis. The *ADRB1*, *ADRB2*, and *ADRB3* genes each harbor common polymorphisms that cause amino acid changes and might be of importance for gene expression or receptor properties. These adrenoreceptor polymorphisms

TABLE I
NONSYNONYMOUS BETA-ADRENORECEPTOR POLYMORPHISMS

Position	AA substitution	Putative regulatory site	Adipocyte phenotype
ADRB1			
145[a]	Ser49Gly	N-terminus	Unknown
1165	Gly389Arg	C-terminus	No impact on lipolysis
ADRB2			
-47	Arg-19Cys	Beta upstream peptide	
46	Arg16Gly	N-terminus	Higher potency of TER in inducing lipolysis
79	Gln27Glu	N-terminus	No impact on lipolysis
491	Thr164Ile	Transmembrane domain	Lower potency of TER in inducing lipolysis
ADRB3			
827	Trp64Arg	Intracellular loop	Unclear if impact on lipolysis

[a]Position relative to start codon. AA, amino acid; TER, terbutaline, that is, beta-2 adrenoreceptor-specific agonist.

could impact cell metabolism and hereby contribute to the pathophysiology of obesity and related metabolic diseases. The common beta-adrenoreceptor polymorphisms are summarized in Table I. There is evidence from *in vitro* studies that some of these receptor variants might be important for catecholamine-induced lipolysis in human adipocytes. A more detailed summary, including nonfat cell *in vitro* phenotypes associated with the variants and additional rare mutations, has been published.[26]

The *ADRB1* gene is encoded by an intronless gene on human chromosome 10q24–q26. Two common polymorphisms have been detected in the gene: in the N-terminus, Ser49Gly with minor allele frequency (MAF) (Gly) about 15% in most populations, and in the C-terminus Gly389Arg with varying MAF in different populations. Ser49Gly and Gly389Arg display linkage disequilibrium (LD). The Gly49 allele is more susceptible to agonist induced downregulation than Ser49, but has to our knowledge not been investigated for impact on fat cell function. As regards Gly389Arg, the Arg389 allele exhibits higher basal and maximal isoprenaline-induced activation of AC than the Gly389 allele. Isoprenaline is a nonselective beta-adrenoreceptor agonist. However, the Gly389Arg polymorphism has no impact on lipolytic response to *ADRB1* stimulation in human adipocyte cultures. As for clinical impact, the *ADRB1* gene Gly389Arg polymorphism was initially reported to be associated with obesity.[27,28] However, a more recent investigation in Danes including ~7600 subjects concluded that the Arg389Gly polymorphism does not have any clinically important impact on the pathogenesis of obesity.[29] Finally, Ser49Gly, but not Gly389Arg, has in one study been reported to predict long-term changes in body weight.[26,30]

The *ADRB2* gene is encoded by an intronless gene on chromosome 5q31–q32. Several polymorphisms have been described in the human *ADRB2* gene. Three common polymorphisms result in the substitution of an amino acid and affect the function of the beta 2 adrenoreceptor *in vitro*. Additional polymorphisms have been detected in the 5′ region of the gene that contains transcriptional regulatory activity. One polymorphism is located at codon 16 and substitutes an arginine for a glycine (Arg16Gly). Another polymorphism is located at codon 27 and substitutes glutamic acid for glutamine (Gln27Glu). Both these variants are common and are located in the extracellular amino-terminal region of the beta 2 adrenoreceptor. *In vitro*, these polymorphisms alter cellular trafficking and influence desensitization of the receptor. Codon 164 of the *ADRB2* gene harbors a rare polymorphism resulting in the substitution of isoleucine for threonine (Thr164Ile). Thr164Ile is located in the receptor transmembrane-spanning domain. This polymorphism influences agonist binding and decreases coupling of the Gs protein to the receptor, which results in reduced basal and agonist induced activation of AC. The 5′ region of the *ADRB2* gene includes a short open reading frame for a leader peptide that affects *ADRB2* gene expression. A polymorphism within the leader peptide region at position -47 from the start codon, which substitutes an arginine for a cystidine, possibly modulates receptor translation and has a clear impact on receptor density.[26]

Among adrenoreceptors, *ADRB2* is believed to be the most important one for regulation of human fat cell lipolysis.[2] As regards the Arg16Gly polymorphism, the Gly16 allele displays increased agonist sensitivity and evokes a stronger lipolytic response *in vitro* in abdominal subcutaneous adipocytes from female donors than the Arg16 allele.[31] According to one study, this polymorphism has no effect on basal and maximum lipolysis or glycerol release.[31] More recently, the Arg16 allele has been shown to be associated with a blunted increase in circulating FFA in both genders following stimulation with the beta-adrenoreceptor agonist isoprenaline.[32] In women only, a blunted increase in circulating glycerol levels and decreased fat oxidation was observed. The cause of the discrepancy between these studies as regards the results for glycerol release is unclear. The Gln27Gly polymorphism has no impact on either basal or beta-2 adrenoreceptor agonist stimulated lipolysis.[31] As regards the Thr164Ile polymorphism, the potency of a beta-2 adrenoreceptor selective agonist to induce lipolysis in human subcutaneous fat cells has been reported to be lower in patients with the Thr164Ile genotype than in those with Thr164Thr.[26] By contrast, in the same cohort, basal and maximal lipolysis is unaltered. Similar results for the Thr164Ile polymorphism have been reported in isolated human mast cells.[26]

Several groups have investigated the common *ADRB2* Arg16Gly and Gln27Glu polymorphisms for association with obesity, hypertension, and T2D with inconsistent results.[26] More recent studies, which include substantially larger patient materials than earlier studies, have provided results that are more informative as regards the importance of these polymorphisms in the etiology of common metabolic diseases in the general population. Thus, in one Danish study including 7800 unrelated middle-aged white people, there was no consistent evidence for association of the *ADRB2* receptor variants Arg16Gly and Gln27Glu with obesity or hypertension.[33] However, both the Gly allele of the Arg16Gly polymorphism and the Glu allele of the Gln27Glu variant displayed nominal association with systolic blood pressure. Furthermore, there was a nominal association between the Arg16 allele frequency and T2D. A meta-analysis for association between Glu27Gln and the Arg16Gly polymorphisms and obesity has been performed in 10,404 (Glu27Gln) and 4,328 (Arg16-Gly) subjects after excluding samples where controls displayed Hardy–Weinberg disequilibrium.[34] The summary odds ratio (OR) from overall meta-analysis showed no association between the *ADRB2* polymorphisms and obesity. However, in ethnic groups with low Glu27 allele frequency, that is, Asians, Pacific Islanders, and American Indians, the ORs revealed a significant obesity risk associated with Glu27Gln. Thus, the Glu27 allele in the *ADRB2* gene might be a significant risk factor for obesity in Asians, Pacific Islanders, and American Indians, but not in Europeans.

Among subjects in a given population, specific alleles at nearby chromosomal polymorphisms tend to occur in specific combinations, so-called haplotypes. Sometimes, haplotypes display phenotypic associations that cannot be deduced to specific polymorphisms. This may either be due to as yet unidentified causative variants, or functional interactions between polymorphisms constituting the haplotype. *ADRB2* gene haplotypes, which are constituted of unique genotype combinations at 13 polymorphisms in the gene, influence *ADRB2* gene expression and receptor density.[35] Haplotypes based on 10 of these polymorphisms have been determined in 141 overweight or obese women and related to subcutaneous fat cell lipolysis in this population.[36] Common *ADRB2* haplotypes differ markedly in sensitivity of lipolysis to catecholamines and ADRB2 agonists, as well as maximum catecholamine-induced lipolysis. There is no influence of *ADRB2* gene haplotypes on lipolytic sensitivity to ADRB1, ADRB3, or ADRA2 agonists, which supports that *ADRB2* haplotypes specifically affect ADRB2 function. One of the *ADRB2* gene haplotypes has been reported to protect against obesity in women, but not in men.[37]

The human *ADRB3* gene is encoded on chromosome 8p12–p11.2. *ADRB3* may, according to *in vitro* results, primarily be of fundamental importance for regulating lipolysis in human omental adipose tissue.[38] A common *ADRB3* variant in the first intracellular loop, Trp64Arg, has been extensively studied in connection with common metabolic traits after it was reported in 1995 that the Arg64 allele associated with abdominal obesity, increased weight-gain, insulin resistance, and age at onset of T2D.[26] The functional properties of the Trp64Arg polymorphisms are unclear.[26] Different results have been reported regarding the impact of Trp64Arg on lipolysis in human subcutaneous and visceral adipocytes; the Arg64Arg genotypes have been associated with no or decreased sensitivity of lipolysis to a beta-3 adrenoreceptor-specific ligand, with no impact on basal or maximal lipolysis.[26] Finally, Trp64Arg has been reported to be associated with basal metabolic rate.[26]

As for clinical associations, the *ADRB3* Trp64Arg polymorphism has previously been examined for association with obesity, insulin resistance, and T2D with various outcomes.[26] As is often the case, small sample sizes make it difficult to draw any definitive conclusions. More importantly, in a recent large study encompassing 7605 Danish subjects, Trp64Arg was not associated with obesity, but the Arg64 allele was nominally associated with T2D, OR 1.15 (CI: 1.01–1.31), with increased insulin resistance as estimated by homeostasis model assessment (HOMA), and higher levels of plasma glucose 2 h after glucose ingestion.[39] Furthermore, the same authors have performed a meta-analysis including a total of 27 studies and 18,891 subjects in which the Arg16 allele was borderline associated with T2D ($P = 0.06$).[39] In addition, another group performed a meta-analysis of 97 studies involving 44,833 individuals and reported a significant association between the Arg64Trp gene variant and BMI; Arg64 allele carriers have 0.24 kg/m^2 higher BMI compared with noncarriers.[40] This meta-analysis also revealed substantial heterogeneity among the studies. The difference in BMI between genotypes in East Asians was almost four times larger than that in Europeans in whom no significant association was observed. The Trp64Arg ADRB3 genetic variant might thus be associated with BMI in East Asians, but not in Europeans. Genetic variation in *ADRA2* is, to our knowledge, not implicated in regulation of metabolic traits including fat cell function.

G proteins relay information from cell surface receptors to intracellular effectors. G proteins are heterotrimers composed of three subunits labeled alpha, beta, and gamma, each having a number of isoforms. Stimulating beta-adrenoreceptors are coupled to the G protein Gs. The guanine nucleotide-binding protein beta 3 (*GNB3*) gene on human chromosome 12p13 encodes one of the G protein beta subunits. A common polymorphism, 825C>T, in the *GNB3* gene induces an alternative splicing of exon 9 that results in a 41 amino acid shorter active peptide called Gbeta3-s.[41] Functional studies suggest that

Gbeta3-s enhances G protein signaling.[41] However, the 825T variant is associated with a decreased amount of Gbeta3 protein and blunted lipolysis in human fat cells.[42] The GNB3 825C>T polymorphism has been investigated for association with hypertension, obesity, and T2D with conflicting results as summarized.[43] In a recent larger study encompassing 7518 white Danish subjects, the 825C>T variant was not associated with hypertension, obesity, T2D, or the metabolic syndrome after correcting for multiple comparisons.[43] A meta-analysis of hypertension has been performed by the authors of the Danish study, which demonstrated a high degree of heterogeneity between study populations of different ethnic origin. In the meta-analysis, there was a tendency toward an increased risk of hypertension among 825T allele carriers, but this result was not statistically significant.

The guanine nucleotide-binding protein alpha stimulating (*GNAS*) gene on chromosome 20q13.2 encodes the alpha subunit of the stimulatory G protein, Gs-alpha. *GNAS* is a complex imprinted locus that produces multiple transcripts through the use of different promoters and alternative splicing. The G-alpha protein system participates in the regulation of lipid mobilization. Thus, a G>A transition at position -1211 in the human *GNAS* promoter that results in altered upstream stimulatory factor 1 (USF1) transcription factor binding has been reported. This variant affects Gs-alpha expression and lipolysis. In a preliminary study, carriers of the -1211GG genotype lost significantly more weight compared with A allele carriers during low-calorie diet.[44]

Next to catecholamines, atrial natriuretic peptide (ANP) is the best described stimulator of human fat cell lipolysis. ANP is released from the myocardium as a prehormone that is cleaved into proANP, which subsequently itself is cleaved into the biologically active ANP. ANP stimulates lipolysis in human fat cells through a cyclic guanosine monophosphate (cGMP) signaling pathway.[45] The physiological role of ANP is not known. It is believed, though, that the molecule is of importance for maintenance of lipolysis during exercise.[2] The *ANP* gene is encoded on chromosome 1p36.2. One common polymorphism in the *ANP* gene promoter region, T-381C (rs198389), has been investigated for association with T2D. A number of large case-control studies have reported a protective effect of the CC genotype against T2D.[46,47] A meta-analysis with a total of 49,279 subjects showed a modest protective effect of the CC genotype on T2D risk, which is consistent with previous studies.[47]

Additional peptides have been shown to regulate fat cell lipolysis *in vitro*; for example, neuropeptides inhibit human adipocyte lipolysis *in vitro*; their role *in vivo* being unknown.[48] The cell surface G protein-coupled receptor 74 (GPR74) displays high affinity for neuropeptides, although the endogenous ligand is unknown. In humans, GPR74 is predominantly expressed in the brain, heart, and adipose tissue. In human fat cells, GPR74 stimulation and inhibition cause a marked decrease and increase, respectively, of lipolysis, which is linked to catecholamine stimulation through beta-adrenoreceptors.[49] We have detected a common haplotype in the *GPR74* gene that in three independent samples is associated with leanness, and with enhanced human adipocyte lipolysis.[49]

Hormone-sensitive lipase (LIPE) is traditionally considered the rate-limiting step for basal and stimulated lipolysis in humans. Following stimulation by catecholamines or ANP, PKA phosphorylates LIPE resulting in its translocation from the cytosol to the lipid droplet. Obese subjects display decreased adipose tissue expression and activity of LIPE, which is in agreement with a blunted response of lipolysis to catecholamines in obesity.[2] The *LIPE* gene is encoded on chromosome 19q13. An intronic repeat in the *LIPE* gene has in human primary fat cell cultures been associated with blunted *in vitro* response of lipolysis to catecholamines.[50] This repeat has no known function. The intronic repeat, as well as a functional polymorphism in the promoter, has been investigated for impact on obesity and related traits with inconsistent results as summarized.[51] ATGL (PNPLA2) is the only other described triglyceride lipase of importance for TAG hydrolysis in human fat cells.[2] ATGL seems more important than LIPE for basal lipolysis. In contrast to LIPE, the adipocyte levels of ATGL are not influenced by adiposity. The *ATGL* gene is encoded on chromosome 11p15.5. Mutations in *ATGL* and its coactivator CGI-58 (*ABHD5*) are causes of neutral lipid storage diseases, which are characterized by the presence of TAG-containing cytoplasmic droplets in many organs but not obesity.[52,53] By contrast, polymorphisms in these genes, which only recently have been shown to be of importance for human adipocyte function, have to our knowledge so far not been investigated for impact on adipose tissue lipolytic regulation or common clinical metabolic disturbances.

Gene variants of transcription factors, which control expression of genes in the lipolytic pathway, could also be of importance for lipolytic regulation. USF1 regulates the expression of several genes involved in lipid and glucose metabolism, among them genes regulating lipolysis. USF1 specifically controls the expression of *LIPE* in adipocytes. The human *USF1* gene is encoded on chromosome 1q22–q23. Interestingly, a common intronic variant in the *USF1* gene is associated with increased response of lipolysis to catecholamines in fat cells.[54] In another study, the same *USF1* variant has been reported to be associated with higher *in vivo* antilipolytic insulin sensitivity.[55] The cause of these seemingly incongruous results is unknown.

Proteins covering lipid droplets in adipocytes participate in regulation of lipolysis. PLIN is the most extensively studied lipid droplet protein[2] and inhibits basal lipolysis. Lipolytic stimuli, including catecholamines and ANP, activate PKA and PKG, which phosphorylate PLIN. Following phosphorylation, PLIN dissociates from the lipid droplet hereby facilitating translocation of LIPE to the droplet in order to initiate hydrolysis of TAGs. PLIN thus coordinates lipid storage and release and is, therefore, a strong candidate for genetic studies. The predominant isoform of PLIN found in mature human adipocytes is PLIN A. Adipocyte levels of PLIN protein are inversely correlated to lipolytic rate. Furthermore, obesity, which is associated with resistance to catecholamine lipolysis, is also associated with increased levels of PLIN in adipose tissue. The *PLIN* gene is encoded on chromosome 15q26. An intronic *PLIN* allele has been associated with higher basal and catecholamine-induced lipolysis, as well as with reduced *PLIN* content in fat cells.[56] However, no function has been ascribed to this polymorphism. Additional common silent *PLIN* gene variants have been reported to be associated with BMI, but the results are preliminary and need to be replicated in additional larger cohorts.[57] Recent results have highlighted the importance of additional lipid droplet proteins in regulation of human lipolysis, for example, the cell death-inducing DNA fragmentation factor A-like effector (CIDE) family members FSP27 and CIDEA.[58–60] CIDEA has additional important metabolic functions in human fat cells including substrate oxidation switch.[61] Common *CIDEA* gene polymorphisms have been investigated for association with obesity phenotypes with varying results in different samples; however, investigated samples were not large.[62,63]

Among other genes of potential importance for lipolysis in humans is aquaporin 7 (*AQP7*) on chromosome 9p13, which encodes a glycerol channel protein primarily expressed in adipose tissue. A functional promoter polymorphism in this gene is associated with reduced adipose tissue *AQP7* mRNA levels and increased BMI in two cohorts.[64]

C. Energy Expenditure

As mentioned earlier, adult humans have functional BAT.[11,12] BAT increases energy expenditure by expression of uncoupling protein 1 (UCP1). UCP1 is a protein located in the inner mitochondrial membrane, which allows protons to reenter the matrix, bypassing the ATP synthase. The dissipation of the proton electrochemical gradient leads to an uncoupled and accelerated respiration and heat production as reviewed.[65] Human BAT in adults is difficult to isolate for protein quantification. There is no easy method to directly measure adipose tissue energy expenditure that can be used in large-scale studies. It is, as far as we know, unknown whether there is a hereditary impact on expression of genes regulating mitochondrial function in human adipose tissue. However, it has been shown that mRNA levels of *UCP1* in WAT are downregulated in morbid obese humans.[66] The *UCP1* gene is encoded on chromosome 4q31. Common genetic variations in the *UCP1* gene have been investigated for impact on obesity, T2D, and related metabolic disturbances with inconclusive results.[65] Most studied cohorts have been small, which probably is an important reason for the different results reported in these studies.

III. Genetics of WAT Endocrine Function

WAT secretes peptide hormones and inflammatory mediators, together termed adipokines, that through para/autocrine and endocrine action participate in regulation of local as well as systemic metabolic homeostasis. For hormones that are uniquely produced in adipose tissue, that is, leptin and adiponectin, levels in the general circulation indirectly reflect adipose tissue production. However, other factors such as plasma clearance are also important determinants of the circulating levels of these hormones. Hormone levels in serum are relatively easy to determine in large cohorts. As for genetic analysis of adipose tissue endocrine function, leptin and adiponectin are, therefore, the most extensively studied. Below is a description of the most important human adipokines, in our opinion, which have been examined for genetic regulation.

The primary function of the adipose tissue-secreted hormone leptin is to give information about nutritional status to the hypothalamus and to limit food intake, but leptin also has peripheral effects.[4] Monogenic defects in the leptin signaling pathway are described elsewhere in this book. Importantly, variation in leptin levels in the population is partly under genetic influence. According to genetic epidemiological studies, heritability estimates for leptin levels in the general circulation range from 0.28 to 0.55.[67,68] Furthermore, a substantial proportion of the genetic effects is common to leptin levels and obesity.[69] Since mutations in the *LEP* gene on chromosome 7q31.3 were identified as a cause of monogenic obesity, common polymorphisms in the *LEP* gene have been extensively studied for impact on adiposity. A number of common polymorphisms in the 5′ regulatory region have shown association with *LEP* levels or obesity in several studies.[70] Most studies of *LEP* gene variants are of limited size and it is, therefore, difficult based on these results to draw any definitive conclusions regarding the impact of common *LEP* variants on circulating leptin levels and adiposity. Surprisingly, there is to our knowledge no more recent large (involving several thousand subjects) genetic studies of the human *LEP* gene, nor are there any recent meta-analyses of published studies.

Adiponectin (APM1, ADIPOQ) is an adipokine with an unclear role in humans, but rodent studies provide clear evidence that ADIPOQ participates in regulation of energy homeostasis and has a beneficial impact on glucose and lipid metabolism.[6] A direct effect of ADIPOQ in humans remains to be firmly established, for example, by *in vivo* administration. In some but not all studies,

ADIPOQ levels are found to be reduced in human obesity and insulin-resistant states including T2D and upregulated following weight loss.[6] Furthermore, ADIPOQ has also been demonstrated in rodent models to have anti-inflammatory and anti-atherogenic properties, and is independently associated with cardiovascular disease.[71] In the circulation, ADIPOQ exists as different multimers. In particular, high molecular weight ADIPOQ tracks correlates of total ADIPOQ.[72] The heritability of circulating ADIPOQ levels has in several genetic epidemiological studies been estimated to be strong, in general between 46% and 88%.[15,73] Importantly, there is also evidence for common quantitative trait loci for ADIPOQ levels and obesity-related phenotypes.[15] The *ADIPOQ* gene is located on chromosome 3q27. Several early studies have investigated association between common *ADIPOQ* polymorphisms and circulating levels of ADIPOQ, as well as obesity, T2D, and cardiovascular disease, but with inconsistent results between different samples.[74] A recent meta-analysis of three GWAS for circulating ADIPOQ levels ($n = 8531$) with validation of the lead associated polymorphisms in additional cohorts ($n = 6202$) confirmed that polymorphisms at the *ADIPOQ* locus demonstrate the strongest associations with ADIPOQ levels.[71] The polymorphism most strongly associated with circulating adiponectin levels lies 30 kb upstream of the *ADIPOQ* gene. None of the polymorphisms at the *ADIPOQ* locus demonstrates a robust relationship with BMI, T2D, and HOMA as a measure of insulin resistance, or cardiovascular disease. The same meta-analysis has identified a variant in a novel gene, ADP-ribosylation factor-like 15 (*ARL15*), which is associated with lower circulating levels of ADIPOQ. This risk allele at ARL15 is also associated with a higher risk of coronary heart disease and, nominally, an increased risk of T2D and several metabolic traits. Expression of *ARL15* was highest in skeletal muscle, which does not synthesize ADIPOQ, but is the main site of insulin-mediated glucose disposal. These findings suggest that *ARL15* may be involved in cellular insulin resistance and/or ADIPOQ trafficking and secretion.

Retinol-binding protein 4 (RBP4) is a hormone secreted primarily by visceral fat, liver, and muscle that in rodents induces insulin resistance in liver and muscle.[6] Elevated plasma RBP4 levels have been observed in humans with insulin resistance and T2D. In a genetic epidemiological study, the heritability for plasma RBP4 levels has been reported to be 63%.[75] However, the inverse correlation between plasma RBP4 and insulin sensitivity disappeared after correction for covariates indicating that elevated plasma RBP4 in T2D is a secondary phenomenon and that plasma RBP4 plays at most a minor role in the development of insulin resistance in humans. Thus, the role of RBP4 in development of metabolic disturbances humans is unclear. However, common genetic variations in the *RBP4* gene have been indicated in susceptibility to T2D and insulin resistance, possibly through an effect on *RBP4* expression.[76]

Adipose tissue releases a number of proinflammatory peptides implicated in regulation of insulin signaling and lipolysis including interleukin 6 (IL6), tumor necrosis factor alpha (TNFA), and monocyte chemoattractrant factor 1 (MCP1).[77–79] Several secreted inflammatory mediators seem to predominantly originate from macrophages in the stroma vascular fraction of adipose tissue. The number of macrophages in adipose tissue correlates with both adipocyte size and body mass and is believed to be of pathogenic importance for development of obesity-associated insulin resistance.[80] Adipose macrophage-secreted factors may either induce insulin resistance by systemic effects, that is, impairment of insulin signaling in liver and muscle, or by local inhibition of expression of adipogenic transcription factors and adipogenesis, which leads to adipose tissue hypertrophy.[81]

Human subcutaneous adipose tissue secretion of the cytokine IL6 correlates with BMI.[78] Human adipocytes express the IL6-receptor system consisting of the IL6 receptor (IL6R) and the signal transducing protein (IL6ST, gp130), which suggests that IL6 has auto- and paracrine effects.[82] It is established that IL6 stimulates lipolysis in human adipocytes and exerts anti-insulin actions.[82] Furthermore, treatment with IL6 downregulates various adipogenic markers, which suggests that IL6 inhibits adipogenesis.[82] The IL6-signaling complex has not been investigated for direct genetic influence on various adipose tissue phenotypes, only on its possible consequences for obesity and T2D. The *IL6* gene is located on chromosome 7p21. A tag-SNP haplotype analysis of the entire *IL6* gene in two large independent cohorts ($n = 3100$) has indicated that variability in the *IL6* gene is significantly associated with adiposity, but not T2D or circulating IL6 levels.[83] Furthermore, according to a meta-analysis of 27,000 subjects, the association between the *IL6* gene and adiposity is unlikely to be caused by the extensively studied *IL6* promoter polymorphism -174G>C.[83] A tag-SNP haplotype analysis of $n = 5900$ subjects has not revealed significant association between the *IL6* gene and T2D.[84] Furthermore, according to a meta-analysis of $n = 17{,}000$, there is no association between the promoter polymorphism -174G>C and diabetes risk.[84] The *IL6R* gene is encoded on chromosome 1q21.3. Inconsistent results regarding the association between *IL6R* gene variants, obesity, and T2D have been reported. *IL6R* Asp358Ala substitution has been associated with obesity in Pima Indians, but this was not confirmed in a large Danish study.[85,86] However, the Danish study has reported that *IL6R* Asp358Ala is associated with T2D risk. A tag-SNP haplotype analysis of the *IL6R* gene in another large cohort has not confirmed an association between *IL6R* gene variants, including Asp358Ala, and T2D.[87] Nominal associations between variants in the *IL6ST* gene and components of the metabolic syndrome have been reported, but these results need to be confirmed in larger studies.[88]

Adipose tissue expression of TNFA is increased in obesity and insulin-resistant states.[77] In humans, in contrast to rodents, the TNFA produced within WAT is not released into the systemic circulation and must, therefore, act locally.[78] Actually, treatment with TNFA *in vitro* has profound impact on adipocyte function including attenuation of insulin signaling, stimulation of lipolysis, and inhibition of fatty acid oxidation and adipogenesis.[2] These effects are primarily mediated via TNF receptor 1 (TNFR1).[2] TNFA stimulates lipolysis by several mechanisms: downregulating expression and phosphorylation of PLIN; inhibiting insulin signaling resulting in attenuation of the antilipolytic action of insulin.[2] The *TNFA* gene is encoded on chromosome 6p21.3 and harbors several common polymorphisms in the proximal promoter. Three of these polymorphisms have been indicated to affect expression on the gene *in vitro*, -863C/A, -308G/A, and -238G/A.[89–91] The A allele at -863C/A, but not other promoter polymorphisms, has in a limited study been reported to be associated with reduced secretion of TNFA from human subcutaneous adipose tissue pieces *in vitro* in lean but not obese subjects.[92] The same -863A allele has also been associated with lower serum TNFA levels.[89] However, this polymorphism has no known impact on clinical phenotypes. The *TNFA* promoter polymorphism -308G/A is the one that has been most extensively evaluated for impact on obesity. The G allele at -308G/A has in a recent meta-analysis been associated with a modest increased risk of developing obesity, OR 1.23 in a total of 3562 individuals from eight studies.[93] In additional meta-analyses, the same promoter polymorphism has been associated with systolic arterial blood pressure in a total of 1624 individuals from four studies, and plasma insulin levels in a total of 3720 subjects from 16 studies.[93] The -238G/A and -857C/A polymorphisms do not associate with obesity.[94,95]

MCP1 is secreted by adipocytes and macrophages and is believed to promote macrophage infiltration into adipose tissue in obese humans.[79,96] The *MCP1* (*CCL2*) gene is located on chromosome 17q11.2–q12. As far as we know, only one study has investigated the impact of *MCP1* gene polymorphisms on metabolic phenotypes. A polymorphism in the distal regulatory region of *MCP1* is associated with circulating MCP1 levels, insulin sensitive and T2DM, but not with obesity.[97]

IV. Genetics of Fat Cell Morphology

Obesity is accompanied by an increase in fat cell size as well as number; their relative importance differs markedly between individuals. Thus, some obese have fat cell hypertrophy, that is, few large fat cells, and others hyperplasia, that is, many small fat cells.[7] A similar distinction is observed in lean adults.[7] Total adipocyte number in adults is a major determinant for fat mass.

In addition, the number of fat cells stays constant in adulthood even after marked weight loss, indicating that the number of adipocytes is set during adolescence.[16] It has recently been shown that around 10% of fat cells are renewed annually and that the absolute number of new adipocytes generated each year is 70% lower ($p < 0.001$) in hypertrophy than in hyperplasia, which suggest that primary defects in adipogenesis could be an important factor for determining fat cell morphology and its complications.[7] Adipose tissue hypertrophy is a risk factor for cardiovascular and metabolic disorders.[7,8] This has been related to altered adipokine secretion and enhanced lipid mobilization, as well as to limited opportunities for adipose tissue expansion, which results in TAG in liver and skeletal muscle and metabolic disturbances in these organs.[9,10,98]

Based on the above observations supporting important clinical effects of the relation between fat cell size and number, it is noteworthy that there is a hereditary impact on fat cell size, which possibly reflects variation in adipogenesis. Thus, in 46 investigated monozygotic twin pairs, there is a significant within pair resemblance in adipocyte diameter.[13] This is the largest published study of human fat cell morphology. The primary genetic regulators of adipose tissue morphology in humans are unknown, but genes regulating adipogenesis are obvious candidates. The transcriptional control of adipogenesis involves the sequential activation of a series of transcription factors; CCAAT/enhancer-binding proteins (CEBP) alpha/beta/gamma stimulate peroxisome proliferator-activated receptor gamma (PPARG) expression. PPARG in turn induces expression of genes giving the fat cell phenotype and promoting TAG storage.[98] Sterol regulatory element-binding protein (SREBP) 1c is another transcription factor that stimulates adipogenesis and TAG storage. A host of other genes have been shown to regulate adipogenesis in experimental models, but their role for human adipogenesis *in vivo* is incompletely known (Fig. 2).[98] Polymorphisms in genes regulating adipogenes have primarily been investigated for impact on adiposity and T2D due to lack of large cohorts with adipose tissue morphology phenotypes. Below follow a summary of key findings from association studies of adipogenesis candidate genes.

Rare monogenic disorders of adipogenesis cause lipodystrophy with severe disturbances in development of some or all WAT depots. The *PPARG* gene is located on chromosome 3p25. Dominant negative mutations in the DNA- and ligand-binding domain of the *PPARG* gene is one cause of partial lipodystrophy. Affected family members manifest loss of distal limb and gluteal fat, although abdominal adipose tissue is relatively preserved.[3] This disorder is commonly associated with severe insulin resistance, early onset T2D, and marked dyslipidemia. The metabolic disturbances in lipodystrophy have been linked to the inability of adipose tissue to trap and store FFA and underscores how important the proper function of WAT is for whole body insulin sensitivity.[99]

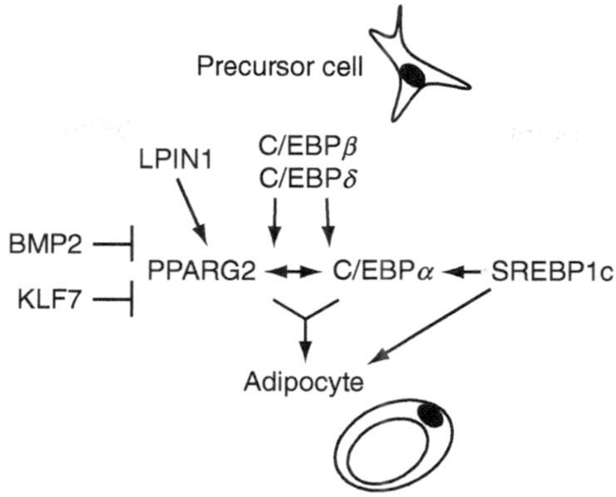

FIG. 2. Regulation of adipogenesis. Only factors discussed in the text are shown.

A rare gain-of-function mutation in *PPARG* was originally associated with extreme obesity (Gln115–Pro115), in agreement with the hypothesis that predisposition to store fat is a cause of obesity.[100] However, this result has not been confirmed in subsequent studies.[3,101]

PPARG exists in three isoforms differing in their 5′ sequence. *PPARG2* is the most abundant isoform in human adipose tissue, and it is the only isoform, that is, exclusively expressed in this organ.[102] BMI has been reported to increase, or to have no impact on, PPARG expression in subcutaneous adipose tissue.[103,104] A common coding gene variant, Pro12Ala which is located in the 5′ activity function region, that is, specific for the PPARG2 isoform is the most frequently found genetic variant of PPARG. Functional studies of Pro12Ala have shown a modest reduction of target gene transactivation by the Ala variant in the absence or presence of ligand due to reduced DNA-binding capacity.[105] Furthermore, the Ala variant has reduced ability to induce adipocyte differentiation *in vitro* compared to its wild-type counterpart.[106] In 78 investigated lean subjects, the 12Alla allele has been associated with increased insulin sensitivity of suppression of lipolysis.[107] It is unknown how this effect is mediated. The Pro12Ala polymorphism has in numerous studies been investigated for impact on common metabolic disorders. In a meta-analysis of 41 published and 2 unpublished studies (a total of 42,910 subjects), Ala12 carriers have been reported to have a modest but significant 19% T2D risk reduction.[108] However, because this allele occurs with a high frequency in certain populations, this modest effect can translate into a large subpopulation at decreased risk for T2D. Furthermore, this association is highly heterogeneous between studies. A great proportion of heterogeneity is explained by BMI, with risk reduction being greater when BMI is lower. In addition, risk reduction is greater in Asia than in Europe and, among Europeans, higher in Northern Europe, barely significant in Central Europe, and nonexistent in Southern Europe. A complementary meta-analysis comprising approximately 32,000 nondiabetic subjects in 57 studies has been performed to assess the effect of the Pro12Ala polymorphism on prediabetic traits.[109] Across all studies, the Pro12Ala polymorphism has no significant effect on diabetes-related traits. Only in selected subgroups, such as Caucasians and obese subjects, there is an association of the Ala allele with greater BMI and greater insulin sensitivity. The PPARG2 isoform is primarily expressed in adipose tissue and suppression of lipolysis in this organ by the Ala12 allele, as described above, is one potential mechanism by which *PPARG* could regulate insulin sensitivity. A previous meta-analysis encompassing 40 data sets comprising approximately 19,000 subjects has also reported that the Pro12Ala Ala allele is associated with higher BMI in the overweight and obese subset.[110]

The *SREBF1* gene on chromosome 17p11.2 encodes the transcription factors SREBP1a and SREBP1c by differential transcription start sites.[111] Whereas SREBP1a is primarily implicated in cholesterol metabolism, SREBP1c stimulates expression of genes promoting lipid storage and adipogenesis.[98] SREBP1c hereby induces both glucose utilization and lipid metabolism, which suggests that low levels of SREBP1c could be a contributing factor in the pathogenesis of insulin resistance and T2D. Levels of SREBP1c are indeed reduced in the adipose tissue of obese and T2D patients.[112,113] Association between common polymorphisms in *SREBP1* and T2D has been replicated in several cohorts.[114–116] A recent large study of 15,734 Danish subjects confirmed association between *SREBP1* gene variants and T2D.[117] According to a meta-analysis of this and previous published studies, there is a significant association between common *SREBP1* gene variants and T2D ($P = 0.001$).[117] Furthermore, the diabetes-associated alleles associate with a modestly increased HbA1c levels in two samples. Because SREBP1c is a mediator of insulin action, this finding is consistent with the presence of a yet undefined subtle loss-of-function in the presence of this *SREBF1* gene variant. In addition, one haplotype in the *SREBP1* gene has been associated with T2D, as well as morbid obesity and dyslipidemia in French Caucasian cohorts.[118] This haplotype carries variants in the 5′ untranslated region that could affect mRNA and protein levels.

Lamin A/C (LMNA) is an element of the nuclear lamina. The *LMNA* gene is encoded on chromosome 1q21.2. Rare mutations in the *LMNA* gene that lead to accumulation of LMNA precursor is one cause of familial partial lipodystrophy, which indicates a crucial role of *LMNA* in human adipogenesis.[119] In partial lipodystrophy caused by mutations in *LMNA*, accumulated protein precursor interacts with SREBP1, thereby decreasing the pool of active

SREBP1c that normally promotes preadipocyte differentiation.[119] The original observation that a silent variant in *LMNA* is associated with obesity in two aboriginal Canadian populations has stimulated other groups to evaluate the relationship between variations in the *LMNA* gene and metabolic disturbances in additional populations.[120,121] A Danish study has reported that the T-allele of the common *LMNA* polymorphism rs4641 is nominally associated with T2D and with elevated fasting plasma glucose levels in a population-based study of 5,395 middle-aged individuals.[122] However, most studies have not been able to reproduce an association between rs4642 and T2D.[123–125] Two meta-analyses of available rs4641 genotype data including 15,591 and 13,694 participants, respectively, have reported *P*-values of 0.054–0.001 for associations with T2D.[124,125] Another *LMNA* variant, a synonymous polymorphism (D446D) in exon 7, has showed nominal evidence of association with T2D in French Europids. A meta-analysis of available D446D genotype data from 7,819 participants provided support for a modest association with T2D.[124] Common LMNA polymorphisms have further been evaluated for impact on T2D-related metabolic phenotypes. Whereas some studies report nominal associations between *LMNA*, obesity, and dyslipidemia phenotypes, others do not.[122–124] In conclusion, the available data do not support a major effect of common *LMNA* variation on T2D susceptibility.

Lipin (LPIN1) acts upstream of PPARG in adipocyte differentiation and promotes fat storage in the mouse. The impact of LPIN1 on human fat cell differentiation is unknown; however, human adipose tissue expression of LPIN1 has recently been shown to be downregulated in subjects with obesity and the metabolic syndrome, and to increase following weight reduction in obese subjects.[126,127] Furthermore, adipose tissue LPIN1 levels are inversely correlated with insulin resistance as measured by HOMA, and positively correlated with glucose transport in human adipocytes.[126,127] The *LPIN1* gene is encoded on chromosome 2p21. Three rare nonsynonymous variants have been detected in the *LPIN1* gene. These variants do not influence protein expression and phosphorylation, and do not segregate with insulin resistance.[128] A number of studies have evaluated the role of common variation in the *LPIN1* gene in relation to adiposity and quantitative metabolic phenotypes. The results have been inconsistent across studies. SNP rs2716610 and a SNP in high LD, rs2716609, have been associated with BMI in Finns and with obesity in the Quebec Family Study, but not in a German cohort.[126,129,130] In a meta-analysis of available studies including 8,504 subjects, one common *LPIN1* variant is associated with BMI with nominal *P*-value 0.042.[128] Moreover, *LPIN1* haplotypes have been reported to have opposite impact on the metabolic syndrome in different cohorts.[130] Together, these results do not support an important role for *LPIN1* in predisposition to obesity and the metabolic syndrome.

Members of the bone morphogenetic protein (BMP) family are important regulators of adipogenesis in experimental models.[131] BMPs are growth and differentiation factors of the transforming growth factor beta superfamily.[132] Genetic variation in genes encoding for BMPs, and their receptors could, therefore, influence human adipose tissue morphology and adiposity. mRNA levels of the BMP receptor 1A (BMPR1A) are increased in both visceral and subcutaneous adipose tissue of obese subjects. Common polymorphisms in the *BMPR1A* gene have been associated with obesity in two independent cohorts.[133] Furthermore, homozygous carriers of the obesity risk alleles have higher BMPR1A mRNA expression in adipose tissue. Another member of the BMP family, *BMP2*, inhibits adipogenesis.[134] A common polymorphism in the *BMP2* gene has been reported to disrupt a putative posttranscriptional regulatory motif.[135] Reporter genes with the C allele of this ultra-conserved sequence are differentially expressed in mesenchymal cells, and the C/C genotype is associated with lower subcutaneous fat volumes. However, another group has found no association between *BMP2* polymorphisms and fat or lean mass.[136]

Some members of the Krüppel-like factor (KLF) family of transcription factors, initially shown to play important roles in cellular development and differentiation, have been reported to be involved in adipogenesis. Krüppel-like factor 7 (*Klf7*) has been shown to inhibit adipogenesis in rodent cells.[137] A subsequent large Danish study has investigated common polymorphisms in *KLF7* for association with T2D in 8,777 individuals, and with obesity in 14,818 individuals.[138] No association between common *KLF7* variants and T2D was observed; the previously reported association in Japanese individuals could not be confirmed. One polymorphism was reported to be associated with protection against obesity, $P = 0.001$. This association has to our knowledge not been confirmed in other populations.

V. Adipocyte Phenotypes of Established Susceptibility Genes for Obesity

Recently, more than 20 genes predisposing to common obesity have been identified.[139] The function of most of these genes in disease pathogenesis is unknown. It has been speculated that susceptibility genes for obesity primarily regulate food intake because most of the genes are highly expressed in the central nervous system, and all known monogenic defects causing obesity interfere with hypothalamic pathways and have a profound impact on food intake and satiety.[139] However, functional studies of susceptibility genes for common obesity support direct peripheral metabolic effects.

The human gene with strongest impact on common obesity is the fat mass and obesity-associated gene (*FTO*). Mice with a mutation in the *Fto* gene, the rodent homologue of human *FTO*, have normal food intake and physical activity, but reduced fat mass and increased energy expenditure.[140] The mutated mice display clear alterations in peripheral metabolism, for example, substrate switch and gene expression profiling reveals increased expression of fat and carbohydrate metabolism genes and an improved inflammatory profile in WAT. Studies on humans also support a functional role of *FTO* in WAT. FTO mRNA in subcutaneous adipose tissue is increased with increasing BMI and is enriched in fat cells versus the stroma vascular fraction of adipose tissue.[141] By contrast, there is no impact of BMI on muscle FTO levels. Another group has reported that in skeletal muscle, FTO mRNA expression is inversely associated to fat and positively to glucose oxidation rates, as well as positively correlated with expression of genes involved in oxidative phosphorylation.[142] In analysis of a sample comprising >300 subjects for which adipocyte phenotypes were available, the *FTO* T-allele was associated with increased basal adipocyte glycerol release *in vitro* and increased plasma glycerol levels *in vivo* compared with carriers of the A allele.[141] By contrast, there were no genotype effects on catecholamine-stimulated lipolysis or lipogenesis. However, according to another study, the heritability of *FTO* gene expression in adipose tissue and skeletal muscle is low and not influenced by obesity-associated *FTO* genotype.[142]

SH2B adaptor protein 1 (*SH2B1*) is another susceptibility gene for common obesity that is expressed in both the central nervous system and peripheral tissues. Neuronal *SH2B1* controls body weight; however, *SH2B1* expressed in peripheral organs may also be important for regulation of body weight. Interestingly, *SH2B1* modulates both insulin sensitivity and adipogenesis.[143,144]

INSIG2 was the first susceptibility gene for obesity that was identified by a genome-wide search.[145] Association of variation in the *INSIG2* gene with obesity has been difficult to replicate in some cohorts. However, a recent large meta-analysis suggests that *INSIG2* is associated with extreme degrees of obesity, and consequently that heterogeneous effects from different study designs may mask an underlying association when unaccounted for.[146] The function of *INSIG2* is unknown; unlike *INSIG1* it does not seem to regulate SREBP1c.[147] However, adipocyte differentiation is associated with a 13-fold increase in expression of INSIG2, which suggests a potential important role in adipogenesis and fat cell metabolism.[147]

Genes with a specific impact on adipose tissue distribution rather than adiposity *per se* may be more likely than other susceptibility genes for obesity to primarily act on metabolic function in peripheral organs, rather than on central pathways regulating food intake. Interestingly, recently three genes primarily associated with abdominal obesity have been identified: lysophospolipase-like 1 (*LYPLAL1*), transcription factor AP2-beta (*TFAP2B*), and peptide methionine sulfoxide reductase (*MSRA*).[148] *LYPLAL1* encodes a lipase; however, the function of LYPLAL1 in metabolically active organs in humans has to our knowledge not been investigated. The *TFAP2B* gene encodes a transcription factor that regulates adipokine secretion. By contrast, the *MSRA* gene has no defined metabolic function. These new susceptibility genes for adiposity are particularly interesting as therapeutic targets, since abdominal obesity is a strong risk factor for the metabolic complications of obesity. However, thorough functional adipose studies of the genes involved in regional fat distribution are hampered by the difficulty in obtaining inner fat depots and lack of relevant mouse models.

VI. Caveats with Genetic Studies of Adipose Function

An important methodological problem with all genetic studies of common traits is lack of statistical power. This is particularly relevant for adipose tissue due to small sample size and few available cohorts for meta-analyses. The techniques for studies of its function are difficult, cumbersome, and expensive. Therefore, few laboratories currently have the capacity to perform even small studies. Large cohorts, that is, several thousand subjects, are as far as we know not existing at the moment. Meta-analyses are also hampered by methodological issues since the ways to study a specific aspect of adipose biology vary between laboratories. Furthermore, only subcutaneous WAT is easily available. It is hard to get sufficient samples to investigate the genetics of BAT or visceral WAT.

Some easy-to-get information about the secretory function of adipose tissue can be obtained by studying specific plasma markers, such as LEPN, ADIPOQ, glycerol, or FFA. However, most markers are subject to diurnal variation and the correlation with true secretion rates from adipose tissue is very poor.

VII. Summary and Future Directions

Disturbances in adipose tissue function are believed to contribute to adiposity and related metabolic disorders, such as insulin resistance. Genetic epidemiological studies have clearly demonstrated that there is a strong hereditary impact on adipocyte lipolysis, adipokine release, and adipose tissue morphology. However, despite extensive functional and genetic investigations of numerous candidate genes, there is limited information as to what genes regulate human fat cell phenotypes, and to what extent these genes are

TABLE II
COMMON POLYMORPHISMS IN ADIPOSE-EXPRESSED GENES ASSOCIATED WITH ADIPOCYTE OR CLINICAL PHENOTYPES

Adipose phenotype	Gene	Polymorphism	Associated phenotype
Lipolysis	ADRB2	Arg16Gly	Sensitivity to catecholamines *in vitro*
	ADRB2	Thr 164Ile	Sensitivity to catecholamines *in vitro*
	GPR74	ATAG haplo	BMI, lipolysis *in vivo* and *in vitro*
Adipogenesis	PPARG	Pro12Ala	Adipogenesis, BMI, T2D
	SREBP1	Several	T2D. morbid obesity
Endocrine function	ADIPOQ	rs266717, 5′	Adiponectin levels
	ARL15	rs4311394, intron	Adiponectin levels

important for susceptibility to obesity and T2D. The genes that we consider supported by the strongest scientific data are summarized in Table II: (a) common functional variants in beta-adrenoreceptor genes regulate adipocyte lipolytic sensitivity to agonists, (b) a *GPR74* haplotype is associated with BMI and lipolysis, (c) common variants in the genes *PPARG* and *SREBP1*, which promote adipogenesis and lipid storage, are associated with T2D and possible adiposity, (d) a common variant in the *ADIPOQ* gene displays the strongest association with circulation levels of ADIPOQ throughout the genome, but is not associated with any clinical phenotype, by contrast, and (e) a common variant in *ARL15* associates with both circulating levels of ADIPOQ and susceptibility to coronary heart disease. Whether ADIPOQ mediates the impact of ARL15 on susceptibility to ischemic heart disease or is a disease marker is unclear.

The limited availability of large human cohorts with adipocyte phenotypes is a major obstacle to unambiguously identify gene variants regulating adipose function. GWAS of common complex clinical traits, such as obesity and T2D, have highlighted that very large sample sizes are needed to establish significant associations between traits and gene variants. The necessity for large samples is partly due to heterogeneity in the pathogenesis of the clinical disorder, that is, different pathways may cause the same clinical disorder in different individuals. From this perspective, intermediate phenotypes are attractive to study since it may be easier to demonstrate significant associations between such traits and gene variants with smaller sample sizes. However, intermediate phenotypes such as fat cell lipolysis are heavily influenced by nongenetic factors such as adiposity, age, gender, and hormones. Therefore, larger sample sizes than are typically available are probably needed to demonstrate significance between gene variants and adipocyte phenotypes in a genome-wide search.

Below are some complementary approaches which we believe are crucial to elucidate the influence of genetic variation on adipose tissue phenotypes in the population.

1. Global transcriptome profiling is anticipated to be important to identify new genes regulating adipose function by highlighting and guiding researchers to candidate genes. Transcriptome profiling has already been used to identify new and unexpected genes and pathways in human adipose tissue implicated in adiposity and T2D, for example, CIDEA, inflammation, and energy expenditure.[149–151]
2. Combining global transcriptome profiling with genome-wide linkage or association analysis in the same subjects can be used to identify expression quantitative trait loci (eQTLs), that is, *cis* and *trans*-acting gene variants that regulate specific gene expression. In a pioneering work, Emilsson *et al.* performed global eQTL mapping on subcutaneous adipose tissue from 673 subjects.[152] Through segregation analysis, these authors showed that there is a hereditary impact on expression of around 70% of genes in adipose tissue. All subjects in the cohort were used for genome-wide linkage analysis, which detected around 1500 *cis*-regulated transcripts. Genes whose expression in adipose tissue is regulated both by a relevant clinical trait such as obesity and by genetic variants are strong candidates for regulating fat cell function. Defining functional polymorphisms in such genes may reduce the need for large cohorts with adipose tissue phenotypes.
3. Large-scale allele-specific expression studies on adipose tissue samples provide a complementary approach to detect genes with mRNA levels in adipose tissue that are *cis*-regulated; in this case, only genes with polymorphisms in the measured transcripts can be detected.[153] This technique needs to be optimized to detect also low-abundance transcripts.
4. Various methods for massive parallel sequencing have scaled up efficiency of sequencing many-fold. With these new techniques, it has become feasible in many laboratories to screen candidate genes or genome regions for rare functional mutations by sequencing hundreds of subjects. Candidate genes may be selected from key pathways or global transcriptome profiles of relevant organs. The aim is to identify rare variants with a large impact on the studied trait. To optimize the search for clinically relevant rare variants, it is an advantage to sequence individuals with an extreme value for the studied trait, which are more likely to carry alleles with a large impact on the phenotype. In analysis of adipose tissue biology, such individuals may include patients with familial partial lipodystrophy, which has been reported to be more common than previously thought, or obese with low leptin levels.[154]

5. siRNA is a technique to evaluate function of a gene by experimentally downregulating-specific mRNA in cells *in vitro*. Large-scale screening of hundreds of genes by siRNA in experimental models is a complementary approach to identify candidate genes with impact on adipocyte function.[155]
6. Natural gene expression is regulated by independently coded miRNA, which regulates levels of other mRNA by degradation or sequestration. Several hundred miRNA have been detected in the human genome. Common polymorphisms regulating miRNA function can hypothetically contribute to variation in adipocyte phenotypes. Tools for large-scale measurement of miRNA and genotyping of regulatory polymorphisms are available.
7. Finally, epigenetics is defined as changes in DNA structure that are inherited by subsequent cell generations and affect gene expression that may influence adipose tissue phenotypes, and can be screened using novel techniques and platforms.

References

1. Lofgren P, Hoffstedt J, Naslund E, Wiren M, Arner P. Prospective and controlled studies of the actions of insulin and catecholamine in fat cells of obese women following weight reduction. *Diabetologia* 2005;**48**:2334–42.
2. Lafontan M, Langin D. Lipolysis and lipid mobilization in human adipose tissue. *Prog Lipid Res* 2009;**48**:275–97.
3. Jeninga EH, Gurnell M, Kalkhoven E. Functional implications of genetic variation in human PPARgamma. *Trends Endocrinol Metab* 2009;**20**:380–7.
4. Farooqi IS. Genetic and hereditary aspects of childhood obesity. *Best Pract Res Clin Endocrinol Metab* 2005;**19**:359–74.
5. Wajchenberg BL. Subcutaneous and visceral adipose tissue: their relation to the metabolic syndrome. *Endocr Rev* 2000;**21**:697–738.
6. Koerner A, Kratzsch J, Kiess W. Adipocytokines: leptin—the classical, resistin—the controversical, adiponectin—the promising, and more to come. *Best Pract Res Clin Endocrinol Metab* 2005;**19**:525–46.
7. Arner E, Westermark PO, Spalding KL, Britton T, Ryden M, Frisen J, et al. Adipocyte turnover: relevance to human adipose tissue morphology. *Diabetes* 2010;**37**:1513–22.
8. Pausova Z. From big fat cells to high blood pressure: a pathway to obesity-associated hypertension. *Curr Opin Nephrol Hypertens* 2006;**15**:173–8.
9. Skurk T, Alberti-Huber C, Herder C, Hauner H. Relationship between adipocyte size and adipokine expression and secretion. *J Clin Endocrinol Metab* 2007;**92**:1023–33.
10. Smith U. Studies of human adipose tissue in culture. I. Incorporation of glucose and release of glycerol. *Anat Rec* 1972;**172**:597–602.
11. Cypess AM, Lehman S, Williams G, Tal I, Rodman D, Goldfine AB, et al. Identification and importance of brown adipose tissue in adult humans. *N Engl J Med* 2009;**360**:1509–17.

12. van Marken Lichtenbelt WD, Vanhommerig JW, Smulders NM, Drossaerts JM, Kemerink GJ, Bouvy ND, et al. Cold-activated brown adipose tissue in healthy men. *N Engl J Med* 2009;**360**:1500–8.
13. Despres JP, Bouchard C. Monozygotic twin resemblance in fatness and fat cell lipolysis. *Acta Genet Med Gemellol (Roma)* 1984;**33**:475–80.
14. Stich V, Harant I, De Glisezinski I, Crampes F, Berlan M, Kunesova M, et al. Adipose tissue lipolysis and hormone-sensitive lipase expression during very-low-calorie diet in obese female identical twins. *J Clin Endocrinol Metab* 1997;**82**:739–44.
15. Comuzzie AG, Funahashi T, Sonnenberg G, Martin LJ, Jacob HJ, Black AE, et al. The genetic basis of plasma variation in adiponectin, a global endophenotype for obesity and the metabolic syndrome. *J Clin Endocrinol Metab* 2001;**86**:4321–5.
16. Lofgren P, Andersson I, Adolfsson B, Leijonhufvud BM, Hertel K, Hoffstedt J, et al. Long-term prospective and controlled studies demonstrate adipose tissue hypercellularity and relative leptin deficiency in the postobese state. *J Clin Endocrinol Metab* 2005;**90**:6207–13.
17. Sugandhan S, Khandpur S, Sharma VK. Familial chylomicronemia syndrome. *Pediatr Dermatol* 2007;**24**:323–5.
18. Poehlman ET, Despres JP, Marcotte M, Tremblay A, Theriault G, Bouchard C. Genotype dependency of adaptation in adipose tissue metabolism after short-term overfeeding. *Am J Physiol* 1986;**250**:E480–5.
19. Goodarzi MO, Wong H, Quinones MJ, Taylor KD, Guo X, Castellani LW, et al. The 3′ untranslated region of the lipoprotein lipase gene: haplotype structure and association with post-heparin plasma lipase activity. *J Clin Endocrinol Metab* 2005;**90**:4816–23.
20. Goodarzi MO, Taylor KD, Guo X, Hokanson JE, Haffner SM, Cui J, et al. Haplotypes in the lipoprotein lipase gene influence fasting insulin and discovery of a new risk haplotype. *J Clin Endocrinol Metab* 2007;**92**:293–6.
21. Tuncman G, Erbay E, Hom X, De Vivo I, Campos H, Rimm EB, et al. A genetic variant at the fatty acid-binding protein aP2 locus reduces the risk for hypertriglyceridemia, type 2 diabetes, and cardiovascular disease. *Proc Natl Acad Sci USA* 2006;**103**:6970–5.
22. Tsuchiya T, Schwarz PE, Bosque-Plata LD, Geoffrey Hayes M, Dina C, Froguel P, et al. Association of the calpain-10 gene with type 2 diabetes in Europeans: results of pooled and meta-analyses. *Mol Genet Metab* 2006;**89**:174–84.
23. Carlsson E, Fredriksson J, Groop L, Ridderstrale M. Variation in the calpain-10 gene is associated with elevated triglyceride levels and reduced adipose tissue messenger ribonucleic acid expression in obese Swedish subjects. *J Clin Endocrinol Metab* 2004;**89**:3601–5.
24. Hoffstedt J, Ryden M, Lofgren P, Orho-Melander M, Groop L, Arner P. Polymorphism in the Calpain 10 gene influences glucose metabolism in human fat cells. *Diabetologia* 2002;**45**:276–82.
25. Mauriege P, Despres JP, Marcotte M, Tremblay A, Nadeau A, Moorjani S, et al. Adipose tissue lipolysis after long-term overfeeding in identical twins. *Int J Obes Relat Metab Disord* 1992;**16**:219–25.
26. Leineweber K, Buscher R, Bruck H, Brodde OE. Beta-adrenoceptor polymorphisms. *Naunyn Schmiedebergs Arch Pharmacol* 2004;**369**:1–22.
27. Mason DA, Moore JD, Green SA, Liggett SB. A gain-of-function polymorphism in a G-protein coupling domain of the human beta1-adrenergic receptor. *J Biol Chem* 1999;**274**:12670–4.
28. Dionne IJ, Garant MJ, Nolan AA, Pollin TI, Lewis DG, Shuldiner AR, et al. Association between obesity and a polymorphism in the beta(1)-adrenoceptor gene (Gly389Arg ADRB1) in Caucasian women. *Int J Obes Relat Metab Disord* 2002;**26**:633–9.

29. Gjesing AP, Andersen G, Albrechtsen A, Glumer C, Borch-Johnsen K, Jorgensen T, et al. Studies of associations between the Arg389Gly polymorphism of the beta1-adrenergic receptor gene (ADRB1) and hypertension and obesity in 7677 Danish white subjects. *Diabet Med* 2007;**24**:392–7.
30. Linne Y, Dahlman I, Hoffstedt J. beta1-Adrenoceptor gene polymorphism predicts long-term changes in body weight. *Int J Obes (Lond)* 2005;**29**:458–62.
31. Large V, Hellstrom L, Reynisdottir S, Lonnqvist F, Eriksson P, Lannfelt L, et al. Human beta-2 adrenoceptor gene polymorphisms are highly frequent in obesity and associate with altered adipocyte beta-2 adrenoceptor function. *J Clin Invest* 1997;**100**:3005–13.
32. Jocken JW, Blaak EE, Schiffelers S, Arner P, van Baak MA, Saris WH. Association of a beta-2 adrenoceptor (ADRB2) gene variant with a blunted in vivo lipolysis and fat oxidation. *Int J Obes (Lond)* 2007;**31**:813–9.
33. Gjesing AP, Andersen G, Burgdorf KS, Borch-Johnsen K, Jorgensen T, Hansen T, et al. Studies of the associations between functional beta2-adrenergic receptor variants and obesity, hypertension and type 2 diabetes in 7,808 white subjects. *Diabetologia* 2007;**50**:563–8.
34. Jalba MS, Rhoads GG, Demissie K. Association of codon 16 and codon 27 beta 2-adrenergic receptor gene polymorphisms with obesity: a meta-analysis. *Obesity (Silver Spring)* 2008;**16**:2096–106.
35. Drysdale CM, McGraw DW, Stack CB, Stephens JC, Judson RS, Nandabalan K, et al. Complex promoter and coding region beta 2-adrenergic receptor haplotypes alter receptor expression and predict in vivo responsiveness. *Proc Natl Acad Sci USA* 2000;**97**:10483–8.
36. Eriksson P, Dahlman I, Ryden M, Hoffstedt J, Arner P. Relationship between beta-2 adrenoceptor gene haplotypes and adipocyte lipolysis in women. *Int J Obes Relat Metab Disord* 2004;**28**:185–90.
37. Jiao H, Dahlman I, Eriksson P, Kere J, Arner P. A common beta2-adrenoceptor gene haplotype protects against obesity in Swedish women. *Obes Res* 2005;**13**:1645–50.
38. Lonnqvist F, Krief S, Strosberg AD, Nyberg S, Emorine LJ, Arner P. Evidence for a functional beta 3-adrenoceptor in man. *Br J Pharmacol* 1993;**110**:929–36.
39. Gjesing AP, Andersen G, Borch-Johnsen K, Jorgensen T, Hansen T, Pedersen O. Association of the beta3-adrenergic receptor Trp64Arg polymorphism with common metabolic traits: studies of 7605 middle-aged white people. *Mol Genet Metab* 2008;**94**:90–7.
40. Kurokawa N, Young EH, Oka Y, Satoh H, Wareham NJ, Sandhu MS, et al. The ADRB3 Trp64Arg variant and BMI: a meta-analysis of 44833 individuals. *Int J Obes (Lond)* 2008;**32**:1240–9.
41. Siffert W, Rosskopf D, Siffert G, Busch S, Moritz A, Erbel R, et al. Association of a human G-protein beta3 subunit variant with hypertension. *Nat Genet* 1998;**18**:45–8.
42. Ryden M, Faulds G, Hoffstedt J, Wennlund A, Arner P. Effect of the (C825T) Gbeta(3) polymorphism on adrenoceptor-mediated lipolysis in human fat cells. *Diabetes* 2002;**51**:1601–8.
43. Andersen G, Overgaard J, Albrechtsen A, Glumer C, Borch-Johnsen K, Jorgensen T, et al. Studies of the association of the GNB3 825C>T polymorphism with components of the metabolic syndrome in white Danes. *Diabetologia* 2006;**49**:75–82.
44. Frey UH, Hauner H, Jockel KH, Manthey I, Brockmeyer N, Siffert W. A novel promoter polymorphism in the human gene GNAS affects binding of transcription factor upstream stimulatory factor 1, Galphas protein expression and body weight regulation. *Pharmacogenet Genomics* 2008;**18**:141–51.
45. Lafontan M, Moro C, Berlan M, Crampes F, Sengenes C, Galitzky J. Control of lipolysis by natriuretic peptides and cyclic GMP. *Trends Endocrinol Metab* 2008;**19**:130–7.

46. Meirhaeghe A, Sandhu MS, McCarthy MI, de Groote P, Cottel D, Arveiler D, et al. Association between the T-381C polymorphism of the brain natriuretic peptide gene and risk of type 2 diabetes in human populations. *Hum Mol Genet* 2007;**16**:1343–50.
47. Choquet H, Cavalcanti-Proenca C, Lecoeur C, Dina C, Cauchi S, Vaxillaire M, et al. The T-381C SNP in BNP gene may be modestly associated with type 2 diabetes: an updated meta-analysis in 49 279 subjects. *Hum Mol Genet* 2009;**18**:2495–501.
48. Valet P, Berlan M, Beauville M, Crampes F, Montastruc JL, Lafontan M. Neuropeptide Y and peptide YY inhibit lipolysis in human and dog fat cells through a pertussis toxin-sensitive G protein. *J Clin Invest* 1990;**85**:291–5.
49. Dahlman I, Dicker A, Jiao H, Kere J, Blomqvist L, van Harmelen V, et al. A common haplotype in the G-protein-coupled receptor gene GPR74 is associated with leanness and increased lipolysis. *Am J Hum Genet* 2007;**80**:1115–24.
50. Hoffstedt J, Arner P, Schalling M, Pedersen NL, Sengul S, Ahlberg S, et al. A common hormone-sensitive lipase i6 gene polymorphism is associated with decreased human adipocyte lipolytic function. *Diabetes* 2001;**50**:2410–3.
51. Carlsson E, Johansson LE, Strom K, Hoffstedt J, Groop L, Holm C, et al. The hormone-sensitive lipase C-60G promoter polymorphism is associated with increased waist circumference in normal-weight subjects. *Int J Obes (Lond)* 2006;**30**:1442–8.
52. Fischer J, Lefevre C, Morava E, Mussini JM, Laforet P, Negre-Salvayre A, et al. The gene encoding adipose triglyceride lipase (PNPLA2) is mutated in neutral lipid storage disease with myopathy. *Nat Genet* 2007;**39**:28–30.
53. Lefevre C, Jobard F, Caux F, Bouadjar B, Karaduman A, Heilig R, et al. Mutations in CGI-58, the gene encoding a new protein of the esterase/lipase/thioesterase subfamily, in Chanarin–Dorfman syndrome. *Am J Hum Genet* 2001;**69**:1002–12.
54. Hoffstedt J, Ryden M, Wahrenberg H, van Harmelen V, Arner P. Upstream transcription factor-1 gene polymorphism is associated with increased adipocyte lipolysis. *J Clin Endocrinol Metab* 2005;**90**:5356–60.
55. Kantartzis K, Fritsche A, Machicao F, Stumvoll M, Machann J, Schick F, et al. Upstream transcription factor 1 gene polymorphisms are associated with high antilipolytic insulin sensitivity and show gene–gene interactions. *J Mol Med* 2007;**85**:55–61.
56. Mottagui-Tabar S, Ryden M, Lofgren P, Faulds G, Hoffstedt J, Brookes AJ, et al. Evidence for an important role of perilipin in the regulation of human adipocyte lipolysis. *Diabetologia* 2003;**46**:789–97.
57. Soenen S, Mariman EC, Vogels N, Bouwman FG, den Hoed M, Brown L, Westerterp-Plantenga MS. Relationship between perilipin gene polymorphisms and body weight and body composition during weight loss and weight maintenance. *Physiol Behav* 2009.
58. Liu K, Zhou S, Kim JY, Tillison K, Majors D, Rearick D, et al. Functional analysis of FSP27 protein regions for lipid droplet localization, caspase-dependent apoptosis, and dimerization with CIDEA. *Am J Physiol Endocrinol Metab* 2009;**297**:E1395–413.
59. Nordstrom EA, Ryden M, Backlund EC, Dahlman I, Kaaman M, Blomqvist L, et al. A human-specific role of cell death-inducing DFFA (DNA fragmentation factor-{alpha})-like effector A (CIDEA) in adipocyte lipolysis and obesity. *Diabetes* 2005;**54**:1726–34.
60. Puri V, Ranjit S, Konda S, Nicoloro SM, Straubhaar J, Chawla A, et al. Cidea is associated with lipid droplets and insulin sensitivity in humans. *Proc Natl Acad Sci USA* 2008;**105**:7833–8.
61. Laurencikiene J, Stenson BM, Arvidsson Nordstrom E, Agustsson T, Langin D, Isaksson B, et al. Evidence for an important role of CIDEA in human cancer cachexia. *Cancer Res* 2008;**68**:9247–54.
62. Dahlman I, Kaaman M, Jiao H, Kere J, Laakso M, Arner P. The CIDEA gene V115F polymorphism is associated with obesity in Swedish subjects. *Diabetes* 2005;**54**:3032–4.

63. Zhang L, Miyaki K, Nakayama T, Muramatsu M. Cell death-inducing DNA fragmentation factor alpha-like effector A (CIDEA) gene V115F (G–>T) polymorphism is associated with phenotypes of metabolic syndrome in Japanese men. *Metabolism* 2008;**57**:502–5.
64. Prudente S, Flex E, Morini E, Turchi F, Capponi D, De Cosmo S, et al. A functional variant of the adipocyte glycerol channel aquaporin 7 gene is associated with obesity and related metabolic abnormalities. *Diabetes* 2007;**56**:1468–74.
65. Jia JJ, Tian YB, Cao ZH, Tao LL, Zhang X, Gao SZ, Ge CR, Lin QY, Jois M. The polymorphisms of UCP1 genes associated with fat metabolism, obesity and diabetes. *Mol Biol Rep* 2010;**37**:1513–22.
66. Esterbauer H, Oberkofler H, Liu YM, Breban D, Hell E, Krempler F, et al. Uncoupling protein-1 mRNA expression in obese human subjects: the role of sequence variations at the uncoupling protein-1 gene locus. *J Lipid Res* 1998;**39**:834–44.
67. Narkiewicz K, Szczech R, Winnicki M, Chrostowska M, Pawlowski R, Lysiak-Szydlowska W, et al. Heritability of plasma leptin levels: a twin study. *J Hypertens* 1999;**17**:27–31.
68. Jordan J, Brabant G, Brinsuk M, Tank J, Horn R, Luft FC, et al. Heritability of free and receptor-bound leptin in normal twins. *Am J Physiol Regul Integr Comp Physiol* 2005;**288**: R1411–6.
69. Kaprio J, Eriksson J, Lehtovirta M, Koskenvuo M, Tuomilehto J. Heritability of leptin levels and the shared genetic effects on body mass index and leptin in adult Finnish twins. *Int J Obes Relat Metab Disord* 2001;**25**:132–7.
70. Jiang Y, Wilk JB, Borecki I, Williamson S, DeStefano AL, Xu G, et al. Common variants in the 5′ region of the leptin gene are associated with body mass index in men from the National Heart, Lung, and Blood Institute Family Heart Study. *Am J Hum Genet* 2004;**75**:220–30.
71. Richards JB, Waterworth D, O'Rahilly S, Hivert MF, Loos RJ, Perry JR, et al. A genome-wide association study reveals variants in ARL15 that influence adiponectin levels. *PLoS Genet* 2009;**5**:e1000768.
72. Liu Y, Retnakaran R, Hanley A, Tungtrongchitr R, Shaw C, Sweeney G. Total and high molecular weight but not trimeric or hexameric forms of adiponectin correlate with markers of the metabolic syndrome and liver injury in Thai subjects. *J Clin Endocrinol Metab* 2007;**92**:4313–8.
73. Cesari M, Narkiewicz K, De Toni R, Aldighieri E, Williams CJ, Rossi GP. Heritability of plasma adiponectin levels and body mass index in twins. *J Clin Endocrinol Metab* 2007;**92**:3082–8.
74. Gable DR, Hurel SJ, Humphries SE. Adiponectin and its gene variants as risk factors for insulin resistance, the metabolic syndrome and cardiovascular disease. *Atherosclerosis* 2006;**188**:231–44.
75. Ribel-Madsen R, Friedrichsen M, Vaag A, Poulsen P. Retinol-binding protein 4 in twins: regulatory mechanisms and impact of circulating and tissue expression levels on insulin secretion and action. *Diabetes* 2009;**58**:54–60.
76. Kovacs P, Geyer M, Berndt J, Kloting N, Graham TE, Bottcher Y, et al. Effects of genetic variation in the human retinol binding protein-4 gene (RBP4) on insulin resistance and fat depot-specific mRNA expression. *Diabetes* 2007;**56**:3095–100.
77. Hotamisligil GS, Arner P, Caro JF, Atkinson RL, Spiegelman BM. Increased adipose tissue expression of tumor necrosis factor-alpha in human obesity and insulin resistance. *J Clin Invest* 1995;**95**:2409–15.
78. Mohamed-Ali V, Goodrick S, Rawesh A, Katz DR, Miles JM, Yudkin JS, et al. Subcutaneous adipose tissue releases interleukin-6, but not tumor necrosis factor-alpha, in vivo. *J Clin Endocrinol Metab* 1997;**82**:4196–200.

79. Dahlman I, Kaaman M, Olsson T, Tan GD, Bickerton AS, Wahlen K, et al. A unique role of monocyte chemoattractant protein 1 among chemokines in adipose tissue of obese subjects. *J Clin Endocrinol Metab* 2005;**90**:5834–40.
80. Weisberg SP, McCann D, Desai M, Rosenbaum M, Leibel RL, Ferrante Jr. AW. Obesity is associated with macrophage accumulation in adipose tissue. *J Clin Invest* 2003;**112**:1796–808.
81. Lacasa D, Taleb S, Keophiphath M, Miranville A, Clement K. Macrophage-secreted factors impair human adipogenesis: involvement of proinflammatory state in preadipocytes. *Endocrinology* 2007;**148**:868–77.
82. Path G, Bornstein SR, Gurniak M, Chrousos GP, Scherbaum WA, Hauner H. Human breast adipocytes express interleukin-6 (IL-6) and its receptor system: increased IL-6 production by beta-adrenergic activation and effects of IL-6 on adipocyte function. *J Clin Endocrinol Metab* 2001;**86**:2281–8.
83. Qi L, Zhang C, van Dam RM, Hu FB. Interleukin-6 genetic variability and adiposity: associations in two prospective cohorts and systematic review in 26, 944 individuals. *J Clin Endocrinol Metab* 2007;**92**:3618–25.
84. Qi L, van Dam RM, Meigs JB, Manson JE, Hunter D, Hu FB. Genetic variation in IL6 gene and type 2 diabetes: tagging-SNP haplotype analysis in large-scale case-control study and meta-analysis. *Hum Mol Genet* 2006;**15**:1914–20.
85. Wolford JK, Colligan PB, Gruber JD, Bogardus C. Variants in the interleukin 6 receptor gene are associated with obesity in Pima Indians. *Mol Genet Metab* 2003;**80**:338–43.
86. Hamid YH, Urhammer SA, Jensen DP, Glumer C, Borch-Johnsen K, Jorgensen T, et al. Variation in the interleukin-6 receptor gene associates with type 2 diabetes in Danish whites. *Diabetes* 2004;**53**:3342–5.
87. Qi L, Rifai N, Hu FB. Interleukin-6 receptor gene variations, plasma interleukin-6 levels, and type 2 diabetes in US women. *Diabetes* 2007;**56**:3075–81.
88. Gottardo L, De Cosmo S, Zhang YY, Powers C, Prudente S, Marescotti MC, et al. A polymorphism at the IL6ST (gp130) locus is associated with traits of the metabolic syndrome. *Obesity (Silver Spring)* 2008;**16**:205–10.
89. Skoog T, van't Hooft FM, Kallin B, Jovinge S, Boquist S, Nilsson J, Eriksson P, Hamsten A. A common functional polymorphism (C–>A substitution at position -863) in the promoter region of the tumour necrosis factor-alpha (TNF-alpha) gene associated with reduced circulating levels of TNF-alpha. *Hum Mol Genet* 1999;**8**:1443–9.
90. Kroeger KM, Carville KS, Abraham LJ. The -308 tumor necrosis factor-alpha promoter polymorphism effects transcription. *Mol Immunol* 1997;**34**:391–9.
91. Kaluza W, Reuss E, Grossmann S, Hug R, Schopf RE, Galle PR, et al. Different transcriptional activity and in vitro TNF-alpha production in psoriasis patients carrying the TNF-alpha 238A promoter polymorphism. *J Invest Dermatol* 2000;**114**:1180–3.
92. Skoog T, Eriksson P, Hoffstedt J, Ryden M, Hamsten A, Armer P. Tumour necrosis factor-alpha (TNF-alpha) polymorphisms-857C/A and -863C/A are associated with TNF-alpha secretion from human adipose tissue. *Diabetologia* 2001;**44**:654–5.
93. Sookoian SC, Gonzalez C, Pirola CJ. Meta-analysis on the G-308A tumor necrosis factor alpha gene variant and phenotypes associated with the metabolic syndrome. *Obes Res* 2005;**13**:2122–31.
94. Herrmann SM, Ricard S, Nicaud V, Mallet C, Arveiler D, Evans A, et al. Polymorphisms of the tumour necrosis factor-alpha gene, coronary heart disease and obesity. *Eur J Clin Invest* 1998;**28**:59–66.
95. Walston J, Seibert M, Yen CJ, Cheskin LJ, Andersen RE. Tumor necrosis factor-alpha-238 and -308 polymorphisms do not associated with traits related to obesity and insulin resistance. *Diabetes* 1999;**48**:2096–8.

96. Weisberg SP, Hunter D, Huber R, Lemieux J, Slaymaker S, Vaddi K, et al. CCR2 modulates inflammatory and metabolic effects of high-fat feeding. *J Clin Invest* 2006;**116**:115–24.
97. Simeoni E, Hoffmann MM, Winkelmann BR, Ruiz J, Fleury S, Boehm BO, et al. Association between the A-2518G polymorphism in the monocyte chemoattractant protein-1 gene and insulin resistance and Type 2 diabetes mellitus. *Diabetologia* 2004;**47**:1574–80.
98. Sethi JK, Vidal-Puig AJ. Thematic review series: adipocyte biology. Adipose tissue function and plasticity orchestrate nutritional adaptation. *J Lipid Res* 2007;**48**:1253–62.
99. Gurnell M. Peroxisome proliferator-activated receptor gamma and the regulation of adipocyte function: lessons from human genetic studies. *Best Pract Res Clin Endocrinol Metab* 2005;**19**:501–23.
100. Ristow M, Muller-Wieland D, Pfeiffer A, Krone W, Kahn CR. Obesity associated with a mutation in a genetic regulator of adipocyte differentiation. *N Engl J Med* 1998;**339**:953–9.
101. Bluher M, Paschke R. Analysis of the relationship between PPAR-gamma 2 gene variants and severe insulin resistance in obese patients with impaired glucose tolerance. *Exp Clin Endocrinol Diabetes* 2003;**111**:85–90.
102. Giusti V, Verdumo C, Suter M, Gaillard RC, Burckhardt P, Pralong F. Expression of peroxisome proliferator-activated receptor-gamma1 and peroxisome proliferator-activated receptor-gamma2 in visceral and subcutaneous adipose tissue of obese women. *Diabetes* 2003;**52**:1673–6.
103. Vidal-Puig AJ, Considine RV, Jimenez-Linan M, Werman A, Pories WJ, Caro JF, et al. Peroxisome proliferator-activated receptor gene expression in human tissues. Effects of obesity, weight loss, and regulation by insulin and glucocorticoids. *J Clin Invest* 1997;**99**:2416–22.
104. Rieusset J, Andreelli F, Auboeuf D, Roques M, Vallier P, Riou JP, et al. Insulin acutely regulates the expression of the peroxisome proliferator-activated receptor-gamma in human adipocytes. *Diabetes* 1999;**48**:699–705.
105. Deeb SS, Fajas L, Nemoto M, Pihlajamaki J, Mykkanen L, Kuusisto J, et al. A Pro12Ala substitution in PPARgamma2 associated with decreased receptor activity, lower body mass index and improved insulin sensitivity. *Nat Genet* 1998;**20**:284–7.
106. Masugi J, Tamori Y, Mori H, Koike T, Kasuga M. Inhibitory effect of a proline-to-alanine substitution at codon 12 of peroxisome proliferator-activated receptor-gamma 2 on thiazolidinedione-induced adipogenesis. *Biochem Biophys Res Commun* 2000;**268**:178–82.
107. Stumvoll M, Wahl HG, Loblein K, Becker R, Machicao F, Jacob S, et al. Pro12Ala polymorphism in the peroxisome proliferator-activated receptor-gamma2 gene is associated with increased antilipolytic insulin sensitivity. *Diabetes* 2001;**50**:876–81.
108. Ludovico O, Pellegrini F, Di Paola R, Minenna A, Mastroianno S, Cardellini M, et al. Heterogeneous effect of peroxisome proliferator-activated receptor gamma2 Ala12 variant on type 2 diabetes risk. *Obesity (Silver Spring)* 2007;**15**:1076–81.
109. Tonjes A, Scholz M, Loeffler M, Stumvoll M. Association of Pro12Ala polymorphism in peroxisome proliferator-activated receptor gamma with Pre-diabetic phenotypes: meta-analysis of 57 studies on nondiabetic individuals. *Diabetes Care* 2006;**29**:2489–97.
110. Masud S, Ye S. Effect of the peroxisome proliferator activated receptor-gamma gene Pro12Ala variant on body mass index: a meta-analysis. *J Med Genet* 2003;**40**:773–80.
111. Hua X, Wu J, Goldstein JL, Brown MS, Hobbs HH. Structure of the human gene encoding sterol regulatory element binding protein-1 (SREBF1) and localization of SREBF1 and SREBF2 to chromosomes 17p11.2 and 22q13. *Genomics* 1995;**25**:667–73.
112. Kolehmainen M, Vidal H, Alhava E, Uusitupa MI. Sterol regulatory element binding protein 1c (SREBP-1c) expression in human obesity. *Obes Res* 2001;**9**:706–12.

113. Sewter C, Berger D, Considine RV, Medina G, Rochford J, Ciaraldi T, et al. Human obesity and type 2 diabetes are associated with alterations in SREBP1 isoform expression that are reproduced ex vivo by tumor necrosis factor-alpha. *Diabetes* 2002;**51**:1035–41.
114. Laudes M, Barroso I, Luan J, Soos MA, Yeo G, Meirhaeghe A, et al. Genetic variants in human sterol regulatory element binding protein-1c in syndromes of severe insulin resistance and type 2 diabetes. *Diabetes* 2004;**53**:842–6.
115. Harding AH, Loos RJ, Luan J, O'Rahilly S, Wareham NJ, Barroso I. Polymorphisms in the gene encoding sterol regulatory element-binding factor-1c are associated with type 2 diabetes. *Diabetologia* 2006;**49**:2642–8.
116. Felder TK, Oberkofler H, Weitgasser R, Mackevics V, Krempler F, Paulweber B, et al. The SREBF-1 locus is associated with type 2 diabetes and plasma adiponectin levels in a middle-aged Austrian population. *Int J Obes (Lond)* 2007;**31**:1099–103.
117. Grarup N, Stender-Petersen KL, Andersson EA, Jorgensen T, Borch-Johnsen K, Sandbaek A, et al. Association of variants in the sterol regulatory element-binding factor 1 (SREBF1) gene with type 2 diabetes, glycemia, and insulin resistance: a study of 15, 734 Danish subjects. *Diabetes* 2008;**57**:1136–42.
118. Eberle D, Clement K, Meyre D, Sahbatou M, Vaxillaire M, Le Gall A, et al. SREBF-1 gene polymorphisms are associated with obesity and type 2 diabetes in French obese and diabetic cohorts. *Diabetes* 2004;**53**:2153–7.
119. Capanni C, Mattioli E, Columbaro M, Lucarelli E, Parnaik VK, Novelli G, et al. Altered pre-lamin A processing is a common mechanism leading to lipodystrophy. *Hum Mol Genet* 2005;**14**:1489–502.
120. Hegele RA, Cao H, Harris SB, Zinman B, Hanley AJ, Anderson CM. Genetic variation in LMNA modulates plasma leptin and indices of obesity in aboriginal Canadians. *Physiol Genomics* 2000;**3**:39–44.
121. Hegele RA, Huff MW, Young TK. Common genomic variation in LMNA modulates indexes of obesity in Inuit. *J Clin Endocrinol Metab* 2001;**86**:2747–51.
122. Wegner L, Andersen G, Sparso T, Grarup N, Glumer C, Borch-Johnsen K, et al. Common variation in LMNA increases susceptibility to type 2 diabetes and associates with elevated fasting glycemia and estimates of body fat and height in the general population: studies of 7, 495 Danish whites. *Diabetes* 2007;**56**:694–8.
123. Mesa JL, Loos RJ, Franks PW, Ong KK, Luan J, O'Rahilly S, et al. Lamin A/C polymorphisms, type 2 diabetes, and the metabolic syndrome: case-control and quantitative trait studies. *Diabetes* 2007;**56**:884–9.
124. Duesing K, Charpentier G, Marre M, Tichet J, Hercberg S, Froguel P, et al. Evaluating the association of common LMNA variants with type 2 diabetes and quantitative metabolic phenotypes in French Europids. *Diabetologia* 2008;**51**:76–81.
125. Owen KR, Groves CJ, Hanson RL, Knowler WC, Shuldiner AR, Elbein SC, et al. Common variation in the LMNA gene (encoding lamin A/C) and type 2 diabetes: association analyses in 9, 518 subjects. *Diabetes* 2007;**56**:879–83.
126. Suviolahti E, Reue K, Cantor RM, Phan J, Gentile M, Naukkarinen J, et al. Cross-species analyses implicate Lipin 1 involvement in human glucose metabolism. *Hum Mol Genet* 2006;**15**:377–86.
127. van Harmelen V, Ryden M, Sjolin E, Hoffstedt J. A role of lipin in human obesity and insulin resistance: relation to adipocyte glucose transport and GLUT4 expression. *J Lipid Res* 2007;**48**:201–6.
128. Fawcett KA, Grimsey N, Loos RJ, Wheeler E, Daly A, Soos M, et al. Evaluating the role of LPIN1 variation in insulin resistance, body weight, and human lipodystrophy in UK population. *Diabetes* 2008;**57**:2527–33.

129. Loos RJ, Rankinen T, Perusse L, Tremblay A, Despres JP, Bouchard C. Association of lipin 1 gene polymorphisms with measures of energy and glucose metabolism. *Obesity (Silver Spring)* 2007;**15**:2723–32.
130. Wiedmann S, Fischer M, Koehler M, Neureuther K, Riegger G, Doering A, et al. Genetic variants within the LPIN1 gene, encoding lipin, are influencing phenotypes of the metabolic syndrome in humans. *Diabetes* 2008;**57**:209–17.
131. Neumann K, Endres M, Ringe J, Flath B, Manz R, Haupl T, et al. BMP7 promotes adipogenic but not osteo-/chondrogenic differentiation of adult human bone marrow-derived stem cells in high-density micro-mass culture. *J Cell Biochem* 2007;**102**:626–37.
132. Soderberg SS, Karlsson G, Karlsson S. Complex and context dependent regulation of hematopoiesis by TGF-beta superfamily signaling. *Ann NY Acad Sci* 2009;**1176**:55–69.
133. Bottcher Y, Unbehauen H, Kloting N, Ruschke K, Korner A, Schleinitz D, et al. Adipose tissue expression and genetic variants of the bone morphogenetic protein receptor 1A gene (BMPR1A) are associated with human obesity. *Diabetes* 2009;**58**:2119–28.
134. Schmitt B, Ringe J, Haupl T, Notter M, Manz R, Burmester GR, et al. BMP2 initiates chondrogenic lineage development of adult human mesenchymal stem cells in high-density culture. *Differentiation* 2003;**71**:567–77.
135. Devaney JM, Tosi LL, Fritz DT, Gordish-Dressman HA, Jiang S, Orkunoglu-Suer FE, et al. Differences in fat and muscle mass associated with a functional human polymorphism in a post-transcriptional BMP2 gene regulatory element. *J Cell Biochem* 2009;**107**:1073–82.
136. McGuigan F, Larzenius E, Callreus M, Gerdhem P, Luthman H, Akesson K. Variation in the bone morphogenetic protein-2 gene: effects on fat and lean body mass in young and elderly women. *Eur J Endocrinol* 2008;**158**:661–8.
137. Kanazawa A, Kawamura Y, Sekine A, Iida A, Tsunoda T, Kashiwagi A, et al. Single nucleotide polymorphisms in the gene encoding Kruppel-like factor 7 are associated with type 2 diabetes. *Diabetologia* 2005;**48**:1315–22.
138. Zobel DP, Andreasen CH, Burgdorf KS, Andersson EA, Sandbaek A, Lauritzen T, et al. Variation in the gene encoding Kruppel-like factor 7 influences body fat: studies of 14 818 Danes. *Eur J Endocrinol* 2009;**160**:603–9.
139. Hofker M, Wijmenga C. A supersized list of obesity genes. *Nat Genet* 2009;**41**:139–40.
140. Church C, Lee S, Bagg EA, McTaggart JS, Deacon R, Gerken T, et al. A mouse model for the metabolic effects of the human fat mass and obesity associated FTO gene. *PLoS Genet* 2009;**5**: e1000599.
141. Wahlen K, Sjolin E, Hoffstedt J. The common rs9939609 gene variant of the fat mass- and obesity-associated gene FTO is related to fat cell lipolysis. *J Lipid Res* 2008;**49**:607–11.
142. Grunnet LG, Nilsson E, Ling C, Hansen T, Pedersen O, Groop L, et al. Regulation and function of FTO mRNA expression in human skeletal muscle and subcutaneous adipose tissue. *Diabetes* 2009;**58**:2402–8.
143. Morris DL, Cho KW, Zhou Y, Rui L. SH2B1 enhances insulin sensitivity by both stimulating the insulin receptor and inhibiting tyrosine dephosphorylation of insulin receptor substrate proteins. *Diabetes* 2009;**58**:2039–47.
144. Yoshiga D, Sato N, Torisu T, Mori H, Yoshida R, Nakamura S, et al. Adaptor protein SH2-B linking receptor-tyrosine kinase and Akt promotes adipocyte differentiation by regulating peroxisome proliferator-activated receptor gamma messenger ribonucleic acid levels. *Mol Endocrinol* 2007;**21**:1120–31.
145. Herbert A, Gerry NP, McQueen MB, Heid IM, Pfeufer A, Illig T, et al. A common genetic variant is associated with adult and childhood obesity. *Science* 2006;**312**:279–83.
146. Heid IM, Huth C, Loos RJ, Kronenberg F, Adamkova V, Anand SS, et al. Meta-analysis of the INSIG2 association with obesity including 74, 345 individuals: does heterogeneity of estimates relate to study design? *PLoS Genet* 2009;**5**:e1000694.

147. Krapivner S, Popov S, Chernogubova E, Hellenius ML, Fisher RM, Hamsten A, van't Hooft FM. Insulin-induced gene 2 involvement in human adipocyte metabolism and body weight regulation. *J Clin Endocrinol Metab* 2008;**93**:1995–2001.
148. Lindgren CM, Heid IM, Randall JC, Lamina C, Steinthorsdottir V, Qi L, et al. Genome-wide association scan meta-analysis identifies three Loci influencing adiposity and fat distribution. *PLoS Genet* 2009;**5**:e1000508.
149. Dahlman I, Linder K, Arvidsson Nordstrom E, Andersson I, Liden J, Verdich C, et al. Changes in adipose tissue gene expression with energy-restricted diets in obese women. *Am J Clin Nutr* 2005;**81**:1275–85.
150. Dahlman I, Forsgren M, Sjogren A, Nordstrom EA, Kaaman M, Naslund E, et al. Downregulation of electron transport chain genes in visceral adipose tissue in type 2 diabetes independent of obesity and possibly involving tumor necrosis factor-{alpha}. *Diabetes* 2006;**55**:1792–9.
151. Clement K, Viguerie N, Poitou C, Carette C, Pelloux V, Curat CA, et al. Weight loss regulates inflammation-related genes in white adipose tissue of obese subjects. *FASEB J* 2004;**18**:1657–69.
152. Emilsson V, Thorleifsson G, Zhang B, Leonardson AS, Zink F, Zhu J, et al. Genetics of gene expression and its effect on disease. *Nature* 2008;**452**:423–8.
153. Heap GA, Yang JH, Downes K, Healy BC, Hunt KA, Bockett N, et al. Genome-wide analysis of allelic expression imbalance in human primary cells by high-throughput transcriptome resequencing. *Hum Mol Genet* 2010;**19**:122–34.
154. Herbst KL, Tannock LR, Deeb SS, Purnell JQ, Brunzell JD, Chait A. Kobberling type of familial partial lipodystrophy: an underrecognized syndrome. *Diabetes Care* 2003;**26**:1819–24.
155. Guo Y, Walther TC, Rao M, Stuurman N, Goshima G, Terayama K, et al. Functional genomic screen reveals genes involved in lipid-droplet formation and utilization. *Nature* 2008;**453**:657–61.

The Genetics of Brown Adipose Tissue

Leslie P. Kozak and Robert A. Koza

Pennington Biomedical Research Center, Baton Rouge, Louisiana, USA

Brown adipose tissue is highly differentiated and has evolved as a mechanism for heat production based upon uncoupling of mitochondrial oxidative phosphorylation. Additionally, large amounts of lipid can be stored in the cells to provide fuel necessary for heat production upon adrenergic stimulation from the central nervous system, and a highly developed vascular system evolved to rapidly deliver heat to vital organs. For unknown reasons, the development of brown adipocytes has two independent pathways: one originates from muscle progenitor cells in the fetus and leads to a fully functional cell at birth (interscapular-type brown fat), while the other transiently emerges in traditional white fat depots at weaning, regresses, and then can be induced in adult

Progress in Molecular Biology
and Translational Science, Vol. 94
DOI: 10.1016/S1877-1173(10)94004-0

mice upon adrenergic stimulation. No genetic variants have been found for interscapular fat, but naturally occurring alleles at eight genetic loci in mice lead to over 100-fold variation for brown adipocytes in white fat upon adrenergic stimulation. The ability to activate this potential for energy expenditure is of great interest in obesity research.

I. Introduction

Recent publications establishing the existence of discrete brown adipose tissue (BAT) depots in humans, which can be activated by reductions in the ambient temperature, have suddenly spurred interest in studies on the biology of brown fat conducted over the past 40 years. Particularly in mice, in which specific genes can be conditionally inactivated, the function of brown fat in the regulation of body temperature and body weight can be rigorously established, and models for the regulation of brown fat differentiation can be tested. A review highlighting the current excitement in brown fat biology has recently been published in a special issue of *Cell Metabolism* (Volume 11, Issue 4, 2010). Despite the progress, there are many unanswered questions on both the function of brown fat and the regulation of its induction. Specifically, the strong data indicating that mice with higher levels of brown adipocytes, particularly those residing in white fat depots, are more able to respond to adrenergic stimulation to burn off fat warrants further investigation. Accordingly, elucidating the differentiation mechanisms for induction of brown fat in mice and applying the technology to obese humans are important goals. In this review, we have examined the studies that address what we consider to be the most critical issues in the brown fat field as opposed to providing a noncritical but comprehensive review of the literature.

II. Genetics of BAT in Mice

Controlled genetic studies on obesity phenotypes in mammals originated with the discovery of the *ob* mutation in the C57BL/6J (B6) inbred mouse.[1] Subsequent investigations on this mouse mutant led to the discovery of a biological mechanistic model for food intake that guided the identification of leptin and its receptor as the underlying molecular mechanism.[2,3] Though less celebrated, a more profound finding for not only obesity, but also for virtually every phenotype that ranges in severity from disease to normality, was the finding by Coleman and Hummel that obesity and diabetes phenotypes associated with the *ob* mutation depended upon the genetic background on which the *ob* gene is expressed.[4] Thus the idea was established that inbred strains themselves were reservoirs of variable alleles that can affect the severity of obesity and diabetes-related traits. Surwit was among the first investigators to recognize that allelic variability among so-called normal inbred strains provided a complex genetic model for type 2 diabetes as a consequence of differential sensitivity to an obesogenic environment created by feeding mice a high-fat diet.[5,6] With this model, Surwit showed that the B6 strain readily developed obesity and eventually diabetes after eight or more weeks on the high-fat diet, whereas the A/J strain was relatively resistant to the development of diet-induced obesity. The genetic basis for this quantitative trait remains unresolved.

Concurrent with the discovery of genetic models for diet-induced obesity in the mouse, the development of agonists to the atypical β-adrenoreceptor of rodents (i.e., β3-adrenergic receptor; Adrb3) by Arch *et al.*[7] initiated experiments in the rat by Himms-Hagen *et al.* showing that the Adrb3 agonist CL-316,243 reduced adiposity, increased whole-body energy expenditure, and induced uncoupling protein 1 (*Ucp1*) expression in interscapular brown adipose tissue (iBAT) and brown adipocytes in white fat depots of rats fed a high-fat diet and in the genetically obese Zucker rat.[8,9] The possibility that the difference in susceptibility to diet-induced obesity between B6 and A/J mice may be due to genetic variation in the adrenergic-mediated induction of *Ucp1* was suggested by Surwit and colleagues, who showed large differences in the induction of *Ucp1* in retroperitoneal fat between B6 and A/J mice.[10] Himms-Hagen *et al.* also showed that the induction of *Ucp1* in both iBAT and the traditional white fat depots occurred in the absence of hyperplasia,[11] yet it was well known that chronic adrenergic stimulation by cold exposure leads to a robust hyperplasia in iBAT.[12] Treatment with Adrb3 agonist suppressed obesity in both *ob/ob* mice and B6 mice fed a high-fat diet, but most interesting was that the expression of *Ucp1* together with brown adipocytes was induced in traditional white fat depots in both mouse models. Since cold exposure stimulates both cell proliferation and induction of *Ucp1* in iBAT, whereas an Adrb3 agonist induces *Ucp1* without cell proliferation, the action of the Adrb3 agonist appears to be restricted to regulating the brown adipocyte differentiation program in the absence of cell proliferation. These studies highlighted the existence of an effective mechanism based upon thermogenesis in brown fat to reduce obesity by manipulation of the environment (i.e., ambient temperature) or by the administration of an adrenergic agonist selectively targeting Adrb3 in the adipocyte.

These initial studies on the robust induction of *Ucp1* by Adrb3 agonists shifted the focus from iBAT to brown adipocytes induced in traditional white fat depots. While this phenotype was predicted from several earlier studies on mice, and in cats and rats exposed to the cold,[13–15] particularly intriguing was the finding that A/J and B6 mice, which varied in diet-induced obesity, also varied in their induction of *Ucp1* in white adipose tissue when administered an Adrb3 agonist.[10] Accordingly, variant alleles for genes associated with *Ucp1* expression and the brown adipocyte differentiation program are extant within the common inbred strains, thereby providing a genetic system to identify genes critical for the induction of brown fat in white fat depots. In contrast, no variation associated with the amount of iBAT among inbred strains has been described; this will be addressed in more depth later.

The investigation into the mechanisms involved in the induction of brown fat has proceeded along two independent lines, one based upon the systematic genetic analysis of the genes controlling the variability in brown fat induction in various stocks of inbred strains of mice and the other on the insights derived from genetically engineered mutations of various genes, several of which were serendipitously shown to produce lean mice with induced numbers of brown adipocytes in white fat. In this review, we will attempt to integrate the two research strategies. While the molecular and transgenic approaches have identified candidate genes with putative physiological and biochemical functions and are often considered more causal, they also suffer from being susceptible to both *in vitro* and *in vivo* artifacts. On the other hand, physiological studies that exploit the genetic variation extant in the mouse population, although being more dependent on correlations, are nonetheless able to provide more novel complementary systemic insights into the brown fat biology.

A. Characteristics of Brown Adipocyte Induction in Mice Important for the Design of Genetic Studies

In the design of a quantitative trait loci (QTL) study to identify genes controlling brown fat induction, the selection of the fat depot becomes critical, because each fat depot has its own unique phenotype.[16] One is faced with the decision to select the white fat depot that one ascertains, with incomplete data, will be most amenable to analysis and informative. First of all, the iBAT in adult mice is not variable between A/J, B6, and 129/SvJ strains.[16,17] However, large variation in levels of *Ucp1* expression are found among white fat depots, and each depot varies independently among strains (Fig. 1). Among the peritoneal depots, gonadal fat showed large variation between strains, but the expression was low compared to that detected in other depots. Mesenteric fat and retroperitoneal fat both have relatively high levels of expression and big differences between strains and are therefore suitable for a QTL analysis. We selected the retroperitoneal fat for the QTL analysis because it is a fat depot also found in humans (the epididymal fat is not) and is a discrete tissue that can be more easily and rapidly dissected from the mouse compared with mesenteric fat.

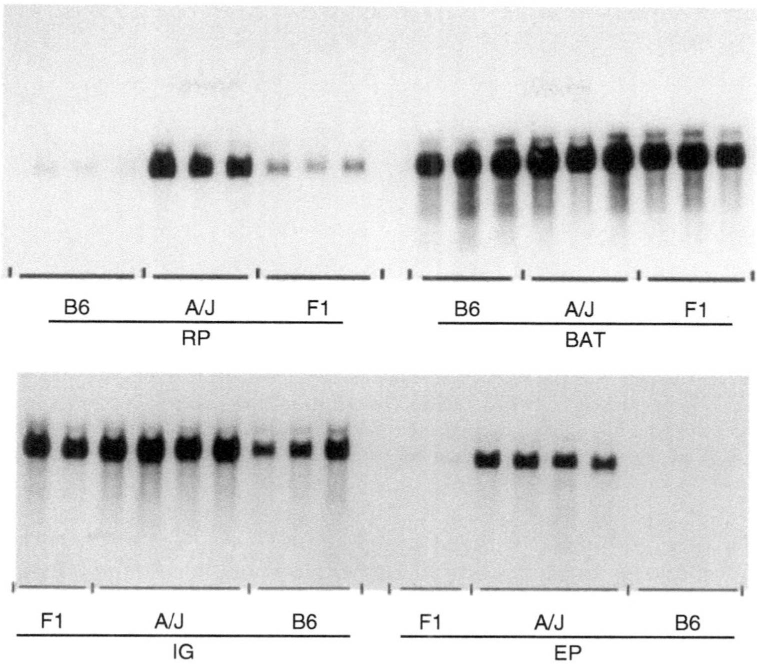

FIG. 1. The levels of *Ucp1* mRNA in the retroperitoneal fat (RP), interscapular brown fat (iBAT), inguinal fat (IG), and epididymal fat (EP) in total RNA isolated from A/J, B6, and (B6 × A/J)F1 mice are shown. Mice were treated with the β3 agonists for 7 days before isolation of RNA, at a time when induction of *Ucp1* mRNA had reached a plateau. 10 mg of RNA from BAT was analyzed, as well as 15 mg from the other fat depots.

Given these features of *Ucp1* expression in different white fat depots, we cannot assume that the QTLs associated with the *Ucp1* mRNA levels in retroperitoneal fat also control *Ucp1* in other depots, especially the subcutaneous fat.

The morphological features of a white fat depot with functional brown adipocytes are shown in Fig. 2. *Ucp1* mRNA induced in the brown adipocytes of the retroperitoneal fat depot of a AXB8 or A/J mouse can reach levels that approach 10–70% of that present in iBAT. These brown adipocytes are located in clusters of cells that show striking resemblance to the morphology of iBAT. Other areas in the tissue retain the morphology typical of a white fat depot.

Recently, several reports of gene knockout (KO) models (e.g., the autophagy-related 7 (yeast) gene [*Atg7*] and the regulatory associated protein of MTOR, complex 1 gene [*Rptor*][18,19]) claim that brown adipocytes are induced in white fat. This claim is based upon reduced whole-body adiposity, small increases in *Ucp1* in white fat depots, and a histology which shows multilocular cells interspersed

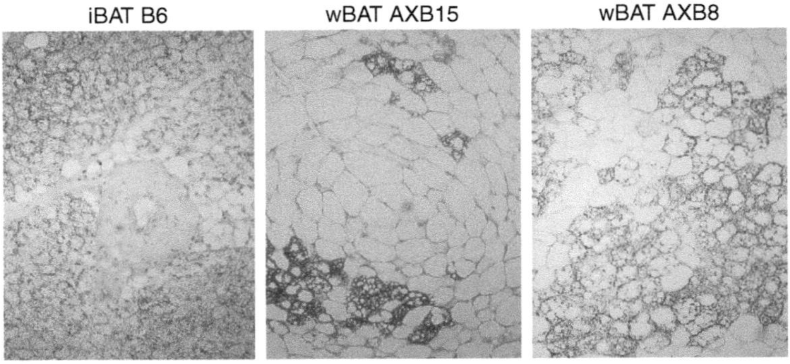

FIG. 2. Ucp1-immunohistology of paraffin-embedded sections from interscapular brown fat (iBAT) and retroperitoneal fat (wBAT) from AXB8 and AXB15 mice demonstrates the organization of brown adipocytes into clusters that resemble fields of densely packed brown adipocytes characteristic of iBAT. The size of these regions in wBAT may be quite variable, depending upon the strength of the induction.

among white adipocytes; however, these cells have not been immunostained for Ucp1 and therefore resemble macrophages as much as brown adipocytes. It is questionable whether the brown fat phenotype in some of these mutants represents a physiologically normal induction of brown adipocytes. One of the rare gene KO models that indeed show a bona fide induction of brown adipocytes in white fat, which resembles the cold induction in selected inbred strains, involves the vitamin D receptor gene (*Vdr*).[20] The relevance of gene KO models to brown adipocyte induction in white fat will be discussed later.

B. Strain Variation and Recombinant Inbred Strains of Mice

The phenotypes of obesity and thermogenesis affected by Adrb3 agonists and ambient temperature in A/J and B6 mice are complex. They involve upregulation of *Ucp1* and other molecular components of thermogenesis in existing iBAT and *de novo* induction of new differentiated brown adipocyte populations in white adipose tissue, both of which are controlled in part by the sympathetic nervous system. In addition, the phenotypes depend on the variation in susceptibility to obesity that exists between A/J and B6 mice.[6,16] To unravel the genetic and molecular complexity of brown adipocyte induction in white fat depots and determine its role in energy balance and obesity, we utilized two genetic tools: the AXB recombinant inbred (RI) strains and progeny from backcross and intercross matings between A/J and B6 mice and the RI strains. The AXB RI lines were formed by crossing A/J and B6 mice and intercrossing the resulting F1 progeny to establish an F2 population that is then brother–sister mated for a minimum of 20 generations to establish inbred lines.[21,22] If a genetic trait is complex, that is, controlled by allelic variation at more than one gene, then alleles from each gene associated with the trait in the A/J and B6 parental strain will be fixed in new combinations in the different RI lines. Thus, each RI line with these novel recombinant chromosome patterns will provide potentially unlimited numbers of mice to establish the phenotypes determined by each gene or combination of genes not present in the original parental lines.

As shown in Fig. 3, the levels of *Ucp1* mRNA in the retroperitoneal fat depot of the parent A/J mice exposed to a 4 °C ambient temperature for 1 week were 40 times higher than that found in B6 mice. However, the novel recombination of alleles in the RI lines generated a continuum of lines of mice with levels of *Ucp1* that at the lower extreme in AXB10 mice were slightly less than that observed in B6 mice and at the other extreme in AXB8 mice were 2.5 times higher than that occurring in A.J mice (Fig. 3).[23] The chromosomal linkage analysis of genes controlling *Ucp1* induction will be discussed later with the QTL analysis of brown fat induction in intercross and backcross progeny.

III. Genetic Variability in the Induction of Brown Fat and the Obese Phenotype

We used a subset of the RI lines to ask two questions essential to the analysis of the relationships between constitutive and inducible levels of BAT and obesity. The first question asked whether the potential for induction of brown adipocytes in white fat determined the development of obesity in the absence of adrenergic stimulation by cold exposure or Adrb3 agonist treatment. In other words, does a constitutive level of *Ucp1* mRNA (which in mice comes from the *Ucp1* expressed in the discrete iBAT depots and does not show strain variation[17,24,25]) affect the development of obesity? Conceivably, this might occur if a cafeteria-like diet directly induced thermogenesis through activation of iBAT Ucp1 as previously proposed by Rothwell and Stock.[26] Parental A/J, B6, and six RI lines with variable levels of *Ucp1* mRNA inducible in white fat were fed a high-fat diet for 18 weeks at an ambient temperature of 23 °C to establish an obese state characteristic for each line. Weight gain varied from 11 g in A/J to 26 g in B6 mice as previously observed.[10] However, AXB8 mice, which had the highest capacity to induce *Ucp1* mRNA in white fat depots upon adrenergic stimulation, gained 24 g of fat, only slightly different than that observed for B6 mice, which had the lowest level of *Ucp1* expression (Fig. 4A). This result indicates that in mice fed a high-fat diet at 23 °C, the capacity for induction of *Ucp1*/BAT in white fat had no impact on the

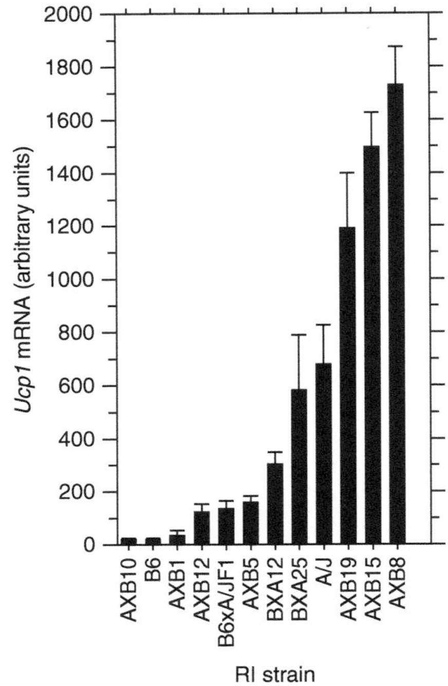

n	7	16	13	12	10	11	13	5	16	5	8	8
Sig.	*	*	*	*	*	*						
							*					
								*	*			
										*	*	
											*	*

FIG. 3. Bar graph shows *Ucp1* mRNA levels in the retroperitoneal fat pad from parental A/J, B6, F1 hybrid, and RI strains of mice derived from the parental mice. Mice were housed at 5 °C in groups of 2–3 mice per pen for 7 days before the isolation of RNA. Induction in the cold is similar to that caused by a β3-adrenergic agonist. Data was presented as the mean ± SE. The number of animals (*n*) of each strain and the results of statistical analyses by ANOVA (Fisher's PLSD test in Statview 4.1) are given in the lower chart. Strains sharing asterisks in a row are not significantly different from each other at $P = 0.05$.

development of diet-induced obesity at normal ambient temperatures. One could argue that despite similarities in the levels of *Ucp1* mRNA or protein in the iBAT of A/J and B6 mice, some unknown form of sympathetically mediated activation of iBAT-dependent, diet-induced thermogenesis exists among A/J mice at 23 °C to reduce diet-induced obesity. However, we have recently emphasized that the absence of increased diet-induced obesity in B6-*Ucp1* KO mice, with no capacity for *Ucp1*-dependent thermogenesis by a brown

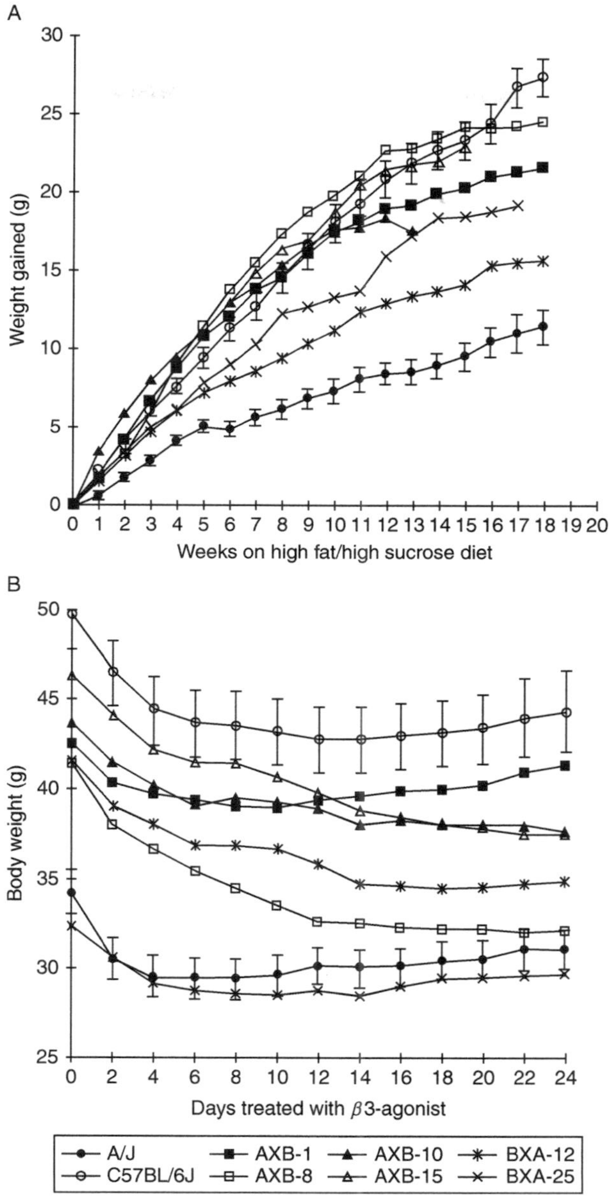

FIG. 4. (A) Body weight gained by parental and RI strains fed a high-fat/high-sucrose diet starting at 7 weeks of age and continuing for 18 weeks. (B) After 18 weeks, mice continued to be fed the high-fat/high-sucrose diet and were also treated with the β3-adrenergic agonist CL 316,243, as described.[16] Changes in body weight are shown. The mean ± SE for the B6 and A/J parental strains are shown, whereas only the means have been plotted for the RI stains. The number of mice in each group ranges from 9 to 12.

adipocyte, indicates that diet-induced thermogenesis by brown fat is more myth than actual fact.[27] The significance of these experiments on brown fat-dependent, diet-induced thermogenesis in mice lies in whether the differences in brown fat recently detected in humans, often living at or near thermoneutrality, is functionally related to the level of obesity and diabetes in these individuals.[28] These experiments in mice suggest that the differences in BAT in adult humans, as recently observed,[29–32] will not be a significant factor in determining variable obesity in humans at thermoneutral ambient temperatures.

In the second part of the experiment, the same eight lines of mice with variable levels of obesity described in Fig. 4A were treated with the Adrb3 agonist CL-316243 for 24 days while continuing to be fed the high-fat diet (Fig. 4B). All mice lost body weight; however, the amount of body weight lost depended upon the levels of *Ucp1* mRNA that were induced in the white fat depots (Fig. 3)[16]; that is, mice with higher levels of adrenergically induced *Ucp1* mRNA in retroperitoneal fat lost more body weight. More recently, similar data was obtained in mice exposed to the cold (4 °C) rather than administered an Adrb3 agonist.[33] Thus, genetically determined variability in the induction of brown adipocytes in white fat depots is a very significant factor in how an individual will respond to adrenergic stimulation by exposure to cold or drugs. The results also suggest that pharmaceuticals that can induce the number of brown adipocytes in discrete BAT depots or white fat depots would be an effective antiobesity strategy; however, this relationship between obesity and the capacity to induce *Ucp1* does not exist under normal room temperature (~ 25 °C).

IV. Chromosomal Mapping of Genes Controlling Brown Fat Induction by QTL Analysis: Rationale and Strategy for a Genetic Analysis

Given that there is probably no antiobesity strategy in mice and rats that is as effective as the induction of brown adipocytes in white fat depots, identifying genes that could enhance the induction of brown adipocytes is an important goal. The analysis of the RI strains provided evidence that the induction of brown adipocytes in white fat upon adrenergic stimulation involved multiple genes; however, neither the number of genes nor their chromosomal location could be determined from analysis of strain distribution patterns alone. To identify the genes controlling brown adipocyte induction, we pursued a QTL analysis in which the primary phenotype was *Ucp1* mRNA levels in retroperitoneal fat depots after 7 days at 5 °C, initially measured by semiquantitative Northern blots[16] and later by quantitative RT-PCR with TaqMan probes using the ABI sequence detection system.[23] Since there was a strong correlation between inducible *Ucp1* mRNA and the content of brown adipocytes determined by immunostaining for Ucp1 in the tissues of mice exposed to the cold,[16] we think that the *Ucp1* mRNA subphenotype is a reliable surrogate for brown adipocyte induction. The conversion of a white fat depot devoid of brown adipocytes to one rich in brown adipocytes requires the activation of brown fat-specific adipogenesis, mitochondrial biogenesis, and increased innervation and vascularization.[34] However, the most specific molecular marker and functionally important biochemical change is the activation of the gene encoding Ucp1, which is located in the inner mitochondrial membrane where it functions as the key mechanism for heat production by uncoupling oxidative phosphorylation. Mice with an inactivated *Ucp1* gene cannot tolerate an acute exposure to the cold because they have lost an important mechanism for thermogenesis.[35] Since *Ucp1* expression phenotype is near the end of a complex pathway from cold receptors on the surface of the skin to the processing of signals in the central nervous system that activate signaling and transcription factors in the adipocyte capable of being converted to a brown adipocyte, variable activity of any gene involved at any step of this complex process can be revealed by QTL analysis, if its effects on the levels of *Ucp1* mRNA are sufficiently robust and precise. As we shall show, gene expression for several genes (*Ppara*, type 2 deiodinase [*Dio2*], and *Ppargc1a*) implicated in *Ucp1* expression are highly variable in white fat of cold-exposed mice and show high correlations with *Ucp1* expression.

Based upon evidence that multilocular brown adipocytes in white fat depots were not labeled with BrdU in rats treated with the Adrb3 agonist CL 316,243,[11] we assumed that cell proliferation of a progenitor brown adipocyte does not occur in white fat depots, unlike the cell proliferation that occurs in iBAT during chronic cold exposure.[12] Genes controlling the proliferation of adipocyte progenitors in white fat are assumed not to be involved in the induction of brown adipocytes in white fat. In addition, gene expression data showing that maximal levels of *Ppargc1a* are reached within 1 day of adrenergic stimulation[24] and changes in adipocyte morphology showing the emergence of a multilocular adipocyte from a unilocular adipocyte are consistent with the hypothesis that mature unilocular white adipocytes are directly converted to multilocular brown adipocytes (Fig. 5A and B). If in fact brown adipocytes arise from mature white adipocytes, then it suggests that the brown adipocyte differentiation program is established during a window of postnatal adipose tissue development between 10 and 21 days of age when brown adipocytes transiently appear and then disappear.[25] Since not all adipocytes look like brown adipocytes, this differentiation process has been implemented in selected mature white adipocytes at 10 days of age. Thus, in adult stages, only adrenergic stimulation of this adipocyte is required to initiate mitochondrial biogenesis and *Ucp1* expression. Such a mechanism may not even involve an

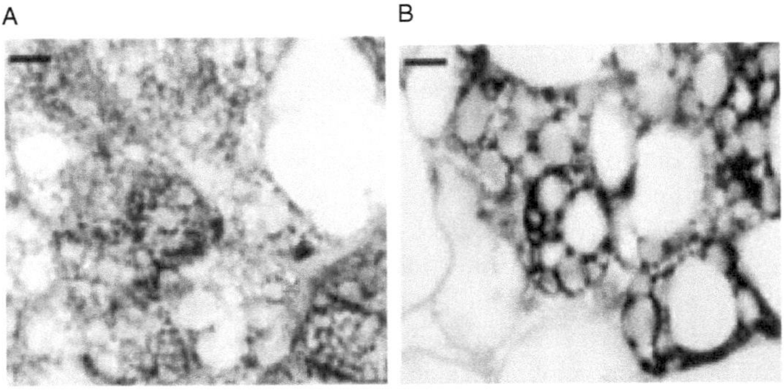

FIG. 5. Comparison of the morphology of brown adipocytes in interscapular brown fat (iBAT) (A) and retroperitoneal fat (B) after immunohistology with anti-*Ucp1* antibody. Note that the unilocular lipid droplet in mature white adipocytes gradually becomes fragmented into multilocular distribution of lipid vesicles as the induction proceeds. The magnification bar measures 10 μm.

epigenetic modification of chromatin structure, but merely activation of protein kinase A (Prka) signaling and other mechanisms controlling cAMP levels. Viewed from this perspective, the search for candidate genes controlling the induction of brown adipocytes becomes focused on genetic mechanisms determining the subpopulation of white adipocytes in a white fat depot with a chromatin structure unique for brown adipocytes; parenthetically, such a chromatin structure remains to be determined. The brown adipocytes in traditional white fat depots may be masquerading as white adipocytes until they are adrenergically stimulated. A research goal will be to identify the molecular basis of the chromatin structure that enables a rapid transcription of brown adipocyte-specific genes. A component of such a mechanism could be nuclear receptor interacting protein 1 (Nrip1).[36,37]

In addition to the effects of the cold environment and/or adrenergic agonist stimulation on induction of *Ucp1*/BAT in the retroperitoneal fat depot, the idea that a high-fat/high-sucrose (cafeteria) diet induces brown fat thermogenesis as a physiological mechanism for the regulation of energy balance has been promoted by several publications (reviewed by Cannon[34]). If a cafeteria diet had a significant role in the induction of *Ucp1* in white fat depots, then the genes controlling this function should be uncovered in a QTL analysis. The strategy for this study employed two independent cohorts of (B6 × A/J) F1 × A/J backcross mice. One cohort of ~400 mice was weaned onto a high-fat diet (58 kcal%, Research Diets 12331) at 3 weeks of age and continued on the diet during 7 days of exposure to the cold at 4 °C beginning at 8 weeks of age, then sacrificed. A second cohort of mice was weaned onto a low-fat chow diet (Picolab 5053) and continued on the diet until they were exposed to the cold at 8 weeks of age. RNA was isolated from the retroperitoneal fat depot for gene expression analysis and DNA from the tail for genotyping.

In summary, the major factors underlying this strategy for the QTL analysis of brown fat inductions were:

1. The developmental features of brown adipocytes in white fat.
2. The focus on identifying QTLs controlling *Ucp1* expression, as the most downstream component of functional brown fat, followed by the integration of QTLs affecting both *Ucp1* transcription and signaling molecules associated with *Ucp1* transcription.
3. The role of a cafeteria diet on brown adipocyte induction in white fat.

An important shortcoming of our strategy is that we completely focused on only one gene integral to the terminally differentiated brown adipocyte, that is, the *Ucp1* gene. In retrospect, analysis of components of the mitochondria electron transport chain or oxidative phosphorylation (i.e., ATP synthase) or of lipid oxidation would have provided additional insights into the genetic control of brown adipocyte induction in white fat. Do QTLs that control *Ucp1* expression also control the expression of the components of the electron transport chain? What we did achieve was a parallel genetic analysis of critical transcription factors implicated in the regulation of *Ucp1*.

V. Transcription Factors Controlling *Ucp1* Expression and QTLs Associated with Brown Adipocyte Induction in Retroperitoneal Fat

Our efforts to identify genes controlling brown adipocyte induction by QTL analysis occurred in parallel with studies from several laboratories on the molecular analysis of the brown adipocyte differentiation program, based upon transgenic studies and transient expression analysis of the *Ucp1* promoter in cell cultures. The rationale behind the molecular approach was predicated on the fact that *Ucp1* expression was restricted to brown adipocytes. Therefore, identification of the signaling and transcription factors determining *Ucp1* expression in brown adipocytes would reveal brown fat-specific regulatory factors that could be used to induce an increase in brown adipocytes. In a transgenic mouse, a 3.7 kb DNA fragment from the 5′ flanking region of the *Ucp1* gene was able to show cold induction of a mini-gene restricted to brown fat, providing the DNA target that was necessary and sufficient to identify DNA regulatory motifs responsible for *Ucp1*-specific expression.[38] Subsequently, transient expression analyses in brown adipocyte cell cultures were used to identify transcription factors interacting with motifs in the 2.5 kb DNA regions (Fig. 6). Two pivotal regions were identified, a 220 bp enhancer located ~2.8 kb upstream of the transcription start site and a basal promoter extending 272 bp upstream of the transcription start site.[39,40] Both regions contained cAMP responsive element-binding protein (Creb) domains (Cre2 and Cre4) that lose their ability to support transcription when mutated.[40] More recently, it has been shown by chromatin immunoprecipitation (ChIP) assays that activating transcription factor 2 (Atf2) binds to Cre2 in the enhancer, whereas Creb binds to the Cre4 basal promoter.[41]

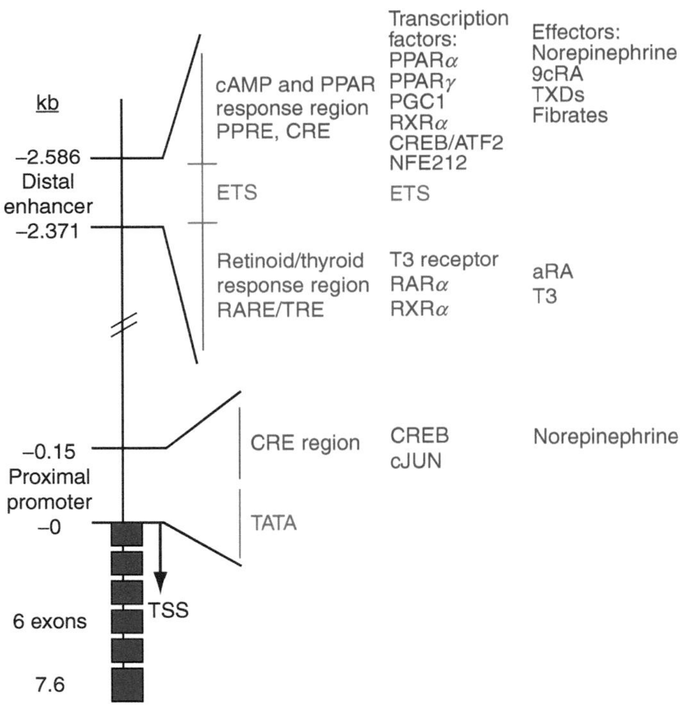

FIG. 6. Schematic representation of the *Ucp1* gene on mouse chromosome 8. A transgene carrying 3 kb of the 5′ flanking region, 0.3 kb of the 3′ flanking region, and exons 1, 2, and 6 carried the regulatory information for inducibility by cold exposure specifically in brown fat.[38] Mutation analysis of selective motifs in a 220 bp segment 2.371 kb from the TTS and a 150 bp segment in the proximal promoter identified CRE motifs essential for inducibility by catecholamine analogs in brown fat cell lines. Further studies showed the location of DNA motifs for transcription factors, listed on the right, that have been strongly implicated in the regulation of *Ucp1*.

A key finding for the understanding of *Ucp1* regulation was the identification of a direct repeat (peroxisome proliferator response element; Ppre) in the enhancer region that interacts with Ppara and Pparg transcription factors.[42] Accordingly, the Cres provided a mechanism for adrenergic (cold-mediated) regulation of *Ucp1*, whereas the Ppre provided a DNA-binding site for transcription factors regulating the adipocyte differentiation pathway. The latter led to the identification of Ppargc1a, the transcriptional coactivator of Pparg.[43] The initial evidence was interpreted to indicate that the brown fat-specific expression and cold inducibility of Ppargc1a was consistent with it being a master transcriptional regulator of the brown fat differentiation program; however, more careful quantitative analysis of *Ppargc1a* expression and the phenotypes of the *Ppargc1a* KO mice indicate that Ppargc1a has a broad regulatory role as a modulator of transcription of *Ucp1* and other genes associated with mitochondria structure and function in both brown fat and skeletal muscle, as well as being a regulator of genes of gluconeogenesis in liver.[44–46] This elucidation of the transcriptional regulation of *Ucp1* identified several transcription factors involved in the brown fat differentiation program, but none that could be considered brown fat-specific master regulators, that is, expressed and functioning only in the brown adipocyte.

With this brief diversion to provide some necessary background on the molecular mechanisms controlling *Ucp1* expression, we can now return to the QTL study with backcross progeny. The molecular analysis of *Ucp1* regulation has not revealed any truly brown fat-specific signaling or transcription molecules; however, a regulatory model in which brown fat-specific expression is determined by a modular molecular mechanism becomes a plausible, though a somewhat unsatisfying, alternative. We reasoned that if an interactive molecular mechanism controls brown adipocyte induction in white fat, perhaps the interactive transgressive mechanism revealed by the analysis of RI lines and backcross and intercross progeny, as described below, in fact is the manifestation of such a mechanism at the gene level. Two new cohorts of ~ 400 (A/J × B6)F1 backcross mice were generated to test the idea that the known transcription factors of *Ucp1* were associated with *Ucp1* induction phenotypes following 7 days in the cold. As described above, mice in cohort 1 were fed a standard low-fat chow (11.9 kcal% fat) diet throughout the protocol, including during the week at 5 °C, whereas mice in cohort 2 were fed a high-fat diet containing 58 kcal% fat (Research Diets 12331) from weaning until the end of the cold exposure (mice were subjected to cold exposure at 5 °C for 7 days at 2 months of age). The rationale behind conducting a QTL analysis of mice fed a high-fat diet was based upon the general belief that *Ucp1* was inducible by a cafeteria diet and protected the animal against diet-induced obesity.[26,34] Accordingly, identification of a pathway or gene controlling the induction of *Ucp1* by high-fat/sucrose was viewed as an important secondary goal of the QTL analysis.

Prior to analysis of gene expression in the backcross mice, we assessed the variation in expression of genes implicated in *Ucp1* transcription between A/J and B6 mice exposed to cold for 7 days. Differences in expression between the mouse strains were too small for analysis in backcross mice (less than twofold) for the following genes: delta-like 1 homolog (Drosophila) (*Dlk1*); nuclear factor, erythroid derived 2, like 2 (*Nfe2l2*); sterol regulatory element-binding transcription factor 1 (*Srebf1*); *Pparg*; CCAAT/enhancer-binding proteins (C/EBP), α (*Cebpa*), β (*Cebpb*), and δ (*Cebpd*); forkhead box C2 (*Foxc2*); and *Adrb3*. Robust differences between B6 and A/J were found for *Ppara* and *Dio2*, and modest variation was found for *Ppargc1a*. In addition, the high-fat diet showed variable increases in the expression of these latter genes (Table I). Based upon the difference in gene expression in parental progeny, we used MIT markers to map chromosomal regions associated with levels of *Ppara*, *Ppargc1a*, and *Dio2* mRNA in backcross progeny. When the chromosomal linkages for *Ucp1* QTLs on chromosomes 2, 3, 8, and 19, previously established in analysis of backcross mice and RI strains (Fig. 7), were compared to the location of QTLs for *Ppara*, *Ppargc1a*, and *Dio2*, a remarkable coincidence was observed for the core QTLs, defined as those controlling *Ucp1*

TABLE I
EFFECT OF DIETARY FAT ON GENES ASSOCIATED WITH BROWN FAT INDUCTION

		Strain			
Gene symbol	Diet	A/J	B6	F1	BC
Ucp1	LF	100	4	17	30
	HF	84	15	25	35
Ppara	LF	100	25	32	42
	HF	138	45	65	71
Ppargc1a	LF	100	44	57	63
	HF	162	75	72	90
Dio2	LF	100	7	17	7
	HF	166	28	65	45

Effect of dietary fat on expression of *Ucp1*, *Ppara*, *Ppargc1a*, and *Dio2* in retroperitoneal fat after 7 days of cold exposure. Mice were fed either a low-fat (LF) chow diet (LabDiet 5053; 13 kcal% fat) or a high-fat diet (Research Diets, D12331; 58 kcal% fat) from weaning (3 weeks) until after cold exposure from weeks 8 to 9. Gene expression was measured via quantitative RT-PCR and normalized to cyclophilin b (*Ppib*). Data is represented as mean RNA levels with values of each gene set at 100% for LF-fed A/J mice. At least seven mice per group were analyzed for A/J, B6, and the (A/J × B6)F1 progeny (F1) whereas a minimum of 384 mice were analyzed for the (A/J × B6)F1 × A/J backcross (BC) progeny.

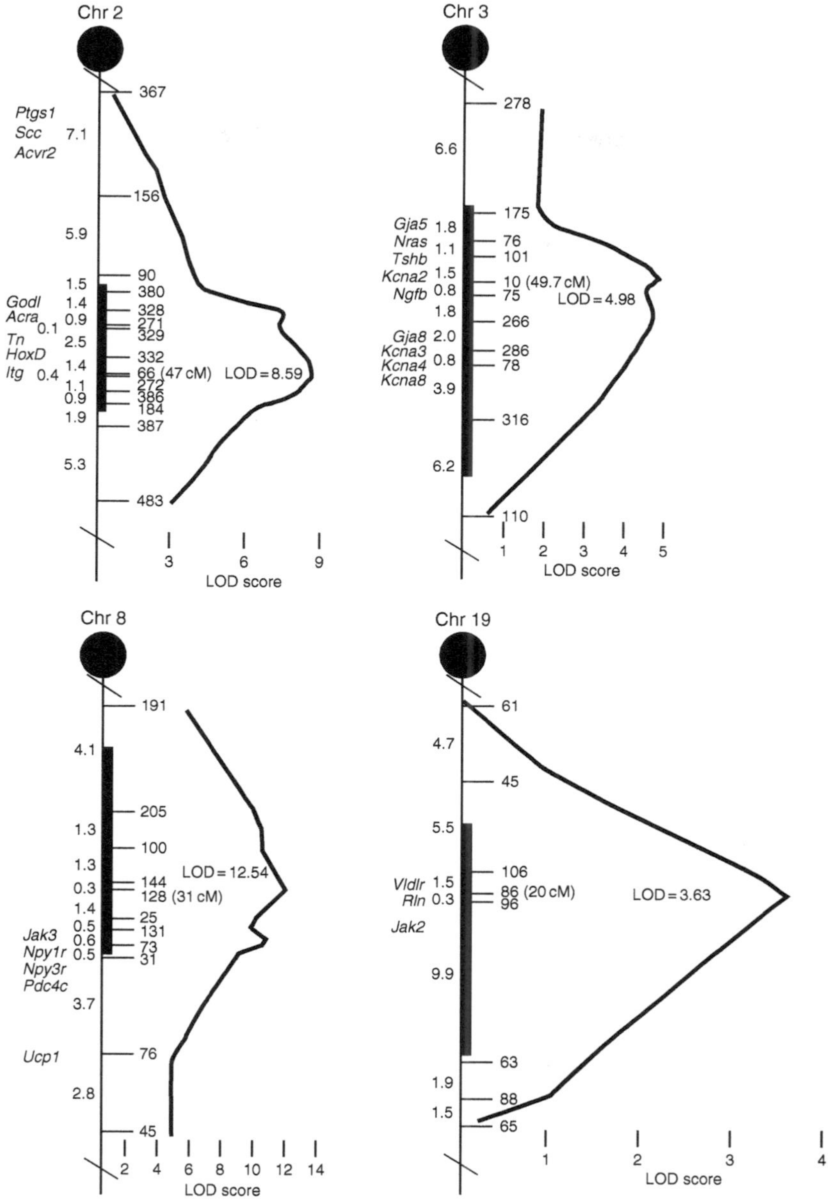

FIG. 7. Interval maps of the *Iba* (*induction of brown adipose tissue*) genes. Interval maps were generated for the four *Iba* QTLs located on chromosomes 2, 3, 8, and 19 based on recombinations between microsatellite markers using 789 backcross mice. The *solid line* represents the LOD scores generated for each interval with the peak LOD score indicated on each plot. The enhanced region within each chromosome is the 1-LOD confidence interval. LOD scores and confidence intervals for the *Iba* loci were generated using MapManager QTb8[139] and procedures described by Lander and Botstein.[140] Centimorgan distances and possible candidate genes are listed on the *left-hand side* of each plot.

expression (Fig. 8). In mice fed a high-fat diet, chromosomes 2, 3, 8, and 19 were associated with levels of *Ucp1* mRNA; however, unlike *Ucp1* expression, not all QTLs were involved with the *Ppara*, *Ppargc1a*, and *Dio2*. These associations for backcross mice fed either the low- or the high-fat diet are shown in Fig. 9. The complexity of QTLs is greater for mice fed the chow diet than the high-fat diet. The identities of the genes in the QTLs that determine the expression phenotypes are unknown. The fact that the peak of the QTL on chromosome 8 is very close to *Ucp1* provides a plausible candidate to account for variation in *Ucp1* mRNA levels, but it is difficult to imagine a mechanism by which Ucp1 regulates *Ppara*. Other genes under the peak of the chromosome 8 QTL appear to be better candidates at this time.

The interpretation of the QTL analysis suggests that, of the many genes implicated in *Ucp1* and brown adipocyte induction in white fat, at least three of these (*Ppara*, *Ppargc1a*, and *Dio2*), which show strong positive correlations with *Ucp1* expression, could play a role in controlling the induction of brown adipocytes. However, given that the mRNA trait is a continuous variable in segregating progeny, these regulatory factors probably do not function as a switch mechanism to initiate a differentiation process for brown fat differentiation. Rather, they are probably modulators that are involved in determining the continuous variable trait. Serendipity has given us many gene KO and transgenic overexpressing models in which inactivation or overexpression of a gene has initiated the activation of brown adipocyte induction in white fat depots. Table II, which provides a list of these genes and their chromosomal locations, will be discussed below.

VI. Transgressive Variation: A Manifestation of Synergistic Interactions Between QTLs and Between Transcription Factors and Signaling Molecules that Control *Ucp1* Expression

Synergy in the expression of *Ucp1* has been observed at two different levels, first at the level of interacting genes and second at the level of transcription and signaling factors implicated in *Ucp1* expression. The strain distribution of *Ucp1* mRNA levels in retroperitoneal fat of cold-exposed AXB and BXA RI mice provided clear evidence for the existence of transgressive variation for *Ucp1* expression, that is, the production of offspring with phenotypes that exceed the parental extremes.[16] The novel combinations of alleles at several loci contributing to *Ucp1* expression in RI strains result in higher levels than that which occur in the A/J strain. The gene interactions underlying transgressive variation are a unique feature of *Ucp1* expression that has not been observed in many other gene systems. The first QTL analysis of the (B6 × A/J)F1 × A/J backcross mice identified chromosomes 2, 3, 8, and 19 to be associated with *Ucp1* expression, and similar to the RI line data, the levels of *Ucp1* mRNA in the backcross progeny had ranges of expression that also showed transgressive variation; that

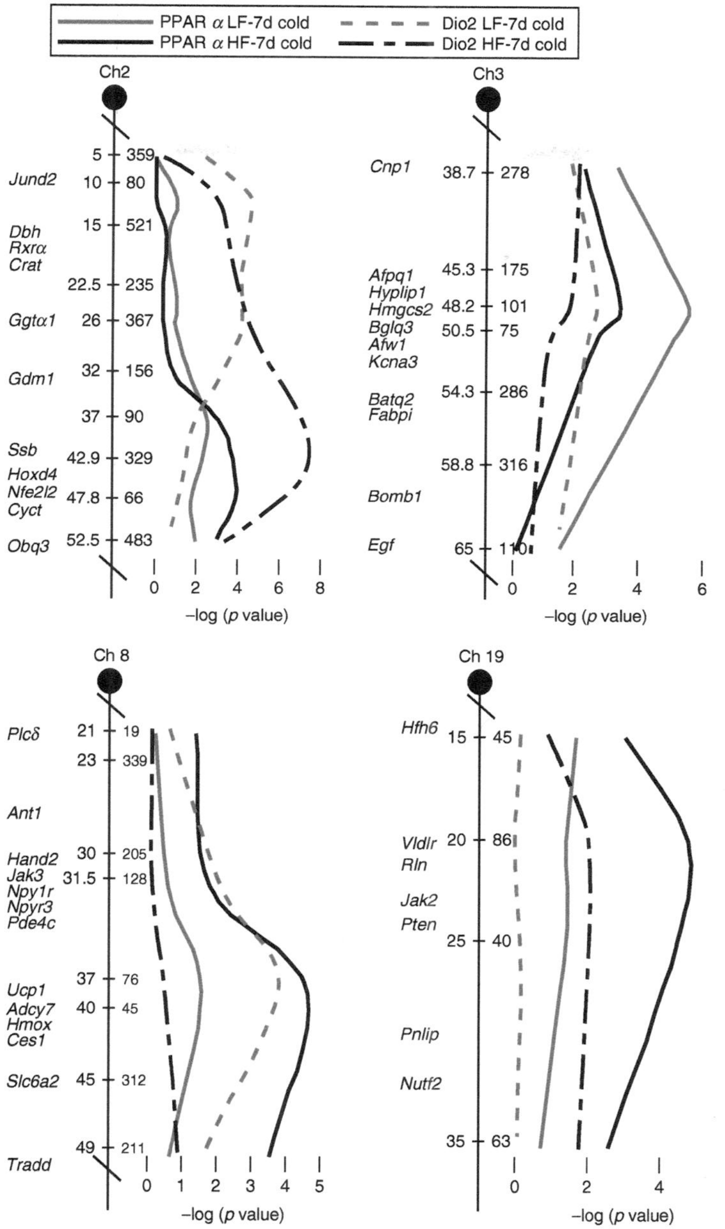

FIG. 8. Interval maps of QTLs controlling *Ppara* and *Dio2* mRNA levels in mice fed a low-fat and high-fat diet based upon recombinations between microsatellite markers using 387 and 386 mice fed a low-fat and high-fat diet. Microsatellite markers are shown on the right-hand side of each chromosome, and centiMorgan (cM) distance and candidate genes are shown on the left-hand side. The statistical significance for the association between mRNA levels and chromosomal markers are given as the negative log of the *p* value. Linkage maps were generated by ANOVA of microsatellite markers with gene expression levels and positioned on the chromosome with the use of the Jackson Laboratory Genome Database.

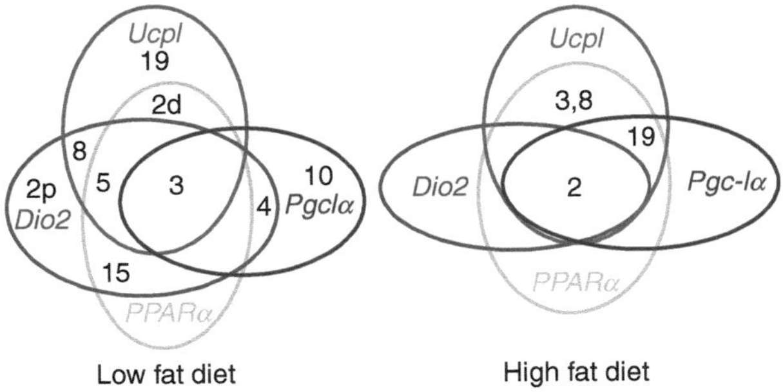

FIG. 9. Venn diagrams for chromosomes and gene expression illustrate the overlap of selective chromosomes in regulating genes associated with the regulatory network controlling *Ucp1* function. Diagram also underscores the profound effects of diet on genes associated with regulatory networks of *Ucp1* expression. Numbers refer to chromosomes carrying QTLs, and those enclosed within a circle indicate those QTLs associated with regulation of the indicated gene.

is, they exceeded the levels of *Ucp1* expression found in the A/J mice, approaching those found in AXB8 mice.[23,24] Positive effects on *Ucp1* mRNA levels are associated with an A/J allele at the peak QTL on chromosomes 2, 3, and 8, whereas the B6 allele on the QTL for chromosome 19 was associated with elevated *Ucp1* mRNA. Accordingly, transgressive variation is based in part upon the interaction of a variant gene encoded by the B6 allele on chromosome 19 with a variant gene on chromosomes 2, 3, and/or 8 encoded by an A/J allele. Sixteen sets of mice with allelic assignments possible at chromosomes 2, 3, 8, and 19 are rank ordered with respect to *Ucp1* mRNA levels (Table III). As predicted from the strain distribution pattern of the RI lines, AXB8 and seven other strains with levels that exceed A/J mice have A/J alleles at QTLs on chromosomes 2, 3, and 8 but a B6 allele on chromosome 19. This interpretation was further tested with an intercross of 2 RI lines, AXB8 and AXB10, that would assess the association of A/J and B6 alleles at the QTLs on chromosomes 8 and 19. As shown in Fig. 10A, strong synergistic interactions elevate *Ucp1* expression in mice homozygous for the A/J allele at DNA segment, Chr 8, Massachusetts Institute of Technology 45 (*D8Mit45*) and homozygous for the B6 allele on chromosome 19 near *D19Mit86* ($\sim$20 cM). Thus, an A/J allele at *D19Mit86* shows dominance in suppressing the effects of an A/J allele near *D8Mit45* at 40.5 cM (Fig. 10B and C). The chromosome 8 QTL affecting *Ucp1* mRNA production is near the *Ucp1* gene itself at 37.7 cM. It is possible that *Ucp1* itself is structurally variable between A/J and B6 mice and that it directly interacts by a *trans*-mechanism with a gene located near *D19Mit86*.

We have shown for the first time in mammals how *trans*-acting genes on known chromosomes in a very specific QTL system controlling *Ucp1* expression are giving rise to transgressive expression at the molecular level. Previously, it was

TABLE II

Transgenic and knockout models of *Ucp1* and brown fat

	Mouse				Human			
Gene	Symbol	Chr	Chr position/cM	Physical location (mb)	Symbol	Chr	Chr position/cM	Physical location (mb)
Uncoupling protein 1[35,89]	*Ucp1*	8	38	85.81	*UCP1*	4	4q28-q31	141.5
β1-Adrenergic receptor[67,100]	*Adrb1*	19	51	56.8	*ADRB1*	10	10q24-q26	115.8
β2-Adrenergic receptor[101]	*Adrb2*	18	34	62.34	*ADRB2*	5	5q31-q32	148.2
β3-Adrenergic receptor[103]	*Adrb3*	8	10	28.34	*ADRB3*	8	8p12-p11.2	37.82
Thyroid hormone receptor α[130]	*Thra*	11	57	98.6	*THRA*	17	17q11.2	38.22
Dopamine β-hydroxylase[108]	*Dbh*	2	15.5	27.02	*DBH*	9	9q34	136.5
Leptin[1]	*Lep*	6	10.5	29.01	*LEP*	7	7q31.3	127.88
Glycerol phosphate dehydrogenase 2, mitochondrial[131]	*Gpd2*	2	33	57.09	*GPD2*	2	2q24.1	157.29
Protein kinase A RII β subunit[68]	*Prkar2b*	12	B1	32.64	*PRKAR2B*	7	7q22	106.68
Forkhead box C2[50]	*Foxc2*	8	65.5	123.64	*FOXC2*	16	16q22-q24	86.6
Retinoblastoma 1[51]	*Rb1*	14	41	73.59	*RB1*	13	13q14.2	48.88
Nuclear receptor coactivator 2[132]	*Ncoa2*	1	1	13.13	*NCOA2*	8	8q13.3	71.02

(*Continues*)

TABLE II (*Continued*)

	Mouse				Human			
Gene	Symbol	Chr	Chr position/cM	Physical location (mb)	Symbol	Chr	Chr position/cM	Physical location (mb)
Nuclear receptor coactivator 1[132]	*Ncoa1*	12	A2-A3	4.25	*NCOA1*	2	2p23	24.81
Regulatory associated protein of MTOR, complex 1[18]	*Rptor*	11	E2	119.47	*RPTOR*	17	17q25.3	78.52
Nuclear receptor coactivator 3[74]	*Ncoa3*	2	H2-H4	165.82	*NCOA3*	20	20q12	46.13
Ribosomal protein S6 kinase, polypeptide 1[76]	*Rps6kb1*	11	C	86.33	*RPS6KB1*	17	17q23.1	57.97
Eukaryotic translation initiation factor 4E-binding protein 1[75,83]	*Eif4ebp1*	8	8	28.37	*EIF4EBP1*	8	8p12	37.89
Nitric oxide synthase 3, endothelial cell[133]	*Nos3*	5	9	23.87	*NOS3*	7	7q36	150.69
CCAAT/enhancer-binding protein (C/EBP), β[134]	*Cebpb*	7	12	35.9	*CEBPB*	19	19q13.1	33.79
Nuclear receptor subfamily 1, group H, member 3[135]	*Nr1h3*	2	40.4	91.02	*NR1H3*	11	11p11.2	47.27
PR domain containing 1[56,57]	*Prdm16*	4	E2	153.69	*PRDM16*	1	1p36.23-p33	2.99

Bone morphogenetic protein 7[136]	*Bmp7*	2	102	172.7	*BMP7*	20	20q13	55.74
Peroxisome proliferative activated receptor, γ, coactivator 1 α[45,46]	*Ppargc1a*	5	C1	51.85	*PPARGC1A*	4	4p15.1	23.79
Peroxisome proliferative activated receptor, γ, coactivator 1 β[137]	*Ppargc1b*	18	E1	61.46	*PPARGC1B*	5	5q32	149.11
Deiodinase, iodothyronine, type II[138]	*Dio2*	12	D3	91.96	*DIO2*	14	14q24.2-q24.3	80.66
Thyroid hormone receptor β	*Thrb*	14	A3	18.81	*THRB*	3	3p24.2	24.16
Vitamin D receptor[20]	*Vdr*	15	F1	97.68	*VDR*	12	12q13.11	48.24
Peroxisome proliferator activated receptor α[52]	*Ppara*	15	48.8	85.57	*PPARA*	22	22q13.31	46.55
Nuclear receptor interacting protein 1[37]	*Nrip1*	16	C3.1	76.29	*NRIP1*	21	21q11.2	16.33

TABLE III

STRAIN DISTRIBUTION PATTERNS IN RI LINES IN RELATIONSHIP TO *UCP1* MRNA LEVELS

	Chromosome 8						Chromosome 19			
	8M128	8M31	Ucp1	8M45	8M242	RP Ucp1	19M106	19M86	19M99	19M30
Strain	31 cM	34 cM	37.7 cM	40.5 cM	47 cM	mRNA	18.5 cM	20 cM	20.3 cM	21.4 cM
A/J	A	A	A	A	A	668	A	A	A	A
C57BL/6J	B	B	B	B	B	16	B	B	B	B
(B6 X A/J)F1	AB	AB	AB	AB	AB	129	AB	AB	AB	AB
A X B 8	A	A	A	A	A	1719	B	B	B	B
A X B 4	A	A	A	A	A	1700	B	B	B	B
A X B 15	A	A	A	A	A	1490	B	B	B	B
B X A 14	A	A	A	A	B	1439	B	B	B	B
A X B 14	A	A	A	A	A	1247	B	B	A	A
A X B 19	A	A	A	A	A	1181	B	B	B	B
A X B 20	A	A	A	A	A	896	B	B	B	B
A X B 18	A	A	A	A	A	630	B	B	B	B
B X A 25	A	A	A	A	B	576	B	B	B	B
A X B 24	B	B	A	A	A	562	B	B	B	B
B X A 11	B	A	A	A	A	387	A	A	A	A
B X A 4	A	A	A	A	B	359	A	B	B	B
B X A 12	A	A	A	A	A	295	B	B	B	B
A X B 6	B	B	B	B	B	226	A	A	A	A
B X A 17	B	A	A	A	A	200	B	B	B	B
B X A 8	B	A	A	A	A	186	B	B	B	B
A X B 5	B	B	B	B	B	153	B	B	B	B
A X B 12	B	B	B	B	B	122	A	A	A	B
B X A 7	B	B	B	B	B	86	A	A	A	A
A X B 2	A	B	B	A	A	77	A	A	A	B
B X A 24	B	B	B	B	B	55	A	A	A	A
B X A 2	B	B	A	A	A	51	A	A	A	A
B X A 13	B	B	B	B	B	51	A	A	A	A
A X B 1	B	B	B	B	A	40	B	B	B	B
B X A 1	B	B	B	B	B	37	B	B	B	B
A X B 10	B	B	B	B	A	13	B	A	A	A

Genotyping data in this table was assembled from the RI strain distribution patterns for AXB and BXA in the Mouse Genome Database (The Jackson Laboratory, Bar Harbor, ME) and by genotyping RI strain DNA. Microsatellite markers and relative positions (cM) for chromosomes 8 and 19 are noted on the top of the table. Boxed genotypes indicate RI strains with recombination between microsatellite markers.

shown that the occurrence of spontaneous ovarian teratomas in recombinant inbred strains was greater than that observed in progenitor strains.[47] A genetic analysis of morphological traits of yeast has revealed a complex QTL system with

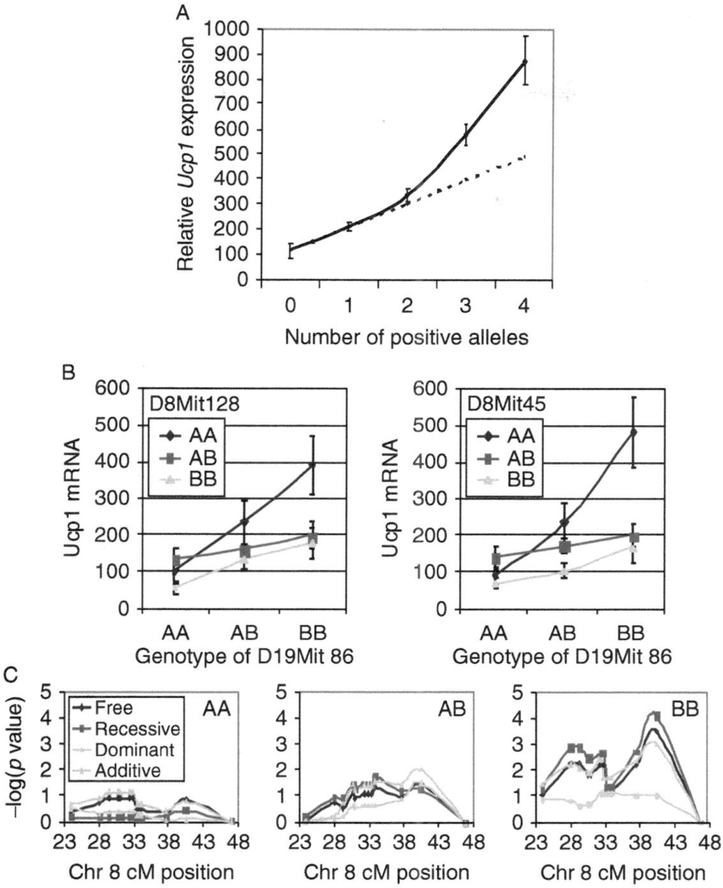

FIG. 10. (A) Synergistic effect of *Iba* QTLs on retroperitoneal *Ucp1* mRNA induction. This is a graphic representation demonstrating the complex epistatic and additive interactions between the four *Iba* loci. The *solid line* represents the experimental retroperitoneal fat *Ucp1* mRNA induction levels in relationship to the number of positive contributing loci. The *dashed line* indicates the theoretical value if interactions between *Iba* loci were only additive. (B) Effect of interactions between loci near *Ucp1* (*D8Mit45*, 40.5 cM; *D8Mit128*, 31 cM) and *D19Mit86* (20 cM) on retroperitoneal fat *Ucp1* mRNA induction in 286 intercross progeny from AXB10 and AXB8. Line interaction plots demonstrate a strong epistatic interaction between loci on chromosomes 8 and 19. This interaction appears to be stronger for *D8Mit45*, a locus nearer to *Ucp1* than for *D8Mit128*, the major QTL defined in the (A/J × B6)F1 × A/J backcross. Interaction plot data was generated using Statview v.4.5. (C) This figure represents the linkage analysis of a QTL on chromosome 8 using 286 intercross progeny from AXB10 and AXB8. The QTL analysis was performed for individual groups of mice having the AA, AB, and BB genotypes for a QTL on chromosome 19 near *D19Mit86*. Results demonstrate a very strong recessive gene effect near a locus at 40 cM on chromosome 8 for retroperitoneal fat *Ucp1* mRNA induction. This effect is observed only in animals with a BB allele on chromosome 19. Data for these plots was generated using MapManager QTb8.[139]

transgressive variation which has been correlated with gene expression analysis to produce results similar to the regulation of *Ucp1* and brown fat induction.[48] Others have conducted a methodological global gene expression analysis of liver and kidney with chromosome substitution strains between A/J and B6 mice to assess allelic effects on gene expression and found that allelic variation at several chromosomes affects expression of 4209 transcripts.[49] Many of the transcripts had levels of expression that exceeded those found in the parental strains, suggestive of transgressive variation.[49]

Evidence from backcross and intercross analyses indicated that interactions between a B6 allele on chromosome 19 with an A/J allele on chromosome 8 and possibly chromosome 3 synergize to maximize expression of *Ucp1*. Since we also showed that these QTLs determined variation in the levels of *Ppara*, *Ppargc1a*, and *Dio2* (Fig. 8), we speculate that these transcription factors and signaling molecules, as well as others, interact synergistically to maximize the expression of *Ucp1*. An overview of the signaling and transcription pathways controlling *Ucp1* expression is presented in Fig. 11. Below each of the factors is a number that indicates the ratio of expression for that factor between A/J and B6 mice. One immediately notices that all increases in expression are found for factors that lie downstream of *Prka*. There were no differences in expression for *Prka* or other factors that are upstream of *Prka*. Therefore, while retinoblastoma (*Rb*) and *Foxc2* have been implicated in the induction of brown adipocytes in transgenic mice and cell cultures,[50,51] variations in their expression are not determining the differences in the induction of *Ucp1* in A/J and B6 mice. However, the key feature of this system for the control

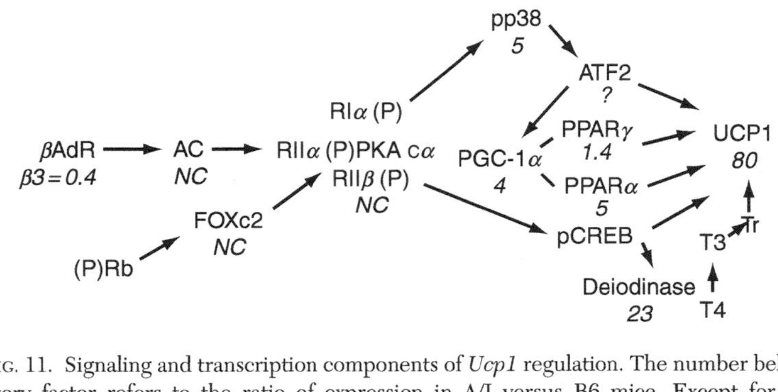

FIG. 11. Signaling and transcription components of *Ucp1* regulation. The number below each regulatory factor refers to the ratio of expression in A/J versus B6 mice. Except for the β3-adrenergic receptor, expression is higher in B6 than A/J at either the mRNA or protein levels; however, these strain-dependent differences all occur downstream of Pka. Most important is that none of the strain-dependent differences for the regulatory factors approaches the difference observed for *Ucp1* at either the mRNA or protein levels. This suggests that synergy among the regulatory pathways occurs with a changing cast of factors depending on the genotypes of *Iba* genes and environmental conditions. NC, no difference in expression between A/J and B6.

of *Ucp1* is that, while many regulatory components known to have essential roles in *Ucp1* expression have elevated expression in A/J mice, relative to *Ucp1* the differences in expression are very modest. For example, there is a 4–5-fold difference in expression of the master regulator *Ppargc1a* and other transcription factors between A/J and B6 mice, whereas *Ucp1* is expressed 80-fold greater in A/J than B6 mice. Interpretation of this expression data is facilitated by a map of the structure of the 5′ regulatory region of *Ucp1* (Fig. 6). The regulation of *Ucp1* is determined by distinct sites for *Creb*, *Atf2*, *Pparg/Ppara*, *Nr1h3*, *Txnrd1*, and interactions among several of the transcription factors with the coactivator *Ppargc1a*. *Ppargc1a* has been shown to be involved in the transcription of *Ucp1* and additional aspects of mitochondrial biogenesis as a coactivator with Pparg/Rxra and Thrb in cell culture models.[43] It has also been found that mice carrying a targeted mutation for *Ppargc1a* have no reduction in *Ucp1* expression.[46] Similarly, mice with a targeted mutation for *Ppara* have no suppression of *Ucp1* expression in liver or retroperitoneal fat.[52,53] Therefore, transcription factors with well-documented functions in *Ucp1* transcription *in vitro* often turn out to be nonessential for its expression when inactivated by gene targeting *in vivo*. This suggests that many of the sites for transcription of *Ucp1* are in fact redundant, and this redundancy involves not only the DNA regulatory motif but extends to the factors themselves. The redundancy facilitates the synergy by which many factors can participate in the determination of the expression of *Ucp1*. It becomes evident that the components of the environment, including diet and ambient temperature, exert their effects on *Ucp1* through their actions on the factors modulating *Ucp1* expression.

How does this molecular information on the structure and transcription of the *Ucp1* gene relate to the readout of the brown fat genetic system as we have studied in A/J and B6 and the associated RI lines derived from these mice? A continuous variability occurs in which mRNA levels are positively correlated with *Ppara*, *Ppargc1a*, and *Dio2*, and this variability is determined by synergistically interacting genes within the QTLs on chromosomes 2, 3, 8, and 19. Thus, the synergy acts on different levels—at one level on the expression of the transcription factors and signaling mechanisms and at another level on the terminal product of the process, that is, on *Ucp1* expression.

VII. The Development of a Brown Adipocyte Lineage

A. The Brown Fat Lineage in White Fat Depots

Our QTL studies to identify genes controlling brown adipocytes in white fat depots revealed additional findings of significant interest. Some of these, such as variable induction among white fat depots, were important from the perspective of selecting the best tissue to conduct the genetic experiment. For example, the differences between A/J and B6 mice were large for the interperitoneal fat depots, although the rank order of expression was retroperitoneal fat = mesenteric fat ≫ epididymal fat; while the difference between A/J and B6 mice was small in inguinal fat, the level of expression was higher than even retroperitoneal fat.[16] However, the most striking finding was that iBAT, while having the highest level of expression, showed no difference between A/J and B6.[24,53] We reasoned that if the induction of brown adipocytes in adult mice recapitulated early development and was based upon the same molecular mechanisms for all brown adipocytes, then if we analyzed gene and protein expression in brown fat during early development in utero and during suckling, *Ucp1* in iBAT would be higher in the fetus of A/J compared to B6 mice. The analysis showed that at no time during development from the fetus *in utero* until 4 months of age were any strain-dependent differences detected in iBAT gene expression.[25] In contrast, expression and histological analysis of retroperitoneal fat showed a transient induction of brown adipocytes between 10 and 30 days of age. Importantly, although brown adipocytes initially appeared in the retroperitoneal fat (we have since found that brown adipocytes are induced in the inguinal fat with essentially the same time course as retroperitoneal fat) of both A/J and B6 mice, the process was aborted in the B6 mice before the peak of brown adipocyte expression occurred at ~24 days of age.[25] By 2 months of age, the brown adipocytes have essentially disappeared from the white fat; however, if mice are exposed to an ambient temperature of 4 °C, the brown adipocytes are reinduced in A/J mice, but not in B6 mice. This genetic variability, characteristic of brown adipocytes found in white fat depots but not in interscapular fat, suggests that the developmental origins of iBAT are fundamentally different from the brown adipocytes that reside in white fat depots.

The absence of genetic variability for a trait in a normal animal suggests that the trait (in this case, the presence of iBAT) is essential for survival of the animal through its function in maintaining a normal body temperature at birth. The newborn mouse, because of its small size and absence of hair, is particularly susceptible to a cold environment. Consequently, functional brown fat in the discrete depots in the interscapular region and around vital organs is fully functional at birth. On the other hand, brown adipocytes in white fat depots, which appear at ~20 days of age, are not as essential for the maintenance of body temperature and therefore more tolerant of genetic variability. Furthermore, the neonatal pig, which has poor thermoregulation and no *Ucp1* in its brown fat due to a deletion of exons 3–5 that occurred about 20 million years ago,[54,55] can manage to survive the cold stress of early development, underscoring the ability of mammals to adapt in the absence of *Ucp1* with alternative mechanisms of thermogenesis.

B. The iBAT Lineage

Recently, the idea from genetic studies that brown adipocytes in interscapular fat and those in white fat tissues have independent origins has been corroborated by an independent molecular cell lineage study[56,57]. Remarkable progress has been made in identifying critical signaling and transcription factors regulating this differentiation pathway in the fetus. The initial clue came from comparisons of gene expression in microarray experiments of RNA from brown preadipocytes to white preadipocytes that showed a molecular signature in iBAT indicating a lineage relationship to muscle progenitors.[58] Concurrently, a global molecular screen of transcription factors led to the identification of PR domain containing 16 (Prdm16), a 140 kD PR (PRD1-BF-RiZI homologous)-domain-containing protein that was preferentially expressed in cells of brown fat but not white fat lineage.[56] *Prdm16* transfected into undifferentiated adipocyte precursor cells causes activation of the brown adipocyte program including mitochondrial biogenesis, but transfection into mature adipocytes fails to initiate the program. Similarly, transgenic mice carrying transcription factor *Fabp4-Prdm16*-cDNA convert white fat to brown fat. There is a clear relationship between expression of *Prdm16* and muscle and brown adipocyte lineages. Introduction of Prdm16 into MEF, muscle, and skin progenitors converts these cells into brown adipocytes, whereas introduction of *Prdm16* siRNA vectors into immortalized brown adipocytes to suppress expression blocks the brown adipocyte program.[56] Prdm16 is part of a transcriptional complex that activates brown adipogenesis in myoblast progenitors and that includes Pparg and C/ebpb, two critical players in white adipogenesis.[59] The transcriptional coactivator Ppargc1a, which is a key regulator for the transcription of genes of mitochondrial structure and function, is included within the Prdm16 transcriptional complex that directs the brown adipogenesis program. In addition, the program for brown adipogenesis includes a Prdm16 complex in which Ppargc1a is displaced by C-terminal-binding proteins 1 and 2 (Ctbp1, Ctbp2), which are transcriptional corepressors that are proposed to suppress the transcription of white fat-specific genes such as resistin.[60] A model for the pathways of brown adipogenesis in interscapular fat and white fat depots is shown in Fig. 12. An important component of the lineage study is the evidence that initiation of the brown adipogenesis program for interscapular-like brown fat depends on bone morphogenetic protein 7 (Bmp7) to trigger commitment of mesenchymal progenitor cells by induction of Prdm16, Pparg, and *Cebpb*.

Although this transcriptional model for the regulation of muscle cell progenitors to the lineage of iBAT has been worked out by cell culture systems, it has been only partially validated in transgenic mice. White adipocytes, as defined by the gene expression experiments of Timmons *et al.*[58] and Seale

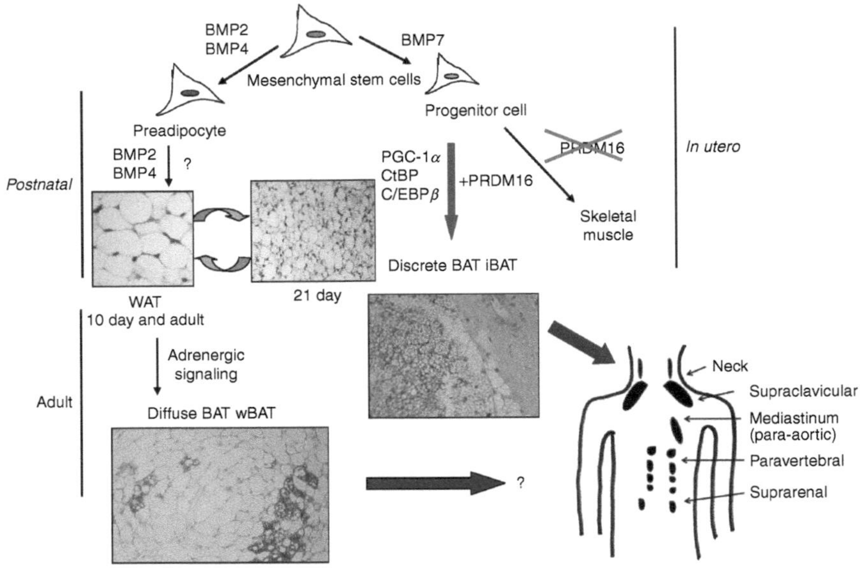

FIG. 12. A model illustrating brown adipocyte development in interscapular brown fat (iBAT) and for the diffuse brown adipocytes found in traditional white fat depots. iBAT originates from mesenchymal stem cells that express Prdm16 and associated transcription factors when stimulated by Bmp7. In the absence of Prdm16, progenitor cells can differentiate into the muscle lineage. On the right-hand side of the illustration, Bmp2 and 4 promote the development of the white adipocyte lineage. Within this lineage, selective white adipocytes have the capacity to differentiate into brown adipocytes at about 10 days of age. This is a transient stage, since these brown adipocytes will spontaneously convert to white adipocytes by about 40 days of age; however, upon adrenergic stimulation in adults, they can convert back to brown adipocytes.

et al.,[56] do not carry the molecular footprints of the skeletal muscle cells, suggesting they have a separated independent lineage from iBAT. However, a transgenic mouse in which *Prdm16* is driven by the fatty acid-binding protein 4, adipocyte (*Fabp4*) promoter had a robust expression of brown adipocytes in white fat depots without the need for adrenergic stimulation.[56] This is expected, since all other components of the Prdm16 transcriptional complex described above are present within the adipocytes. The result also suggests that this program can be initiated quite late in white progenitors, since Fabp4 is one of the late markers of white adipocytes.[61] However, as we have discussed previously, white fat depots can form brown adipocytes in adult animals when adrenergically stimulated, but do these brown adipocytes also originate from muscle cell progenitors? A clever experiment was designed to test whether the brown adipocytes that are normally induced in white fat depots are derived from the Prdm16 myoblast lineage.[57] The expression of *Myf5* and other muscle-specific genes in the differentiated brown fat cell constitutes molecular evidence of the muscle lineage.[58] Furthermore, a *Myf5* promoter driving a *Cre* recombinase transgene can activate a *Yfp* reporter in skeletal muscle and interscapular fat of adult mice. However, the brown adipocytes that can be induced in mature white fat depots by treating mature adult mice with the Adrb3 agonist CL 316,243 failed to express Yfp.[57] This failure of Myf5 to activate *Cre* and *Yfp* expression in brown adipocytes of white fat supports the genetic evidence that its origins are not from a skeletal muscle lineage.

C. Elucidating the Developmental Pathway for Brown Adipocytes in White Fat Depots: Evidence from Transgenic and Gene KO Models

The elucidation of a molecular basis for development of the brown adipocyte lineage represented by the interscapular fat depot has progressed enormously over the past 3 years, largely through the work of Spiegelman and his colleagues.[62] Through our traditional genetic studies and the work of others, we know that the brown adipocytes induced in white fat depots have a different lineage of origin than iBAT. Listed here are features of wBAT (brown adipocytes of white fat) that make them distinct from iBAT cells;

1. Under normal physiological conditions, wBAT is induced by adrenergic signaling generally from reduced ambient temperature.[13–16] (This does not exclude the possibility that drugs, such as thiazolidinediones, might also be able to activate the wBAT program.)
2. The induction of *Ucp1*, *Ppargc1a*, and other genes of thermogenesis in white adipose tissue takes minutes, and brown adipocytes appear in a day or two.[24] In iBAT, however, although cold exposure will also acutely induce expression of *Ucp1* and *Ppargc1a*, new brown adipocytes appear only after several days and require chronic exposure to cold.[12]
3. wBAT emerges from resident white adipocytes and does not depend upon proliferation of precursor cells or differentiation of a progenitor cell.[8] iBAT arises after proliferation from a reservoir of precursor cells.[63]
4. Differentiation of iBAT begins in the fetus at ~ 17 days gestation, and it is fully functional at birth and remains at a high level in mice throughout life.[64,65] wBAT begins to appear, both molecularly and morphologically, at about 12 days of age and reaches a maximum at ~ 21 days of age, and then the brown adipocyte phenotype diminishes and is essentially absent by 6 weeks of age[25]; however, wBAT reappears in adult mice upon adrenergic stimulation.

A model for the differentiation pathway for wBAT must account for the initial induction during early postnatal development, which is independent of a cold environment, and an adrenergic-dependent mechanism that occurs in adult mice. While exposing mice to a cold environment or treatment with Adrb3 agonists led to increases of wBAT and suggested that adrenergic signaling was critical for its induction, the mechanism that gives selective white adipocytes the capacity to respond to an adrenergic signal remains unknown. It is an important question that is relevant to the apparent failure of Adrb3 agonists to stimulate energy expenditure sufficiently in humans.[66] Several transgenic models consistent with a key role for adrenergic signaling in the induction of wBAT are the *Fabp4-Adrb1* transgenic mouse,[67] the *Prka-RIIb* KO[68] and the *Fabp4-FoxC2*[50] transgenic mouse, each of which increases the Prka signaling pathway (Table III). The *Fabp4* promoter leads to constitutive expression of the adrenergic receptor or *Foxc2*, which induces the Prka RIα regulatory subunit with higher affinity for cAMP. The *Prka-RIIb* KO compensates for the loss of RIIβ with increased RIα. Thus, constitutively higher Prka activity bypasses the need for adrenergic stimulation and induces wBAT. All three models show an increase in the number of brown adipocytes that increases the capacity to resist diet-induced obesity. There is an additional pathway for brown adipocyte induction, which is closely related to those just described, that has its origins in the use of the SV40 T-antigen to induce brown fat tumors (hibernomas) for the purpose of generating immortalized brown adipocyte cell lines.[69,70] SV40 T-antigen stimulates the expression of the brown adipocyte program by suppression of the retinoblastoma 1 protein (Rb1).[51] Evidence indicates that suppression of Rb1 increases expression of *Foxc2*, effectively increasing the efficacy of the Prka pathway by upregulating the RIα subunit of Prka.

As shown in Table II, several gene KOs with resistance to diet-induced obesity phenotypes have been interpreted as being a consequence of increasing energy expenditure (oxygen consumption by indirect calorimetry) through induction of brown fat and *Ucp1* expression in white fat depots. These genetic models are of great interest because they may help us to identify regulatory pathways that function independently of adrenergic and Prka signaling. These genes may function in the initial induction of brown adipogenesis that occurs in the developing white fat depots between 10 and 21 days of age in a manner analogous to Prdm16-mediated induction of iBAT in the fetus. This expectation assumes that the critical period of wBAT differentiation is during the postnatal period, just as the mid- to late-fetal period is critical for iBAT development. The genes can be grouped into seemingly unrelated pathways that include: TNF-receptor (*Traf*)[71] and protein kinase IKKε (*Ikbke*) in the inflammatory pathway[72]; *Vdr*[20] and nuclear receptor subfamily 0, group B, member 2 (*Nr0b2*) (an orphan nuclear receptor)[73]; nuclear receptor coactivator 3

(*Ncoa3*; aka *p/CIP* and *SRC-1 p160*) (coactivator of nuclear hormone receptors)[74] and four genes, *Rptor*,[18] *Atg7*,[19] eukaryotic translation initiation factor 4E-binding protein 1 (*Eif4ebp1*)[75] and ribosomal protein S6 kinase, polypeptide 1 (*Rps6kb1*),[76] associated with the mTOR pathway.

With the exception of *Vdr*,[20] the energy expenditure phenotypes of several of these genes are problematic in that the calculation for oxygen consumption has normalized the total oxygen consumption per mouse to body weight. The consequence of this calculation is that the lean KO mice have increased energy expenditure simply because the large mass of fat is assumed to have the same contribution to whole-body energy expenditure as lean mass. We have discussed this problem in detail recently, as did Jean Himms-Hagen about 12 years ago.[77,78] It is highly improbable that any of these mice have energy expenditure that is different from the wild-type mice. Other phenotypes are also suspect in several of these KO models. It is certainly clear that diet-induced obesity is significantly reduced in these gene KO models; however, whether a significant level of induction of brown adipocytes and *Ucp1* occurs is uncertain without more quantitative analyses of *Ucp1* expression and brown adipocyte content than were presented in these studies. When a robust induction of wBAT occurs, clusters of brown adipocytes are observed as a mosaic pattern with white adipocytes (Fig. 2).[16,20] This was not observed, for example, in the *Rptor* KO mouse; rather, cells with a multilocular morphology were interspersed between white adipocytes, and no areas of brown fat were shown. Also, in mice with a robust induction of wBAT, *Ucp1* is induced to levels between 10% and 70% of that seen in iBAT.[25] Such data was not presented in these studies. Granneman *et al.*[79] published an informative study on induction of gene expression in adipose tissue in mice treated with the Adrb3 agonist CL-316, 243. Whereas many studies of mice treated with Adrb3 agonists have observed a strong induction of *Ucp1* and the mosaic pattern of brown adipocytes, Granneman found multilocular cells dispersed randomly among the white adipocytes and strong induction of genes of mitochondrial β-oxidation and electron transport, but very little *Ucp1* expression that did not cluster with the mitochondrial genes. It was concluded that the major process occurring in the white fat tissue involved induction of the inflammatory response and genes associated with the catabolic activity. While it is clear that diet-induced obesity is suppressed in these KO models, the conclusion that this is caused by increased energy expenditure from brown adipocytes needs to be reevaluated. It is important to keep in mind that *Ucp1−/−* and wild-type mice at normal room temperature show no difference in oxygen consumption when fed either a low- or high-fat diet.[80] *Ucp1−/−* mice can show higher oxygen consumption than wild-type mice, but only at temperatures below 20 °C (higher levels of oxygen consumption by wild-type mice is never observed). Therefore, it seems implausible that a mouse with only a slight increase in *Ucp1* mRNA levels as observed in the KO models of genes in the mTOR pathway will show a highly significant increase in oxygen consumption.

An additional peculiarity of a KO gene from the mTOR pathway concerns Eif4ebp1, which is a translational inhibitor that loses its capacity to inhibit translation when it is hyperphosphorylated, presumably by the protein kinase mTOR.[81] Initially, inactivation of *Eif4ebp1* was reported to increase wBAT, suggesting that the increased translational activity resulting from the absence of Eif4ebp1 to inhibit Eif4e somehow stimulated brown adipogenesis in white fat. On the other hand, the *Rptor* KO, which should increase the hypophosphorylated state of Eif4ebp1 by reducing the capacity of mTOR to bind to Eif4ebp1 and thereby increasing binding of Eif4ebp1 to Eif4e to inhibit translation,[82] also stimulates brown adipogenesis in white fat and reduces diet-induced obesity.[18] Thus, opposing effects on Eif4ebp and its effects on translation have the same phenotype. It turns out that inactivation of *Rps6kb1*, which is also a substrate protein kinase of mTOR, also increases brown adipogenesis. To add more confusion to this story implicating the mTOR/Eif4ebp pathway in brown adipogenesis, it has been reported that the *Eif4ebp1/Eif4ebp2* double mutant is not lean with increased brown adipocytes, but it has increased diet-induced obesity and insulin resistance.[83] The authors argue that the switch in phenotypes is a consequence of differences in the genetic background of the original and more recent KO models. In support of this explanation, they cite the changes in phenotypes on *ob/ob* mice. These include the effects of the BALB/cJ and FVB/N background compared to the C57BL/6J background on the severity of the obese phenotype in *ob/ob* mice.[84,85] However, these effects of genetic background are on the degree of the obesity, not like the *Eif4ebp* KOs in which a different background completely switches phenotypes. Given the serious caveats about the energy expenditure and *Ucp1*/brown adipocyte phenotypes of the mTOR-associated gene KOs alluded to above, a reevaluation of a role for mTOR in the brown adipogenesis program needs to be undertaken.

Two of the transgenic models are particularly interesting and potentially informative. One is the *Vdr* KO mouse which shows all the features of a genuine induction of wBAT[20]; however, the underlying mechanism for brown adipocyte induction is unclear. The other is the *Nrip1* KO mouse in which *Ucp1* and brown adipocytes are induced in white fat. Nrip1 is a corepressor which has been shown by ChIP assays to bind to the *Ucp1* enhancer region.[36,37] Consequently, elevated levels of Nrip1 would normally suppress *Ucp1* expression in white adipose tissue, and its inactivation by gene targeting or downregulation by adrenergic signaling would lead to upregulation of *Ucp1*. The induction of the complete brown adipogenesis program would suggest that additional genes of mitochondrial structure and function are also regulated by Nrip1.

VIII. Transgenic Models to Explore the Role of Brown Fat in Thermogenesis and Body Weight Regulation

In 1979, with the publication of a study in which rats fed a cafeteria diet were less obese than expected from caloric intake, the concept was born that diet-induced thermogenesis was determined by brown fat.[26] Tests of this concept have been conducted with several transgenic and KO models of brown fat. The underlying assumption for the hypothesis of diet-induced thermogenesis has been that the brown fat system is a highly efficient physiological system for heat production; it is centered upon the existence of a protein (Ucp1) located in the inner mitochondria that utilizes the membrane potential to produce heat by uncoupling respiration and oxidative phosphorylation, rather than to drive ATP production. It was rationalized that this system to produce heat, while primarily necessary to protect body temperature, could also be used to burn off excess calories to maintain the organism (mice or humans) in energy balance and free of insulin resistance. It should also be emphasized that the intricate, complex regulation of heat production by the sympathetic nervous system and the extensive vascularization to carry heat to vital organs optimize the efficacy and efficiency of heat production by BAT.[86] Therefore, in the initial development of transgenic models, it was predicted that reducing capacity for brown fat thermogenesis by destroying brown adipocytes or inactivating the *Ucp1* gene would lead to obesity because the capacity of diet-induced thermogenesis would be reduced.[87]

Indeed, this prediction appeared to be reached when *Ucp1*-DTA transgenic mice in which the *Ucp1* promoter was used to drive expression of the cytotoxic diphtheria toxin A chain (DTA), specifically in brown adipocytes, became obese under standard conditions in which mice were fed a chow diet.[88] Accordingly, transgenic mice with a 50–60% reduction in the interscapular fat depot were obese and possessed many of the phenotypes of the metabolic syndrome. The opposing strategy increased Ucp1-based thermogenesis in order to test the hypothesis that increased thermogenesis from brown fat suppresses obesity phenotypes. In a transgenic mouse, the fat-specific *Fabp4* promoter caused ectopic expression of *Ucp1*[89] in white fat depots and high constitutive expression in brown adipocytes.[89] In this model, it was anticipated that the enhanced production of *Ucp1* from a constitutively active promoter in brown adipocytes and novel expression of *Ucp1* in white adipocytes would reduce fat stores because of increased heat production, and indeed, the suppression of obesity and insulin resistance from both diet-induced and genetic obesity was observed in the *Fabp4-Ucp1* transgenic mice.[89,90] The suppression of genetic and diet-induced obesity also occurred in myosin light chain 2 (*Mlc2*)-*Ucp1* transgenic mice, presumably from ectopic expression of *Ucp1* in skeletal muscle.[91]

Although the ectopic overexpression of *Ucp1* in adipocytes and skeletal muscle had positive effects by suppression of obesity and amelioration of insulin resistance, ectopic overexpression of *Ucp1*[89] is not without its negative phenotypes, as evidenced by the increase in hypertension, dietary atherosclerosis, and oxidative stress that occurs when *Ucp1* is ectopically expressed in aorta smooth muscle cells.[92] Accordingly, inefficient metabolism can cause vascular disease, as well as prevent type 2 diabetes. There are also limits to overexpression of *Ucp1*, as evidenced by the consequences of increasing the level of *Ucp1* in the brown adipocytes by generating mice homozygous for the *Fabp4-Ucp1* transgene. Simply doubling the amount of *Ucp1* from the transgene caused the level of *Ucp1* in the mitochondria to reach toxic levels that destroyed the brown adipocytes.[93] The iBAT in the *Fabp4-Ucp1* tg/tg mice lost over 95% of their brown adipocytes; however, unlike the obese *Ucp1*-DTA mice that had lost only 50% of their brown adipocytes, the *Fabp4-Ucp1* tg/tg mice with almost complete absence of *Ucp1* expression were lean and resistant to diet-induced obesity and had normal levels of food intake.[93] Consistent with this loss of brown adipocytes, the *Fabp4-Ucp1* tg/tg mice were not able to survive when exposed to an ambient temperature of 4 °C (body temperature dropped 10 degrees within 3 h), whereas the *Ucp1*-DTA mice were within a degree of normal body temperature after 50 h at 4 °C. Furthermore, it was also clearly shown that *Ucp1*-DTA mice were hyperphagic,[94,95] possibly from the insertion of the transgene into a normal gene that also causes cataracts.[96]

Accordingly, the initial interpretation of the *Ucp*-DTA transgenic mice that a 50% reduction in brown adipocytes suppressed diet-induced thermogenesis and also caused hyperphagia was not supported by the phenotypes of the *Fabp4-Ucp1* transgenic model. Although these conflicting studies of *Ucp1*-DTA and *Fabp4-Ucp1* transgenic mice should have questioned the conclusion that the *Ucp1*-DTA mice supported the idea of diet-induced thermogenesis, it failed to do so, possibly because the concept is so appealing that variation in the amount of brown fat accounted for part of the obesity epidemic through brown-fat-dependent, diet-induced thermogenesis.[97] Therefore, by inducing brown fat levels in humans, obesity would be suppressed because the obesogenic diet could induce thermogenesis, independent of cold or β-adrenergic stimulation.

Three additional transgenic lines of mice are critical to the issue of diet-induced thermogenesis by brown fat. The first of these is the *Ucp1−/−* mice which established the role of Ucp1 in heat production by brown fat because the mice deficient for *Ucp1* were unable to tolerate the cold at 4 °C, similar to the *Fabp4-Ucp1* tg/tg mice.[35] In addition, like the *Fabp4-Ucp1* tg/tg mice, *Ucp1−/−* mice were resistant to diet-induced obesity. Thus, mice with the loss of the major thermogenic system did not become more metabolically efficient and obese; rather, they are more metabolically inefficient. Consistent with this interpretation, mice without the capacity to synthesize norepinephrine due to inactivation of the dopamine β-hydroxylase gene are cold-sensitive and resistant to diet-induced obesity just like *Ucp1−/−* mice. The interpretation of the phenotypes of *Ucp1* KO mice is that *Ucp1* and brown fat are not associated with diet-induced thermogenesis. Rather, they regulate body temperature in response to the external temperature, and if the level of *Ucp1* is changed, it will affect the efficiency of heat production or its distribution, with secondary consequences on body weight.

How then do we interpret the original studies of Rothwell and Stock[26]? Studies on the physiology of thermogenesis by brown fat following norepinephrine administration to cold-acclimated rats provided the critical insights. Through the application of radiolabeled microspheres to measure blood flow in BAT and other tissues, Foster and Frydman showed that blood flow through BAT accounted for 33% of cardiac output, even though BAT mass constitutes only 1.3% of total body mass. In the same animals, skeletal muscle constituted 38% of body mass yet accounted for only 8.4% of cardiac output.[98] Foster and colleagues subsequently applied the microsphere technique for blood flow measurements, together with determinations of arteriovenous oxygen concentrations across the interscapular fat depot to determine whether oxygen consumption was increased in BAT of rats fed a cafeteria diet.[99] While oxygen consumption increased in BAT of rats shifted from 28 to 24 °C, there was no increase in oxygen consumption by BAT of rats fed a cafeteria diet compared to chow, even though at the whole-animal level, cafeteria-fed rats had higher rates of oxygen consumption than chow-fed rats. Clearly, this data does not support the idea that a cafeteria diet stimulates brown fat thermogenesis. Although diet-induced thermogenesis may exist at the whole-animal level by some unknown mechanism, it is not a function of brown fat thermogenesis. This conclusion is also consistent with the finding that there is no difference in oxygen consumption between *Ucp1−/−* and wild-type mice fed a high-fat diet.[80]

Gene KO models for the role of adrenergic signaling in brown fat thermogenesis and body weight regulation have not provided a consistent picture of the function of the adrenergic receptors. None of the mice carrying KO alleles for β1-, β2-, β1/β2-, or β3-receptors were cold-sensitive or overtly obese.[100–104] Finally, a mouse lacking β1-, β2-, and β3-adrenergic receptors, the β-less mouse, was cold-sensitive, devoid of *Ucp1* expression, had brown fat cells that resembled small white adipocytes, and was sensitive to diet-induced obesity.[105,106] These phenotypes were interpreted as indicating that lack of functional brown fat prevented the mice from controlling fat stores by diet-induced thermogenesis. The problem with this interpretation is that mice with an inactive *Ucp1* gene are cold-sensitive and highly resistant to obesity,[35,107] as are dopamine β-hydroxylase-deficient mice that cannot synthesize norepinephrine.[108,109] The fact that transgenic mice that overexpress an α-adrenergic receptor are obese suggests that the balance between α- and β-signaling pathways is critically important for the development of obesity,[110] and mice without β-adrenergic receptors would be severely imbalanced. An additional argument to consider is that the idea that primitive humans, who were either running to catch their food or running to avoid becoming a meal, evolved a thermogenic mechanism to burn off excess calories, which they rarely had, seems incompatible with evolution. UCP1 and brown fat evolved to maintain body temperature, a physiological function essential for survival.[111,112] To propose that UCP1 evolved to protect the individual from obesity is tantamount to saying that the ability to run evolved to protect against obesity.

IX. BAT in Humans

Although Wahl and colleagues showed that adult humans possessed BAT from the analysis of PET/CT images of cancer patients and that its appearance depended on the environmental temperature,[113] it took another 6 years before the rest of the field took notice.[29–32] In addition to the association between brown fat content and environmental temperature, less brown fat is present in older people and individuals with elevated BMI. This has led to the speculation that increased obesity may be causally linked to the reduced content of brown fat. It may also mean that obesity is more causally associated with the ambient temperature. As we have described, mice with increased wBAT respond to β-adrenergic stimulation to actively reduce adipose stores. Therefore, the effectiveness of the thermogenic mechanism will depend on a genetically determined capacity to induce brown fat or other mechanisms of thermogenesis, such as those present in skeletal muscle. Evaluating whether the capacity for humans to induce brown fat by a procedure as simple as reducing the ambient temperature or by treatment with a drug to stimulate brown fat induction is certainly an important and feasible research effort.

Cold-induced thermogenesis, as evidenced by increased energy expenditure assessed by indirect calorimetry, can be detected in individuals maintained at an ambient temperature of 16 °C for 84 h.[114] Total energy expenditure increased on average 5.5%; however, among a group of 13 lean male subjects, the increase in energy expenditure ranged from 0% to 14% over basal levels of energy expenditure. Energy expenditure also increased to a similar degree in the same cohort of individuals from the feeding of a high-fat diet. Since thermogenesis from both cold exposure and a high-fat diet were correlated, and both are associated via an increase in sympathetic activity (although not in this study), the authors speculated that the same thermogenic mechanism determines both cold- and diet-induced thermogenesis. This same group of investigators has recently provided evidence with an independent cohort of lean male individuals that cold adaptation increased energy expenditure 2.8%, and this change correlated with increased state 4 respiration in mitochondria from skeletal biopsies.[115] It was concluded that cold adaptation is associated with increased mitochondrial uncoupling in skeletal muscle.

The studies by Wijers and colleagues are some of the first to show an increase in cold-induced, nonshivering energy expenditure in humans. An earlier study failed to find evidence for increased nonshivering adaptive thermogenesis.[116] However, two earlier studies showed that reducing the ambient temperature from 28 to 20 °C increased energy expenditure from 4% to 12%.[117,118] These studies conclude that mild reductions in ambient temperature could increase energy expenditure whether or not individuals were clothed. The individual variation observed by the latter was similar to that of Wijers and colleagues, who also found a large individual variation among lean young males in the study. Additional studies to establish whether this variation in cold-inducible energy expenditure has physiological effects could be very informative and could lead to genetic studies on thermogenesis in humans. Studies with mice and rats have established that stimulation of thermogenesis by cold exposure or β-adrenergic agonists has profound effects on body adiposity that are related to the capacity for brown fat thermogenesis.[8,10,33,119] It has also been established through the analysis of energy balance in mice deficient for brown fat and *Ucp1* that, even in the absence of *Ucp1*, thermogenic mechanisms in both the fat depot and skeletal muscle are able to dissipate energy stores.[90,107] In fact, the nonbrown fat sites of thermogenesis, by being less efficient metabolically in generating heat for the maintenance of body temperature, lead to a greater expenditure of energy stores.[80] Accordingly, reduction of body fat stores by increasing energy expenditure to maintain body temperature at reduced ambient temperature is a feasible strategy for weight reduction. It needs to be emphasized that, in mice, an efficient, powerful method for thermogenesis and maintenance of body temperature is vital to survival because of their small size. It has now been well-documented that when mutations occur that inactivate thermogenic pathways, the mouse quickly adapts to activate an alternative pathway. There are now in excess of six such pathways that have been implicated in the control of body temperature.[27]

X. The Genetics of BAT and *UCP1* in Humans

The genetic and molecular studies of *Ucp1* and brown fat in mice indicate that brown fat in discrete fat depots that are established and functional at birth have a different cell lineage than those brown adipocytes that emerge in white fat depots at weaning, disappear shortly thereafter, and then can be induced again in adult mice exposed to β-adrenergic stimulation. The relevance of this developmental story about brown fat in mice to that in humans is that the brown fat in discrete depots present from birth do not vary genetically like those found in the white fat depots, and that this variation in the development of wBAT has a huge effect on the resistance of mice to obesity when they are adrenergically stimulated. Therefore, a key question is what do the BAT depots currently being detected in humans by PET/CT correspond to? Will humans have both types of brown adipocytes? Van Marken Lichtenbelt assayed biopsy samples of fat depots detected by PET/CT, and those expected to be brown fat had high levels of UCP1, whereas the white fat depots showed no UCP1 signal.[29] On the other hand, in a careful quantitative assay of *UCP1* mRNA in both intraperitoneal and extraperitoneal fat depots of control versus obese human subjects, *UCP1* mRNA was uniformly present in fat from both sources, was present at almost an order of magnitude higher in intraperitoneal fat than extraperitoneal fat (different from the mouse), and was higher in the intraperitoneal fat of lean compared to obese subjects (approximately twofold).[120] There have been no reports of induction of *UCP1* expression in humans, nor did Oberkofler *et al.* note whether significant individual variability was present in their subject population.[120]

The availability of polymorphisms in the structure of the *UCP1* gene has been used to link UCP1 to abnormal energy balance phenotypes including obesity. Several genetic studies have shown significant associations between A and G (−3826) *UCP1* variant in the 5′ regulatory region[121] and obesity in French Canadian,[122] French,[123] Czech,[124] Polish,[125] Japanese,[126] and Danish[127] populations. In addition, similar positive associations have been detected between obesity phenotypes and an amino acid substitution within the *ADRB3* gene[128] The fact that UCP1 and ADRB3 are both critical components of brown fat thermogenesis suggests that additive effects of the variants associated with obesity phenotypes may be present in individuals carrying both A to G (−3826) *UCP1* and Trp/Arg^{64} *ADRB3* variants; one study suggested the presence of such interactions,[123] whereas a second study failed to show interactions.[129] The effects of these mutations to genes of brown fat on obesity phenotypes have been interpreted to suggest that brown fat metabolism plays a role in the regulation of body weight. However, without supportive data such as functional evidence that the A to G (−3826) variant of *UCP1* modulates *UCP1* transcription or the Trp/Arg^{64} variant of *ADRB3* affects ligand binding, it is just as probable that these variants are chromosomal markers linking phenotypes to other neighboring variant candidate genes on the chromosome.

XI. Summary and Perspectives

The intent of this review is to provide a perspective on how to apply the discoveries and lessons of the study of the biology of brown fat in rodents, especially in mice where most of the genetic studies have been performed, to the upcoming investigations on brown fat in humans and how they can be applied to the problem of human obesity. Given the convincing data on the presence of brown fat in humans and the fact that therapies for human obesity have been slow in coming, a major effort can be anticipated in the near future to find ways to enhance/activate brown fat thermogenesis.

In mice, the discrete BAT depots that are present at birth to protect the animal from the cold are derived from muscle progenitor cells. Given the rapidly emerging information on the molecular pathways controlling their development, it will not be long before small molecules are found that can induce these cells. However, these tissues are relatively fixed and do not vary in amount unless the mouse is chronically exposed to cold. Consequently, we do not know how they are involved in burning off excess calories when adrenergically stimulated. On the other hand, the wBAT shows a rich potential for variable expression, and this variation can be exploited to burn off excess calories. Although we do not have the same depth of knowledge of the transcription and signaling pathway that we have for iBAT, we nevertheless know that they can be induced from mature white adipocytes by stimulation of adrenergic signaling. In other words, simply conditioning the wBAT system by implementing a training program consisting of exposure to a reduced ambient temperature, analogous to physical training, will be sufficient to increase daily energy expenditure by $\sim$ 12% or 240 kcal per day[114,117]

Acknowledgments

The research described in this review was supported by R01-DK089641.

References

1. Ingalls M, Dickie MM, Snell GD. Obese, a new mutation in the house mouse. *J Hered* 1950;**41**:317–8.
2. Coleman DL. Effects of parabiosis of obese with diabetes and normal mice. *Diabetologia* 1973;**9**:294–8.
3. Zhang Y, Proenca R, Maffei M, Barone M, Leopold L, Friedman JM. Positional cloning of the mouse obese gene and its human homologue. *Nature* 1994;**372**:425–32.
4. Coleman DL, Hummel KP. The influence of genetic background on the expression of the obese (ob) gene in the mouse. *Diabetologia* 1973;**9**:287–93.

5. Surwit RS, Kuhn CM, Cochrane C, McCubbin JA, Feinglos MN. Diet-induced type II diabetes in C57BL/6J mice. *Diabetes* 1988;**37**:1163–7.
6. Seldin MF, Mott D, Bhat D, Petro A, Kuhn CM, Kingsmore SF, et al. Glycogen synthase: a putative locus for diet-induced hyperglycemia. *J Clin Invest* 1994;**94**:269–76.
7. Arch JR, Ainsworth AT, Cawthorne MA, Piercy V, Sennitt MV, Thody VE, et al. Atypical beta-adrenoceptor on brown adipocytes as target for anti-obesity drugs. *Nature* 1984;**309**:163–5.
8. Himms-Hagen J, Cui J, Danforth Jr. E, Taatjes DJ, Lang SS, Waters BL, et al. Effect of CL-316, 243, a thermogenic beta 3-agonist, on energy balance and brown and white adipose tissues in rats. *Am J Physiol* 1994;**266**:R1371–82.
9. Ghorbani M, Himms-Hagen J. Appearnace of brown adipocytes in white adipose tissue during CL 316, 243-induced reversal of obesity and diabetes in Zucker fa/fa rats. *Int J Obes* 1997;**21**:465–75.
10. Collins S, Daniel KW, Petro AE, Surwit RS. Strain-specific response to beta 3-adrenergic receptor agonist treatment of diet-induced obesity in mice. *Endocrinology* 1997;**138**:405–13.
11. Himms-Hagen J, Melnyk A, Zingaretti MC, Ceresi E, Barbatelli G, Cinti S. Multilocular fat cells in WAT of CL-316243-treated rats derive directly from white adipocytes. *Am J Physiol Cell Physiol* 2000;**279**:C670–81.
12. Bukowiecki L, Collet AJ, Follea N, Guay G, Jahjah L. Brown adipose tissue hyperplasia: a fundamental mechanism of adaptation to cold and hyperphagia. *Am J Physiol* 1982;**242**: E353–9.
13. Young P, Arch JR, Ashwell M. Brown adipose tissue in the parametrial fat pad of the mouse. *FEBS Lett* 1984;**167**:10–4.
14. Loncar D, Bedrica L, Mayer J, Cannon B, Nedergaard J, Afzelius BA, et al. The effect of intermittent cold treatment on the adipose tissue of the cat. Apparent transformation from white to brown adipose tissue. *J Ultrastruct Mol Struct Res* 1986;**97**:119–29.
15. Cousin B, Cinti S, Morroni M, Raimbault S, Ricquier D, Penicaud L, et al. Occurrence of brown adipocytes in rat white adipose tissue: molecular and morphological characterization. *J Cell Sci* 1992;**103**(Pt 4):931–42.
16. Guerra C, Koza RA, Yamashita H, Walsh K, Kozak LP. Emergence of brown adipocytes in white fat in mice is under genetic control. Effects on body weight and adiposity. *J Clin Invest.* 1998;**102**:412–20.
17. Almind K, Manieri M, Sivitz WI, Cinti S, Kahn CR. Ectopic brown adipose tissue in muscle provides a mechanism for differences in risk of metabolic syndrome in mice. *Proc Natl Acad Sci USA* 2007;**104**:2366–71.
18. Polak P, Cybulski N, Feige JN, Auwerx J, Ruegg MA, Hall MN. Adipose-specific knockout of raptor results in lean mice with enhanced mitochondrial respiration. *Cell Metab* 2008;**8**:399–410.
19. Singh R, Xiang Y, Wang Y, Baikati K, Cuervo AM, Luu YK, et al. Autophagy regulates adipose mass and differentiation in mice. *J Clin Invest* 2009;**119**:3329–39.
20. Narvaez CJ, Matthews D, Broun E, Chan M, Welsh J. Lean phenotype and resistance to diet-induced obesity in vitamin D receptor knockout mice correlates with induction of uncoupling protein-1 in white adipose tissue. *Endocrinology* 2009;**150**:651–61.
21. Bailey DW. Recombinant Inbred Strains. *Transplantation* 1971;**11**:325–7.
22. Taylor BA. Recombinant inbred strains. In: Green MC, editor. *Genetic Variants and Strains of the Laboratory Mouse*. New York: Gustav Fischer Verlag; 1981. p. 397–407.
23. Koza RA, Hohmann SM, Guerra C, Rossmeisl M, Kozak LP. Synergistic gene interactions control the induction of the mitochondrial uncoupling protein (Ucp1) gene in white fat tissue. *J Biol Chem* 2000;**275**:34486–92.

24. Coulter AA, Bearden CM, Liu X, Koza RA, Kozak LP. Dietary fat interacts with QTLs controlling induction of Pgc-1 alpha and Ucp1 during conversion of white to brown fat. *Physiol Genomics* 2003;**14**:139–47.
25. Xue B, Rim JS, Hogan JC, Coulter AA, Koza RA, Kozak LP. Genetic variability affects the development of brown adipocytes in white fat but not in interscapular brown fat. *J Lipid Res* 2007;**48**:41–51.
26. Rothwell NJ, Stock MJ. A role for brown adipose tissue in diet-induced thermogenesis. *Nature* 1979;**281**:31–5.
27. Kozak LP. Brown fat and the myth of diet-induced thermogenesis. *Cell Metab* 2010;**11**:263–7.
28. Timmons JA, Pedersen BK. The importance of brown adipose tissue. *N Engl J Med* 2009;**361**:415–6, author reply 418–21.
29. van Marken Lichtenbelt WD, Vanhommerig JW, Smulders NM, Drossaerts JM, Kemerink GJ, Bouvy ND, et al. Cold-activated brown adipose tissue in healthy men. *N Engl J Med* 2009;**360**:1500–8.
30. Virtanen KA, Lidell ME, Orava J, Heglind M, Westergren R, Niemi T, et al. Functional brown adipose tissue in healthy adults. *N Engl J Med* 2009;**360**:1518–25.
31. Cypess AM, Lehman S, Williams G, Tal I, Rodman D, Goldfine AB, et al. Identification and importance of brown adipose tissue in adult humans. *N Engl J Med* 2009;**360**:1509–17.
32. Saito M, Okamatsu-Ogura Y, Matsushita M, Watanabe K, Yoneshiro T, Nio-Kobayashi J, et al. High incidence of metabolically active brown adipose tissue in healthy adult humans: effects of cold exposure and adiposity. *Diabetes* 2009;**58**:1526–31.
33. Nikonova L, Koza RA, Mendoza T, Chao PM, Curley JP, Kozak LP. Mesoderm-specific transcript is associated with fat mass expansion in response to a positive energy balance. *FASEB J* 2008;**22**:3925–37.
34. Cannon B, Nedergaard J. Brown adipose tissue: function and physiological significance. *Physiol Rev* 2004;**84**:277–359.
35. Enerback S, Jacobsson A, Simpson EM, Guerra C, Yamashita H, Harper ME, et al. Mice lacking mitochondrial uncoupling protein are cold-sensitive but not obese. *Nature* 1997;**387**:90–4.
36. Christian M, Kiskinis E, Debevec D, Leonardsson G, White R, Parker MG. RIP140-targeted repression of gene expression in adipocytes. *Mol Cell Biol* 2005;**25**:9383–91.
37. Leonardsson G, Steel JH, Christian M, Pocock V, Milligan S, Bell J, et al. Nuclear receptor corepressor RIP140 regulates fat accumulation. *Proc Natl Acad Sci USA* 2004;**101**:8437–42.
38. Boyer BB, Kozak LP. The mitochondrial uncoupling protein gene in brown fat: correlation between DNase I hypersensitivity and expression in transgenic mice. *Mol Cell Biol* 1991;**11**:4147–56.
39. Cassard-Doulcier AM, Gelly C, Fox N, Schrementi J, Raimbault S, Klaus S, et al. Tissue-specific and beta-adrenergic regulation of the mitochondrial uncoupling protein gene: control by cis-acting elements in the 5'- flanking region. *Mol Endocrinol* 1993;**7**:497–506.
40. Kozak UC, Kopecky J, Teisinger J, Enerback S, Boyer B, Kozak LP. An upstream enhancer regulating brown-fat-specific expression of the mitochondrial uncoupling protein gene. *Mol Cell Biol* 1994;**14**:59–67.
41. Cao W, Daniel KW, Robidoux J, Puigserver P, Medvedev AV, Bai X, et al. p38 mitogen-activated protein kinase is the central regulator of cyclic AMP-dependent transcription of the brown fat uncoupling protein 1 gene. *Mol Cell Biol* 2004;**24**:3057–67.
42. Sears IB, MacGinnitie MA, Kovacs LG, Graves RA. Differentiation-dependent expression of the brown adipocyte uncoupling protein gene: regulation by peroxisome proliferator-activated receptor gamma. *Mol Cell Biol* 1996;**16**:3410–9.
43. Puigserver P, Wu Z, Park CW, Graves R, Wright M, Spiegelman BM. A cold-inducible coactivator of nuclear receptors linked to adaptive thermogenesis. *Cell* 1998;**92**:829–39.

44. Yoon JC, Puigserver P, Chen G, Donovan J, Wu Z, Rhee J, et al. Control of hepatic gluconeogenesis through the transcriptional coactivator PGC-1. *Nature* 2001;**413**:131–8.
45. Lin J, Wu PH, Tarr PT, Lindenberg KS, St-Pierre J, Zhang CY, et al. Defects in adaptive energy metabolism with CNS-linked hyperactivity in PGC-1alpha null mice. *Cell* 2004;**119**:121–35.
46. Leone TC, Lehman JJ, Finck BN, Schaeffer PJ, Wende AR, Boudina S, et al. PGC-1alpha deficiency causes multi-system energy metabolic derangements: muscle dysfunction, abnormal weight control and hepatic steatosis. *PLoS Biol* 2005;**3**:e101.
47. Eppig JJ, Wigglesworth K, Varnum DS, Nadeau JH. Genetic regulation of traits essential for spontaneous ovarian teratocarcinogenesis in strain LT/Sv mice: aberrant meiotic cell cycle, oocyte activation, and parthenogenetic development. *Cancer Res* 1996;**56**:5047–54.
48. Nogami S, Ohya Y, Yvert G. Genetic complexity and quantitative trait loci mapping of yeast morphological traits. *PLoS Genet* 2007;**3**:e31.
49. Shockley KR, Churchill GA. Gene expression analysis of mouse chromosome substitution strains. *Mamm Genome* 2006;**17**:598–614.
50. Cederberg A, Gronning LM, Ahren B, Tasken K, Carlsson P, Enerback S. FOXC2 is a winged helix gene that counteracts obesity, hypertriglyceridemia, and diet-induced insulin resistance. *Cell* 2001;**106**:563–73.
51. Hansen JB, Jorgensen C, Petersen RK, Hallenborg P, De Matteis R, Boye HA, et al. Retinoblastoma protein functions as a molecular switch determining white versus brown adipocyte differentiation. *Proc Natl Acad Sci USA* 2004;**101**:4112–7.
52. Kersten S, Seydoux J, Peters JM, Gonzalez FJ, Desvergne B, Wahli W. Peroxisome proliferator-activated receptor alpha mediates the adaptive response to fasting. *J Clin Invest* 1999;**103**:1489–98.
53. Xue B, Coulter A, Rim JS, Koza RA, Kozak LP. Transcriptional synergy and the regulation of Ucp1 during brown adipocyte induction in white fat depots. *Mol Cell Biol* 2005;**25**:8311–22.
54. Trayhurn P, Temple NJ, Van Aerde J. Evidence from immunoblotting studies on uncoupling protein that brown adipose tissue is not present in the domestic pig. *Can J Physiol Pharmacol* 1989;**67**:1480–5.
55. Berg F, Gustafson U, Andersson L. The uncoupling protein 1 gene (UCP1) is disrupted in the pig lineage: a genetic explanation for poor thermoregulation in piglets. *PLoS Genet* 2006;**2**: e129.
56. Seale P, Kajimura S, Yang W, Chin S, Rohas LM, Uldry M, et al. Transcriptional control of brown fat determination by PRDM16. *Cell Metab* 2007;**6**:38–54.
57. Seale P, Bjork B, Yang W, Kajimura S, Chin S, Kuang S, et al. PRDM16 controls a brown fat/skeletal muscle switch. *Nature* 2008;**454**:961–7.
58. Timmons JA, Wennmalm K, Larsson O, Walden TB, Lassmann T, Petrovic N, et al. Myogenic gene expression signature establishes that brown and white adipocytes originate from distinct cell lineages. *Proc Natl Acad Sci USA* 2007;**104**:4401–6.
59. Kajimura S, Seale P, Kubota K, Lunsford E, Frangioni JV, Gygi SP, et al. Initiation of myoblast to brown fat switch by a PRDM16-C/EBP-beta transcriptional complex. *Nature* 2009;**460**:1154–8.
60. Kajimura S, Seale P, Tomaru T, Erdjument-Bromage H, Cooper MP, Ruas JL, et al. Regulation of the brown and white fat gene programs through a PRDM16/CtBP transcriptional complex. *Genes Dev* 2008;**22**:1397–409.
61. MacDougald OA, Lane MD. Transcriptional regulation of gene expression during adipocyte differentiation. *Annu Rev Biochem* 1995;**64**:345–73.
62. Kajimura S, Seale P, Spiegelman BM. Transcriptional control of brown fat development. *Cell Metab* 2010;**11**:257–62.

63. Bukowiecki LJ, Géloën A, Collet AJ. Proliferation and differentiation of brown adipocytes from interstitial cells during cold acclimation. *Am J Physiol* 1986;**250**:C880–7.
64. Houstek J, Kopecky J, Rychter Z, Soukup T. Uncoupling protein in embryonic brown adipose tissue–existence of nonthermogenic and thermogenic mitochondria. *Biochim Biophys Acta* 1988;**935**:19–25.
65. Giralt M, Martin I, Iglesias R, Vinas O, Villarroya F, Mampel T. Ontogeny and perinatal modulation of gene expression in rat brown adipose tissue. Unaltered iodothyronine 5'-deiodinase activity is necessary for the response to environmental temperature at birth. *Eur J Biochem* 1990;**193**:297–302.
66. Arch JR. beta(3)-Adrenoceptor agonists: potential, pitfalls and progress. *Eur J Pharmacol* 2002;**440**:99–107.
67. Soloveva V, Graves RA, Rasenick MM, Spiegelman BM, Ross SR. Transgenic mice over-expressing the beta 1-adrenergic receptor in adipose tissue are resistant to obesity. *Mol Endocrinol* 1997;**11**:27–38.
68. Cummings DE, Brandon EP, Planas JV, Motamed K, Idzerda RL, McKnight GS. Genetically lean mice result from targeted disruption of the RII beta subunit of protein kinase A [see comments]. *Nature* 1996;**382**:622–6.
69. Ross SR, Choy L, Graves RA, Fox N, Solevjeva V, Klaus S, et al. Hibernoma formation in transgenic mice and isolation of a brown adipocyte cell line expressing the uncoupling protein gene. *Proc Natl Acad Sci USA* 1992;**89**:7561–5.
70. Kozak UC, Held W, Kreutter D, Kozak LP. Adrenergic regulation of the mitochondrial uncoupling protein gene in brown fat tumor cells. *Mol Endocrinol* 1992;**6**:763–72.
71. Romanatto T, Roman EA, Arruda AP, Denis RG, Solon C, Milanski M, et al. Deletion of tumor necrosis factor-alpha receptor 1 (TNFR1) protects against diet-induced obesity by means of increased thermogenesis. *J Biol Chem* 2009;**284**:36213–22.
72. Chiang SH, Bazuine M, Lumeng CN, Geletka LM, Mowers J, White NM, et al. The protein kinase IKKepsilon regulates energy balance in obese mice. *Cell* 2009;**138**:961–75.
73. Wang L, Liu J, Saha P, Huang J, Chan L, Spiegelman B, et al. The orphan nuclear receptor SHP regulates PGC-1alpha expression and energy production in brown adipocytes. *Cell Metab* 2005;**2**:227–38.
74. Wang Z, Qi C, Krones A, Woodring P, Zhu X, Reddy JK, et al. Critical roles of the p160 transcriptional coactivators p/CIP and SRC-1 in energy balance. *Cell Metab* 2006;**3**:111–22.
75. Tsukiyama-Kohara K, Poulin F, Kohara M, DeMaria CT, Cheng A, Wu Z, et al. Adipose tissue reduction in mice lacking the translational inhibitor 4E- BP1. *Nat Med* 2001;**7**:1128–32.
76. Um SH, Frigerio F, Watanabe M, Picard F, Joaquin M, Sticker M, et al. Absence of S6K1 protects against age- and diet-induced obesity while enhancing insulin sensitivity. *Nature* 2004;**431**:200–5.
77. Himms-Hagen J. On raising energy expenditure in ob/ob mice. *Science* 1997;**276**:1132–3.
78. Butler AA, Kozak LP. A recurring problem with the analysis of energy expenditure in genetic models expressing lean and obese phenotypes. *Diabetes* 2010;**59**:323–9.
79. Granneman JG, Li P, Zhu Z, Lu Y. Metabolic and cellular plasticity in white adipose tissue I: effects of beta3-adrenergic receptor activation. *Am J Physiol Endocrinol Metab* 2005;**289**: E608–16.
80. Anunciado-Koza R, Ukropec J, Koza RA, Kozak LP. Inactivation of UCP1 and the glycerol phosphate cycle synergistically increases energy expenditure to resist diet-induced obesity. *J Biol Chem* 2008;**283**:27688–97.
81. Hara K, Yonezawa K, Kozlowski MT, Sugimoto T, Andrabi K, Weng QP, et al. Regulation of eIF-4E BP1 phosphorylation by mTOR. *J Biol Chem* 1997;**272**:26457–63.
82. Hara K, Maruki Y, Long X, Yoshino K, Oshiro N, Hidayat S, et al. Raptor, a binding partner of target of rapamycin (TOR), mediates TOR action. *Cell* 2002;**110**:177–89.

83. Le Bacquer O, Petroulakis E, Paglialunga S, Poulin F, Richard D, Cianflone K, et al. Elevated sensitivity to diet-induced obesity and insulin resistance in mice lacking 4E-BP1 and 4E-BP2. *J Clin Invest* 2007;**117**:387–96.
84. Qiu J, Ogus S, Mounzih K, Ewart-Toland A, Chehab FF. Leptin-deficient mice backcrossed to the BALB/cJ genetic background have reduced adiposity, enhanced fertility, normal body temperature, and severe diabetes. *Endocrinology* 2001;**142**:3421–5.
85. Haluzik M, Colombo C, Gavrilova O, Chua S, Wolf N, Chen M, et al. Genetic background (C57BL/6J versus FVB/N) strongly influences the severity of diabetes and insulin resistance in ob/ob mice. *Endocrinology* 2004;**145**:3258–64.
86. Nicholls DG, Locke RM. Thermogenic mechanisms in brown fat. *Physiol Rev* 1984;**64**:1–64.
87. Hirsch J. Obesity. Some heat but not enough light [news; comment]. *Nature* 1997;**387**:27–8.
88. Lowell BB, Susulic VS, Hamann A, Lawitts JA, Himms-Hagen J, Boyer BB, Kozak LP, Flier JS. Development of obesity in transgenic mice after genetic ablation of brown adipose tissue. *Nature* 1993;**366**:740–2.
89. Kopecky J, Clarke G, Enerback S, Spiegelman B, Kozak LP. Expression of the mitochondrial uncoupling protein gene from the aP2 gene promoter prevents genetic obesity. *J Clin Invest* 1995;**96**:2914–23.
90. Kopecky J, Hodny Z, Rossmeisl M, Syrovy I, Kozak LP. Reduction of dietary obesity in aP2-Ucp transgenic mice: physiology and adipose tissue distribution. *Am J Physiol* 1996;**270**: E768–75.
91. Li B, Nolte LA, Ju JS, Han DH, Coleman T, Holloszy JO, et al. Skeletal muscle respiratory uncoupling prevents diet-induced obesity and insulin resistance in mice. *Nat Med* 2000;**6**:1115–20.
92. Bernal-Mizrachi C, Gates AC, Weng S, Imamura T, Knutsen RH, DeSantis P, et al. Vascular respiratory uncoupling increases blood pressure and atherosclerosis. *Nature* 2005;**435**:502–6.
93. Stefl B, Janovska A, Hodny Z, Rossmeisl M, Horakova M, Syrovy I, et al. Brown fat is essential for cold-induced thermogenesis but not for obesity resistance in aP2-Ucp mice. *Am J Physiol* 1998;**274**:E527–33.
94. Hamann A, Benecke H, Le Marchand-Brustel Y, Susulic VS, Lowell BB, Flier JS. Characterization of insulin resistance and NIDDM in transgenic mice with reduced brown fat. *Diabetes* 1995;**44**:1266–73.
95. Hamann A, Flier JS, Lowell BB. Decreased brown fat markedly enhances susceptibility to diet-induced obesity, diabetes, and hyperlipidemia. *Endocrinology* 1996;**137**:21–9.
96. Chang B, Hawes NL, Davisson MT, Hechenlively JR. A new microphthalmia and cataract condition caused by a transgene integration. *Invest Ophthalmol Vis Sci* 1997;**38**:S586.
97. Lowell BB, Bachman ES. {beta}-Adrenergic Receptors, Diet-induced Thermogenesis, and Obesity. *J Biol Chem* 2003;**278**:29385–8.
98. Foster DO, Frydman ML. Nonshivering thermogenesis in the rat. II Measurements of blood flow with microspheres point to brown adipose tissue as the dominant site of the calorigenesis induced by noradrenaline. *Can J Physiol Pharmacol* 1978;**56**:110–22.
99. Ma SW, Foster DO, Nadeau BE, Triandafillou J. Absence of increased oxygen consumption in brown adipose tissue of rats exhibiting "cafeteria" diet-induced thermogenesis. *Can J Physiol Pharmacol* 1988;**66**:1347–54.
100. Rohrer DK, Desai KH, Jasper JR, Stevens ME, Regula Jr. DP, Barsh GS, et al. Targeted disruption of the mouse beta1-adrenergic receptor gene: developmental and cardiovascular effects. *Proc Natl Acad Sci USA* 1996;**93**:7375–80.
101. Chruscinski AJ, Rohrer DK, Schauble E, Desai KH, Bernstein D, Kobilka BK. Targeted disruption of the beta2 adrenergic receptor gene. *J Biol Chem* 1999;**274**:16694–700.

102. Rohrer DK, Chruscinski A, Schauble EH, Bernstein D, Kobilka BK. Cardiovascular and metabolic alterations in mice lacking both beta1- and beta2-adrenergic receptors. *J Biol Chem* 1999;**274**:16701–8.

103. Susulic VS, Frederich RC, Lawitts J, Tozzo E, Kahn BB, Harper M-E, et al. Targeted disruption of the ß3-adrenergic receptor gene. *J Biol Chem* 1995;**270**:29483–92.

104. Revelli JP, Preitner F, Samec S, Muniesa P, Kuehne F, Boss O, et al. Targeted gene disruption reveals a leptin-independent role for the mouse beta3-adrenoceptor in the regulation of body composition. *J Clin Invest* 1997;**100**:1098–106.

105. Bachman ES, Dhillon H, Zhang CY, Cinti S, Bianco AC, Kobilka BK, et al. betaAR signaling required for diet-induced thermogenesis and obesity resistance. *Science* 2002;**297**:843–5.

106. Jimenez M, Leger B, Canola K, Lehr L, Arboit P, Seydoux J, et al. Beta(1)/beta(2)/beta(3)-adrenoceptor knockout mice are obese and cold-sensitive but have normal lipolytic responses to fasting. *FEBS Lett* 2002;**530**:37–40.

107. Liu X, Rossmeisl M, McClaine J, Riachi M, Harper ME, Kozak LP. Paradoxical resistance to diet-induced obesity in UCP1-deficient mice. *J Clin Invest* 2003;**111**:399–407.

108. Thomas SA, Palmiter RD. Thermoregulatory and metabolic phenotypes of mice lacking noradrenaline and adrenaline. *Nature* 1997;**387**:94–7.

109. Ste Marie L, Luquet S, Curtis W, Palmiter RD. Norepinephrine- and epinephrine-deficient mice gain weight normally on a high-fat diet. *Obes Res* 2005;**13**:1518–22.

110. Valet P, Grujic D, Wade J, Ito M, Zingaretti MC, Soloveva V, et al. Expression of human alpha 2-adrenergic receptors in adipose tissue of beta 3-adrenergic receptor-deficient mice promotes diet-induced obesity. *J Biol Chem* 2000;**275**:34797–802.

111. Saito S, Saito CT, Shingai R. Adaptive evolution of the uncoupling protein 1 gene contributed to the acquisition of novel nonshivering thermogenesis in ancestral eutherian mammals. *Gene* 2008;**408**:37–44.

112. Hughes DA, Jastroch M, Stoneking M, Klingenspor M. Molecular evolution of UCP1 and the evolutionary history of mammalian non-shivering thermogenesis. *BMC Evol Biol* 2009;**9**:4.

113. Cohade C, Mourtzikos KA, Wahl RL. "USA-Fat": prevalence is related to ambient outdoor temperature-evaluation with 18F-FDG PET/CT. *J Nucl Med* 2003;**44**:1267–70.

114. Wijers SL, Saris WH, van Marken Lichtenbelt WD. Individual thermogenic responses to mild cold and overfeeding are closely related. *J Clin Endocrinol Metab* 2007;**92**:4299–305.

115. Wijers SL, Schrauwen P, Saris WH, van Marken Lichtenbelt WD. Human skeletal muscle mitochondrial uncoupling is associated with cold induced adaptive thermogenesis. *PLoS ONE* 2008;**3**:e1777.

116. Bruck K, Baum E, Schwennicke HP. Cold-adaptive modifications in man induced by repeated short-term cold-exposures and during a 10-day and-night cold-exposure. *Pflugers Arch* 1976;**363**:125–33.

117. Dauncey MJ. Influence of mild cold on 24hr energy expenditure, resting metabolism and diet-induced thermogenesis. *Br J Nutr* 1981;**45**:257–67.

118. Warwick P, Busby R. Influence of mild cold on 24 h energy expenditure in 'normally' clothed adults. *Br J Nutr* 1990;**63**:481–8.

119. Inokuma K, Okamatsu-Ogura Y, Omachi A, Matsushita Y, Kimura K, Yamashita H, et al. Indispensable role of mitochondrial UCP1 for antiobesity effect of beta3-adrenergic stimulation. *Am J Physiol Endocrinol Metab* 2006;**290**:E1014–21.

120. Oberkofler H, Dallinger G, Liu YM, Hell E, Krempler F, Patsch W. Uncoupling protein gene: quantification of expression levels in adipose tissues of obese and non-obese humans. *J Lipid Res* 1997;**38**:2125–33.

121. Cassard-Doulcier AM, Bouillaud F, Chagnon M, Gelly C, Dionne FT, Oppert JM, et al. The Bcl I polymorphism of the human uncoupling protein (ucp) gene is due to a point mutation in the 5'-flanking region. *Int J Obes Relat Metab Disord* 1996;**20**:278–9.

122. Oppert JM, Vohl MC, Chagnon M, Dionne FT, Cassard-Doulcier AM, Ricquier D, et al. DNA polymorphism in the uncoupling protein (UCP) gene and human body fat. *Int J Obes Relat Metab Disord* 1994;**18**:526–31.
123. Clement K, Ruiz J, Cassard-Doulcier AM, Bouillaud F, Ricquier D, Basdevant A, et al. Additive effect of A–>G (-3826) variant of the uncoupling protein gene and the Trp64Arg mutation of the beta 3-adrenergic receptor gene on weight gain in morbid obesity. *Int J Obes Relat Metab Disord* 1996;**20**:1062–6.
124. Sramkova D, Krejbichova S, Vcelak J, Vankova M, Samalikova P, Hill M, et al. The UCP1 gene polymorphism A-3826G in relation to DM2 and body composition in Czech population. *Exp Clin Endocrinol Diabetes* 2007;**115**:303–7.
125. Kiec-Wilk B, Wybranska I, Malczewska-Malec M, Leszczynska-Golabek L, Partyka L, Niedbal S, et al. Correlation of the -3826A > G polymorphism in the promoter of the uncoupling protein 1 gene with obesity and metabolic disorders in obese families from southern Poland. *J Physiol Pharmacol* 2002;**53**:477–90.
126. Mori H, Okazawa H, Iwamoto K, Maeda E, Hashiramoto M, Kasuga M. A polymorphism in the 5' untranslated region and a Met229–>Leu variant in exon 5 of the human UCP1 gene are associated with susceptibility to type II diabetes mellitus. *Diabetologia* 2001;**44**:373–6.
127. Urhammer SA, Hansen T, Borch-Johnsen K, Pedersen O. Studies of the synergistic effect of the Trp/Arg64 polymorphism of the beta3-adrenergic receptor gene and the -3826 A–>G variant of the uncoupling protein-1 gene on features of obesity and insulin resistance in a population-based sample of 379 young Danish subjects. *J Clin Endocrinol Metab* 2000;**85**:3151–4.
128. Clement K, Vaisse C, Manning BS, Basdevant A, Guy-Grand B, Ruiz J, et al. Genetic variation in the beta 3-adrenergic receptor and an increased capacity to gain weight in patients with morbid obesity. *N Engl J Med* 1995;**333**:352–4.
129. Ramis JM, Gonzalez-Sanchez JL, Proenza AM, Martinez-Larrad MT, Fernandez-Perez C, Palou A, et al. The Arg64 allele of the beta 3-adrenoceptor gene but not the -3826G allele of the uncoupling protein 1 gene is associated with increased leptin levels in the Spanish population. *Metabolism* 2004;**53**:1411–6.
130. Gauthier K, Plateroti M, Harvey CB, Williams GR, Weiss RE, Refetoff S, et al. Genetic analysis reveals different functions for the products of the thyroid hormone receptor alpha locus. *Mol Cell Biol* 2001;**21**:4748–60.
131. Brown LJ, Koza RA, Everett C, Reitman ML, Marshall L, Fahien LA, et al. Normal thyroid thermogenesis but reduced viability and adiposity in mice lacking the mitochondrial glycerol phosphate dehydrogenase. *J Biol Chem* 2002;**277**:32892–8.
132. Picard F, Gehin M, Annicotte J, Rocchi S, Champy MF, O'Malley BW, et al. SRC-1 and TIF2 control energy balance between white and brown adipose tissues. *Cell* 2002;**111**:931–41.
133. Nisoli E, Clementi E, Paolucci C, Cozzi V, Tonello C, Sciorati C, et al. Mitochondrial biogenesis in mammals: the role of endogenous nitric oxide. *Science* 2003;**299**:896–9.
134. Carmona MC, Hondares E, Rodriguez de la Concepcion ML, Rodriguez-Sureda V, Peinado-Onsurbe J, Poli V, et al. Defective thermoregulation, impaired lipid metabolism, but preserved adrenergic induction of gene expression in brown fat of mice lacking C/EBPbeta. *Biochem J* 2005;**389**:47–56.
135. Wang H, Zhang Y, Yehuda-Shnaidman E, Medvedev AV, Kumar N, Daniel KW, et al. Liver X receptor alpha is a transcriptional repressor of the uncoupling protein 1 gene and the brown fat phenotype. *Mol Cell Biol* 2008;**28**:2187–200.
136. Tseng YH, Kokkotou E, Schulz TJ, Huang TL, Winnay JN, Taniguchi CM, et al. New role of bone morphogenetic protein 7 in brown adipogenesis and energy expenditure. *Nature* 2008;**454**:1000–4.

137. Lelliott CJ, Medina-Gomez G, Petrovic N, Kis A, Feldmann HM, Bjursell M, et al. Ablation of PGC-1beta results in defective mitochondrial activity, thermogenesis, hepatic function, and cardiac performance. *PLoS Biol* 2006;**4**:e369.
138. Christoffolete MA, Linardi CC, de Jesus L, Ebina KN, Carvalho SD, Ribeiro MO, et al. Mice with targeted disruption of the Dio2 gene have cold-induced overexpression of the uncoupling protein 1 gene but fail to increase brown adipose tissue lipogenesis and adaptive thermogenesis. *Diabetes* 2004;**53**:577–84.
139. Manly KF, Olson JM. Overview of QTL mapping software and introduction to map manager QT. *Mamm Genome* 1999;**10**:327–34.
140. Lander ES, Botstein D. Mapping Mendelian factors underlying quantitative traits using RFLP linkage maps. *Genetics* 1989;**121**:185–99.

Obesity in Single Gene Disorders

Philip L. Beales

Molecular Medicine Unit, UCL Institute of Child Health and Great Ormond Street Hospital for Children, London, United Kingdom

In the 1980s, the heritability of obesity has been calculated to be around 0.80, implying a significant genetic origin and sparking a worldwide "gold-rush" for causative genes. Since the discovery of leptin and the body weight homeostasis pathway it regulates, we have come to a greater understanding of the factors that now control it. Nevertheless, we are still some way off reaching the goal of effectively managing the consequences of dysregulation: obesity, a disease-related state that threatens the well-being of mankind. An unprecedented plethora of data has informed our understanding of obesogenic processes; this chapter examines the evidence, focusing on the substantial role of monogenic diseases.

Progress in Molecular Biology
and Translational Science, Vol. 94
DOI: 10.1016/S1877-1173(10)94005-2

I. Introduction

The pandemic of rising obesity prevalence and related medical problems is significantly impacting the healthcare resources of many developed and now developing countries.[1] One approach to tackling obesity is to understand the underlying factors that control body weight and their genetic determinants. Genome-wide association studies have recently identified common sequence variations or single nucleotide polymorphisms (SNPs) associated with increased body mass index (BMI)[2–4]; however, together, these account for only a small percentage of the inherited variation in BMI. Studies in mouse mutants and patients with severe early-onset obesity have led to the detection of rare variants, many of which have an impact on the leptin–melanocortin system involved in energy homeostasis.[5–7] Herein, the background to our current understanding of this fundamental and highly conserved genetic, physiological, and biochemical network will be explored. Examples of rare obesity-related disorders, in which the causative genes influence more than BMI alone, will also be discussed.

II. Nonsyndromic Obesity

A. Leptin and the Leptin Receptor

Our understanding of the factors controlling energy homeostasis accelerated in the 1990s due to the discovery of leptin and its receptor. The existence of the severely obese *ob/ob* mouse had been known for several decades but identification of the underlying mutated gene revealed the deficiency of the encoded leptin protein.[8] Furthermore, administration of exogenous leptin reduced appetite and body weight in *ob* homozygous mice, and corrected the associated metabolic and endocrine abnormalities.[9–11] Leptin acts through the long isoform of its receptor (LEPRb) to activate the JAK-STAT pathway phosphorylating STAT3 and 5, which translocate to the nucleus to facilitate target gene transcription (Fig. 1).[12,13] Chen and colleagues discovered that the signaling form of LEPR is deleted in the *db/db* mouse, another severely obese mutant.[14]

LEPR proteins are abundant in the hypothalamus and, upon activation, serve to stimulate anorexigenic pathways mediated by pro-opiomelanocortin (POMC) and cocaine and amphetamine-related transcript (CART) containing neurons. Concomitant inhibition of orexigenic pathways occurs through neurons expressing neuropeptide Y (NPY) and agouti-related protein (AgRP). These pathways, therefore, coordinate appetite centrally and metabolism peripherally, such that leptin levels regulate adiposity, food intake, and energy

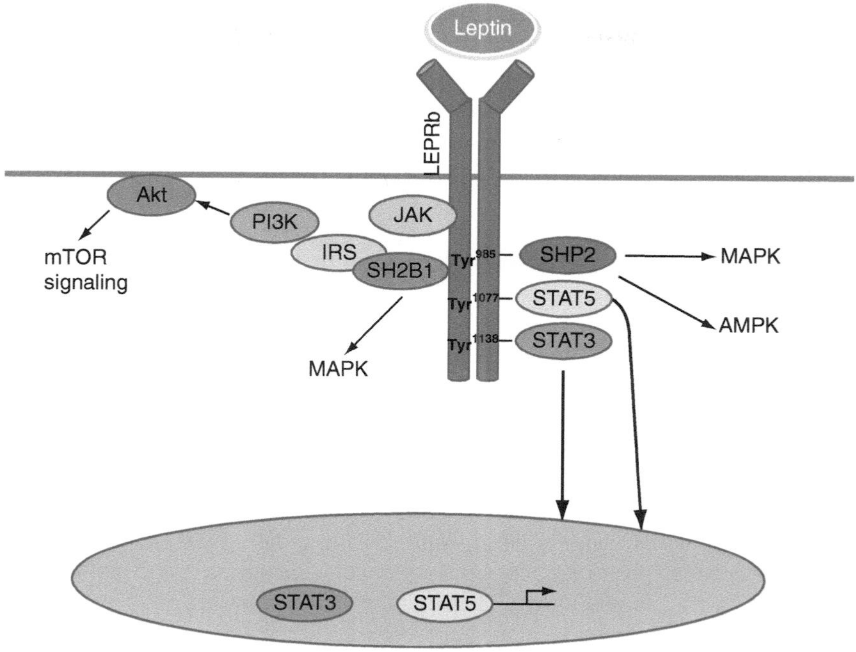

FIG. 1. Model of intracellular leptin signaling. Several key pathways act together to mediate leptin signals. Leptin binds to the long isoform of leptin receptor (LEPRb) which then activates JAK2 which in turn phosphorylates Tyr residues 985, 1077, and 1138. SH2-containing protein tyrosine phosphatase 2 (SHP2) binds to phosphorylated Tyr-985 activating the MAPK pathway. STAT5 and STAT3 bind to Tyr-1077 and -1138, respectively, both of which are then phosphorylated by JAK2. When activated STAT5 and STAT3 translocate to the nucleus to facilitate transcription of target genes. By binding to IRS-1 and IRS-2, SHB2B1 recruits IRS proteins to LEPRb/JAK2 complex resulting in their phosphorylation and subsequent activation of the phosphoinositide 3-kinase (PI3K) pathway. Leptin also regulates the AMP-activated protein kinase (AMPK) and the mammalian target of rapamycin (mTOR) pathways.

expenditure as part of a negative feedback loop. At times of low nutritional availability, low leptin levels will conserve energy, increase food intake, and in severe shortage, shut down reproductive and endocrine activities.[15]

B. Human Leptin and Receptor Mutations

Shortly following the discovery of leptin, the first human mutations emerged in two severely obese cousins.[16] Both patients harbored a homozygous frameshift mutation in *LEP* culminating in a truncated transcript and undetectable leptin levels. At least six further mutations have been found in other families (I. S. Farooqi, personal communication).[17] In 1998, the first mutations in the *LEPR* gene were found in severely obese siblings.[18] The effect of this mutation was to generate an aberrant receptor lacking both the intracellular and transmembrane domain, thus rendering it inactive. Serum leptin levels in these patients were found to be extremely high. Farooqi and colleagues recently identified *LEPR* mutations in eight patients with severe, early-onset obesity.[6] Three missense mutations (A409E, W664R, and H684P) resulted in complete loss of receptor activity, while a further missense (R612H) retained residual receptor activity as measured by the ability to phosphorylate STAT3. In conclusion, approximately 1% of severely obese patients will harbor mutations in *LEP* compared with 3% in *LEPR*. Furthermore, very low or absent serum leptin levels are indications to seek out *LEP* mutations. Leptin receptor deficiency should be considered in patients with severe early-onset obesity.

C. Clinical Phenotypes

Patients with two mutations in either *LEPR* or *LEP* usually have normal birth weights with ensuing rapid weight gain within their first year. The distribution of adipose tissue is similar between the groups and tends to be widespread with the proportion of body fat ranging from 52% to 58%.[6] These patients all experience hyperphagia and constantly crave food, often displaying aggressive behavior when denied.[19] Increased caloric consumption appears to be the basis for the majority of weight gain in both leptin and leptin-receptor-deficient patients as their resting metabolic rates (RMRs) are normal. There is, however, some evidence for defective sympathetic nervous system function in the absence of leptin, possibly explaining the defects in thermogenesis.[20]

Dual leptin and leptin-receptor deficiency is associated with disturbance of the hypothalamic–pituitary–adrenal axis and many patients will have hypothalamic hypothyroidism and hypogonadotrophic hypogonadism.[19] Leptin appears to be necessary for biosynthesis and secretion of thyrotropin-releasing hormone. Free thyroxine levels are low in leptin-deficient mice but usually normal in young patients; however, in some, the thyrotropin levels were high and the secretory patterns abnormal. Thyroid hormone replacement therapy is typically effective.[19]

Onset of puberty in leptin or leptin receptor-deficient patients is delayed or abnormal, consistent with evidence of hypogonadotrophic hypogonadism.[6] These patients do have a reduced adult height if untreated in spite of normal linear growth and IGF-1 levels. It is believed this is caused by an absence of the pubertal growth spurt.[6]

Farooqi and colleagues also found that leptin-deficient children have defective T-cell-mediated immunity, thus explaining the high rates of childhood infections and mortality witnessed in developing countries.[6]

In leptin-deficient patients, daily subcutaneous injections of recombinant human leptin suppressed hyperphagia (reduced calorie intake by up to 84%), reduced fat mass, hyperinsulinemia, and corrected immune deficiency.[19,21] Improvements in pubertal development depended on age at administration. Successful responses have been achieved with doses that raise the plasma leptin levels to just 10% of normal.[19] Weight loss was sustained with continued therapy; however, refractory periods of gain were observed. For those patients with truncated leptin production, exposure to full-length leptin elicited the development of anti-leptin antibodies which may be capable of neutralizing leptin and producing refractory periods.

Serum leptin levels in heterozygous *LEP* mutation carriers are significantly lower than expected for their percentage adiposity, and the subjects are consistently more obese.[22] A small percentage of persons with common obesity have low levels of leptin, suggesting that this subgroup may respond to therapeutic intervention. A large study by Heymsfield and colleagues in which high-dose leptin was given to subjects with common obesity for 28 weeks failed to have a significant and sustained impact on weight loss; however, these subjects were not preselected for hypoleptinemia.[23]

D. POMC and Obesity

The peptide POMC is an important prohormone which when cleaved by prohormone convertase 1 (PC1) results in three peptides of about 50 amino acids including corticotropin (ACTH).[24] Further cleavage of these intermediaries by PC2 yields three smaller melanocyte-stimulating hormone (MSH) peptides—α-MSH, β-MSH, and γ-MSH as well as β-endorphin (Fig. 2). The MSH peptides together with ACTH are termed the melanocortin peptides. β-Endorphin is encoded by the carboxy terminal 30 amino acids residues. It is the differential expression of the two prohormone convertases that generates tissue specificity for POMC-derived peptides. For example, the corticotrophs in the pituitary express mainly PC1, thus generating the longer fragment ACTH.[25] In contrast, skin tissue expresses both PC1 and PC2 giving rise to all types of peptide products in varying proportions. A further level of POMC tissue expression variability is achieved through posttranslational modifications of melanocortin peptides.[26]

The clue to unearthing the role of POMC in weight regulation was revealed following the discovery of the melanocortin receptors (MCRs). Binding of α-MSH to the melanocortin-1 receptor (MC1R) has been shown to stimulate the production of black eumelanin (in mammals and birds) and thus is a key regulator of skin pigmentation.[27] MC2R is predominantly expressed in the zona fasciculata of the adrenal gland and is important for regulation of cortisol production. Mice deficient for MC3R and MC4R revealed novel roles for the melanocortins in weight regulation. MC4R null mice develop extreme obesity

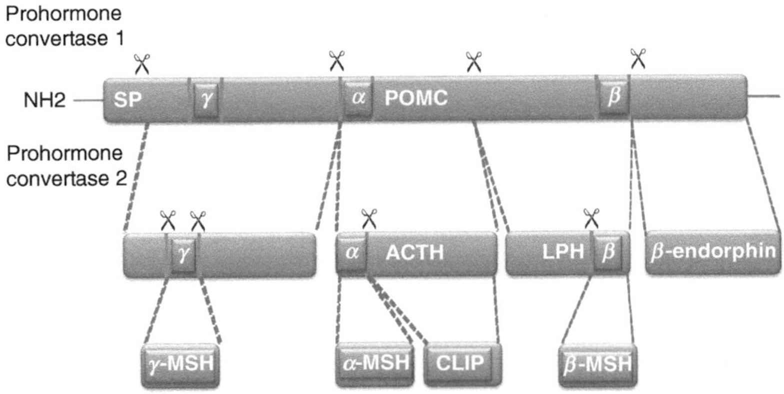

FIG. 2. Pro-opiomelanocortin cleavage. POMC is sequentially cleaved into bioactive peptides by tissue-specific prohormone convertases yielding key products such as α-melanoctye-stimulating hormone (α-MSH), β-melanoctye-stimulating hormone (β-MSH), γ-melanoctye-stimulating hormone (γ-MSH), adrenocorticotrophic hormone (ACTH), corticotropin-like intermediate lobe peptide (CLIP), and β-lipotrophin (LPH). SP, signal peptide.

compared with MC3R-deficient mice which are moderately overweight.[28,29] These and other pharmacological studies have added immeasurably to our overall understanding of the hypothalamic circuits that govern appetite and weight regulation. Administration of leptin, for example, leads to an increase in expression of POMC in the arcuate nucleus helping to piece together an anorexigenic pathway.[30] Following increased adiposity, promotion of leptin secretion stimulates leptin receptor activity in the arcuate nucleus to generate POMC-derived MSH peptides which then signal to their receptors such as MC4R in the paraventricular nuclei (Fig. 3).

In 1998, the first recessive mutations in the *POMC* gene were found in obese patients with a complex phenotype.[31–33] As one might predict, in addition to obesity, the probands suffered from hypocortisolism as well as hair and skin hypopigmentation consistent with absent or inactive cleavage products. Most of the original cases developed neonatal hypoglycemia, seizures, and cholestasis, which are symptoms of adrenal insufficiency and were reversible with glucocorticoid therapy. The underlying pathology was confirmed at postmortem in one child in which the adrenal glands were small and the zona fasciculata absent. By contrast, the zona glomerulosa (aldosterone production) and medulla (catecholamine production) were preserved, indicating a specific lack of ACTH exposure. In POMC-deficient mice, the medulla is, however, structurally abnormal.[34] In the European children with POMC deficiency, all were reported to have red-hair in their first year reminiscent of persons with

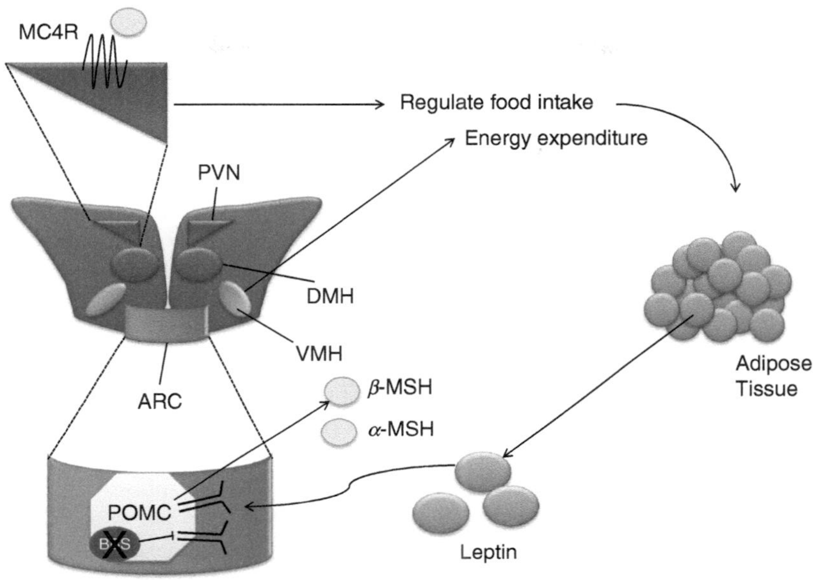

FIG. 3. The leptin regulatory pathway. Leptin is secreted from peripheral adipose tissue, which enters the circulation to relay information to the hypothalamus regarding energy stores. Leptin regulates nerve activity in key regions of the hypothalamus; the arcuate nucleus (ARC), the ventromedial hypothalamus (VMH), and the paraventricular nucleus (PVN). Upon activation of the leptin receptor, appetite is suppressed and energy expenditure promoted. In the ARC, POMC containing neurons bear leptin receptors, activation of which leads to production and secretion of POMC-derived peptides (e.g., MSH). MSH peptides signal to the PVH and specifically to neurons bearing the transmembrane MC4 receptors. BBS proteins are required for LEPRb signaling.

inactivating mutations in MC1R.[27,35] Initially, it was surprising to observe obesity in glucocorticoids-deficient states in which severe weight loss is usually the norm. In fact, in mutant *POMC* mice, administration of corrective doses of glucocortisol exacerbated the obesity.[36] In all cases of *POMC* deficiency, the patients had voracious appetites. Obligate carriers all have normal cortisol levels and pigmentation; however, they do have a propensity to develop mild obesity.[32,33]

Overall, mutations in *POMC* are extremely rare in the general population. The position of *POMC* mutations appears to be important though at least three reported missense mutations are believed to directly disrupt the melanocortin peptide-encoding regions of *POMC*. Two further missense mutations appear to alter the peptide-receptor binding affinity. R236G disrupts a cleavage site in between β-MSH and β-endorphin culminating in an aberrant fusion protein that can bind to MC4R but with reduced activity.[37] Y221C, a rare nonsynonymous polymorphism found to be associated with obesity in general European populations, produces a variant β-MSH peptide with reduced MC4R-binding capability.[38,39]

E. PC1 and Obesity

There exist at least seven endoproteases responsible for residue-specific cleavage of inactive hormone precursors. Some PCs are ubiquitously expressed; however, PC1 and PC2 expression is restricted to neuroendocrine cells.[40,41] Here, they function within secretory vesicles to process bioactive hormones and neuropeptides such as POMC (see above), insulin, and glucagon (Fig. 2). It is the expression of combinations of PCs and their peptide sequence specificity that confers cellular and ultimately differential tissue-specific functions such as the production and secretion of predominantly glucagon from α-cells or insulin from β-cells of the endocrine pancreas.

Mutations in PC1 have now been found in a handful of patients presenting variably with the combination of hyperphagia, early-onset obesity, diarrhea, and abnormal glucose homeostasis. In the first case of a female reported with postprandial hypoglycemia, early-onset obesity and hypogonadotrophic hypogonadism, ovulation was induced with gonatrophin administration but she developed insulin-requiring gestational diabetes.[42] Further investigations revealed virtually absent plasma levels of mature insulin but very high levels of proinsulin and intermediate split proinsulin. She also had high levels of POMC and proglucagon consistent with impaired/absent PC1 activity. The patients were subsequently determined to harbor compound heterozygous mutations in *PC1*; a missense (Gly593Arg) shown to alter the catalytic activity of the peptide and halt its release from the endoplasmic reticulum, and an intron 5 splice donor site base change resulting in exon skipping and premature termination of translation.[43,44]

A second patient, a girl presenting with neonatal hypoglycemia and profuse diarrhea precipitated by any feeds, only recovered on total parental nutrition.[45] She had patchy villous atrophy on small intestinal biopsy but the extent of this was deemed insufficient to cause the severe degree of malabsorption. Even on reduced calorie total parenteral nutrition, the patient became severely obese. Further biochemical investigation revealed high levels of proinsulin and split proinsulin (des-64,65), whereas insulin and des-31,32 proinsulin were near normal, suggestive of PC1 cleavage dysfunction. She was subsequently shown to harbor compound heterozygous mutations; a nonsense and a single alanine (213) deletion which was demonstrated to disrupt catalytic activity.

A third reported child presented with watery diarrhea, duodenal villous atrophy, and a requirement for TPN.[46] Later, he was able to tolerate some formula feeds but rapidly gained weight with hyperphagia, becoming severely obese by 6 years. He had levels of high proinsulin similar to the other two patients although there was no history of hypoglycemia. He was homozygous for a missense mutation in *PC1* shown to impair maturation and catalytic activity of PC1.

How are these expected and unexpected phenotypes to be accounted for? The hypoglycemia probably arises from compensation of insulin deficiency by hyper-secretion of proinsulin, which has 5% of the activity of insulin but five times its plasma half-life.[47] The insulin-like effect of proinsulin is therefore sustained, giving rise to relative hypoglycemia, especially between meals.

The early onset of diarrhea and the inability to absorb basic nutrients in the absence of functioning PC1 is similar to patients with defective neurogenin-3.[48] Despite grossly normal epithelial histology, these patients have malabsorptive diarrhea and poorly developed enteroendocrine cells, which are important for efficient absorption.

PC1 deficiency-induced obesity is most likely a consequence of defective MSH signaling in the hypothalamus as previously described. This is consistent with the high plasma levels of uncleaved POMC seen in PC1-deficient patients. Surprisingly, ACTH levels are relatively normal in these patients suggesting that there may be an alternative biosynthetic pathway.

F. Animal Models of PC1 Deficiency

Three different mouse models disrupting the *Pc1* locus have been reported. In the first two, the phenotype was inconsistent with that reported in human patients.[49] In the first of these models, the whole *Pc1* gene was deleted and in the second, the promoter and exon 1 were targeted. Both these mutants resulted in significant embryonic lethality with zero null offspring present for the full gene deletion compared with 40% lethality in the 5′ deleted nulls. In the latter surviving mutants, there was mild chronic diarrhea and stunted growth with decreased levels of circulating growth hormone (GH) peptide probably owing to impaired processing of the GH prohormone.[49] A third murine knock-in model homozygous for the Asn222Asp missense mutation sited within the catalytic domain developed a phenotype more akin to humans.[50] The homozygotes are hyperphagic, obese, and have high circulating proinsulin levels. There was, however, residual Pc1 catalytic activity and reduced hypothalamic (~45%) α-MSH levels. Heterozygous mice displayed an intermediate phenotype in contrast with human heterozygous mutation carriers.

G. Melanocortin-4 Receptor Deficiency

As previously described, there are five known melanocortin peptide receptors designated MC1R–MC5R. Of this family of G-protein coupled, 7-transmembrane domain receptors, MC3R and MC4R are highly expressed in the brain and especially the hypothalamus. They are most abundant in two distinct subgroups of neurons of the hypothalamic arcuate nucleus responsible for sensing energy stores and controlling appetite. While the melanocortin peptides and CART promote anorexia, NPY and AGRP (a potent antagonist of MC3R and MC4R) are orexigenic, increasing food intake and thus obesity if unchecked. Leptin is also expressed in the arcuate nucleus and regulates both these neuronal populations in a reciprocal manner.

Huszar and colleagues (1997) generated a murine model deleted for *Mc4r* in which the null mutants developed maturity-onset obesity with hyperphagia, hyperinsulinemia, and hyperglycemia.[28] Interestingly, these mice also had accelerated linear growth over their wild-type littermates. Heterozygous mice display an intermediate phenotype indicative of a gene dosage effect.

A year later, the first human *MC4R* heterozygous mutations were reported in obese patients.[51,52] Present in various ethnic populations, *MC4R* mutations represent the commonest form of monogenic obesity accounting for 6% of individuals with early-onset severe obesity and 1–2% of unselected obese adults.[53] The majority of mutations were found in heterozygous subjects, and generally, carriers of homozygous mutations have a greater BMI than single mutation carriers. Of the mutations in *MC4R*, 70% are missense, with the remainder made up of nonsense and frameshift mutations.[54] Among heterozygous carriers, the penetrance of obesity is reduced with some individuals being nonobese. The inheritance pattern of *MC4R* for the obesity trait has been best described as codominant with modulation of the expressivity and penetrance.[5]

H. The MC4R-Deficient Phenotype

In addition to obesity, MC4R-deficient children have hyperinsulinemia and increased linear growth achieving a height standard deviation score (SDS) + 2 compared with the general population expected height.[55] These findings are consistent with the mouse Mc4r models. In addition to an increase in adiposity, patients also have an increase in their lean body mass not observed in other forms of monogenic obesity, such as leptin deficiency. The degree of hyperphagia is not as severe as that experienced by leptin-deficient patients and is proportional to the level of receptor dysfunction.[5]

It has been reported that MC4R deficiency might protect against hypertension as blood pressure values are lower than BMI-matched control (MC4R intact) subjects.[47] These observations are also consistent with Mc4r-deficient mice, which have lower mean arterial pressures than their wild-type littermates.[56] It was postulated that loss of sympathetic nervous activity may be responsible for lowering blood pressure. In support of this hypothesis is the discovery that the increase in heart rate upon waking, normally driven by SNS activation, was attenuated in MC4R-deficient subjects. Moreover, urinary noradrenaline excretion was significantly reduced in the same patients, reflecting the effect of MC4R on synaptic terminal release of noradrenaline. These results highlight the important role of MC4R in blood pressure regulation.

I. BDNF/TrkB

Recent evidence implicates the brain-derived neurotrophic factor (BDNF) and its receptor, tyrosine receptor kinase B (TrkB, encoded by the neurotrophic tyrosine kinase, receptor, type 2 (*NTRK2*) gene), in the regulation of appetite and energy balance.[57] Mice haploinsufficient for Bdnf or with a brain-specific conditional deletion of Bdnf develop hyperphagia and obesity.[58,59]

It has been shown that BDNF is expressed at high levels in the ventromedial hypothalamus, where its expression is regulated by nutritional state and by MC4R signaling.[58] Similar to *MC4R* mutants, mice hypomorphic ($\sim$25% expression levels) for the BDNF receptor TrkB showed hyperphagia and excessive weight gain on high-fat diets. Furthermore, BDNF infusion into the brain suppresses the hyperphagia and excessive weight gain observed in MC4R-deficient mice, indicating that TrkB may be a downstream effector of MC4R signaling by which it controls energy balance. *BDNF* and *NTRK2* are therefore good candidates for early-onset obesity. Yeo *et al.* screened both these genes in a boy with early-onset obesity, hyperphagia, developmental delay, specific impairment of short-term memory, stereotyped behaviors, and impaired nociception, a phenotype closely resembling the *Ntrk2* mouse mutant, revealing a heterozygous *de novo* missense mutation in *NTRK2*.[60] Further analysis of cells expressing the highly conserved but mutated tyrosine (Y722C) residue showed markedly impaired BDNF-stimulated mitogen-activated protein kinase phosphorylation, as predicted. The extended phenotype in the index case, which encompasses developmental and neurological deficits, is in keeping with the widespread expression of TrkB throughout the central nervous system, where a role has been proposed in the promotion of neuronal survival and differentiation and the regulation of synaptic function. The nociception deficit in this patient is reminiscent of mutations in the related receptor TrkA, responsible for congenital insensitivity to pain with anhydrosis.[61] In another case, a girl with a *de novo* chromosomal inversion and loss of one functional copy of *BDNF* presented with hyperphagia, severe obesity, cognitive impairment, and hyperactivity.[62] Mice with complete deficiency of BDNF die in the early postnatal period; however, *BDNF* heterozygotes survive but develop aggressiveness, hyperactivity, and hyperphagia.[57,59,63]

J. Nesfatin

In 2006, a new anorexigenic secreted protein, nucleobindin-2 (NUCB2) was shown to affect satiety.[64] Rat cerebrospinal fluid contains nesfatin-1, an amino-terminal fragment derived from NUCB2, and its expression is decreased in the hypothalamic paraventricular nucleus (PVN) under starved conditions. Intracerebroventricular and peripheral injection of nesfatin-1 decreases food intake in a dose-dependent manner, in contrast to antibody neutralization of nesfatin-1 which stimulates appetite.[65] Oh and colleagues went on to show that nesfatin-1 operates through the melanocortin signaling pathway.[64]

Nesfatin-1 has recently been reported to act as an antihyperglycemic agent.[66] Intravenous injection of nesfatin-1 significantly reduced blood glucose in hyperglycemic db/db mice. NUCB2 is also expressed in human and rat islet beta cells, but not in other islets cell types or the exocrine pancreas suggesting a role for nesfatin in beta cell function and glycemic control.[67] Research on the role of nesfatin holds great promise for the design of interventions for obesity and diabetes but no mutations in man have yet been reported.

K. Single-Minded Homolog 1 (*SIM1*) Gene

Holder *et al.* (2000) reported a girl with early-onset obesity and a *de novo* balanced translocation between 1p22.1 and 6q16.2.[68] Subsequent mapping of the translocation breakpoints revealed disruption to the *SIM1* gene on 6q, culminating in separation of the 5′ promoter region and the basic helix–loop–helix (bHLH) domain from the 3′ period, aryl hydrocarbon receptor, single-minded (PAS), and putative transcriptional regulation domains. *SIM1* is the human homologue of the *Drosophila* transcription factor single-minded, containing bHLH and PAS domains. Dimerization of bHLH–PAS proteins is a prerequisite for nuclear translocation, and the specificity of dimerization is partly facilitated by the PAS domains. Thus, separation as in this translocation is likely to lead to a loss of function. The authors postulated that haploinsufficiency of SIM1, possibly acting upstream or downstream of the melanocortin 4 receptor (MC4R) in the PVN, was responsible for severe obesity in their patient. Supporting evidence for the role of SIM1 in obesity came from the report of Faivre and coworkers, in which a boy with a Prader-Willi-like phenotype carried a deletion of chromosome 6q16.1–q21, resulting in deletion of *SIM1*.[69]

Sim1 is expressed in the developing kidney and central nervous system and is essential for formation of the supraoptic nucleus and PVN of the hypothalamus.[70] Previous neuroanatomical and pharmacological studies have implicated the PVN in the regulation of body weight: PVN neurons express MC4R and appear to be physiological targets of α-MSH, which inhibits food intake.[71–73] *Sim1* homozygous mice die perinatally, but haploinsufficiency, as in the case described by Holder *et al.* (2000),[68] causes early-onset obesity, increased linear growth, hyperleptinemia, and hyperinsulinemia.[74] Hyperphagia appears to be the cause of the weight gain without any discernible impact on thermogenesis, in contrast with MC4R- and leptin-deficient mice.

III. Syndromic Obesity

A. Albright Hereditary Osteodystrophy

Albright hereditary osteodystrophy (AHO) is a complex disorder that is often associated with obesity. AHO describes a range of physical features, including round face, short adult stature, obesity, brachydactyly, and ectopic ossifications.[75] Hypoparathyroidism secondary to end-organ resistance to parathyroid hormone may also be present, termed "pseudohypoparathyroidism" (PHP).[76] Several cases of AHO without end-organ resistance have also been reported and termed "pseudopseudohypoparathyroidism" (PPHP). PHP is further subdivided into types Ia, Ib, Ic, and II. Type Ia and PPHP forms of AHO are caused by inactivating mutations in the tissue specifically imprinted gene *GNAS1* resulting in reduction of the encoded Gsα protein.[77]

Typically, patients with AHO have normal birth weights, but later the majority develop generalized obesity, some with hyperphagia. Although uncertain, it is possible that underlying etiology of the obesity lies in the role of Gsα as a transducer of G-protein coupled transmembrane receptors that mediate anorexigenic signals. As discussed above, the melanocortin receptor MC4R is also transduced by Gsα but its activity has not been widely investigated in AHO.[78]

Ubiquitously expressed G-proteins couple many hormone receptors to the adenylate cyclase second messenger system. Because patients with PHP1a may have reduced cAMP responses to β-adrenergic stimulation in fat cell membranes, reduced lipolysis and sympathetic activity may also contribute to obesity in AHO.[78,79]

B. Alström Syndrome

Alström syndrome (AS) is a rare recessive disorder with around 300 known cases. Patients typically present with early-onset obesity, hyperinsulinemia (frequently with acanthosis nigricans) which often progresses to type 2 diabetes mellitus, dilated cardiomyopathy, short stature, renal failure, and male hypogonadism. Infants usually display nystagmus and photophobia, eventually progressing to cone and rod photoreceptor degeneration, making it a key differential diagnosis with the Bardet–Biedl syndrome (BBS).[80,81] AS may be associated with hepatic dysfunction, hepatic steatosis, and hyperlipidemia, especially hypertriglyceridemia. In addition, 89% patients develop mild to profound bilateral sensorineural hearing loss at an average age of 5 years, owing to cochlear neuronal degeneration.[81]

Rapid weight gain occurs during infancy but tends to plateau in adolescence with a truncal distribution (BMI ranges from 21 to 53).[81] This is often accompanied by rapid growth with height above the 50th centile before puberty but at a cost of achieving a final adult height below 5th centile. Despite the improvement in adiposity and BMI with age, the insulin resistance continues to increase.[82] There are numerous associated endocrine disturbances, including GH deficiency and hyper or hypogonadotrophic hypogonadism in males.[83,84] Among females, hirsutism, precocious puberty, and amenorrhoea have been reported.

Postmortem and tissue biopsy analyses have revealed widespread organ fibrosis. This includes glomerular and renal interstitial fibrosis, pancreatic interstitial fibrosis, micronodular cirrhosis of the liver with severe portal fibrosis, and fatty changes. In three patients with dilated cardiomyopathy, moderate to severe myocardial fibrosis was observed. Fibrosis extended to the gonads with testicular atrophy and obliterating fibrosis of seminiferous tubules; ovaries were densely packed with fibrotic tissue. The lungs and bladder were not spared either.[81]

The underlying gene, *ALMS1* was identified on 2p13.[85,86] Most mutations in *ALMS1* reported so far are nonsense or frameshift mutations leading to premature termination codons.[47] Mutations appear to cluster in the largest exons; 8, 10, and 16. A large deletion involving intron 8 and exon 9 has been described. The most common mutation, 10775delC, accounts for the $>15\%$ of the sequence changes reported. Neither the predicted gene nor protein sequence has similarity to any other genes although there are several conserved sequence motifs of limited functional significance. Of interest are the presence of a large 8 kb exon containing a tandem-repeat domain and in exon 1, a polyglutamic acid/polyalanine tract, the length of which does not appear to impact on the AS phenotype. There do not appear to be any phenotype–genotype correlations and there is no strong evidence for locus heterogeneity.

ALMS1 is ubiquitously expressed throughout all organ tissues but especially in brain, lung, heart, kidney, large intestine, spleen, eye, ovary, and testis.[85] However, in a proteomic study of the centrosome, ALMS1 was identified as a component protein. In 2005, Hearn and coworkers demonstrated centrosomal and basal body localization suggesting involvement in ciliary function and perhaps explaining the phenotypic overlap with BBS (see below).[87]

Common variations in the *ALMS1* gene were not associated with type 2 diabetes mellitus in two studies of Dutch and UK populations.[88,89]

Three mouse models of AS have been reported; one a genetrap, $Alms1^{-/-}$, a spontaneous mutation, *fat aussie* (*Foz/Foz*) and a chemically induced nonsense mutation, $Alms1^{L213X/L213X}$—each recapitulates many aspects of the human phenotype.[90,91] They are of normal weight when young but, between 8 and 17 weeks of age, develop obesity and hyperinsulinaemia.[92] Later, diabetes develops, accompanied by pancreatic islet hyperplasia and islet cysts. On a normal chow diet, *foz/foz* mice developed simple steatosis, whereas on a high-fat diet, they developed steatohepatitis and failed to induce fatty acid oxidation and microsomal triglyceride transfer protein.[91] Female mice are fertile but male mice are sterile due to a progressive germ cell loss leading to failure of spermatogenesis. *Alms1*$^{-/-}$ mice develop retinal dysfunction and late-onset hearing loss. In the photoreceptors, there was mislocalisation of rhodopsin to the outer nuclear layer suggesting that ALMS1 has a role in intracellular trafficking.[90] The third model (*Alms1*$^{L213X/L213X}$), harboring an ethyl nitrosourea-induced, premature truncation mutation in exon 10 of *Alms1*, also showed a similar phenotype to the previously described models and furthermore displayed normal ciliogenesis.[93] The normal development of the cilia in these models may in part be due to the fact that, while the mutation in *Alms1*$^{L213X/L213X}$, like that in *Alms1*$^{-/-}$ mice, is predicted to lead to an early truncation of the protein, both models produce relatively normal expression of *Alms1* mRNA. *In vitro* studies by Li *et al.* (2007) showed that depletion of *Alms1* mRNA using RNA interference gave rise to defective cilia in renal medullary collecting duct (mIMCD3) cells.[93] Moreover, knockdown of *Alms1* in primary kidney cells from *Alms1*$^{L213X/L213X}$ mutants also disrupted ciliogenesis. This suggests that in all three mouse models, residual Alms1 function is present and that the ALMS phenotype in humans and mice is a consequence of partial ciliary dysfunction.

C. Bardet–Biedl Syndrome

BBS is a highly heterogeneous disorder inherited in a mainly recessive manner. Clinical features include retinal degeneration, cognitive impairment, obesity, renal cystic disease, polydactyly, and genital hypoplasia/malformation.[94] There are numerous secondary craniofacial, endocrine, neurological, and behavioral features which can assist in early diagnosis and overlap with AS. Although most infants with BBS are born with normal birth weight, by one year most are showing signs of significant weight gain. There may be few other signs of the syndrome during infancy as up to one-third of cases do not have polydactyly and signs of visual impairment do not typically emerge until 6–8 years of age (night blindness). The majority of adults have a BMI > 30 kg/m^2 often accompanied by hypertension, dyslipidemia, and type 2 diabetes mellitus.[94]

Mutations in 15 genes have now been identified in BBS patients for whom there is little evidence of any phenotype–genotype correlation (reviewed in Ref. 95). These account for approximately 70% of all cases, suggesting there remain further causative gene(s) yet to be cloned or alternative disease mechanisms to uncover. The BBS genes (*BBS1-12*, *MKS1*, *MKS2*, and *Cep290*) have few sequence similarities to each other or other protein groups. Three, *BBS6*, *BBS10*, and *BBS12* have strong homology with the type II group of chaperones and account for around 30% of all mutations.[96–99] Only *BBS3/ARL6* and *BBS11/TRIM32* encode known proteins (a member of the Ras superfamily of small GTP-binding proteins and an E3 ubiquitin ligase, respectively).[100–102] Mutations in *MKS1*, *MKS3*, and *Cep290*, traditionally cause phenotypes consistent with Meckel and Joubert syndromes, respectively (reviewed in Ref. 103).

Recent evidence suggests that BBS is probably caused, at least in part, by dysfunction of primary cilia and the evolutionarily conserved intraflagellar transport (IFT) process which facilitates movement of proteins into and out of the cilium/flagellum (synonymous structures) (see Ref. 104 for review). This is based on the following observations. All BBS proteins studied thus far localize to the cilium/basal body/centrosome complex (Fig. 4). In mammalian

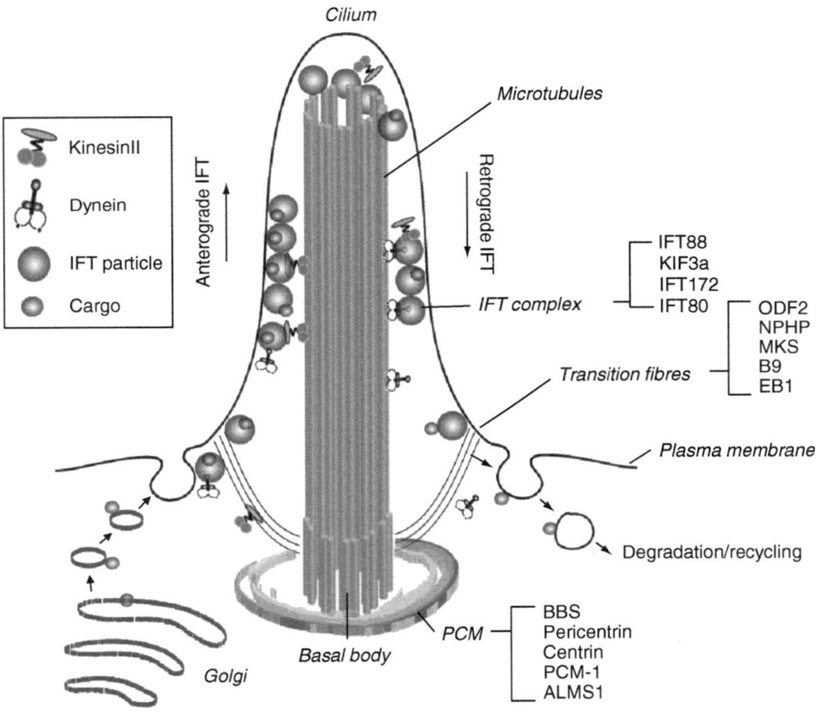

FIG. 4. The structure of the primary cilium. This diagram illustrates the concept of intraflagellar transport (IFT) and the component parts therein. The protein cargo is manufactured in the Golgi apparatus and carried by vesicles to the cell membrane where receptor proteins may also be picked up and translocated to the IFT particles docked at the transition zone. Under the influence of the kinesin motors, the IFT/cargo assembly is transported along the microtubular axoneme (anterograde transport). At the tip, cargo may be unloaded; further proteins/molecules and kinesin itself are loaded and carried back down the axoneme (retrograde transport) facilitated by the dynein molecular motor. In this way, proteins are destined for degradation or recycling. Loss of function of several components proteins, for example, Kif3a, Ift88, Bbs2, 4, 6, and Alms1, results in hyperphagia-associated obesity in animal models.

cultured cells, BBS1, BBS2, BBS4, BBS5, BBS6, and BBS8 localize to the basal body and pericentriolar region, whereas BBS4 and BBS5 have been observed in the ciliary axoneme.[105–109]

Studies indicate that many BBS proteins function in microtubular processes. For example, BBS4 appears to function as an adaptor protein to facilitate IFT cargo loading. Further evidence comes from studies of several *Bbs* mouse mutants, in which each develops severe retinal degeneration similar to patients.[108,110–113] In photoreceptors, rhodopsin relies on IFT for transport to the outer segment—in *Bbs* mutants, rhodopsin accumulates in the cell body, triggering apoptosis.[108,113] Anosmia was recently reported in *Bbs1* and *Bbs4* mutants arising from depletion of olfactory proteins in the ciliary layer of olfactory neurones.[111] Subsequently, anosmia was demonstrated in BBS patients, a novel feature of the syndrome. Finally, repression of *Bbs* proteins in zebrafish delays IFT-dependent movement of melanosomes, confirming their role in microtubule-based transport.[114]

1. Obesity and BBS

Obesity is a cardinal feature of the BBS phenotype, beginning in early childhood and progressing with age; it is usually distributed about the trunk and proximal limbs. A survey of BBS patients in the UK identified 72% of adults as overweight (BMI > 25kg/m^2) and 52% as obese (BMI > 30kg/m^2).[94]

At present, the physiological and biochemical abnormalities underlying obesity in BBS are poorly understood; however, a number of recent studies are illuminating our understanding. A case-control study showed no significant differences between RMR between obese BBS and controls suggesting no underlying defect in metabolism.[115] *Bbs*-deficient mouse models (*Bbs2*, *Bbs4*, and *Bbs6*) are initially runty at birth but display progressive weight gain associated with increased food intake, culminating in obesity at about 12 weeks.[116] This suggests that disruption of appetite regulation leading to hyperphagia underlies weight gain in BBS but may be coupled with lower activity levels.

Leptin, secreted by adipocytes, acts on its receptor in the hypothalamus signaling a decrease in food intake and an increase in energy expenditure, to limit weight gain. *Bbs6*$^{-/-}$ mice have elevated serum leptin levels, with leptin resistance developing before the onset of obesity.[112] It is perhaps noteworthy that primary cilia are present on all cells in the mammalian central nervous system, including leptin-responsive hypothalamic neurons.

Caenorhabditis elegans bbs genes and *tub-1* (the orthologue of the gene mutated in tubby mice, which display a BBS-like phenotype with late-onset obesity, retinal degeneration, infertility, and hearing loss) are expressed exclusively in ciliated sensory neurons and undergo IFT-like transport within the ciliary axoneme. One study suggested that proteins encoded by *bbs-1* and *tub-1* form part of an ancient, conserved neuroendocrine signaling axis from ciliated neurons to fat storage tissue, as worms with mutations in *bbs-1* or *tub-1* and *kat-1* (a fatty acid thiolase expressed in *C. elegans* fat storage tissue) display a synergistic increase in fat deposition.[117] In this model, bbs-1 protein acts in a subset of ciliated neurons poised to sense external and internal nutrient levels, co-coordinating the regulation of fat storage. It is likely that such a pathway is conserved in mammals, as Tubby and *Bbs* mutant mice and human BBS patients display obesity. In mice, the tub protein is neuronal-specific and prominent in the arcuate, PVN, VMH, and DMH nuclei. Hypothalami from tubby mutant mice display increased expression of orexigenic NPY in the VMH and DMH, and, in the arcuate nucleus, reduced expression of anorexigenic POMC, orexigenic NPY and AGRP, and abnormal cholinergic and GABAergic vascular innervation.[118,119] Together, these data suggest that disruption of Bbs or Tub protein function in ciliated hypothalamic neurons might predispose to obesity.

2. Primary Cilia and Satiety

The possible role of cilia in regulation of weight and in particular satiation has been suggested following the study of two conditional *IFT* mutant mouse models; *Kif3a* and *Ift88/Tg737*.[120] To generate conditional null animals, the researchers systemically disrupted IFT using a tamoxifen-inducible Cre recombinase (CAGG-creER^{TM}). Four months after tamoxifen treatment, mice developed obesity and concomitant hyperinsulinemia, hyperglycemia, and hyperleptinemia compared with their wild-type littermates. Loss of primary cilia was confirmed in null mice. They were also hyperphagic and remained normally active. To confirm that this increased appetite was directly due to loss of cilia in the hypothalamus, the conditional mice were crossed with a synapsin1-cre (Syn1-cre) line to specifically target the central nervous system. These mice also became obese. To determine which neuronal type might be responsible, the conditional *Kif3a* mutant was crossed with a *POMC*-cre deletor line. These mice also gained significant amounts of weight but to a much lesser extent than the Kif3a-CAGG-creER or Syn1-cre lines indicating that there are likely additional factors outside of POMC neuronal activity involved in weight regulation but for which cilia play a central role.

More recently, Ramouni and colleagues showed that *Bbs2*, *Bbs4*, and *Bbs6* mutant mice are leptin resistant with high circulating leptin levels.[121] They were able to show that Pomc expression in the hypothalamus was reduced in null *Bbs* mice similar to *Ift88* mutants suggesting that defective POMC signaling likely accounts for obesity in BBS.

In *Bbs* mutant mice which were calorie-restricted to adjust their weights and serum leptin levels to that of control animals, intracerebral injections of leptin did not induce further weight loss or reduce food intake.[122] However, downstream MCR signaling was shown to be unaffected, indicating that leptin receptor signaling is specifically impaired in Bbs mice. Activation of the "b" isoform of the leptin receptor (LEPRb) by leptin leads to phosphorylation of signal transducer and activator of transcription-3 (STAT3) which in turn promotes POMC transcription (see above). Bbs mice have reduced STAT3 phosphorylation upon administration of exogenous leptin thus confirming that BBS proteins are required for LEPRb-mediated signaling (Fig. 3).[122] It has been suggested that defective LEPRb trafficking from the golgi apparatus underlies this defect.[122]

3. Cilia on Adipocytes

In contrast to the neurocentric model of obesity described above, there is emerging support for a peripheral or adipocentric pathway leading to obesity. One recent study has revealed the presence of cilia on differentiating preadipocytes, the number of which are reduced in the absence of BBS10 and BBS12.[123] Furthermore, in these knockdown adipocytes, perixisome proliferator-activated receptor γ (PPARG) and activated glycogen synthase kinase 3β (GSK3β) levels were significantly elevated, suggesting the possibility that cilia signaling may influence adipogenesis. PPARG is the master regulator of preadipocyte differentiation. Increased canonical Wnt signaling inhibits PPARG activity,[108] whereas Hedgehog signaling reduces expression of PPARG in 3T3-L1 preadipocyte cell lines.[124] GSK3β is a central player in both the canonical (β-catenin dependent) Wnt pathway—where its inactivation results in dephosphorylation of β-catenin and its nuclear accumulation—and the Sonic hedgehog (Shh) pathway whereby inhibition of GSK3β stabilizes Gli2 and Gli3, permitting transcription of target genes. Intact cilia function has been shown to be essential for hedgehog signal transduction in mammals[125,126] and perhaps for Wnt signaling[127,128] thus providing two possible complementary mechanisms that promote adipogenesis in ciliopathies such as BBS.

4. BBS Association Studies

A study by Croft and colleagues first suggested that heterozygous carriers were at risk of obesity.[129,130] Attempts to show that *BBS* gene sequence variants may be associated with general non-syndromic obesity have met with mixed success. Reed and colleagues investigated 17 genetic markers spanning chromosomal regions implicated in five different obesity syndromes including BBS and BMI in 44 families segregating for non-syndromic morbid obesity.[131] Sib-pair analyses failed to reveal evidence of linkage between any of the markers and obesity in these families. Among 60 Danish white men with juvenile-onset obesity who were screened for five variants in *MKKS/BBS6*, no significant association was found.[132] Another study did not find any association of the common M390R mutation in BBS1 with obesity among Newfoundlanders.[133] A recent large population study, however, suggests that variations at *BBS* genes are associated with risk of common obesity. Benzinou and colleagues genotyped 12 variants from the coding and conserved regions of *BBS1*, *BBS2*, *BBS4*, and *BBS6* in 1943 French-Caucasian obese subjects and 1299 French-Caucasian nonobese, nondiabetic controls.[134] A *BBS2* SNP was associated with common adult obesity, whereas the *BBS4* and *BBS6* SNPs were associated with common early-onset childhood obesity and common adult morbid obesity, respectively.

D. Carpenter Syndrome

Carpenter syndrome, or acrocephalopolysyndactyly type II, is characterized by preaxial polydactyly of the feet together with craniosynostosis and generalized or truncal obesity in older patients.[135,136] In the hands, there is often brachydactyly and syndactyly. Dental anomalies are common, with retention of milk teeth and hypodontia.

Mutations in *RAB23* were identified in Carpenter syndrome patients.[137] Rab proteins are small guanosine triphosphatases (GTPases) of the Ras superfamily involved in the regulation of intracellular membrane trafficking. RAB23 is an essential negative regulator of the Shh pathway. The related protein Indian hedgehog is important for bone development, but whether the Shh pathway activity is the sole cause of obesity (see Section III.C) remains to be determined. RAB23 is also an important regulator of vesicular transport and perturbation of this process may also impact weight regulatory pathways.

E. Cohen Syndrome

Cohen syndrome (CS) patients characteristically have a history of developmental delay, severe cognitive impairment, and maladaptive behavior in addition to a typical facial appearance. They usually have down-sloping palpebral fissures, mild maxillary hypoplasia, a prominent nasal root, micrognathia, high arching palate, thick hair, and an open mouth expression where the upper lip barely covers the upper incisors giving the appearance of incisoral prominence.[138–140] Many have microcephaly at birth.[141,142] CS babies often have low birthweights and failure to thrive owing to feeding difficulties.[138,139] Sometimes during mid-childhood, patients gain weight and develop truncal adiposity although it is rarely severe. Short stature is common. Delayed puberty is commonly encountered and cases with GH, testosterone deficiency, hypogonadotrophic hypogonadism, and insulin resistance were reported.[138,143–145]

CS patients typically develop a progressive retinopathy with early-onset myopia.[138,145,146] There is a wide range of additional ocular defects with pigmentary changes around the macular (giving rise to the typical "Bull's Eye" maculopathy) occurring as young as 3 years of age.[138] A chorioretinal dystrophy with accompanying electroretinographic (ERG) changes is usually evident by 5 years.[147] Progressive visual field loss with night blindness is present by 10 years.

Neutropenia is an additional and characteristic association with CS but is not always present. The total white count may appear normal but an aberrant differential count is sometimes revealing. Despite resting low neutrophil counts, patients appear to be able to mount relevant responses to bacterial infection and bone marrow analysis has shown normal cellularity.[148,149] The mechanism of this neutropenia is unknown but increased neutrophil adhesion has been documented in CS.[149]

In view of the associated obesity, cognitive impairment and retinal degeneration, the differential diagnosis of CS must be considered alongside BBS and AS. Some patients with 1p36 deletions have a similar facial appearance to CS, and thus cytogenetic analysis should be considered in each case.[150]

Mutations in the responsible gene, vacuolar protein sorting 13 homolog B (*VPS13B*) (originally named *COH1*), were first identified in Finnish families in whom CS is most commonly found.[54,151] *VPS13B* is a large gene, spanning 864 kb of genome with a transcript of 14 kb and an open-reading frame of 4022 codons. Although common founder mutations have been observed in the Finnish and Amish CS population, the positions of mutations in other cases are variable and without any phenotype correlation. Most of the 70 plus mutations described so far are nonsense.

The function of VPS13B remains unknown although homologs such as Vps13p are involved in intracellular vesicular trafficking.[152] Expression of Vps13B in the mouse is widespread among neurons of the postnatal brain, but has low level embryonic expression, suggesting a role in neuronal differentiation, but not in proliferation.[153] This may explain the postnatal microcephaly seen in CS patients.

F. MORM

In 2006, a new autosomal recessive disorder, MORM syndrome (mental retardation, truncal obesity, retinal dystrophy, and micropenis) was described and mapped to 9q34.[154] Fourteen affected individuals from a consanguineous family presented with static moderate mental retardation, truncal obesity, a congenital nonprogressive retinal dystrophy, and hypogonadism in males. There was no polydactyly, and the retinal dystrophy did not progress, thus distinguishing the condition from BBS. In any case, linkage excluded all the *BBS* loci known at that time.

Recently, the same group identified a mutation in inositol polyphosphate-5-phosphatase, (*INPP5E*), a cilia-localized protein important for maintaining cilium stability, in the same family with MORM syndrome.[155] *Inpp5e* homozygous mutant mice die during embryonic development or soon after birth. These mice had bilateral anophthalmos (100%) where eye development ceased at the optic vesicle stage, just before the appearance of the optic cup, postaxial hexadactyly, renal cysts, bifid sternum, delayed ossification of metacarpals and phalanges, and cleft palate. In these mice, 30% of $Inpp5e^{\Delta/\Delta}$ embryos developed anencephaly or exencephaly. Mutations in *INPP5E* have also been reported to cause another form of ciliopathy, Joubert syndrome, linking phosphotidylinositol (PtdIns) signaling with the primary cilium.[156]

G. Ulnar-Mammary Syndrome

Inheritance of Ulnar-mammary syndrome (UMS) is autosomal dominant with variable penetrance. UMS, sometimes referred to as Pallister or Schinzel syndrome, is characterized by ulnar ray defects with missing digits or postaxial polydactyly, anal and genitourinary abnormalities, and hypoplasia of the apocrine glands (with reduced sweating) and breasts.[157,158] These are usually accompanied by generalized obesity. Males may have small genitalia, delayed puberty, and reduced fertility, whereas females may display small or absent breasts with hypoplastic nipples. Anal atresia, hypoplastic kidneys, uterine malformations, laryngeal stenosis and/or web, ectopic canine teeth, and delayed growth have also been reported.[158–160]

The accompanying limb abnormality is variable, as illustrated in a report of an affected mother and son.[161] Both had postaxial polydactyly of one hand, and the son had unilateral oligodactyly with absent ulna and hypoplasia of the ipsilateral shoulder girdle. The mother also had a single kidney.

A heterozygous mutation in the causative gene, T-box 3 (*TBX3*) encoding the T-box transcription factor, was positionally cloned in a large Utah family with UMS consisting of 33 affected individuals.[162,163]

Endocrine studies on two affected brothers reveal that the hypogonadism is likely secondary to gonadotropin deficiency.[164] A recent report describes a child with bilateral ulnar defects, inverted nipples, short stature with associated GH deficiency, and cryptorchidism. This boy also had a hypoplastic anterior pituitary and an ectopic posterior pituitary gland, ventricular septal defect, and Wolff–Parkinson–White (WPW) syndrome (cardiac conduction defects).[165] This result is consistent with recent studies indicating a requirement for Tbx3 in murine cardiac outflow tract development and specification of the atrioventricular conduction system.[166,167]

Although the cause of obesity in UMS is unknown, one recent study has established that TBX3 is produced in three discrete neuronal populations of the adult mouse brain including NPY-expressing neurons in the arcuate nucleus. It is conceivable that loss of TBX3 might impact on satiety regulatory pathways.[168]

IV. Copy Number Variants in Obesity

In 2005, the discovery of the existence of wide ranging and common copy number polymorphisms (CNVariants or CNV) in the human genome has changed our view of genome architecture and disease.[169]

A recent study of copy number variation in 300 white obese subjects with developmental delay found several large (> 500 kb) deletions commonly present in over 7000 controls.[170] Five patients shared overlapping deletions on 16p11.2 segregating with severe obesity and present in only 2 of the 7366 controls. One gene of many lying in the deleted interval, SH2B adaptor protein 1 (*SH2B1*) is involved in leptin and insulin signaling (Fig. 1). Deletion carriers exhibited hyperphagia and severe insulin resistance disproportionate for the degree of obesity.

Other studies have also suggested an association with a CNV at 10q11.22 associated with BMI.[171] The CNV contributed 1.6% of BMI variation and covered a potentially important obesity gene-pancreatic polypeptide receptor 1 (*PPYR1*), a key regulator of energy homeostasis. It remains to be determined whether these genes contribute to the observed obesity phenotype attributable to the deletions described.

V. Closing Remarks

It has become very clear in the last 3 years that no single genetics approach will suffice to understand the complexity surrounding the mechanisms controlling homeostasis of body weight and the etiology of obesity, fast becoming the scourge of modern society. More recent genome-wide association studies have revealed the fat mass and obesity associated (*FTO*) gene as the major genetic player in complex forms of obesity.[172] Nonetheless, this gene alone does represent only a small contribution to susceptibility toward obesity, and it is fast emerging that simple loss of function of FTO does not account for obesity.[173] It is, therefore, likely that most disease gene associations involve subtle regulatory variations that will take time to evaluate and integrate into a regulatory network that will undoubtedly consist of epigenetic, epigenomic, and environmental variables. Contrast this with the study of monogenic obesity disorders, some of which have been described here, in which key regulatory pathways have been unveiled that directly control body weight homeostasis, energy expenditure, and even fertility. If we are to fully understand and conquer the causes of obesity, we will have to combine all these approaches especially if we are to target therapies and improve management, the holy grail of obesity research.

References

1. Drewnowski A, Popkin BM. The nutrition transition: new trends in the global diet. *Nutr Rev* 1997;**55**(2):31–43.
2. Dina C, Meyre D, Gallina S, Durand E, Korner A, Jacobson P, et al. Variation in FTO contributes to childhood obesity and severe adult obesity. *Nat Genet* 2007;**39**(6):724–6.
3. Thorleifsson G, Walters GB, Gudbjartsson DF, Steinthorsdottir V, Sulem P, Helgadottir A, et al. Genome-wide association yields new sequence variants at seven loci that associate with measures of obesity. *Nat Genet* 2009;**41**(1):18–24.
4. Willer CJ, Speliotes EK, Loos RJ, Li S, Lindgren CM, Heid IM, et al. Six new loci associated with body mass index highlight a neuronal influence on body weight regulation. *Nat Genet* 2009;**41**(1):25–34.
5. Farooqi IS, Keogh JM, Yeo GS, Lank EJ, Cheetham T, O'Rahilly S. Clinical spectrum of obesity and mutations in the melanocortin 4 receptor gene. *N Engl J Med* 2003;**348**(12):1085–95.
6. Farooqi IS, Wangensteen T, Collins S, Kimber W, Matarese G, Keogh JM, et al. Clinical and molecular genetic spectrum of congenital deficiency of the leptin receptor. *N Engl J Med* 2007;**356**(3):237–47.
7. Walley AJ, Asher JE, Froguel P. The genetic contribution to non-syndromic human obesity. *Nat Rev Genet* 2009;**10**(7):431–42.
8. Zhang Y, Proenca R, Maffei M, Barone M, Leopold L, Friedman JM. Positional cloning of the mouse obese gene and its human homologue. *Nature* 1994;**372**(6505):425–32.
9. Pelleymounter MA, Cullen MJ, Baker MB, Hecht R, Winters D, Boone T, et al. Effects of the obese gene product on body weight regulation in ob/ob mice. *Science* 1995;**269**(5223):540–3.
10. Halaas JL, Gajiwala KS, Maffei M, Cohen SL, Chait BT, Rabinowitz D, et al. Weight-reducing effects of the plasma protein encoded by the obese gene. *Science* 1995;**269**(5223):543–6.
11. Campfield LA, Smith FJ, Guisez Y, Devos R, Burn P. Recombinant mouse OB protein: evidence for a peripheral signal linking adiposity and central neural networks. *Science* 1995;**269**(5223):546–9.
12. Vaisse C, Halaas JL, Horvath CM, Darnell Jr. JE, Stoffel M, Friedman JM. Leptin activation of Stat3 in the hypothalamus of wild-type and ob/ob mice but not db/db mice. *Nat Genet* 1996;**14**(1):95–7.
13. Gong Y, Ishida-Takahashi R, Villanueva EC, Fingar DC, Munzberg H, Myers Jr. MG. The long form of the leptin receptor regulates STAT5 and ribosomal protein S6 via alternate mechanisms. *J Biol Chem* 2007;**282**(42):31019–27.
14. Chen H, Charlat O, Tartaglia LA, Woolf EA, Weng X, Ellis SJ, et al. Evidence that the diabetes gene encodes the leptin receptor: identification of a mutation in the leptin receptor gene in db/db mice. *Cell* 1996;**84**(3):491–5.
15. Flier JS. Clinical review 94: what's in a name? In search of leptin's physiologic role. *J Clin Endocrinol Metab* 1998;**83**(5):1407–13.
16. Montague CT, Farooqi IS, Whitehead JP, Soos MA, Rau H, Wareham NJ, et al. Congenital leptin deficiency is associated with severe early-onset obesity in humans. *Nature* 1997;**387**(6636):903–8.
17. Strobel A, Issad T, Camoin L, Ozata M, Strosberg AD. A leptin missense mutation associated with hypogonadism and morbid obesity. *Nat Genet* 1998;**18**(3):213–5.
18. Clement K, Vaisse C, Lahlou N, Cabrol S, Pelloux V, Cassuto D, et al. A mutation in the human leptin receptor gene causes obesity and pituitary dysfunction. *Nature* 1998;**392**(6674):398–401.
19. Farooqi IS, Matarese G, Lord GM, Keogh JM, Lawrence E, Agwu C, et al. Beneficial effects of leptin on obesity, T cell hyporesponsiveness, and neuroendocrine/metabolic dysfunction of human congenital leptin deficiency. *J Clin Invest* 2002;**110**(8):1093–103.

20. Ozata M, Ozdemir IC, Licinio J. Human leptin deficiency caused by a missense mutation: multiple endocrine defects, decreased sympathetic tone, and immune system dysfunction indicate new targets for leptin action, greater central than peripheral resistance to the effects of leptin, and spontaneous correction of leptin-mediated defects. *J Clin Endocrinol Metab* 1999;**84**(10):3686–95.
21. Licinio J, Caglayan S, Ozata M, Yildiz BO, de Miranda PB, O'Kirwan F, Whitby R, Liang L, Cohen P, Bhasin S, Krauss RM, Veldhuis JD, Wagner AJ, DePaoli AM, McCann SM, Wong ML. Phenotypic effects of leptin replacement on morbid obesity, diabetes mellitus, hypogonadism, and behavior in leptin-deficient adults. *Proc Natl Acad Sci USA* 2004;**101** (13):4531–6.
22. Farooqi IS, Keogh JM, Kamath S, Jones S, Gibson WT, Trussell R, et al. Partial leptin deficiency and human adiposity. *Nature* 2001;**414**(6859):34–5.
23. Heymsfield SB, Greenberg AS, Fujioka K, Dixon RM, Kushner R, Hunt T, et al. Recombinant leptin for weight loss in obese and lean adults: a randomized, controlled, dose-escalation trial. *JAMA* 1999;**282**(16):1568–75.
24. Seidah NG, Benjannet S, Hamelin J, Mamarbachi AM, Basak A, Marcinkiewicz J, et al. The subtilisin/kexin family of precursor convertases. Emphasis on PC1, PC2/7B2, POMC and the novel enzyme SKI-1. *Ann NY Acad Sci* 1999;**885**:57–74.
25. Kato H, Kuwako K, Suzuki M, Tanaka S. Gene expression patterns of pro-opiomelanocortin-processing enzymes PC1 and PC2 during postnatal development of rat corticotrophs. *J Histochem Cytochem* 2004;**52**(7):943–57.
26. Guo L, Munzberg H, Stuart RC, Nillni EA, Bjorbaek C. *N*-acetylation of hypothalamic alpha-melanocyte-stimulating hormone and regulation by leptin. *Proc Natl Acad Sci USA* 2004;**101** (32):11797–802.
27. Valverde P, Healy E, Jackson I, Rees JL, Thody AJ. Variants of the melanocyte-stimulating hormone receptor gene are associated with red hair and fair skin in humans. *Nat Genet* 1995;**11**(3):328–30.
28. Huszar D, Lynch CA, Fairchild-Huntress V, Dunmore JH, Fang Q, Berkemeier LR, et al. Targeted disruption of the melanocortin-4 receptor results in obesity in mice. *Cell* 1997;**88** (1):131–41.
29. Chen AS, Marsh DJ, Trumbauer ME, Frazier EG, Guan XM, Yu H, et al. Inactivation of the mouse melanocortin-3 receptor results in increased fat mass and reduced lean body mass. *Nat Genet* 2000;**26**(1):97–102.
30. Fan W, Boston BA, Kesterson RA, Hruby VJ, Cone RD. Role of melanocortinergic neurons in feeding and the agouti obesity syndrome. *Nature* 1997;**385**(6612):165–8.
31. Krude H, Biebermann H, Luck W, Horn R, Brabant G, Gruters A. Severe early-onset obesity, adrenal insufficiency and red hair pigmentation caused by POMC mutations in humans. *Nat Genet* 1998;**19**(2):155–7.
32. Krude H, Biebermann H, Schnabel D, Tansek MZ, Theunissen P, Mullis PE, et al. Obesity due to proopiomelanocortin deficiency: three new cases and treatment trials with thyroid hormone and ACTH4-10. *J Clin Endocrinol Metab* 2003;**88**(10):4633–40.
33. Farooqi IS, Drop S, Clements A, Keogh JM, Biernacka J, Lowenbein S, et al. Heterozygosity for a POMC-null mutation and increased obesity risk in humans. *Diabetes* 2006;**55** (9):2549–53.
34. Krude H, Gruters A. Implications of proopiomelanocortin (POMC) mutations in humans: the POMC deficiency syndrome. *Trends Endocrinol Metab* 2000;**11**(1):15–22.
35. Rees JL. Genetics of hair and skin color. *Annu Rev Genet* 2003;**37**:67–90.
36. Smart JL, Tolle V, Low MJ. Glucocorticoids exacerbate obesity and insulin resistance in neuron-specific proopiomelanocortin-deficient mice. *J Clin Invest* 2006;**116**(2):495–505.

37. Challis BG, Pritchard LE, Creemers JW, Delplanque J, Keogh JM, Luan J, et al. A missense mutation disrupting a dibasic prohormone processing site in pro-opiomelanocortin (POMC) increases susceptibility to early-onset obesity through a novel molecular mechanism. *Hum Mol Genet* 2002;**11**(17):1997–2004.
38. Biebermann H, Castaneda TR, van Landeghem F, von Deimling A, Escher F, Brabant G, et al. A role for beta-melanocyte-stimulating hormone in human body-weight regulation. *Cell Metab* 2006;**3**(2):141–6.
39. Lee YS, Challis BG, Thompson DA, Yeo GS, Keogh JM, Madonna ME, et al. A POMC variant implicates beta-melanocyte-stimulating hormone in the control of human energy balance. *Cell Metab* 2006;**3**(2):135–40.
40. Hutton JC. Insulin secretory granule biogenesis and the proinsulin-processing endopeptidases. *Diabetologia* 1994;**37**(Suppl. 2):S48–56.
41. Scopsi L, Gullo M, Rilke F, Martin S, Steiner DF. Proprotein convertases (PC1/PC3 and PC2) in normal and neoplastic human tissues: their use as markers of neuroendocrine differentiation. *J Clin Endocrinol Metab* 1995;**80**(1):294–301.
42. O'Rahilly S, Gray H, Humphreys PJ, Krook A, Polonsky KS, White A, et al. Brief report: impaired processing of prohormones associated with abnormalities of glucose homeostasis and adrenal function. *N Engl J Med* 1995;**333**(21):1386–90.
43. Jackson RS, Creemers JW, Ohagi S, Raffin-Sanson ML, Sanders L, Montague CT, et al. Obesity and impaired prohormone processing associated with mutations in the human prohormone convertase 1 gene. *Nat Genet* 1997;**16**(3):303–6.
44. Jackson RS, O'Rahilly S, Brain C, Nussey SS. Proopiomelanocortin products and human early-onset obesity. *J Clin Endocrinol Metab* 1999;**84**(2):819–20.
45. Jackson RS, Creemers JW, Farooqi IS, Raffin-Sanson ML, Varro A, Dockray GJ, et al. Small-intestinal dysfunction accompanies the complex endocrinopathy of human proprotein convertase 1 deficiency. *J Clin Invest* 2003;**112**(10):1550–60.
46. Farooqi IS, Volders K, Stanhope R, Heuschkel R, White A, Lank E, et al. Hyperphagia and early-onset obesity due to a novel homozygous missense mutation in prohormone convertase 1/3. *J Clin Endocrinol Metab* 2007;**92**(9):3369–73.
47. Oxford Monographs on Medical Genetics. In: Beales P, Farooqi IS, O'Rahilly S, editors. *Genetics of obesity syndromes*. 1st ed. New York: Oxford University Press, 2009.
48. Wang J, Cortina G, Wu SV, Tran R, Cho JH, Tsai MJ, et al. Mutant neurogenin-3 in congenital malabsorptive diarrhea. *N Engl J Med* 2006;**355**(3):270–80.
49. Zhu X, Zhou A, Dey A, Norrbom C, Carroll R, Zhang C, et al. Disruption of PC1/3 expression in mice causes dwarfism and multiple neuroendocrine peptide processing defects. *Proc Natl Acad Sci USA* 2002;**99**(16):10293–8.
50. Lloyd DJ, Bohan S, Gekakis N. Obesity, hyperphagia and increased metabolic efficiency in Pc1 mutant mice. *Hum Mol Genet* 2006;**15**(11):1884–93.
51. Vaisse C, Clement K, Guy-Grand B, Froguel P. A frameshift mutation in human MC4R is associated with a dominant form of obesity. *Nat Genet* 1998;**20**(2):113–4.
52. Yeo GS, Farooqi IS, Aminian S, Halsall DJ, Stanhope RG, O'Rahilly S. A frameshift mutation in MC4R associated with dominantly inherited human obesity. *Nat Genet* 1998;**20**(2):111–1112.
53. Alharbi KK, Spanakis E, Tan K, Smith MJ, Aldahmesh MA, O'Dell SD, et al. Prevalence and functionality of paucimorphic and private MC4R mutations in a large, unselected European British population, scanned by meltMADGE. *Hum Mutat* 2007;**28**(3):294–302.
54. Yeo GS, Lank EJ, Farooqi IS, Keogh J, Challis BG, O'Rahilly S. Mutations in the human melanocortin-4 receptor gene associated with severe familial obesity disrupts receptor function through multiple molecular mechanisms. *Hum Mol Genet* 2003;**12**(5):561–74.

55. Farooqi IS, Yeo GS, Keogh JM, Aminian S, Jebb SA, Butler G, et al. Dominant and recessive inheritance of morbid obesity associated with melanocortin 4 receptor deficiency. *J Clin Invest* 2000;**106**(2):271–9.
56. Ni XP, Butler AA, Cone RD, Humphreys MH. Central receptors mediating the cardiovascular actions of melanocyte stimulating hormones. *J Hypertens* 2006;**24**(11):2239–46.
57. Kernie SG, Liebl DJ, Parada LF. BDNF regulates eating behavior and locomotor activity in mice. *EMBO J* 2000;**19**(6):1290–300.
58. Xu B, Goulding EH, Zang K, Cepoi D, Cone RD, Jones KR, et al. Brain-derived neurotrophic factor regulates energy balance downstream of melanocortin-4 receptor. *Nat Neurosci* 2003;**6**(7):736–42.
59. Rios M, Fan G, Fekete C, Kelly J, Bates B, Kuehn R, et al. Conditional deletion of brain-derived neurotrophic factor in the postnatal brain leads to obesity and hyperactivity. *Mol Endocrinol* 2001;**15**(10):1748–57.
60. Yeo GS, Connie Hung CC, Rochford J, Keogh J, Gray J, Sivaramakrishnan S, et al. A de novo mutation affecting human TrkB associated with severe obesity and developmental delay. *Nat Neurosci* 2004;**7**(11):1187–9.
61. Indo Y, Tsuruta M, Hayashida Y, Karim MA, Ohta K, Kawano T, et al. Mutations in the TRKA/NGF receptor gene in patients with congenital insensitivity to pain with anhidrosis. *Nat Genet* 1996;**13**(4):485–8.
62. Gray J, Yeo GS, Cox JJ, Morton J, Adlam AL, Keogh JM, et al. Hyperphagia, severe obesity, impaired cognitive function, and hyperactivity associated with functional loss of one copy of the brain-derived neurotrophic factor (BDNF) gene. *Diabetes* 2006;**55**(12):3366–71.
63. Lyons WE, Mamounas LA, Ricaurte GA, Coppola V, Reid SW, Bora SH, et al. Brain-derived neurotrophic factor-deficient mice develop aggressiveness and hyperphagia in conjunction with brain serotonergic abnormalities. *Proc Natl Acad Sci USA* 1999;**96**(26):15239–44.
64. Oh IS, Shimizu H, Satoh T, Okada S, Adachi S, Inoue K, et al. Identification of nesfatin-1 as a satiety molecule in the hypothalamus. *Nature* 2006;**443**(7112):709–12.
65. Shimizu H, Oh IS, Hashimoto K, Nakata M, Yamamoto S, Yoshida N, et al. Peripheral administration of nesfatin-1 reduces food intake in mice: the leptin-independent mechanism. *Endocrinology* 2009;**150**(2):662–71.
66. Su Y, Zhang J, Tang Y, Bi F, Liu JN. The novel function of nesfatin-1: anti-hyperglycemia. *Biochem Biophys Res Commun* 2010;**391**(1):1039–42.
67. Foo K, Brauner H, Ostenson CG, Broberger C. Nucleobindin-2/Nesfatin in the endocrine pancreas: distribution and relationship to glycaemic state. *J Endocrinol* 2010; **204**(3):255–63.
68. Holder Jr. JL, Butte NF, Zinn AR. Profound obesity associated with a balanced translocation that disrupts the SIM1 gene. *Hum Mol Genet* 2000;**9**(1):101–8.
69. Faivre L, Cormier-Daire V, Lapierre JM, Colleaux L, Jacquemont S, Genevieve D, et al. Deletion of the SIM1 gene (6q16.2) in a patient with a Prader-Willi-like phenotype. *J Med Genet* 2002;**39**(8):594–6.
70. Michaud JL, DeRossi C, May NR, Holdener BC, Fan CM. ARNT2 acts as the dimerization partner of SIM1 for the development of the hypothalamus. *Mech Dev* 2000;**90**(2):253–61.
71. Harris M, Aschkenasi C, Elias CF, Chandrankunnel A, Nillni EA, Bjoorbaek C, et al. Transcriptional regulation of the thyrotropin-releasing hormone gene by leptin and melanocortin signaling. *J Clin Invest* 2001;**107**(1):111–1120.
72. Kishi T, Aschkenasi CJ, Lee CE, Mountjoy KG, Saper CB, Elmquist JK. Expression of melanocortin 4 receptor mRNA in the central nervous system of the rat. *J Comp Neurol* 2003;**457**(3):213–35.
73. Liu H, Kishi T, Roseberry AG, Cai X, Lee CE, Montez JM, et al. Transgenic mice expressing green fluorescent protein under the control of the melanocortin-4 receptor promoter. *J Neurosci* 2003;**23**(18):7143–54.

74. Michaud JL, Boucher F, Melnyk A, Gauthier F, Goshu E, Levy E, et al. Sim1 haploinsufficiency causes hyperphagia, obesity and reduction of the paraventricular nucleus of the hypothalamus. *Hum Mol Genet* 2001;**10**(14):1465–73.
75. Albright F, Ellsworth R. Studies on the physiology of the parathyroid glands: I. Calcium and phosphorus studies on a case of idiopathic hypoparathyroidism. *J Clin Invest* 1929;**7**(2):183–201.
76. Albright F, Burnett C, Smith P, Parson W. Pseudohypoparathyroidism: an example of "Seabright-Bantam syndrome. *Endocrinology* 1942;**30**:922–32.
77. Weinstein LS, Yu S, Warner DR, Liu J. Endocrine manifestations of stimulatory G protein alpha-subunit mutations and the role of genomic imprinting. *Endocr Rev* 2001;**22**(5):675–705.
78. Wang H, Johnson GL, Liu X, Malbon CC. Repression of adipogenesis by adenylyl cyclase stimulatory G-protein alpha subunit is expressed within region 146–220. *J Biol Chem* 1996;**271**(36):22022–9.
79. Carel JC, Le Stunff C, Condamine L, Mallet E, Chaussain JL, Adnot P, et al. Resistance to the lipolytic action of epinephrine: a new feature of protein Gs deficiency. *J Clin Endocrinol Metab* 1999;**84**(11):4127–31.
80. Russell-Eggitt IM, Clayton PT, Coffey R, Kriss A, Taylor DS, Taylor JF. Alstrom syndrome. Report of 22 cases and literature review. *Ophthalmology* 1998;**105**(7):1274–80.
81. Marshall JD, Bronson RT, Collin GB, Nordstrom AD, Maffei P, Paisey RB, et al. New Alstrom syndrome phenotypes based on the evaluation of 182 cases. *Arch Intern Med* 2005;**165**(6):675–83.
82. Minton JA, Owen KR, Ricketts CJ, Crabtree N, Shaikh G, Ehtisham S, et al. Syndromic obesity and diabetes: changes in body composition with age and mutation analysis of ALMS1 in 12 United Kingdom kindreds with Alstrom syndrome. *J Clin Endocrinol Metab* 2006;**91**(8):3110–6.
83. Maffei P, Munno V, Marshall JD, Scandellari C, Sicolo N. The Alstrom syndrome: is it a rare or unknown disease? *Ann Ital Med Int* 2002;**17**(4):221–8.
84. Alter CA, Moshang Jr. T. Growth hormone deficiency in two siblings with Alstrom syndrome. *Am J Dis Child* 1993;**147**(1):97–9.
85. Collin GB, Marshall JD, Ikeda A, So WV, Russell-Eggitt I, Maffei P, et al. Mutations in ALMS1 cause obesity, type 2 diabetes and neurosensory degeneration in Alstrom syndrome. *Nat Genet* 2002;**31**(1):74–8.
86. Hearn T, Renforth GL, Spalluto C, Hanley NA, Piper K, Brickwood S, et al. Mutation of ALMS1, a large gene with a tandem repeat encoding 47 amino acids, causes Alstrom syndrome. *Nat Genet* 2002;**31**(1):79–83.
87. Hearn T, Spalluto C, Phillips VJ, Renforth GL, Copin N, Hanley NA, et al. Subcellular localization of ALMS1 supports involvement of centrosome and basal body dysfunction in the pathogenesis of obesity, insulin resistance, and type 2 diabetes. *Diabetes* 2005;**54**(5):1581–7.
88. Patel S, Minton JA, Weedon MN, Frayling TM, Ricketts C, Hitman GA, et al. Common variations in the ALMS1 gene do not contribute to susceptibility to type 2 diabetes in a large white UK population. *Diabetologia* 2006;**49**(6):1209–13.
89. t Hart LM, Maassen JA, Dekker JM, Heine RJ. Lack of association between gene variants in the ALMS1 gene and Type 2 diabetes mellitus. *Diabetologia* 2003;**7**:1023–4.
90. Collin GB, Cyr E, Bronson R, Marshall JD, Gifford EJ, Hicks W, et al. Alms1-disrupted mice recapitulate human Alstrom syndrome. *Hum Mol Genet* 2005;**14**(16):2323–33.
91. Arsov T, Silva DG, O'Bryan MK, Sainsbury A, Lee NJ, Kennedy C, et al. Fat aussie—a new Alstrom syndrome mouse showing a critical role for ALMS1 in obesity, diabetes, and spermatogenesis. *Mol Endocrinol* 2006;**20**(7):1610–22.

92. Arsov T, Larter CZ, Nolan CJ, Petrovsky N, Goodnow CC, Teoh NC, et al. Adaptive failure to high-fat diet characterizes steatohepatitis in Alms1 mutant mice. *Biochem Biophys Res Commun* 2006;**342**(4):1152–9.
93. Li G, Vega R, Nelms K, Gekakis N, Goodnow C, McNamara P, et al. A role for Alstrom syndrome protein, alms1, in kidney ciliogenesis and cellular quiescence. *PLoS Genet* 2007;**3**(1):e8.
94. Beales PL, Elcioglu N, Woolf AS, Parker D, Flinter FA. New criteria for improved diagnosis of Bardet–Biedl syndrome: results of a population survey. *J Med Genet* 1999;**36**(6):437–46.
95. Baker K, Beales PL. Making sense of cilia in disease: the human ciliopathies. *Am J Med Genet C Semin Med Genet* 2009;**151C**(4):281–95.
96. Katsanis N, Beales PL, Woods MO, Lewis RA, Green JS, Parfrey PS, et al. Mutations in MKKS cause obesity, retinal dystrophy and renal malformations associated with Bardet–Biedl syndrome. *Nat Genet* 2000;**26**(1):67–70.
97. Slavotinek AM, Stone EM, Mykytyn K, Heckenlively JR, Green JS, Heon E, et al. Mutations in MKKS cause Bardet–Biedl syndrome. *Nat Genet* 2000;**26**(1):15–6.
98. Stoetzel C, Laurier V, Davis EE, Muller J, Rix S, Badano JL, et al. BBS10 encodes a vertebrate-specific chaperonin-like protein and is a major BBS locus. *Nat Genet* 2006;**38**(5):521–4.
99. Stoetzel C, Muller J, Laurier V, Davis EE, Zaghloul NA, Vicaire S, et al. Identification of a novel BBS gene (BBS12) highlights the major role of a vertebrate-specific branch of chaperonin-related proteins in Bardet–Biedl syndrome. *Am J Hum Genet* 2007;**80**(1):1–11.
100. Chiang AP, Beck JS, Yen HJ, Tayeh MK, Scheetz TE, Swiderski RE, et al. Homozygosity mapping with SNP arrays identifies TRIM32, an E3 ubiquitin ligase, as a Bardet–Biedl syndrome gene (BBS11). *Proc Natl Acad Sci USA* 2006;**103**(16):6287–92.
101. Chiang AP, Nishimura D, Searby C, Elbedour K, Carmi R, Ferguson AL, et al. Comparative genomic analysis identifies an ADP-ribosylation factor-like gene as the cause of Bardet–Biedl syndrome (BBS3). *Am J Hum Genet* 2004;**75**(3):475–84.
102. Fan Y, Esmail MA, Ansley SJ, Blacque OE, Boroevich K, Ross AJ, et al. Mutations in a member of the Ras superfamily of small GTP-binding proteins causes Bardet–Biedl syndrome. *Nat Genet* 2004;**36**(9):989–93.
103. Parisi MA. Clinical and molecular features of Joubert syndrome and related disorders. *Am J Med Genet C Semin Med Genet* 2009;**151C**(4):326–40.
104. Zaghloul NA, Katsanis N. Mechanistic insights into Bardet–Biedl syndrome, a model ciliopathy. *J Clin Invest* 2009;**119**(3):428–37.
105. Li JB, Gerdes JM, Haycraft CJ, Fan Y, Teslovich TM, May-Simera H, et al. Comparative genomics identifies a flagellar and basal body proteome that includes the BBS5 human disease gene. *Cell* 2004;**117**(4):541–52.
106. Kim JC, Badano JL, Sibold S, Esmail MA, Hill J, Hoskins BE, et al. The Bardet–Biedl protein BBS4 targets cargo to the pericentriolar region and is required for microtubule anchoring and cell cycle progression. *Nat Genet* 2004;**36**(5):462–70.
107. Kim JC, Ou YY, Badano JL, Esmail MA, Leitch CC, Fiedrich E, et al. MKKS/BBS6, a divergent chaperonin-like protein linked to the obesity disorder Bardet–Biedl syndrome, is a novel centrosomal component required for cytokinesis. *J Cell Sci* 2005;**118**(Pt 5):1007–20.
108. Ross AJ, May-Simera H, Eichers ER, Kai M, Hill J, Jagger DJ, et al. Disruption of Bardet–Biedl syndrome ciliary proteins perturbs planar cell polarity in vertebrates. *Nat Genet* 2005;**37**(10):1135–40.
109. Ansley SJ, Badano JL, Blacque OE, Hill J, Hoskins BE, Leitch CC, et al. Basal body dysfunction is a likely cause of pleiotropic Bardet–Biedl syndrome. *Nature* 2003;**425**(6958):628–33.

110. Mykytyn K, Mullins RF, Andrews M, Chiang AP, Swiderski RE, Yang B, et al. Bardet–Biedl syndrome type 4 (BBS4)-null mice implicate Bbs4 in flagella formation but not global cilia assembly. *Proc Natl Acad Sci USA* 2004;**101**(23):8664–9.
111. Kulaga HM, Leitch CC, Eichers ER, Badano JL, Lesemann A, Hoskins BE, et al. Loss of BBS proteins causes anosmia in humans and defects in olfactory cilia structure and function in the mouse. *Nat Genet* 2004;**36**(9):994–8.
112. Fath MA, Mullins RF, Searby C, Nishimura DY, Wei J, Rahmouni K, et al. Mkks-null mice have a phenotype resembling Bardet–Biedl syndrome. *Hum Mol Genet* 2005;**14**(9):1109–18.
113. Nishimura DY, Fath M, Mullins RF, Searby C, Andrews M, Davis R, et al. Bbs2-null mice have neurosensory deficits, a defect in social dominance, and retinopathy associated with mislocalization of rhodopsin. *Proc Natl Acad Sci USA* 2004;**101**(47):16588–93.
114. Yen HJ, Tayeh MK, Mullins RF, Stone EM, Sheffield VC, Slusarski DC. Bardet–Biedl syndrome genes are important in retrograde intracellular trafficking and Kupffer's vesicle cilia function. *Hum Mol Genet* 2006;**15**(5):667–77.
115. Grace C, Beales P, Summerbell C, Jebb SA, Wright A, Parker D, et al. Energy metabolism in Bardet–Biedl syndrome. *Int J Obes Relat Metab Disord* 2003;**27**(11):1319–24.
116. Eichers ER, Abd-El-Barr MM, Paylor R, Lewis RA, Bi W, Lin X, et al. Phenotypic characterization of Bbs4 null mice reveals age-dependent penetrance and variable expressivity. *Hum Genet* 2006;**120**(2):211–26.
117. Mak HY, Nelson LS, Basson M, Johnson CD, Ruvkun G. Polygenic control of *Caenorhabditis elegans* fat storage. *Nat Genet* 2006;**38**(3):363–8.
118. Guan XM, Yu H, Van der Ploeg LH. Evidence of altered hypothalamic pro-opiomelanocortin/neuropeptide Y mRNA expression in tubby mice. *Brain Res Mol Brain Res* 1998;**59**(2):273–9.
119. Backberg M, Meister B. Abnormal cholinergic and GABAergic vascular innervation in the hypothalamic arcuate nucleus of obese tub/tub mice. *Synapse* 2004;**52**(4):245–57.
120. Davenport JR, Watts AJ, Roper VC, Croyle MJ, van Groen T, Wyss JM, et al. Disruption of intraflagellar transport in adult mice leads to obesity and slow-onset cystic kidney disease. *Curr Biol* 2007;**17**(18):1586–94.
121. Rahmouni K, Fath MA, Seo S, Thedens DR, Berry CJ, Weiss R, et al. Leptin resistance contributes to obesity and hypertension in mouse models of Bardet–Biedl syndrome. *J Clin Invest* 2008;**118**(4):1458–67.
122. Seo S, Guo DF, Bugge K, Morgan DA, Rahmouni K, Sheffield VC. Requirement of Bardet–Biedl syndrome proteins for leptin receptor signaling. *Hum Mol Genet* 2009;**18**(7):1323–31.
123. Marion V, Stoetzel C, Schlicht D, Messaddeq N, Koch M, Flori E, et al. Transient ciliogenesis involving Bardet–Biedl syndrome proteins is a fundamental characteristic of adipogenic differentiation. *Proc Natl Acad Sci USA* 2009;**106**(6):1820–5.
124. Suh JM, Gao X, McKay J, McKay R, Salo Z, Graff JM. Hedgehog signaling plays a conserved role in inhibiting fat formation. *Cell Metab* 2006;**3**(1):25–34.
125. Huangfu D, Liu A, Rakeman AS, Murcia NS, Niswander L, Anderson KV. Hedgehog signalling in the mouse requires intraflagellar transport proteins. *Nature* 2003;**426**(6962):83–7.
126. Haycraft CJ, Banizs B, Aydin-Son Y, Zhang Q, Michaud EJ, Yoder BK. Gli2 and Gli3 localize to cilia and require the intraflagellar transport protein polaris for processing and function. *PLoS Genet* 2005;**1**(4):e53.
127. Gerdes JM, Liu Y, Zaghloul NA, Leitch CC, Lawson SS, Kato M, et al. Disruption of the basal body compromises proteasomal function and perturbs intracellular Wnt response. *Nat Genet* 2007;**39**(11):1350–60.
128. Corbit KC, Shyer AE, Dowdle WE, Gaulden J, Singla V, Chen MH, et al. Kif3a constrains beta-catenin-dependent Wnt signalling through dual ciliary and non-ciliary mechanisms. *Nat Cell Biol* 2008;**10**(1):70–6.

129. Croft JB, Morrell D, Chase CL, Swift M. Obesity in heterozygous carriers of the gene for the Bardet–Biedl syndrome. *Am J Med Genet* 1995;**55**(1):12–5.
130. Croft JB, Swift M. Obesity, hypertension, and renal disease in relatives of Bardet–Biedl syndrome sibs. *Am J Med Genet* 1990;**36**(1):37–42.
131. Reed DR, Ding Y, Xu W, Cather C, Price RA. Human obesity does not segregate with the chromosomal regions of Prader-Willi, Bardet–Biedl, Cohen, Borjeson or Wilson-Turner syndromes. *Int J Obes Relat Metab Disord* 1995;**19**(9):599–603.
132. Andersen KL, Echwald SM, Larsen LH, Hamid YH, Glumer C, Jorgensen T, et al. Variation of the McKusick-Kaufman gene and studies of relationships with common forms of obesity. *J Clin Endocrinol Metab* 2005;**90**(1):225–30.
133. Fan Y, Rahman P, Peddle L, Hefferton D, Gladney N, Moore SJ, et al. Bardet–Biedl syndrome 1 genotype and obesity in the Newfoundland population. *Int J Obes Relat Metab Disord* 2004;**28**(5):680–4.
134. Benzinou M, Walley A, Lobbens S, Charles MA, Jouret B, Fumeron F, et al. Bardet–Biedl syndrome gene variants are associated with both childhood and adult common obesity in French Caucasians. *Diabetes* 2006;**55**(10):2876–82.
135. Carpenter G. Case of acrocephaly with other congenital malformations. *Proc R Soc Med* 1909;**2**:45–53, 199–201.
136. Temtamy SA. Carpenter's syndrome: acrocephalopolysyndactyly. An autosomal recessive syndrome. *J Pediatr* 1966;**69**(1):111–1120.
137. Jenkins D, Seelow D, Jehee FS, Perlyn CA, Alonso LG, Bueno DF, et al. RAB23 mutations in Carpenter syndrome imply an unexpected role for hedgehog signaling in cranial-suture development and obesity. *Am J Hum Genet* 2007;**80**(6):1162–70.
138. Chandler KE, Kidd A, Al-Gazali L, Kolehmainen J, Lehesjoki AE, Black GC, et al. Diagnostic criteria, clinical characteristics, and natural history of Cohen syndrome. *J Med Genet* 2003;**40**(4):233–41.
139. Kivitie-Kallio S, Norio R. Cohen syndrome: essential features, natural history, and heterogeneity. *Am J Med Genet* 2001;**102**(2):125–35.
140. Cohen Jr. MM, Hall BD, Smith DW, Graham CB, Lampert KJ. A new syndrome with hypotonia, obesity, mental deficiency, and facial, oral, ocular, and limb anomalies. *J Pediatr* 1973;**83**(2):280–4.
141. Fryns JP, Legius E, Devriendt K, Meire F, Standaert L, Baten E, et al. Cohen syndrome: the clinical symptoms and stigmata at a young age. *Clin Genet* 1996;**49**(5):237–41.
142. Partington M, Anderson D. Mild growth retardation and developmental delay, microcephaly, and a distinctive facial appearance. *Am J Med Genet* 1994;**49**(2):247–50.
143. Seifert W, Holder-Espinasse M, Spranger S, Hoeltzenbein M, Rossier E, Dollfus H, et al. Mutational spectrum of COH1 and clinical heterogeneity in Cohen syndrome. *J Med Genet* 2006;**43**(5):e22.
144. Carey JC, Hall BD. Confirmation of the Cohen syndrome. *J Pediatr* 1978;**93**(2):239–44.
145. Resnick K, Zuckerman J, Cotlier E. Cohen syndrome with bull's eye macular lesion. *Ophthalmic Paediatr Genet* 1986;**7**(1):1–8.
146. Norio R, Raitta C, Lindahl E. Further delineation of the Cohen syndrome; report on chorioretinal dystrophy, leukopenia and consanguinity. *Clin Genet* 1984;**25**(1):1–14.
147. Kivitie-Kallio S, Summanen P, Raitta C, Norio R. Ophthalmologic findings in Cohen syndrome. A long-term follow-up. *Ophthalmology* 2000;**107**(9):1737–45.
148. Kivitie-Kallio S, Rajantie J, Juvonen E, Norio R. Granulocytopenia in Cohen syndrome. *Br J Haematol* 1997;**98**(2):308–11.
149. Olivieri O, Lombardi S, Russo C, Corrocher R. Increased neutrophil adhesive capability in Cohen syndrome, an autosomal recessive disorder associated with granulocytopenia. *Haematologica* 1998;**83**(9):778–82.

150. Slavotinek A, Shaffer LG, Shapira SK. Monosomy 1p36. *J Med Genet* 1999;**36**(9):657–63.
151. Kolehmainen J, Black GC, Saarinen A, Chandler K, Clayton-Smith J, Traskelin AL, et al. Cohen syndrome is caused by mutations in a novel gene, COH1, encoding a transmembrane protein with a presumed role in vesicle-mediated sorting and intracellular protein transport. *Am J Hum Genet* 2003;**72**(6):1359–69.
152. Velayos-Baeza A, Vettori A, Copley RR, Dobson-Stone C, Monaco AP. Analysis of the human VPS13 gene family. *Genomics* 2004;**84**(3):536–49.
153. Mochida GH, Rajab A, Eyaid W, Lu A, Al-Nouri D, Kosaki K, et al. Broader geographical spectrum of Cohen syndrome due to COH1 mutations. *J Med Genet* 2004;**41**(6):e87.
154. Hampshire DJ, Ayub M, Springell K, Roberts E, Jafri H, Rashid Y, et al. MORM syndrome (mental retardation, truncal obesity, retinal dystrophy and micropenis), a new autosomal recessive disorder, links to 9q34. *Eur J Hum Genet* 2006;**14**(5):543–8.
155. Jacoby M, Cox JJ, Gayral S, Hampshire DJ, Ayub M, Blockmans M, et al. INPP5E mutations cause primary cilium signaling defects, ciliary instability and ciliopathies in human and mouse. *Nat Genet* 2009;**41**(9):1027–31.
156. Bielas SL, Silhavy JL, Brancati F, Kisseleva MV, Al-Gazali L, Sztriha L, et al. Mutations in INPP5E, encoding inositol polyphosphate-5-phosphatase E, link phosphatidyl inositol signaling to the ciliopathies. *Nat Genet* 2009;**41**(9):1032–6.
157. Pallister PD, Herrmann J, Opitz JM. Studies of malformation syndromes in man XXXXII: a pleiotropic dominant mutation affecting skeletal, sexual and apocrine-mammary development. *Birth Defects Orig Artic Ser* 1976;**12**(5):247–54.
158. Schinzel A, Illig R, Prader A. The ulnar-mammary syndrome: an autosomal dominant pleiotropic gene. *Clin Genet* 1987;**32**(3):160–8.
159. Gay I, Feinmesser R, Cohen T. Laryngeal web, congenital heart disease and low stature. A syndrome? *Arch Otolaryngol* 1981;**107**(8):510–2.
160. Franceschini P, Vardeu MP, Dalforno L, Signorile F, Franceschini D, Lala R, et al. Possible relationship between ulnar-mammary syndrome and split hand with aplasia of the ulna syndrome. *Am J Med Genet* 1992;**44**(6):807–12.
161. Gonzalez CH, Herrmann J, Opitz JM. Studies of malformation syndromes of man XXXXIIB: mother and son affected with the ulnar-mammary syndrome type Pallister. *Eur J Pediatr* 1976;**123**(4):225–35.
162. Bamshad M, Krakowiak PA, Watkins WS, Root S, Carey JC, Jorde LB. A gene for ulnar-mammary syndrome maps to 12q23–q24.1. *Hum Mol Genet* 1995;**4**(10):1973–7.
163. Bamshad M, Lin RC, Law DJ, Watkins WC, Krakowiak PA, Moore ME, et al. Mutations in human TBX3 alter limb, apocrine and genital development in ulnar-mammary syndrome. *Nat Genet* 1997;**16**(3):311–5.
164. Sasaki G, Ogata T, Ishii T, Hasegawa T, Sato S, Matsuo N. Novel mutation of TBX3 in a Japanese family with ulnar-mammary syndrome: implication for impaired sex development. *Am J Med Genet* 2002;**110**(4):365–9.
165. Linden H, Williams R, King J, Blair E, Kini U. Ulnar Mammary syndrome and TBX3: expanding the phenotype. *Am J Med Genet A* 2009;**149A**(12):2809–12.
166. Mesbah K, Harrelson Z, Theveniau-Ruissy M, Papaioannou VE, Kelly RG. Tbx3 is required for outflow tract development. *Circ Res* 2008;**103**(7):743–50.
167. Bakker ML, Boukens BJ, Mommersteeg MT, Brons JF, Wakker V, Moorman AF, et al. Transcription factor Tbx3 is required for the specification of the atrioventricular conduction system. *Circ Res* 2008;**102**(11):1340–9.
168. Eriksson KS, Mignot E. T-box 3 is expressed in the adult mouse hypothalamus and medulla. *Brain Res* 2009;**1302**:233–9.

169. Sharp AJ, Locke DP, McGrath SD, Cheng Z, Bailey JA, Vallente RU, et al. Segmental duplications and copy-number variation in the human genome. *Am J Hum Genet* 2005;**77** (1):78–88.

170. Bochukova EG, Huang N, Keogh J, Henning E, Purmann C, Blaszczyk K, Saeed S, Hamilton-Shield J, Clayton-Smith J, O'Rahilly S, Hurles ME, Farooqi IS. Large, rare chromosomal deletions associated with severe early-onset obesity. *Nature* 2010;**463**:666–70.

171. Sha BY, Yang TL, Zhao LJ, Chen XD, Guo Y, Chen Y, et al. Genome-wide association study suggested copy number variation may be associated with body mass index in the Chinese population. *J Hum Genet* 2009;**54**(4):199–202.

172. Frayling TM, Timpson NJ, Weedon MN, Zeggini E, Freathy RM, Lindgren CM, et al. A common variant in the FTO gene is associated with body mass index and predisposes to childhood and adult obesity. *Science* 2007;**316**(5826):889–94.

173. Boissel S, Reish O, Proulx K, Kawagoe-Takaki H, Sedgwick B, Yeo GS, et al. Loss-of-function mutation in the dioxygenase-encoding FTO gene causes severe growth retardation and multiple malformations. *Am J Hum Genet* 2009;**85**(1):106–11.

A Translational View of the Genetics of Lipodystrophy and Ectopic Fat Deposition

Matthew B. Lanktree,[*,†]
Christopher T. Johansen,[*,†]
Tisha R. Joy,[*] and
Robert A. Hegele[*,†]

[*]*Departments of Medicine and Biochemistry, Schulich School of Medicine & Dentistry, Robarts Research Institute, University of Western Ontario, London, Ontario, Canada*

[†]*Blackburn Cardiovascular Genetics Laboratory, Robarts Research Institute, University of Western Ontario, London, Ontario, Canada*

Progress in Molecular Biology
and Translational Science, Vol. 94
DOI: 10.1016/S1877-1173(10)94006-4

A wide range of lipodystrophy syndromes exist, each with varying clinical presentations, and yet cumulatively they underscore the importance of adipocyte biology in human metabolism. Loss of the ability to retain excess lipids in "classical" adipose tissue stores can lead to the overdevelopment of ectopic fat stores, often creating severe perturbations of both glucose and lipid homeostasis. Linkage analysis and candidate sequencing efforts have successfully identified responsible mutations for multiple forms of lipodystrophy. Recently, the reduction in the cost of DNA sequencing has resulted in discovery of many novel mutations within both known and novel loci. In this review, we present the steps involved in clinical characterization of a suspected lipodystrophy case, an overview of the clinical manifestations, molecular findings, and pathogenic basis of different forms of lipodystrophy, a discussion of therapeutic options for lipodystrophy patients, and an examination of genetic advances that will be used to identify additional pathogenic mechanisms.

I. Introduction

The term lipodystrophy refers to a heterogeneous group of disorders characterized by selective or generalized atrophy of anatomical adipose tissue stores (Table I). Both selective and generalized forms of lipodystrophy can be broadly categorized as either acquired or inherited (also often referred to as congenital or familial).[1] Loss of the ability to retain excess lipids in "classical" adipose tissue stores can lead to the overdevelopment of ectopic fat stores, particularly certain subcutaneous depots, but also within and around skeletal muscle, heart, liver, pancreas, and kidneys and within the arterial wall presenting as atherosclerosis.[2] In some women with lipodystrophy, polycystic ovary syndrome (PCOS) is seen. The severity of associated clinical and metabolic manifestations is often, but not always, proportional to the extent of adipose tissue loss. Development of the constellation of symptoms referred to as the

TABLE I
CLASSIFICATION OF LIPODYSTROPHY

Disease	OMIM #
Generalized	
Congenital generalized lipodystrophy (CGL) (Berardinelli–Seip)	
CGL1—caused by mutations in *AGPAT2*	608594
CGL2—caused by mutations in *BSCL2*	269700
CGL3—caused by mutations in *CAV1, PTRF*	612526
Acquired generalized lipodystrophy (AGL)	
Partial	
Familial partial lipodystrophy (FPLD)	
FPLD1 (Kobberling)	
FPLD2 (Dunnigan)—caused by mutations in *LMNA*	151660
FPLD3—caused by mutations in *PPARG*	604367
FPLD caused by newly discovered mutations—*CAV1, AKT2*	
Acquired partial lipodystrophy (APL) (Barraquer–Simmons)	
APL—some cases associated with *LMNB2* mutations	608709
HIV-associated lipodystrophy—partially due to inhibition of *ZMPSTE24*	
Syndromes that include lipodystrophy as a component	
Mandibuloacral dysplasia (MAD)	
MADA—caused by mutations in *LMNA*	248370
MADB—caused by mutations in *ZMPSTE24*	608612
SHORT syndrome—potentially caused by mutations in *PITX2*	269880
Neonatal progeroid syndrome	264090
Hutchinson–Gilford progeria syndrome (HGPS)—caused by mutations in *LMNA*	176670
Werner syndrome (WRN)—caused by mutations in *RECQL2, LMNA*	277700

metabolic syndrome (MetS), including insulin resistance, glucose intolerance, lipid disturbances (hypertriglyceridemia and reduced high-density lipoprotein (HDL) cholesterol), and hypertension, is common, creating the additional concern of increased risk of cardiovascular disease (Table II).

Inherited forms of lipodystrophy can segregate in families, often following a Mendelian pattern of inheritance. Acquired lipodystrophy represents a complex phenotype that can result from preexisting genetic susceptibility with the compounding influence of nongenetic triggers, such as autoimmune responses, infections, or certain medications. Lipodystrophy may also be observed as a component of rare multisystem syndromes such as mandibuloacral dysplasia (MAD), Hutchinson–Gilford progeria syndrome (HGPS), and Werner syndrome (WRN). Genetic mapping efforts, starting near the end of the twentieth century, have successfully identified the genetic cause of approximately half of lipodystrophy patients who present to the clinic.[3] Recently, reduction in cost and improvement in access to DNA sequencing has resulted in the discovery of many responsible mutations residing within previously known loci. Additionally, sequencing of novel candidate loci, such as caveolin-1 (*CAV1*) and polymerase I

TABLE II
COMMON CLINICAL FEATURES OF LIPODYSTROPHY

Feature	Characteristics
Fat atrophy (lipoatrophy)	– Face: sunken cheeks, hollow temples, sunken eyes, prominent zygomatic arch – Extremities: prominent veins, skinny or muscular appearance – Buttocks: loss of contour, loose skin folds
Fat accumulation (lipohypertrophy)	– Face and neck: "moon facies" and "fat neck syndrome" – Abdomen: increased abdominal girth with visceral fat accumulation – Dorsocervical or supraclavicular fat pad – Fat accumulation in ectopic stores: muscle, heart, liver, pancreas and kidneys and within the arterial wall
Dysglycemia	– Insulin resistance, progressing to glucose intolerance and type 2 diabetes – Acanthosis nigricans
Dyslipidemia	– Hypertriglyceridemia, usually with depressed high-density lipoprotein (HDL) cholesterol – Hypercholesterolemia
Reproductive anomalies	– Menstrual irregularities, polycystic ovarian syndrome – Androgen deficiency
Atherosclerosis	– Elevated cardiovascular and cerebrovascular risk

and transcript release factor (*PTRF*), has uncovered additional causal variants, each resulting in an incremental increase in the proportion of lipodystrophy cases that can be accounted for at the molecular level. In this review, we present a thorough description of the investigations required for clinical characterization of a patient with suspected lipodystrophy, an overview of the clinical manifestations of different forms of lipodystrophy, the molecular basis associated with each of the lipodystrophy phenotypes, and an examination of genetic advances that are currently being applied to identify additional pathogenic mechanisms.

II. Components of a Thorough Investigation of a Suspected Lipodystrophy Case

A. Medical History

The heterogeneity of the lipodystrophies means that no single combination of clinical and biochemical investigations would provide definitive evidence favoring or ruling out a diagnosis. Even molecular analysis—DNA sequencing of known candidate genes—fails to identify a mutation in about half of cases referred to our specialized quaternary center; in such cases, clinical diagnosis becomes more pertinent. Clinical assessment of the patient should include (a) history of attributes directly related to lipodystrophy, such as changes in anthropometry, appetite, or appearance; (b) historical features of insulin resistance, such as acanthosis nigricans; (c) history of features of related metabolic states, such as PCOS (hirsutism, menstrual irregularities); (d) careful assessment of clinical features that can affect other organ systems in some complex and extended syndromes that contain lipodystrophy as a component; (e) history of potential triggering factors or associated conditions, such as autoimmune disorders, infection, trauma, or exposure to medications or xenobiotics; and (f) family history of similar anthropometry, metabolic characteristics, or related conditions (Table III). Physical examination includes (a) careful qualitative or semi-quantitative anthropometric assessment and description; (b) dermatologic evaluation (e.g., acanthosis nigricans, lipomata); (c) cardiovascular system evaluation (blood pressure, peripheral pulses, arterial bruits, cardiac assessment); (d) abdominal evaluation for possible hepatosplenomegaly; (e) assessment for possible micro- and macrovascular complications of diabetes; and (f) assessment of other organ systems as indicated, such as neurological, musculoskeletal, and genitourinary systems.

B. Biochemical Investigations

Biochemical assessment includes documentation of associated dysglycemia (fasting blood glucose, glycated hemoglobin), dyslipidemia (plasma lipid profile), renal function (creatinine, urinary albumin:creatinine ratio), and liver function (serum transaminases). A fasting serum insulin and C-peptide level might help document the presence of hyperinsulinemia. Serum levels of C-reactive protein (CRP) and free fatty acids (FFA) are often elevated. Other measurements such as plasma concentrations of leptin or adiponectin are not indicated as part of routine clinical care, but the determinations might have some academic interest, since they can be dramatically reduced in some patients with lipodystrophy.

C. Adipose Tissue Characterization

Clinical research indicates characteristic changes in skin-fold thickness at particular sites among individuals with certain forms of lipodystrophy, but these vary over a wide range and may not be definitive or pathognomonic in specific patients. Similarly, noninvasive imaging using computed tomography or magnetic resonance imaging (MRI) and quantification of subcutaneous and visceral fat depots show significant mean differences between individuals with particular molecular subtypes of lipodystrophy, but there are no clinical standards for these measurements, and the range of individual values is wide even among individuals with the same molecular lesion. We have found that the mid-thigh is the most specific site for depletion of adipose tissue across several forms of

TABLE III

COMPONENTS OF A COMPREHENSIVE INVESTIGATION OF SUSPECTED LIPODYSTROPHY PATIENT

Modality	Investigation
History	– Directly related to lipodystrophy: onset, changes in weight, appearance, appetite – History of potentially triggering events – Family history of similar characteristic anthropometry or metabolic features – Thorough review of systems with emphasis on organs and systems often affected in lipodystrophy, including polycystic ovarian syndrome
Clinical	– Thorough physical examination: cardiovascular, abdominal, musculoskeletal, neurological, dermatological, genitourinary systems – Specific examination for micro- and macrovascular complications – Qualitative or semi-quantitative anthropometric assessment and description (bioimpedence, skin-fold thickness, hip-to-waist ratio)
Biochemical	– Dysglycemia (fasting blood glucose, hemoglobin A1C) – Dyslipidemia (fasting plasma lipid profile) – Renal function (creatinine, urinary albumin: creatinine ratio) – Liver function tests (serum transaminases) – Insulin resistance (fasting insulin, C-peptide) – Inflammation (CRP)
Imaging	– Magnetic resonance imaging (MRI) for evaluation of ectopic fat stores – Ultrasound measurement of hepatosteatosis, carotid intima-media thickness
Genetic	– Candidate gene sequencing of known lipodystrophy-causing loci

lipodystrophy; this could be the single most useful site for noninvasive visualization of subcutaneous fat. However, noninvasive imaging is not currently recommended for lipodystrophy diagnosis, although selected images—when compared to appropriate controls—can be of academic interest and, if lipoatrophy is obvious radiologically, might be helpful for diagnosis. We recently showed that among female patients with partial lipodystrophy, no simple clinical anthropometric measure correlates well with visceral adipose tissue mass quantified on MRI scanning, while BMI correlates well with subcutaneous adipose stores.[4]

Abdominal ultrasound examination can sometimes document gross hepatosteatosis, which is a common feature in many lipodystrophy types. More advanced noninvasive imaging tools include spectroscopic evaluation to quantify hepatic and myocellular accumulations of adipose tissue.[5]

III. Generalized Lipodystrophies

A. Congenital Generalized (Berardinelli–Seip) Lipodystrophy

1. Clinical Features

Congenital generalized lipodystrophy (CGL) was first described by Berardinelli from Brazil and Seip from Norway in the 1950s[6,7] (Table IV). Infants affected by CGL are often diagnosed soon after birth due to the generalized muscular appearance resulting from a near-absence of adipose tissue.[3] In childhood, patients affected by CGL can be recognized by what has been described as "anabolic syndrome," including a voracious appetite, muscular hypertrophy, accelerated linear growth, advanced bone age, and precocious genitalia development.[1,8,9] Additionally, marked acanthosis nigricans over the neck, axilla, groin, and trunk and acromegaloid features, including enlarged jaw, feet, and hands, are also often present. Umbilical hernia, hepatomegaly secondary to hepatic steatosis, and focal lytic lesions of the appendicular bones are often present.[8–10] Pancreatitis, cardiomyopathy, renal failure, hirsutisim, and intellectual impairment may occur. Approximately 250 patients of multiple ethnicities have been reported in the literature, and the estimated prevalence has been reported to be 1 in 10 million.[1] CGL subtypes have been defined at the molecular level (see below), and phenotypic distinctions can be recognized between groups. CGL type 1 appears to have a less severe phenotype than CGL type 2, with fewer severe biochemical manifestations, less adipose tissue involvement, less cardiovascular disease risk, and later onset of diabetes.[3]

CGL patients often display features of MetS in preadolescence, including overt type 2 diabetes, hyperinsulinemia, insulin resistance, and dyslipidemia (markedly elevated triglycerides (TG) and depressed HDL cholesterol), though hypertension is rare. Serum leptin and adiponectin levels are also drastically reduced.[11] In women with CGL, hirsutism, PCOS, and menstrual abnormalities are common, while in men with CGL, reproduction is typically unaffected.[12] Advanced imaging of CGL patients using MRI confirms the drastic reduction of subcutaneous adipose tissue stores, including supraclavicular, gluteal, mid-thigh, and mid-calf stores. When examining visceral fat stores using MRI, CGL patients show variable quantities compared to controls, or compared to patients with familial partial lipodystrophy (FPLD), acquired partial lipodystrophy (APL), and human immunodeficiency virus (HIV)-related lipodystrophy.[13]

2. Genetics and Pathophysiology

All identified forms of CGL have followed an autosomal recessive pattern of inheritance. The responsible mutation for CGL type 1 (OMIM: 608594) was identified in the 1-acylglycerol-3-phosphate-*O*-acyltransferase 2 (*AGPAT2*)

TABLE IV

Clinical Features of Generalized Lipodystrophies

Variable	CGL1 (*AGPAT2*)	CGL2 (*BSCL2*)	CGL3 (*CAV1*)	AGL
Age of onset	Soon after birth	Soon after birth	3 months after birth	Usually < 20 years
Body mass index	↓↓ to ↔	↓↓ to ↔	↓↓	↓↓ to ↔
Facial fat	↓↓↓↓	↓↓↓↓	↓↓↓↓	↓↓↓
Mechanical fat	↔	↓↓↓↓	↓↓	↓↓ to ↔
Limb fat	↓↓↓↓	↓↓↓↓	↓↓↓↓	↓↓↓↓
Trunk fat	↓↓↓↓	↓↓↓↓	↓↓↓↓	↓↓↓↓ to ↔
Gluteal fat	↓↓↓↓	↓↓↓↓	↓↓↓↓	↓↓↓↓ to ↓↓
Bone marrow fat	↓↓↓↓	↓↓↓↓	↔	↓↓↓↓
Hepatic steatosis	↑↑↑	↑↑↑	↑↑↑↑	↑↑↑↑
Diabetes (DM)	Common	Common	Present	Common
DM age of onset	Adolescence	< 10 years	8 years	Variable
Hypertension	Rare	Rare	Not present	Can be present
Acanthosis nigricans	Present	Present	Present	Very common
Hirsutism	Common	Common	Present	Common
Sexual function	No change	External genitalia pseudohypertrophy often present	Primary amenorrhea	Amenorrhea or PCOS can be present
Fasting insulin	↑↑↑↑	↑↑↑	↑↑↑↑	↑↑↑
Triglyceride	↑↑↑↑	↑↑↑	↑↑↑↑	↑↑ to ↑↑↑
HDL	↓↓ to ↔	↓↓ to ↔	–	↓ to ↔
Leptin	↓↓↓	↓↓↓↓	↓↓↓↓	↓↓
Adiponectin	↓↓↓↓	↓↓↓↓	↓↓↓↓	↓↓↓
Other features	Cardiomyopathy and bone cysts may be present	Mental retardation possible; higher risk of cardiomyopathy than CGL1; bone cysts may be present	–	Subgroups: idiopathic; assoc. w/autoimmune conditions; assoc. w/panniculitis

Missing entries have not been reported in the literature.

(OMIM: 603100) gene found on chromosome 9p34.[14] AGPAT2 catalyzes reactions necessary for the metabolism of lysophosphatidic acid, an important part of TG synthesis, and is associated with increased expression of inflammatory cytokines such as interleukin-6 (IL-6) and tumor necrosis factor alpha (TNF-α).[15] Despite 25 reported mutations in *AGPAT2* (Fig. 1),[16] there appears to be no correlation between mutation location and CGL severity.

A gene on chromosome 11q13 was identified to contain mutations causative for CGL type 2 (OMIM: 269700) and named Bernardinelli–Seip congenital lipodystrophy type 2 (*BSCL2*) (OMIM: 606158).[17] *BSCL2* encodes the "seipin" protein, whose function in CGL is yet to be determined. Seipin is located in the endoplasmic reticulum membrane and expressed primarily in the brain and testes.[17,18] A total of 28 mutations have been reported in *BSCL2*, including splicing and missense mutations, and similar to as seen in *AGPAT2*, no relationship has been described between mutation location and disease severity (Fig. 1).[3]

The discovery of a mutation in the gene encoding *CAV1* (OMIM: 601047) led to defining a third CGL subtype (OMIM: 612526). The only reported CGL type 3 patient thus far is a homozygous carrier of a nonsense mutation (E38X), resulting from a consanguineous mating, leading to a near complete absence of both subcutaneous and visceral adipose tissue (Fig. 1).[19] In addition to fairly classic manifestations of CGL, the patient also had mild hypocalcemia, likely due to vitamin D resistance. Contrary to *AGPAT2* and *BSCL2*, mutations in *CAV1* appear to have additional heterogeneous effects, as *CAV1* has also been described in atypical partial lipodystrophy.[20] *CAV1* is both necessary and sufficient for caveolae formation, making it an important regulator of normal lipid homeostasis and insulin-regulated glucose uptake.[21]

Most recently, homozygous mutations in *PTRF* (OMIM: 603198) have been identified as causative for a subset of CGL patients.[22] The PTRF protein (also known as cavin) is a caveolin-associated protein, essential for the formation and stabilization of caveolae in human cells.[22] The clinical features of *PTRF* mutations were observed in five nonconsanguineous patients who presented with both generalized lipodystrophy and muscular dystrophy. Muscle hypertrophy, muscle mounding, mild metabolic complications, and elevated serum creatine kinase levels were observed in these patients. Skeletal muscle biopsies revealed chronic dystrophic changes, deficiency and mislocalization of all three caveolin family members, and a reduction of caveolae. A survey of our own clinical population has revealed that ~2% of patients with CGL that was previously uncharacterized at the molecular genetic level carry a rare heterozygous mutation in *PTRF* (Hegele, unpublished observations). The lipodystrophy and related clinical features observed in the patients with *PTRF* mutations are likely associated with a secondary deficiency of caveolins. Thus, a primary mutation in *CAV1* or in the associated *PTRF* can both lead to lipodystrophy, highlighting the importance of caveolin biology in the development of lipodystrophy.

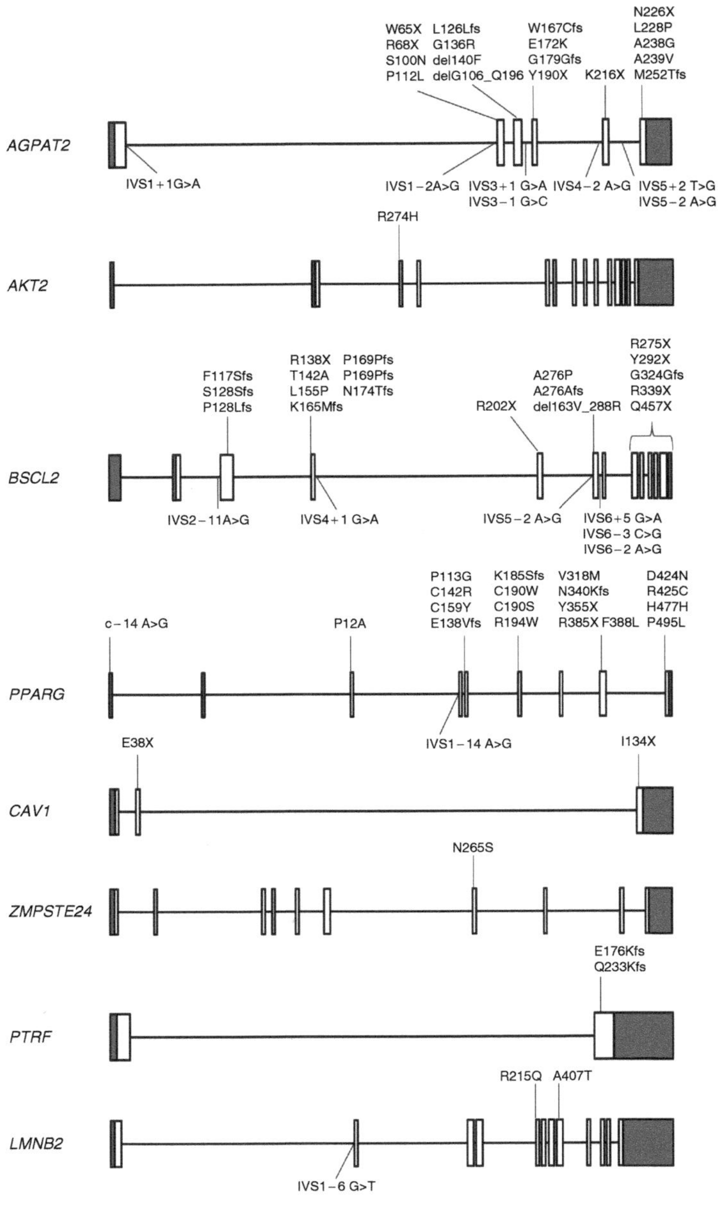

FIG. 1. Lipodystrophy causing mutations. Schematic diagrams of genes mutated in lipodystrophy and syndromes in which lipodystrophy is a component. Exons are drawn as boxes along the horizontal line approximately to scale. Non-synonymous substitution and frameshift and truncation mutations are listed above the corresponding exons, while splicing variants are below the exons. See Fig. 4 for *LMNA* mutations. Abbreviations: del, deletion; fs, frameshift; X, truncation; IVS, intervening sequence.

B. Acquired Generalized Lipodystrophy (AGL)

1. Clinical Features

In 1928, Ziegler first described AGL.[23] AGL is typically diagnosed in childhood to early teens, typically beginning with loss of adipose tissue of the face and upper extremities, progressing to large areas of the body.[1,3] Similar to CGL, patients often have voracious appetites, acanthosis nigricans, and hepatic steatosis. Retro-orbital and intramarrow fat are preserved, and patients may or may not have loss of fat in the palms and soles of the feet.[24] The degree of visceral fat loss is variable. Biochemical findings in AGL include low levels of leptin and adiponectin, hyperinsulinemia, diabetes, and dyslipidemia.

Misra and Garg suggested a classification of AGL patients into three categories based upon presumed etiology: AGL associated with autoimmune disorders, AGL associated with panniculitis, and idiopathic AGL.[24] They found that the prevalence of diabetes and hypertriglyceridemia was higher in the autoimmune and idiopathic groups compared to the panniculitis group.[24] AGL has been reported in the literature in approximately 80 patients, most of whom were white with a 3:1 female–male ratio,[24] though the exact prevalence is unknown.

2. Genetics and Pathophysiology

As of yet, no putative causative or associated mutations have been reported. Candidate gene sequencing of known lipodystrophy genes (*AGPAT2*, *BSCL2*, lamin A/C (*LMNA*), peroxisome proliferator-activated receptor gamma (*PPARG*), lamin B2 (*LMNB2*), *CAV1*, and *PTRF*) and genes encoding additional nuclear envelope proteins (lamin B receptor (*LBR*), lamin B1 (*LMNB1*), and emerin (*EMD*)) has been performed with no obvious variants identified (Hegele, unpublished observations).

IV. Lipodystrophies Affecting Selected Lipid Depots (Partial)

A. Familial Partial Lipodystrophies

1. Clinical Features

FPLD is the most heterogeneous lipodystrophy subtype, including several distinct entities: FPLD1, also known as Kobberling-type lipodystrophy; FPLD2 (OMIM: 151660), also known as Dunnigan-type lipodystrophy and caused by *LMNA* mutations; FPLD3 (OMIM: 604367), caused by *PPARG* mutations; and recently FPLD from mutations in *CAV1*, v-akt murine thymoma viral oncogene homolog 2 (*AKT2*), *PTRF*, and zinc metallopeptidase STE24 (*ZMPSTE24*). Familial forms of lipodystrophy went unreported in the literature until the 1970s[25,26] (Table V). While still rare, FPLD2 is the most

TABLE V

Clinical Features of Partial Lipodystrophies

Variable	Diabetic FPLD2	Nondiabetic FPLD2	FPLD3	APL	HIV-associated
Age of onset	Puberty	Puberty	Puberty to adulthood	Usually < 20 years	Usually < 20 years
Body mass index	↔	↔	↔ to ↑↑	↓↓ to ↔	Typically normal
Facial fat	↔	↔	↓↓ to ↔	↓↓↓ to ↔	↓↓↓ to ↓↓
Mechanical fat	↔	↔	↔	↓↓ to ↔	↔
Upper limb fat	↓↓↓↓	↓↓↓↓	↓↓↓↓	↓↓ to ↓↓↓	↓↓↓↓ to ↔
Lower limb fat	↓↓↓↓	↓↓↓↓	↓↓↓↓	↔ to ↑↑	↓↓↓↓ to ↔
Trunk fat	↑↑↑↑	↑↑↑↑	↑↑↑↑	↓↓↓ to ↓↓	↓↓↓↓ to ↑↑
Gluteal fat	↓↓↓↓	↓↓↓↓	↓↓ to ↔	↔ to ↑↑	↓↓↓↓ to 0
Hepatic steatosis	↔ to ↑↑	↔ to ↑↑	↑↑ to ↑↑↑	Rare	↔ to ↑↑↑
Diabetes (DM)	Present	Absent	Typically present	Uncommon	Common
DM age of onset	~ 35–45 years		~ 13–45 years	Adulthood	Adulthood
Hypertension	Present	Present	Present	Can be present	Can be present
Acanthosis nigricans	Present	Present	Typically present	Rare	Rare
Hirsutism	Rare	Rare	Common	Uncommon	Rare
Sexual function	External genitalia pseudohypertrophy often present	Rarely PCOS	Often menstrual anomalies or PCOS	Rarely PCOS	Often androgen deficiency in ♂; PCOS rarely in ♀
Fasting insulin	↑↑	↑	↑↑↑↑	↔ to ↑↑	↑↑ to ↑↑↑↑
Triglyceride	↑↑↑	↑↑	↑↑ to ↑↑↑↑	↔ to ↑↑	↑↑ to ↑↑↑↑
HDL	↓ to ↔	↓ to ↔	↓↓ to ↔	↔	↑ to ↑↑↑↑
Free fatty acids	↔ to ↑	↔ to ↑	↔ to ↑	–	↑↑↑ to ↑↑↑↑
Leptin	↓↓	↓↓	↓↓↓↓ to ↔	↓	↔ to ↑↑
Adiponectin	↓↓↓	↓↓↓	↓↓↓↓	↓↓	↓↓↓
C-reactive protein	↑	↑↑	↔ to ↑	Unknown	↔ to ↑↑
Other features	–	–	–	Associated with low C3, autoimmune disorders, membranoproliferatiive glomerulonephritis	–

common form of inherited lipodystrophy.[1] In general, patients with FPLD have normal fat distribution during childhood, but during or shortly after puberty there is a progressive and gradual loss of subcutaneous adipose tissue, a triggering event for the metabolic complications that follow. A variable degree of subcutaneous adipose tissue loss over the abdomen, trunk, and chest may occur. In FPLD1, patients typically maintain normal fat stores in the face, neck, and trunk. In FPLD2, fat deposition in the neck and labia increases. In FPLD3, a decrease to total absence of facial fat occurs. MRI imaging has identified increases in ectopic fat depots, mainly within intramyocellular, intraabdominal, and hepatic stores.[27–29] Acanthosis nigricans, a marker of significant insulin resistance, is common. Decaudain et al. proposed an additional diagnosis of metabolic laminopathy (MLP) for patients carrying *LMNA* mutations and associated biochemical findings but with less severe lipoatrophy.[30]

Biochemical manifestations include elevated plasma concentrations of TG, FFA, CRP, insulin, and C-peptide with depressed plasma concentrations of HDL cholesterol, leptin, and adiponectin. Circulating FFA may deposit within the liver and muscles, leading to increased hepatic and peripheral insulin resistance. Lipotoxicity affecting the pancreatic beta-cell with impairment of insulin secretion may hasten loss of glycemic control. Progressing to type 2 diabetes typically occurs in the fourth to fifth decades of life.

In women, hirsutism, PCOS, and menstrual irregularities are common.[4] A higher prevalence of infertility, gestational diabetes, miscarriages, and obstetrical complications has also been reported.[31] Women also have a greater risk of developing diabetes compared to men, particularly if multiparous and if possessing increased visceral adiposity.[32] Elevated risk of atherosclerotic disease is of concern in both sexes, especially in diabetic FPLD2 patients.[33]

Using MRI imaging to quantify adipose stores among patients affected with different FPLD subtypes, significant disparities have been observed.[13] FPLD2 patients experience a greater loss of subcutaneous fat, particularly in the gluteal, thigh, and calf regions, and develop larger ectopic fat stores when compared to FPLD3 patients.[13] Despite the less significant adipose tissue involvement observed in FPLD3, the metabolic complications are more severe. Though a molecular diagnosis can definitely differentiate FPLD2 from FPLD3, advanced imaging could be clinically useful for distinguishing FPLD3 from the more common FPLD2.[3]

2. Genetics and Pathophysiology

All forms of FPLD except FPLD1 have been observed to follow an autosomal dominant inheritance pattern. FPLD1 follows an unclear inheritance pattern and the molecular cause remains unknown. After linkage analysis of families affected with FPLD2 narrowed the causative locus to chromosome 1q21–22,[34] mutations in the gene encoding nuclear lamins A and C were identified in a Canadian kindred.[35] *LMNA* was hypothesized to harbor mutations as it was already known to contain mutations causative for Emery–Dreifuss muscular dystrophy, characterized by progressive muscle wasting with an anatomical arrangement similar to the adipocyte degeneration in FPLD2. DNA sequencing of patients continues to identify novel disease-causing mutations.[36] Currently, over 200 mutations have been identified in *LMNA*,[16] causing 13 different diseases together termed laminopathies, including FPLD2, Emery–Dreifuss muscular dystrophy, limb-girdle muscular dystrophy type 1B, dilated cardiomyopathy type 1A, Charcot-Marie-Tooth, HGPS, atypical WRN, and a range of overlapping syndromes. *LMNA* encodes an intermediate filament protein vital for the structural integrity of the nuclear envelope, transcriptional regulation, nuclear pore functioning, and heterochromatin organization. It remains unknown how the nonsense mutations observed in FPLD2 specifically alter *LMNA* functioning.

PPARG was selected as a candidate for sequencing in FPLD patients due to its important role as a ligand-inducible transcription factor regulating adipogenesis (Fig. 1).[37,38] Moreover, *PPARG* is bound and activated by the thiazolidinedione (TZD) class of drugs, which often results in a repartitioning of fat stores. Mutations in *PPARG* have been identified to work through both dominant negative mechanisms, in which the mutant receptor is able to inhibit the action of the wild-type receptor,[39] and a haploinsufficiency mechanism, in which the 50% reduction in wild-type expression was sufficient to create the phenotype.[38]

After the discovery of *LMNA* mutations as the molecular cause of FPLD2 and *PPARG* as the molecular cause of FPLD3, the molecular source of approximately 50% of patients had been uncovered.[3] Rare sequence variants in *AKT2* were discovered in a family with autosomal dominant insulin resistance and lipodystrophy,[40] but *AKT2* was later found to only account for a very small proportion of lipodystrophy patients.[41] More recently, after the discovery that *Cav1* deficient mice had depleted and abnormal adipocytes, *CAV1* was sequenced in a cohort of lipodystrophy patients with no known disease-causing mutation. Heterozygous *CAV1* frameshift mutations were identified in 3 of 60 patients.[20] As mentioned above, homozygous *CAV1* mutations have been identified in patients with CGL. Interestingly, *CAV3* mutations cause limb and girdle muscular dystrophy (OMIM: 607801) and rippling muscle disease (OMIM: 606072), two disorders that affect large muscle groups but for which adipose tissue appears to be spared from dystrophy or hypertrophy.

B. Acquired Partial Lipodystrophy

1. Clinical Features

APL (OMIM: 608709), also known as Barraquer–Simons syndrome, was the subject of separate reports by Barraquer and Simons early in the twentieth century.[42,43] Patients affected with APL develop adipose tissue loss in the face, neck, arms, thorax, and upper abdomen, progressing in a cephalocaudal order, with onset typically starting in childhood or adolescence. Excess fat is often stored in the hips and legs, particularly in affected women.[1] MRI imaging of female APL patients identified increased subcutaneous calf adiposity compared to FPLD, CGL, or HIV-associated lipodystrophy, making calf adiposity a feature that can be used to distinguish APL from other lipodystrophies.[13]

A relationship between APL and other autoimmune diseases appears to exist, as comorbid systemic lupus erythematosus, juvenile dermatomyositis, and membranoproliferative glomerulonephritis have been reported.[44,45] Additionally, low complement component 3 is often observed.[44] Insulin resistance and overt diabetes are relatively uncommon compared to other forms of lipodystrophy, typically around 10% of patients.[46,47]

2. Genetics and Pathophysiology

The appearance of APL is sporadic, but a wide range of associated secondary factors, such as systemic lupus erythematosis, dermatomyositis, hypocomplementemia, and membranoproliferative glomerulonephritis, are often present and precede the onset of lipodystrophy.[47] Since there is a lack of multigenerational families with multiple affected members, traditional gene mapping strategies are not possible. However, candidate sequencing of *LMNB2* uncovered three different mutations in four of nine APL patients. Comparing the mutation frequency in APL patients to a multiethnic sample of 1100 subjects, the relative risk for the associated mutations was 110 (95% confidence interval 36–271; $P < 1 \times 10^{-5}$). Two of the mutations were missense variants, while the third interrupted a putative splice site in intron 1 (Fig. 1).[47] Other than its localization and homology with nuclear envelope proteins, little is known about *LMNB2*, and the exact mechanism by which mutated *LMNB2* results in lipodystrophy remains to be elucidated. There are no other human disorders so far that have been attributed to mutations in *LMNB2*.

C. HIV-Associated Lipodystrophy

1. Clinical Features

Lipodystrophy was first recognized as a metabolic complication of HIV in the late 1990s.[48] Treatment of HIV using either protease inhibitors (PI) or nucleoside reverse transcription inhibitors (NRTI), alone or as part of highly active antiretroviral therapy (HAART), can exacerbate HIV-associated lipodystrophy.[49] Patients affected with HIV-associated lipodystrophy were initially thought to have Cushing's syndrome due to the presence of a prominent dorsocervical fat pad. Careful endocrinologic examination ruled out altered corticosteroid metabolism as responsible for the adipose redistribution.[50–52] HIV-associated lipodystrophy is by far the most common type of lipodystrophy, and affects males and females equally.[1] Approximately 30–50% of HIV-infected patients experience abnormal fat redistribution,[53] but subtle lipodystrophy may be even more prevalent in HIV than suggested by clinical ascertainment without diagnostic standards or application of sensitive noninvasive diagnostic tools.

HIV-associated lipodystrophy typically presents with peripheral lipoatrophy, especially of the face, with variable lipoatrophy to lipohypertrophy of subcutaneous fat stores.[54–57] MRI imaging of patients with HIV-associated lipodystrophy revealed remarkably similar body-wide adipose distribution as FPLD3 patients.[13] The scanned FPLD3 and HIV-associated lipodystrophy patients displayed increased supraclavicular and visceral adipose deposition, with decreased thigh and mid-calf adiposity.[13] The largest discrepancy between the FPLD3 and HIV-associated lipodystrophy patients was the increased loss of facial fat in HIV-associated lipodystrophy.

Metabolic complications of HIV-associated lipodystrophy include dyslipidemia, insulin resistance, impaired glucose tolerance, and androgen deficiency.[58,59] The typical dyslipidemia pattern associated with HIV-associated lipodystrophy consists of hypertriglyceridemia, which can be severe, together with depressed concentrations of HDL cholesterol and increased total cholesterol, with or without increased low-density lipoprotein (LDL) cholesterol. Despite significant insulin resistance, acanthosis nigricans and PCOS are rarely observed.[60] Hepatic steatosis has been associated with HIV-associated lipodystrophy.[61] HIV-infection may be an independent risk factor for atherosclerosis,[62] and the duration of HIV-infection, treatment with HAART, severity of lipid disturbance, and the presence of lipodystrophy—are all associated with the presence of ultrasound measured carotid atherosclerosis, a strong predictor of poor cardiovascular outcomes.[63]

2. Genetics and Pathophysiology

The pathogenesis of HIV-associated lipodystrophy in treatment naive patients remains unknown.[64] As not all HIV-infected patients who are treated with HAART develop lipodystrophy, a possible genetic susceptibility or environmental interaction can be postulated,[3] possibly representing a complex phenotype similar to what is seen with APL. No associated mutations have been discovered, but perhaps prediction of side effects associated with HAART could be a future application of pharmacogenetics. Interestingly, PI treatment of human and mouse fibroblasts leads to an inhibition of *ZMPSTE24*, the protein product of the gene mutated in MAD type B.[65] Similar to observations of *ZMPSTE24*-mutated patients, PI treatment of human fibroblasts was found to result in an accumulation of farnesyl-prelamin A and nuclear dysmorphology.[65]

Whether the insulin resistance and dyslipidemia that complicate HAART are a direct consequence of the lipodystrophic fat redistribution remains unknown. However, dyslipidemia can occur without clinically ascertained lipodystrophy, suggesting that it is either mechanistically independent or perhaps that it is expressed as a sensitive early marker when mild lipodystrophy is present but not clinically obvious.

V. Syndromes Including Lipodystrophy

A. Mandibuloacral Dysplasia

1. Clinical Features

MAD is a rare syndrome first described in two teenage males with hypoplastic mandibles and clavicles, persistent cranial sutures, acroosteolysis (bone resorption of the distal phalanges), stiff joints, atrophy of the skin over the hands and feet, alopecia, and short stature[66] (Table VI). Other common features of MAD are progeroid features, such as bird-like facies and high-pitched voice, and ectodermal defects such as mottled hyperpigmentation and nail dysplasia.[67,68] Irregularities in adipose tissue deposition are apparent in childhood and are the defining characteristic of the two MAD subtypes: in MAD type A (OMIM: 248370), a partial loss of subcutaneous adipose tissue from the extremities is observed, while in MAD type B (OMIM: 608612), a generalized loss of subcutaneous adipose tissue in the face, trunk, and extremities is observed.[3] Metabolic complications include insulin resistance, dyslipidemia (hypertriglyceridemia and low HDL cholesterol), and impaired glucose tolerance.[19,67–70]

2. Genetics and Pathophysiology

MAD type A is inherited in an autosomal recessive manner and is caused by homozygous mutation of *LMNA*.[16] The responsible mutations lie within the C terminal globular domain (R527H, A529T, A529V) (Fig. 1).[70–72] The R527H mutation was identified on a common haplotype in nine affected individuals in five consanguineous families in Italy, indicating a founder effect.[72] Functional characterization of the R527H mutation indicated that at the cellular level, MAD type A is characterized by the accumulation of unprocessed prelamin type A, alteration of the nuclear architecture, chromatin defects, and increased apoptotic rate, largely due to deficits in DNA damage response and cell cycle control.[73] In MAD type B, also inherited in an autosomal recessive manner, homozygous or compound heterozygous mutations to *ZMPSTE24* are responsible (Fig. 1).[71,74] *ZMPSTE24* encodes a zinc metalloproteinase vital for the

TABLE VI

Clinical Features of Syndromes Including Lipodystrophy

Variable	MAD	SHORT	Neonatal progeroid	HGPS	WRN
Inheritance	Autosomal recessive	Autosomal dominant & recessive	Autosomal recessive	Autosomal dominant & recessive	Autosomal recessive
Age of onset	Childhood	Childhood	Birth	Childhood	Young adulthood
Hair abnormalities	Alopecia	–	Hair loss	Total loss	Graying, male pattern
Dermatological	Skin atrophy, hyperpigmentation	–	–	–	Hyperkeratosis, calcification
Ophthalmological	–	Rieger anomaly	–	–	Cataracts
Facies	Bird-like	Triangular-shaped	Triangular-shaped, aged	–	–
Mandibular hypoplasia	Present	Variable	–	Present	Absent
Stature	Short, sloped shoulders	Short	Short	Short	Short
Hypogonadism	–	–	–	Absent	Present
Bone involvement	Osteoporosis, persistent sutures	Delayed bone age, teething delay	Large fontanel	Osteopenia/ osteoporosis	Osteopenia/ osteoporosis
Joint involvement	Stiff joints	Hyperextensibility	–	–	–
Cognitive development	–	Variable	Impaired	Normal	–
Diabetes	Present	Present	Present	Absent	Variable
Dyslipidemia	Present	–	Present	–	–
Adipose tissue loss	Partial to generalized	Face, trunk, gluteal	Generalized loss	–	Face, trunk, extremities
Cancer susceptibility	–	–	–	Normal	High
Atherosclerosis susceptibility	–	–	–	High	High

processing of prelamin type A, and thus, similarly to MAD type A, an accumulation of unprocessed prelamin type A is observable, leading to nuclear dysmorphology and altered cellular expression.[75]

B. SHORT Syndrome

1. Clinical Features

SHORT syndrome (OMIM: 269880) is an acronym for its typically occurring features: *S*hort stature, *H*yperextensibility of joints, *O*cular depression, *R*ieger anomaly, and *T*eething delay.[76,77] Reiger anomaly refers to a defective development of the anterior chamber of the eye, including the cornea and iris, often producing pupillary anomalies and congenital glaucoma.[78] Patients with SHORT syndrome also often present with lipoatrophy, variably affecting the face, upper extremities, trunk, and gluteal region.[76–78]

2. Genetics and Pathophysiology

SHORT syndrome typically segregates through families in what appears to be a Mendelian fashion, though following both autosomal recessive and dominant inheritance patterns. No responsible mutation has been identified, though a cytogenetic abnormality has been reported in an affected mother and child on chromosome 4q25, which includes the paired-like homeodomain 2 (*PITX2*) gene that is mutated in Rieger syndrome.[79]

C. Neonatal Progeroid Syndrome

1. Clinical Features

Patients affected with neonatal progeroid syndrome, also known as Wiedemann–Rautenstrauch syndrome (OMIM: 264090), are identified at birth by their progeroid features including enlarged triangular-shaped, older-looking face, and minimal scalp hair.[80,81] A large anterior fontanel with prominent veins on the scalp is also often noted, as may be mild to moderate mental retardation. Neonatal mortality may occur and is typically attributed to cerebral hemorrhage or respiratory causes.[3] A generalized absence of subcutaneous fat is present, typically with sparing of the gluteal and sacral regions. Finally, hyperinsulinemia and dyslipidemia have also been reported.[82]

2. Genetics and Pathophysiology

Neonatal progeroid syndrome segregates through families with an autosomal recessive inheritance pattern. A molecular cause for neonatal progeroid syndrome is yet to be elucidated. Sequencing of candidate nuclear envelope protein genes in affected patients has not uncovered any potentially responsible mutations (Hegele, unpublished observations).

D. Hutchinson–Gilford Progeria Syndrome

1. Clinical Features

HGPS was the first progeria syndrome to be described over 100 years ago.[83,84] For children with HGPS, evidence of accelerated aging often becomes apparent within the first year of life. The faces of children with HGPS become wrinkled and aged looking due to a lack of subcutaneous fat; and hair loss, prominent eyes and scalp veins, and mandibular hypoplasia are also present.[85] The children are physically underdeveloped, with below average weight and extremely short stature. Weak, high-pitched voices, delayed dentition, osteolysis, decreased joint mobility, a pyriform (pear-shaped) thorax, and "horse riding" stance are other common features.[86] Importantly, cognitive development is unaffected. Children with HGPS appear to age a decade for every year of life, and advanced respiratory, cardiovascular, and arthritic conditions are prevalent by the time the children are 10 years of age.[86] Advanced coronary or cerebrovascular disease has been reported to be the cause of death of at least 90% of patients.[86]

Lipodystrophic features of HGPS can be observed as early as 6 months, typically following a pattern of progressive subcutaneous adipose tissue loss beginning with the extremities, followed by the trunk and finally the face.[87] In a less severe form of HGPS, termed nonclassical HGPS, patients have lived into the third decade of life.

2. Genetics and Pathophysiology

Most cases of classical HGPS are sporadic and appear to follow an autosomal dominant inheritance pattern,[88] while nonclassical HGPS has been observed in consanguineous families, suggesting an autosomal recessive mode of inheritance.[89] HGPS is a laminopathy, with all responsible mutations identified within the *LMNA* gene. Most frequently reported is a silent mutation (G608G) that activates a cryptic splice site within exon 11, deleting the second half of the exon, which encompasses 150 bp and 50 codons. The mutation is a recurrent *de novo* C > T mutation, as the methylated C can be deaminated to T and then miscopied.[85] A "parent of origin" effect appears to be present, as seven of the reported mutated alleles have been transmitted to the affected offspring from the father,[88,90] suggesting that genomic imprinting, or epigenetics, may be important in HGPS. Reports from cell-based studies of fibroblasts from HGPS patients describe accumulation of mutated lamin A as the primary pathogenic process,[91] though not always, as failure of the mutated lamin A to associate with lamin B1 has also been suggested.[92] Regardless, striking structural abnormalities of the nucleus, including chromatin extrusion from the nuclear envelope, are always observed.[91,92]

E. Werner Syndrome

1. Clinical Features

WRN (OMIM: 277700) is a progeroid syndrome, displaying clinical characteristics of premature aging, including aged facies, premature graying, cataracts, osteoporosis, diabetes mellitus, and increased risk for early atherosclerotic disease and cancer.[93,94] Interestingly, the sarcoma to carcinoma ratio is around 1:1, compared to 1:10 in the general population.[95] Patients with WRN are of short stature, often have weak high-pitched voices, and have severe scleroderma-like dermatological involvement including telangiectasia, hyperkeratosis, and predisposition to trophic ulcers. Calcification of both subcutaneous tissue and arteries are also common features. With regard to adipose tissue, a decrease in subcutaneous deposition in the face, trunk, and extremities is noted. Associated metabolic features of WRN include insulin resistance and hypogonadism, leading to reduced fertility.[96] An atypical form of WRN, with a more severe phenotype than typical WRN and resembling HGPS, has also been described.[97]

2. Genetics and Pathophysiology

Typically, WRN follows an autosomal recessive mode of inheritance, resulting from homozygous or compound heterozygous mutations in the *RECQL2/WRN* locus. To date, a total of 58 mutations have been reported in the *RECQL2/WRN* locus, including splice mutations (resulting in the exclusion of necessary exons), frameshift mutations (resulting in premature truncation of the transcript), and nonsense mutations (changing the amino acid sequence of the protein).[16] The *RECQL2/WRN* locus encodes a RecQ-type DNA helicase vital for DNA repair, replication, transcription, and maintenance of the telomeres, though it remains to be further demonstrated if any or all of these roles are lost in WRN.[96] Recent work indicated that even short-term inhibition of WRN helicase resulted in impaired DNA replication and differential expression patterns, including genes involved in adipocyte differentiation and IL-6 pathways.[98] This evidence supports a link to clinically observed lipodystrophy.

Atypical WRN follows an autosomal dominant inheritance pattern and three heterozygous mutations (A57P, R133L, and L140R), and deletion of a single codon 168 in the *LMNA* locus accounts for a portion of patients.[97,99] As discussed above, *LMNA* is a remarkably pleiotropic gene, with mutations causing a wide variety of phenotypes. The clinical features of the described atypical Werner patients with *LMNA* mutations have many similarities with MAD, partial lipodystrophy, and Hutchinson–Gilford syndrome, perhaps suggesting that a laminopathy, as opposed to a helicase defect, is responsible.[100] DNA sequencing to identify the responsible genetic mutations may be of limited clinical value as of yet, but differentiating patients with late-onset HGPS and early-onset WRN may be important for the development and targeting of future therapeutics.[100]

VI. Overview of Molecular Pathways Affected in Lipodystrophy Syndromes

A. Pathogenesis of Metabolic Disturbances

Despite the heterogeneity in clinical presentation, the similarities in the predominant metabolic abnormalities observed in different forms of lipoatrophy and lipodystrophy suggest that the lack of fat may be the primary cause of insulin resistance and dyslipidemia. An alternative hypothesis is that the lipodystrophy is an epiphenomenon that follows from the insulin resistance, resulting from a genetic or environmental insult. Several factors appear to fail to support this latter proposed mechanism. First, careful anecdotal clinical observations suggest that the loss of adipose tissue usually precedes the development of detectable insulin resistance and associated metabolic disturbances. Second, the spectrum of causative genes in inherited lipodystrophies is so broad—including genes governing basic cellular, structural, and functional properties that are not obviously related mechanistically—that it is difficult to conceive how each one can independently lead to insulin resistance, in turn giving rise to disturbed fat distribution. Third, human genetic syndromes of pure insulin resistance from infancy, such as homozygous insulin receptor defects, are typically not associated with lipodystrophy. Finally, in animal models, genetic ablation of adipose tissue results in metabolic phenotypes that are very similar to those observed in human lipodystrophies.[101–103] Furthermore, transplantation of subcutaneous fat into one of the animal models remarkably reverses the metabolic abnormalities.[104] Additionally, severe loss of subcutaneous fat stores is often associated with especially severe metabolic abnormalities. While it remains possible that severity of metabolic disturbance can be further modulated by the specific etiology of the lipodystrophy, it is very unlikely that the metabolic disturbances—particularly insulin resistance—develop initially and independently of the redistribution of adipose tissue.

B. Pathogenesis of Adipose Tissue Loss

Genes mutated in lipodystrophy syndromes are involved in at least five pathways necessary for regular cellular function: (a) TG synthesis, (b) nuclear hormone receptor activation of adipocyte-specific expression, (c) maintenance of nuclear structure and function via nuclear envelope proteins, (d) DNA unwinding and repair, and (e) endocytosis via caveolin formation (Fig. 2).

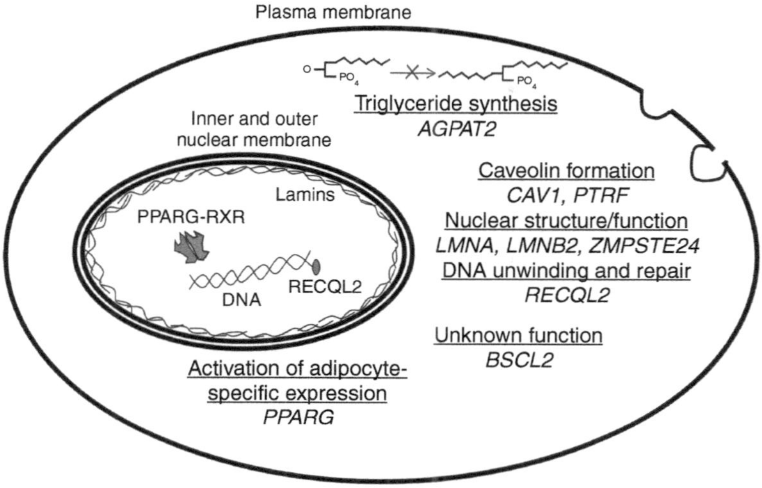

FIG. 2. Function of genes mutated in lipodystrophy. Schematic diagram of the cellular mechanisms interrupted in lipodystrophy and syndromes in which lipodystrophy is a component.

The way in which the loss of these functions directly results in the clinical spectrum observed has not been fully elucidated. Using fine resolution phenotype definition (see phenomics below), differences can often be discerned between individuals with similar clinical diagnoses but disparate underlying molecular mechanisms. On the other hand, it is remarkable that a deficiency of seemingly unrelated proteins, such as a hormone receptor and a nuclear structural protein, produces phenotypes as similar as they are.

VII. Therapy for Lipodystrophy Patients

Standard therapeutic options for clinicians treating lipodystrophy patients are directed toward management of downstream metabolic complications, postponement of the onset of microvascular complications, and primary prevention of macrovascular complications (Table VII). However, recent evidence suggests that leptin replacement therapy will revolutionize the standard care, and several novel therapies are currently being studied in cell- and animal-based studies.

A. Glycemic Control

The principles of glycemic control in FPLD are similar to those in typical type 2 diabetes and are essential to prevent or delay downstream complications. Thus, therapy should include grounding of the patient's knowledge with

TABLE VII
MANAGEMENT STRATEGIES IN LIPODYSTROPHY

Intervention	Comments
Diet and lifestyle	– Trial of nondrug therapies, although efficacy may be limited – Increased exercise; resistance training – Prudent diet, reduced total calories, reduced calories from fat, reduced saturated and trans fats, low glycemic index foods – Competing dietary needs may warrant dietician consultation
Glycemic control	– Similar to treatment of type 2 diabetes – Beginning with oral agents: sulphonylureas, biguanides (metformin), dipeptidyl peptidase-IV inhibitors, thiazolidinediones – Insulin therapy after type 2 diabetes onset – Combination therapy may be required

Dyslipidemia treatment	– Statins directed to LDL cholesterol (LDL-C) and baseline triglycerides (TG) < 5 mmol/L
	  – Generally well tolerated with modest to good efficacy
	  – Unlikely to achieve target lipid levels as monotherapy
	– Fibrates are well tolerated with modest efficacy; appropriate to consider when baseline TG > 5 mmol/L
	  – Unlikely to achieve target lipid levels
	  – Fibrate-statin combination requires periodic monitoring of creatine kinase and liver transaminases
	– Niacin to be used cautiously, side effects include flushing, skin rash, and pruritis and worsened glycemia
	  – Consider longer acting forms (e.g., Niaspan™)

Further atherosclerosis prevention	– Angiotensin converting enzyme (ACE) inhibitors – Angiotensin 2 receptor blocker (ARB)
Leptin replacement	– Works as insulin-sensitizing agent, dramatic improvement of metabolic parameters

diabetes education and dietary counseling, followed as required by administration of oral agents either as monotherapy or in combinations, including sulphonylureas, biguanides (such as metformin), dipeptidyl peptidase-IV inhibitors, and TZDs (such as pioglitazone and rosiglitazone). Among patients with lipoatrophy or lipodystrophy syndromes, troglitazone given over 6 months improved metabolic indices, including reductions of glycated hemoglobin, TG, and FFA together with a modest increase in subcutaneous adipose tissue.[105] However, troglitazone conferred an especially high risk of hepatotoxicity in FPLD, and is no longer available. Some anecdotal data exists regarding administration of rosiglitazone to patients with both *PPARG* and *LMNA* forms of FPLD. Despite the limited data, subjects with the *LMNA* form of FPLD appeared to respond better to TZDs than subjects with the *PPARG* form; one patient with the *LMNA* form had quite a dramatic and sustained improvement over the long term.[106,107] *In vitro* experiments showed that the small molecule tyrosine-based receptor agonist (TA) GI262570 (farglitazar) was superior to TZDs as an inducer of *PPARG* expression in cells with the *PPARG* P467L mutation, suggesting that TA may be a better therapeutic approach in subjects with FPLD3 (the *PPARG* form of FPLD),[108] but further clinical testing is required. In our experience, initiation of both metformin and rosiglitazone administered at usual staring doses to FPLD patients is associated with improved lipid parameters, reduced insulin and C-peptide concentrations, as well as reduced pigmentation of acanthotic skin without hypoglycemia (Hegele, unpublished observations).

As insulin resistance can become severe in FPLD, extremely large doses of insulin are often required, necessitating the use of concentrated preparations. Also, there may be some benefit to coadministration of insulin with an oral insulin-sensitizing agent, potentially reducing the insulin requirement when treating FPLD subjects (Hegele, unpublished observations).

B. Hypertriglyceridemia

Severe hypertriglyceridemia due to chylomicronemia is a complication of lipodystrophy that most commonly presents after diabetes onset. Plasma triglyceride concentrations > 12 mmol/L are associated with a greatly increased risk of acute pancreatitis. Optimization of diet and glycemic control together with use of fibric acid derivatives (fibrates such as gemfibrozil or fenofibrate) can help control plasma TG and avert pancreatitis. Chronic hypertriglyceridemia can also sometimes be associated with hepatosplenomegaly and hepatosteatosis, secondary to nonalcoholic steatohepatitis, possibly progressing to impairment of hepatic function and cirrhosis.[109]

C. Further Treatment to Prevent Atherosclerosis

For lipodystrophy patients, treatment of both elevated blood pressure and dyslipidemia should follow local guidelines for primary prevention of vascular disease in individuals with high atherosclerosis risk. Pharmacological options for lipid lowering include fibric acid derivatives, HMG CoA reductase inhibitors (statins), and sterol absorption inhibitors, depending on the lipid and lipoprotein profile. Options for blood pressure management include agents such as angiotensin converting enzyme inhibitors and angiotensin 2 receptor blockers. The observed CRP elevation observed in lipodystrophies is consistent with the proinflammatory nature of insulin resistance, warranting consideration of aspirin administration for primary prevention of vascular disease.

D. Subcutaneous Leptin Replacement

Leptin is an adipocyte-secreted hormone with vital roles in energy metabolism, and in accordance with lipoatrophy, deficiency of circulating leptin is often observed in lipodystrophy.[110] In a recent report, 48 patients with a variety of lipodystrophies, including AGL, CGL1, CGL2, FPLD2, FPLD3, and lipodystrophy of unknown etiology, were treated with subcutaneous injections of a recombinant form of human methionyl leptin.[111] Across all forms of lipodystrophy, metabolic parameters, including hypertriglyceridemia and glycemic control in particular, were dramatically improved. Further, decreased energy intake, improved insulin sensitivity, and reductions of ectopic fat stores have been reported with leptin replacement.[112,113] Leptin has also been suggested to improve insulin sensitivity, dyslipidemia, and adipose tissue distribution HIV-associated lipodystrophy.[114] Evidence from a randomized control trial, dosing guidelines, and further evaluation of the potential for development of leptin resistance, or tachyphylaxis, are required, but leptin replacement therapy represents a significant improvement for the care of patients affected with a broad range of lipodystrophies.[111]

E. Theoretical Approaches for Future Research

Future treatment strategies at this stage might best be directed toward either preservation of subcutaneous fat depots before irreversible loss occurs in at-risk individuals. Since TZD drug treatment of certain lipodystrophic patients has resulted in modest increases in subcutaneous fat mass measurements in later disease stages, TZDs might in theory be used very early to prevent fat loss. As discussed above, leptin improves both the ectopic fat distribution and metabolic profile among patients with severe complete lipodystrophy. Thus, both leptin and TZDs may qualify as theoretical treatments at an earlier phase of disease. However, better cellular or animal models are needed to define and evaluate the basis of the progressive and selective adipose tissue loss in lipodystrophy before these treatments might be considered in at-risk carrier children.

Fat transplantation is another theoretical approach to improve the metabolic state. In mutant mice with induced lipodystrophy, transplantation of white fat improved insulin resistance in a dose-dependent and leptin-dependent manner, for reasons that are not fully understood.[104] In humans, fat transplantation at any stage of the disease would be complicated by technical issues, as well as the requirement for long-term immunosuppression.

As for attempting to specifically target the molecular mechanism leading to disease, evidence from both cell-based and animal-based studies suggests that the inhibition of farnesylated prelamin A accumulation can abrogate nuclear dysmorphology and subsequent lipodystrophy.[115,116] It has been proposed that either a reduction in *LMNA* transcription or farnesyltransferase inhibitors may be therapeutic strategies.[115,117] In a conditional transgenic mouse, in which the expression of the mutated *LMNA* could be turned on or off, mice developed a lipodystrophy phenotype with expression of the mutant, but reversion to a wild-type phenotype was present 13 weeks after mutant expression was turned off.[118] The farnesyltransferase inhibitor tipifarnib significantly improved the phenotype of HGPS mice.[119] However, further investigation of these novel therapeutic approaches is required before application to human patients.

VIII. Uncovering Additional Responsible Genetic Variants

A. Phenomics

Phenomics is the systematic measurement and analysis of qualitative and quantitative traits, including clinical, biochemical, and imaging methodologies, for the refinement and characterization of a phenotype.[120] Phenomics requires both "deep phenotyping," referring to a thorough collection of a wide breadth of accurate and precise phenotypes, and "phenomic analysis," referring to evaluation of patterns and relationships between genotype–phenotype associations and between individuals with related phenotypes.[121] Multidimensional data, including clinical, biochemical, genetic, and pharmacological response data, can be visualized using heat maps and analyzed using clustering algorithms (Fig. 3). Phenomic evaluation of individuals carrying novel mutations represents a path of investigation to further elucidate the function of the mutated protein.[122] Further, using phenomics, a relationship between *LMNA* mutation location and organ involvement was identified, although the molecular mechanism behind the relationship remains uncharacterized (updated in Fig. 4).[123]

B. Linkage and Autozygosity Mapping

Linkage analysis is a genetic technique that searches for chromosomal segments that cosegregate with the disease phenotype through families and is the analysis technique that has been used to identify the bulk of lipodystrophy genes. The primary limitations of linkage analysis are the requirement of multigenerational pedigrees of affected individuals and the dramatic loss of power in the presence of heterogeneity. Autozygosity mapping is applicable only to recessive conditions and is especially powerful in the context of consanguinity, as it seeks stretches of homozygous alleles along the chromosome that are longer than expected by chance and are common among affected individuals. Genome-wide association studies have become widespread as the focus of many geneticists has moved to complex diseases, but both linkage and

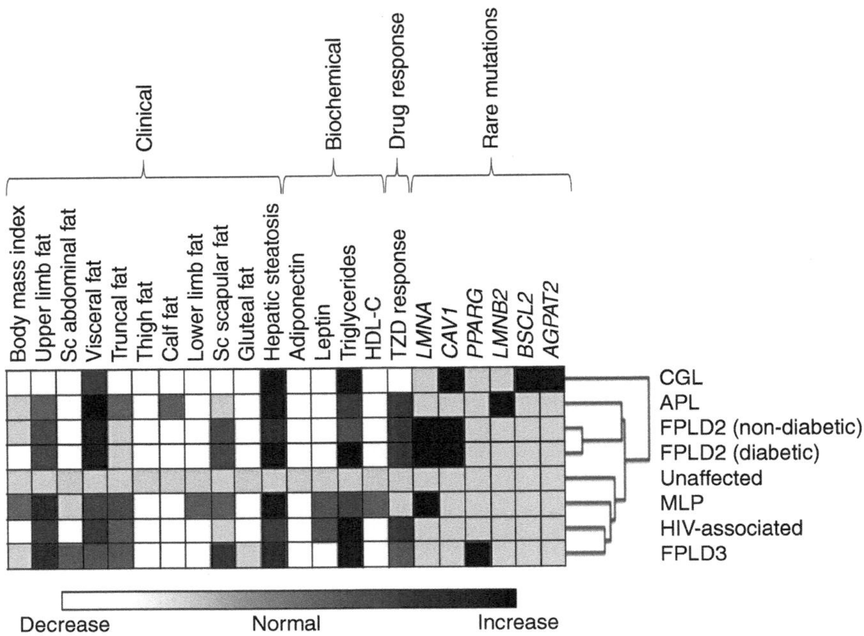

FIG. 3. Phenomic analysis of lipodystrophy subtypes. Although clinically similar, careful phenomic analysis can identify phenotypic differences that may be the result of different molecular etiologies in various lipodystrophy syndromes. A gray-scale heat map of human lipodystrophy syndromes was produced, in which gray indicates the normal reference, darkness toward black indicates an increased severity or frequency of the feature compared to normal reference range, and lightness toward white indicates decreased severity or prevalence of the feature compared to normal reference range. The included syndromes are congenital generalized lipodystrophy (CGL), acquired partial lipodystrophy (APL), metabolic laminopathy (MLP), human immunodeficiency virus-associated lipodystrophy (HIV-associated), and familial partial lipodystrophy types 2 (FPLD2) and 3 (FPLD3). Hierarchical cluster analysis produced a dendrogram, which depicts the similarity between syndromes, and indicates that FPLD2 and FPLD3, caused by mutations within the nuclear lamin A/C gene (*LMNA*) and peroxisome proliferator-activated receptor gamma gene (*PPARG*) respectively, are clearly different by phenomic analysis. Interestingly, due to the small loss of adiposity in MLP, it groups more closely with FPLD3 than FPLD2 despite being caused by mutations in *LMNA*. Additional abbreviations: high-density lipoprotein cholesterol, HDL-C; thiazolidinedione, TZD.

autozygosity mapping, which were traditionally performed using microsatellite marker maps, are also possible using modern high-density genotyping microarrays. Both techniques can be used to search for novel disease-causing mechanisms in families who do not appear to be carrying previously identified mutations.

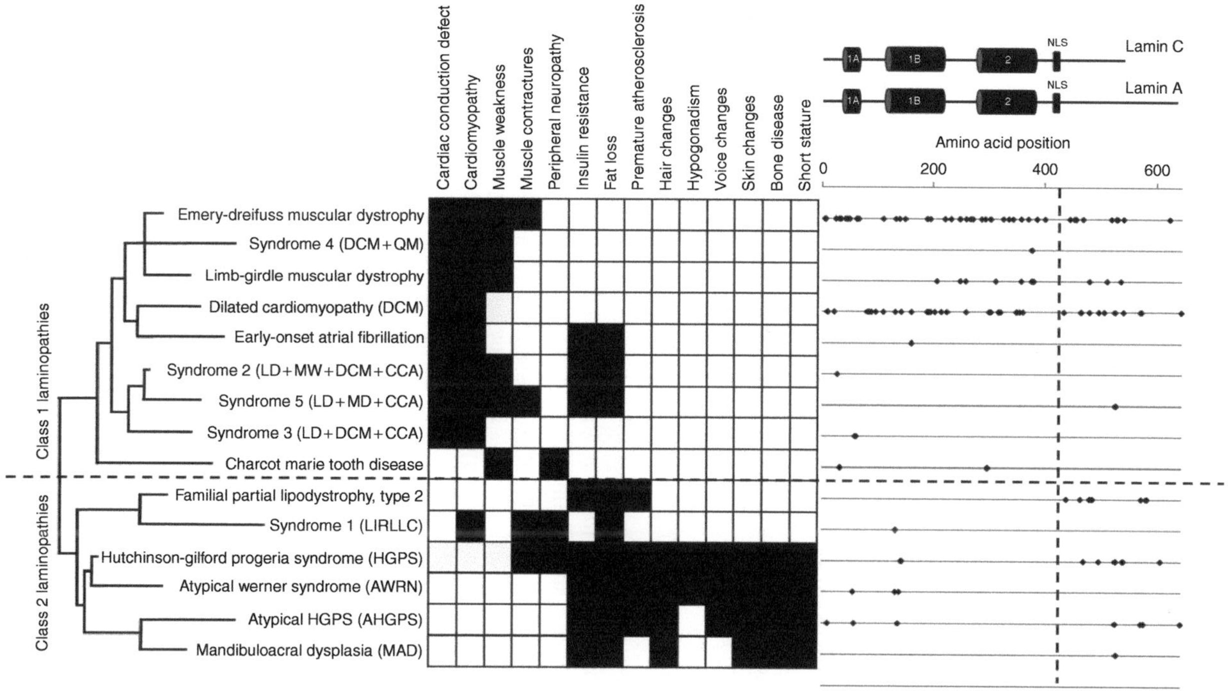

FIG. 4. Examination of the relationship between *LMNA* mutation position and laminopathy phenotype. Hierarchical cluster analysis (HCA) ordered the clinical features across the top of the grid in the center of the figure, as well as the laminopathies across the left side of the grid. Within the center grid, dark shading indicates involvement of the clinical feature in the relevant laminopathy. The clinical similarity between the relevant laminopathies is represented by the cladigram on the extreme left of the figure. On the top right side of the figure lies a schematic representing the linearized *LMNA* cDNA including the location of the major rod domains and nuclear localization signal (NLS). The diamonds indicate reported mutations, with their vertical position corresponding to the associated laminopathy and the horizontal position corresponding to their physical position. The dotted vertical line indicates the location of the NLS, and the horizontal dotted line separates class 1 and class 2 laminopathies defined by HCA. The number of mutations in the four quadrants is 90, 27, 11, 23 in the top left, top right, bottom left, bottom right respectively, corresponding to a 6.9 times higher likelihood of a mutation 5′ of the NLS to be causative of a class 1 laminopathy ($P = 3 \times 10^{-6}$).

C. Copy Number Variation

Copy number variation (CNV) represents duplications or deletions of genomic sequence greater than 1 kb in size. They are generally unidentifiable using capillary-based sequencing, but represent the greatest source of variation among human genomes.[124,125] CNVs have been identified that are causative for Mendelian disorders of metabolism, such as familial hypercholesterolemia[126], but both genome-wide investigations and efforts targeted to *LMNA* have not thus far identified CNVs likely to be responsible for lipodystrophy.[127] Nevertheless, the resolution of array-based techniques for CNV detection is continually improving, and it is quite feasible that CNV represents the disease-causing mechanism for a portion of lipodystrophy patients. Of note, during a genome-wide investigation into CNV in lipodystrophy patients, a patient with triple X syndrome was identified.[128] Cytogenetic evaluation of lipodystrophy patients is not standard care, and thus a diagnosis of triple X syndrome would not normally be identified, but this identification poses an interesting question regarding the effect that triple X syndrome may have on women with other underlying lipodystrophy susceptibilities.

D. Next-Generation Sequencing

The future of genetic investigations attempting to identify rare genetic variants responsible for Mendelian disorders is undoubtedly next-generation sequencing. Whole exome studies, in which the transcribed portion of the genome is sequenced, have already been successfully identifying variants responsible for Mendelian disorders.[129] Large-scale sequencing efforts, such as the 1000 genomes project (www.1000genomes.org), are cataloging human variation such that it can be confidently concluded whether a variant is in less than 1% of the population. A filtering strategy, selecting out all low-frequency DNA variants causing missense, nonsense, or frameshift mutations, creates a pool of potentially causative mutations. If the disorder is recessive, further filtering is possible because either a homozygous change or two potentially deleterious changes (representing compound heterozygosity) must be present. Further elimination of potentially responsible mutations could be obtained through pathway analysis, an *in silico* strategy to evaluate the function of the gene and its interactions. Computational analyses, evaluating the evolutionary conservation of the residue and the potential biochemical impact of the observed amino acid change, may also be helpful. Once a subset of potentially responsible variants is identified, verification of the identified gene through targeted capillary-based sequencing of additional cases can be performed. The final step is to perform *in vitro* and *in vivo* analyses to examine the causal nature of the identified variants. Unfortunately, such a filtering strategy would be unsuccessful if the disease-causing mutation is either in a noncoding region, such as a promoter, splice acceptor, donor, or enhancer, or if it causes a silent change in the coding sequence. HGPS provides a perfect example of how an apparently silent mutation can cause a devastating disease. Nevertheless, whole-exome and whole-genome sequencing will undoubtedly change the way investigators search for the molecular causes of disease.

IX. Inherited Partial Lipodystrophies as a Model for Common Metabolic Syndrome (MetS)

MetS, the proatherogenic clustering of central obesity, dyslipidemia, hypertension, and insulin resistance, affects nearly 30% of individuals in Western societies.[130] This proatherogenic biochemical pattern is largely consistent with the pattern observed in FPLD patients, as a high proportion of FPLD patients have elevated blood pressure ($>$ 130/85 mmHg) and dyslipidemia (TG $>$ 90th percentile and HDL cholesterol $<$ 10th percentile for age and sex) and insulin resistance. The characteristic progressive fat redistribution to central depots rather than outright global fat loss, evolving relatively slowly over years, means that FPLD patients recapitulate many clinical and biochemical attributes of the more common MetS. The common form of MetS is a complicated entity, representing the complex interplay of genetics and environment. FPLD on the other hand, is highly heritable, making FPLD patients an excellent source for insight into biochemical pathways that may be altered in the more common MetS. While *PPARG* mutations were identified in FPLD3 due to the previously known role of PPARγ and the effects of its agonist, the TZD class of drugs, FPLD3 patients clearly demonstrate what happens to an individual with only 50% of functioning PPARγ. Future findings in FPLD patients carrying mutations in yet undiscovered genes may uncover additional important biochemical pathways or may represent novel pharmaceutical targets.

X. Conclusions

A wide spectrum of lipodystrophy syndromes exists, each with varying clinical presentations, and yet cumulatively they underscore the importance of adipocyte biology in human metabolism. In general, the loss of "normal" adipocyte distribution results in severe perturbations of both glucose and lipid homeostasis. Classification of lipodystrophy was traditionally based upon clinical presentation, but with advances in molecular genetics, we are beginning to understand the overlapping pathological processes underlying some of the lipodystrophy subtypes, and to learn from the disparate mechanisms that culminate in similar phenotypes, suggesting that identifying the responsible genetic mutation may assist in the diagnosis and treatment of individual patients. Further study of lipodystrophy cases holds great potential for further characterization of metabolic pathways, the discovery of novel therapies, and improvement in the understanding of rare forms of lipodystrophy, HIV-associated lipodystrophy, and the common MetS.

Acknowledgments

Mr. Lanktree is supported by the Canadian Institutes of Health Research (CIHR) MD/PhD Studentship Award, the University of Western Ontario MD/PhD Program, and is a CIHR Fellow in Vascular Research. Mr. Johansen is supported by CIHR Banting and Best Canada graduate scholarship and is also a CIHR Fellow in Vascular Research. Dr. Hegele is a Career Investigator of the Heart and Stroke Foundation of Ontario, holds the Edith Schulich Vinet Canada Research Chair (Tier I) in Human Genetics, the Martha G. Blackburn Chair in Cardiovascular Research, and the Jacob J. Wolfe Distinguished Medical Research Chair at the University of Western Ontario. This work was supported by CIHR (MOP-13430, MOP-79523, CTP-79853), the Heart and Stroke Foundation of Ontario (NA-6059, T-6018, PRG-4854), Genome Canada through Ontario Genomics Institute, and the Pfizer Jean Davignon Distinguished Cardiovascular and Metabolic Research Award.

References

1. Garg A. Acquired and inherited lipodystrophies. *N Engl J Med* 2004;**350**:1220–34.
2. Dulloo AG, Antic V, Montani JP. Ectopic fat stores: housekeepers that can overspill into weapons of lean body mass destruction. *Int J Obes Relat Metab Disord* 2004;**28**(Suppl. 4): S1–S2.
3. Joy T, Hegele RA. Overview of acquired and genetic lipodystrophies. *Adipose Tissue Health Dis* 2010;**1**:385–402.
4. Joy TR, Hegele RA. Prevalence of reproductive abnormalities among women with familial partial lipodystrophy. *Endocr Pract* 2008;**14**:1126–32.
5. Costa DN, Pedrosa I, McKenzie C, Reeder SB, Rofsky NM. Body MRI using IDEAL. *AJR Am J Roentgenol* 2008;**190**:1076–84.
6. Berardinelli W. An undiagnosed endocrinometabolic syndrome: report of 2 cases. *J Clin Endocrinol Metab* 1954;**14**:193–204.
7. Seip M. Lipodystrophy and gigantism with associated endocrine manifestations. A new diencephalic syndrome? *Acta Paediatr* 1959;**48**:555–74.
8. Seip M, Trygstad O. Generalized lipodystrophy, congenital and acquired (lipoatrophy). *Acta Paediatr Suppl* 1996;**413**:2–28.
9. Westvik J. Radiological features in generalized lipodystrophy. *Acta Paediatr Suppl* 1996;**413**:44–51.
10. Fleckenstein JL, Garg A, Bonte FJ, Vuitch MF, Peshock RM. The skeleton in congenital, generalized lipodystrophy: evaluation using whole-body radiographic surveys, magnetic resonance imaging and technetium-99m bone scintigraphy. *Skeletal Radiol* 1992;**21**:381–8.

11. Haque WA, Shimomura I, Matsuzawa Y, Garg A. Serum adiponectin and leptin levels in patients with lipodystrophies. *J Clin Endocrinol Metab* 2002;**87**:2395.
12. Agarwal AK, Simha V, Oral EA, Moran SA, Gorden P, O'Rahilly S, et al. Phenotypic and genetic heterogeneity in congenital generalized lipodystrophy. *J Clin Endocrinol Metab* 2003;**88**:4840–7.
13. Al-Attar SA, Pollex RL, Robinson JF, Miskie BA, Walcarius R, Little CH, et al. Quantitative and qualitative differences in subcutaneous adipose tissue stores across lipodystrophy types shown by magnetic resonance imaging. *BMC Med Imaging* 2007;**7**:3.
14. Agarwal AK, Arioglu E, De Almeida S, Akkoc N, Taylor SI, Bowcock AM, et al. AGPAT2 is mutated in congenital generalized lipodystrophy linked to chromosome 9q34. *Nat Genet* 2002;**31**:21–3.
15. Agarwal AK, Garg A. Congenital generalized lipodystrophy: significance of triglyceride biosynthetic pathways. *Trends Endocrinol Metab* 2003;**14**:214–21.
16. Stenson PD, Mort M, Ball EV, Howells K, Phillips AD, Thomas NS, et al. The human gene mutation database: 2008 update. *Genome Med* 2009;**1**:13.
17. Magre J, Delepine M, Khallouf E, Gedde-Dahl Jr. T, Van Maldergem L, Sobel E, et al. Identification of the gene altered in Berardinelli–Seip congenital lipodystrophy on chromosome 11q13. *Nat Genet* 2001;**28**:365–70.
18. Lundin C, Nordstrom R, Wagner K, Windpassinger C, Andersson H, von Heijne G, et al. Membrane topology of the human seipin protein. *FEBS Lett* 2006;**580**:2281–4.
19. Kim CA, Delepine M, Boutet E, El Mourabit H, Le Lay S, Meier M, et al. Association of a homozygous nonsense caveolin-1 mutation with Berardinelli–Seip congenital lipodystrophy. *J Clin Endocrinol Metab* 2008;**93**:1129–34.
20. Cao H, Alston L, Ruschman J, Hegele RA. Heterozygous CAV1 frameshift mutations (MIM 601047) in patients with atypical partial lipodystrophy and hypertriglyceridemia. *Lipids Health Dis* 2008;**7**:3.
21. Fruhbeck G, Lopez M, Dieguez C. Role of caveolins in body weight and insulin resistance regulation. *Trends Endocrinol Metab* 2007;**18**:177–82.
22. Hayashi YK, Matsuda C, Ogawa M, Goto K, Tominaga K, Mitsuhashi S, et al. Human PTRF mutations cause secondary deficiency of caveolins resulting in muscular dystrophy with generalized lipodystrophy. *J Clin Invest* 2009;**119**:2623–33.
23. Ziegler LH. Lipodystrophies: report of seven cases. *Brain* 1928;**51**:147–67.
24. Misra A, Garg A. Clinical features and metabolic derangements in acquired generalized lipodystrophy: case reports and review of the literature. *Medicine (Baltimore)* 2003;**82**:129–46.
25. Dunnigan MG, Cochrane MA, Kelly A, Scott JW. Familial lipoatrophic diabetes with dominant transmission. A new syndrome. *Q J Med* 1974;**43**:33–48.
26. Kobberling J, Willms B, Kattermann R, Creutzfeldt W. Lipodystrophy of the extremities. A dominantly inherited syndrome associated with lipatrophic diabetes. *Humangenetik* 1975;**29**:111–20.
27. Garg A, Peshock RM, Fleckenstein JL. Adipose tissue distribution pattern in patients with familial partial lipodystrophy (Dunnigan variety). *J Clin Endocrinol Metab* 1999;**84**:170–4.
28. Garg A, Vinaitheerthan M, Weatherall PT, Bowcock AM. Phenotypic heterogeneity in patients with familial partial lipodystrophy (dunnigan variety) related to the site of missense mutations in lamin a/c gene. *J Clin Endocrinol Metab* 2001;**86**:59–65.
29. Ludtke A, Genschel J, Brabant G, Bauditz J, Taupitz M, Koch M, et al. Hepatic steatosis in Dunnigan-type familial partial lipodystrophy. *Am J Gastroenterol* 2005;**100**:2218–24.
30. Decaudain A, Vantyghem MC, Guerci B, Hecart AC, Auclair M, Reznik Y, et al. New metabolic phenotypes in laminopathies: LMNA mutations in patients with severe metabolic syndrome. *J Clin Endocrinol Metab* 2007;**92**:4835–44.

31. Vantyghem MC, Vincent-Desplanques D, Defrance-Faivre F, Capeau J, Fermon C, Valat AS, et al. Fertility and obstetrical complications in women with LMNA-related familial partial lipodystrophy. *J Clin Endocrinol Metab* 2008;**93**:2223–9.
32. Garg A. Gender differences in the prevalence of metabolic complications in familial partial lipodystrophy (Dunnigan variety). *J Clin Endocrinol Metab* 2000;**85**:1776–82.
33. Vantyghem MC, Pigny P, Maurage CA, Rouaix-Emery N, Stojkovic T, Cuisset JM, et al. Patients with familial partial lipodystrophy of the Dunnigan type due to a LMNA R482W mutation show muscular and cardiac abnormalities. *J Clin Endocrinol Metab* 2004;**89**:5337–46.
34. Peters JM, Barnes R, Bennett L, Gitomer WM, Bowcock AM, Garg A. Localization of the gene for familial partial lipodystrophy (Dunnigan variety) to chromosome 1q21–22. *Nat Genet* 1998;**18**:292–5.
35. Cao H, Hegele RA. Nuclear lamin A/C R482Q mutation in canadian kindreds with Dunnigan-type familial partial lipodystrophy. *Hum Mol Genet* 2000;**9**:109–12.
36. Lanktree M, Cao H, Rabkin SW, Hanna A, Hegele RA. Novel LMNA mutations seen in patients with familial partial lipodystrophy subtype 2 (FPLD2; MIM 151660). *Clin Genet* 2007;**71**:183–6.
37. Agarwal AK, Garg A. A novel heterozygous mutation in peroxisome proliferator-activated receptor-gamma gene in a patient with familial partial lipodystrophy. *J Clin Endocrinol Metab* 2002;**87**:408–11.
38. Hegele RA, Cao H, Frankowski C, Mathews ST, Leff T. PPARG F388L, a transactivation-deficient mutant, in familial partial lipodystrophy. *Diabetes* 2002;**51**:3586–90.
39. Barroso I, Gurnell M, Crowley VE, Agostini M, Schwabe JW, Soos MA, et al. Dominant negative mutations in human PPARgamma associated with severe insulin resistance, diabetes mellitus and hypertension. *Nature* 1999;**402**:880–3.
40. George S, Rochford JJ, Wolfrum C, Gray SL, Schinner S, Wilson JC, et al. A family with severe insulin resistance and diabetes due to a mutation in AKT2. *Science* 2004;**304**:1325–8.
41. Tan K, Kimber WA, Luan J, Soos MA, Semple RK, Wareham NJ, et al. Analysis of genetic variation in Akt2/PKB-beta in severe insulin resistance, lipodystrophy, type 2 diabetes, and related metabolic phenotypes. *Diabetes* 2007;**56**:714–9.
42. Barraquer L. Histoire clinique d'un cas d'atrophie du tissue celluloadipeux. *Neurolog Zentralblatt* 1907;**26**:1072.
43. Simons A. Lipodystrophia progressive. *Z Ges Neurol Psychiatr* 1911;**5**:29–38.
44. Jasin HE. Systemic lupus erythematosus, partial lipodystrophy and hypocomplementemia. *J Rheumatol* 1979;**6**:43–50.
45. Torrelo A, Espana A, Boixeda P, Ledo A. Partial lipodystrophy and dermatomyositis. *Arch Dermatol* 1991;**127**:1846–7.
46. Misra A, Peethambaram A, Garg A. Clinical features and metabolic and autoimmune derangements in acquired partial lipodystrophy: report of 35 cases and review of the literature. *Medicine (Baltimore)* 2004;**83**:18–34.
47. Hegele RA, Cao H, Liu DM, Costain GA, Charlton-Menys V, Rodger NW, et al. Sequencing of the reannotated LMNB2 gene reveals novel mutations in patients with acquired partial lipodystrophy. *Am J Hum Genet* 2006;**79**:383–9.
48. Carr A, Samaras K, Burton S, Law M, Freund J, Chisholm DJ, et al. A syndrome of peripheral lipodystrophy, hyperlipidaemia and insulin resistance in patients receiving HIV protease inhibitors. *AIDS* 1998;**12**:F51–8.
49. Martinez E, Mocroft A, Garcia-Viejo MA, Perez-Cuevas JB, Blanco JL, Mallolas J, et al. Risk of lipodystrophy in HIV-1-infected patients treated with protease inhibitors: a prospective cohort study. *Lancet* 2001;**357**:592–8.

50. Lo JC, Mulligan K, Tai VW, Algren H, Schambelan M. "Buffalo hump" in men with HIV-1 infection. *Lancet* 1998;**351**:867–70.
51. Miller KK, Daly PA, Sentochnik D, Doweiko J, Samore M, Basgoz NO, et al. Pseudo-Cushing's syndrome in human immunodeficiency virus-infected patients. *Clin Infect Dis* 1998;**27**:68–72.
52. Roth VR, Kravcik S, Angel JB. Development of cervical fat pads following therapy with human immunodeficiency virus type 1 protease inhibitors. *Clin Infect Dis* 1998;**27**:65–7.
53. Chen D, Misra A, Garg A. Lipodystrophy in human immunodeficiency virus-infected patients. *J Clin Endocrinol Metab* 2002;**87**:4845–56.
54. Bacchetti P, Gripshover B, Grunfeld C, Heymsfield S, McCreath H, Osmond D, et al. Fat distribution in men with HIV infection. *J Acquir Immune Defic Syndr* 2005;**40**:121–31.
55. FRAM Investigators. Fat distribution in women with HIV infection. *J Acquir Immune Defic Syndr* 2006;**42**:562–71.
56. Tien PC, Cole SR, Williams CM, Li R, Justman JE, Cohen MH, et al. Incidence of lipoatrophy and lipohypertrophy in the women's interagency HIV study. *J Acquir Immune Defic Syndr* 2003;**34**:461–6.
57. Tien PC, Barron Y, Justman JE, Hyman C, Cohen MH, Young M, et al. Antiretroviral therapies associated with lipoatrophy in HIV-infected women. *AIDS Patient Care STDs* 2007;**21**:297–305.
58. Grinspoon S. Androgen deficiency and HIV infection. *Clin Infect Dis* 2005;**41**:1804–5.
59. Grinspoon S, Carr A. Cardiovascular risk and body-fat abnormalities in HIV-infected adults. *N Engl J Med* 2005;**352**:48–62.
60. Johnsen S, Dolan SE, Fitch KV, Killilea KM, Shifren JL, Grinspoon SK. Absence of polycystic ovary syndrome features in human immunodeficiency virus-infected women despite significant hyperinsulinemia and truncal adiposity. *J Clin Endocrinol Metab* 2005;**90**:5596–604.
61. Ristig M, Drechsler H, Powderly WG. Hepatic steatosis and HIV infection. *AIDS Patient Care STDs* 2005;**19**:356–65.
62. Farrugia PM, Lucariello R, Coppola JT. Human immunodeficiency virus and atherosclerosis. *Cardiol Rev* 2009;**17**:211–5.
63. Calza L, Verucchi G, Pocaterra D, Pavoni M, Alfieri A, Cicognani A, et al. Cardiovascular risk factors and ultrasound evaluation of carotid atherosclerosis in patients with HIV-1 infection. *Int J STD AIDS* 2009;**20**:683–9.
64. Carr A. HIV lipodystrophy: risk factors, pathogenesis, diagnosis and management. *AIDS* 2003;**17**(Suppl. 1):S141–8.
65. Coffinier C, Hudon SE, Farber EA, Chang SY, Hrycyna CA, Young SG, et al. HIV protease inhibitors block the zinc metalloproteinase ZMPSTE24 and lead to an accumulation of prelamin A in cells. *Proc Natl Acad Sci USA* 2007;**104**:13432–7.
66. Young LW, Radebaugh JF, Rubin P, Sensenbrenner JA, Fiorelli G, McKusick VA. New syndrome manifested by mandibular hypoplasia, acroosteolysis, stiff joints and cutaneous atrophy (mandibuloacral dysplasia) in two unrelated boys. *Birth Defects Orig Artic Ser* 1971;**7**:291–7.
67. Lombardi F, Gullotta F, Columbaro M, Filareto A, D'Adamo M, Vielle A, et al. Compound heterozygosity for mutations in LMNA in a patient with a myopathic and lipodystrophic mandibuloacral dysplasia type A phenotype. *J Clin Endocrinol Metab* 2007;**92**:4467–71.
68. Miyoshi Y, Akagi M, Agarwal AK, Namba N, Kato-Nishimura K, Mohri I, et al. Severe mandibuloacral dysplasia caused by novel compound heterozygous ZMPSTE24 mutations in two Japanese siblings. *Clin Genet* 2008;**73**:535–44.
69. Cutler DL, Kaufmann S, Freidenberg GR. Insulin-resistant diabetes mellitus and hypermetabolism in mandibuloacral dysplasia: a newly recognized form of partial lipodystrophy. *J Clin Endocrinol Metab* 1991;**73**:1056–61.

70. Simha V, Agarwal AK, Oral EA, Fryns JP, Garg A. Genetic and phenotypic heterogeneity in patients with mandibuloacral dysplasia-associated lipodystrophy. *J Clin Endocrinol Metab* 2003;**88**:2821–4.
71. Garg A, Cogulu O, Ozkinay F, Onay H, Agarwal AK. A novel homozygous Ala529Val LMNA mutation in Turkish patients with mandibuloacral dysplasia. *J Clin Endocrinol Metab* 2005;**90**:5259–64.
72. Novelli G, Muchir A, Sangiuolo F, Helbling-Leclerc A, D'Apice MR, Massart C, et al. Mandibuloacral dysplasia is caused by a mutation in LMNA-encoding lamin A/C. *Am J Hum Genet* 2002;**71**:426–31.
73. di Masi A, D'Apice MR, Ricordy R, Tanzarella C, Novelli G. The R527H mutation in LMNA gene causes an increased sensitivity to ionizing radiation. *Cell Cycle* 2008;**7**:2030–7.
74. Agarwal AK, Fryns JP, Auchus RJ, Garg A. Zinc metalloproteinase, ZMPSTE24, is mutated in mandibuloacral dysplasia. *Hum Mol Genet* 2003;**12**:1995–2001.
75. Shackleton S, Smallwood DT, Clayton P, Wilson LC, Agarwal AK, Garg A, et al. Compound heterozygous ZMPSTE24 mutations reduce prelamin A processing and result in a severe progeroid phenotype. *J Med Genet* 2005;**42**:e36.
76. Aarskog D, Ose L, Pande H, Eide N. Autosomal dominant partial lipodystrophy associated with Rieger anomaly, short stature, and insulinopenic diabetes. *Am J Med Genet* 1983;**15**:29–38.
77. Sorge G, Ruggieri M, Polizzi A, Scuderi A, Di Pietro M. SHORT syndrome: a new case with probable autosomal dominant inheritance. *Am J Med Genet* 1996;**61**:178–81.
78. Brodsky MC, Whiteside-Michel J, Merin LM. Rieger anomaly and congenital glaucoma in the SHORT syndrome. *Arch Ophthalmol* 1996;**114**:1146–7.
79. Karadeniz NN, Kocak-Midillioglu I, Erdogan D, Bokesoy I. Is SHORT syndrome another phenotypic variation of PITX2? *Am J Med Genet A* 2004;**130A**:406–9.
80. Rautenstrauch T, Snigula F. Progeria: a cell culture study and clinical report of familial incidence. *Eur J Pediatr* 1977;**124**:101–11.
81. Wiedemann HR. An unidentified neonatal progeroid syndrome: follow-up report. *Eur J Pediatr* 1979;**130**:65–70.
82. Pivnick EK, Angle B, Kaufman RA, Hall BD, Pitukcheewanont P, Hersh JH, et al. Neonatal progeroid (Wiedemann–Rautenstrauch) syndrome: report of five new cases and review. *Am J Med Genet* 2000;**90**:131–40.
83. Gilford H. Ateleiosis and progeria: continuous youth and premature old age. *Br Med J* 1904;**2**:914–8.
84. Hutchinson J. Case of congenital absence of hair, with atrophic condition of the skin and its appendages, in a boy whose mother had been almost wholly balk from alopecia areata from the age of six. *Lancet* 1886;**1**:923.
85. Pollex RL, Hegele RA. Hutchinson–Gilford progeria syndrome. *Clin Genet* 2004;**66**:375–81.
86. DeBusk FL. The Hutchinson–Gilford progeria syndrome. Report of 4 cases and review of the literature. *J Pediatr* 1972;**80**:697–724.
87. Hennekam RC. Hutchinson–Gilford progeria syndrome: review of the phenotype. *Am J Med Genet A* 2006;**140**:2603–24.
88. Eriksson M, Brown WT, Gordon LB, Glynn MW, Singer J, Scott L, et al. Recurrent de novo point mutations in lamin A cause Hutchinson–Gilford progeria syndrome. *Nature* 2003;**423**:293–8.
89. Khalifa MM. Hutchinson–Gilford progeria syndrome: report of a Libyan family and evidence of autosomal recessive inheritance. *Clin Genet* 1989;**35**:125–32.
90. D'Apice MR, Tenconi R, Mammi I, van den Ende J, Novelli G. Paternal origin of LMNA mutations in Hutchinson–Gilford progeria. *Clin Genet* 2004;**65**:52–4.
91. Goldman RD, Shumaker DK, Erdos MR, Eriksson M, Goldman AE, Gordon LB, et al. Accumulation of mutant lamin A causes progressive changes in nuclear architecture in Hutchinson–Gilford progeria syndrome. *Proc Natl Acad Sci USA* 2004;**101**:8963–8.

92. De Sandre-Giovannoli A, Bernard R, Cau P, Navarro C, Amiel J, Boccaccio I, et al. Lamin a truncation in Hutchinson–Gilford progeria. *Science* 2003;**300**:2055.
93. Lebel M. Werner syndrome: genetic and molecular basis of a premature aging disorder. *Cell Mol Life Sci* 2001;**58**:857–67.
94. McKusick VA. Medical Genetics 1962. *J Chronic Dis* 1963;**16**:457–634.
95. Goto M, Miller RW, Ishikawa Y, Sugano H. Excess of rare cancers in Werner syndrome (adult progeria). *Cancer Epidemiol Biomarkers Prev* 1996;**5**:239–46.
96. Martin GM, Oshima J. Lessons from human progeroid syndromes. *Nature* 2000;**408**:263–6.
97. Chen L, Lee L, Kudlow BA, Dos Santos HG, Sletvold O, Shafeghati Y, et al. LMNA mutations in atypical Werner's syndrome. *Lancet* 2003;**362**:440–5.
98. Turaga RV, Paquet ER, Sild M, Vignard J, Garand C, Johnson FB, et al. The Werner syndrome protein affects the expression of genes involved in adipogenesis and inflammation in addition to cell cycle and DNA damage responses. *Cell Cycle* 2009;**8**:2080–92.
99. Doh YJ, Kim HK, Jung ED, Choi SH, Kim JG, Kim BW, et al. Novel LMNA gene mutation in a patient with Atypical Werner's Syndrome. *Korean J Intern Med* 2009;**24**:68–72.
100. Hegele RA. Drawing the line in progeria syndromes. *Lancet* 2003;**362**:416–7.
101. Burant CF, Sreenan S, Hirano K, Tai TA, Lohmiller J, Lukens J, et al. Troglitazone action is independent of adipose tissue. *J Clin Invest* 1997;**100**:2900–8.
102. Moitra J, Mason MM, Olive M, Krylov D, Gavrilova O, Marcus-Samuels B, et al. Life without white fat: a transgenic mouse. *Genes Dev* 1998;**12**:3168–81.
103. Shimomura I, Hammer RE, Richardson JA, Ikemoto S, Bashmakov Y, Goldstein JL, et al. Insulin resistance and diabetes mellitus in transgenic mice expressing nuclear SREBP-1c in adipose tissue: model for congenital generalized lipodystrophy. *Genes Dev* 1998;**12**:3182–94.
104. Gavrilova O, Marcus-Samuels B, Graham D, Kim JK, Shulman GI, Castle AL, et al. Surgical implantation of adipose tissue reverses diabetes in lipoatrophic mice. *J Clin Invest* 2000;**105**:271–8.
105. Arioglu E, Duncan-Morin J, Sebring N, Rother KI, Gottlieb N, Lieberman J, et al. Efficacy and safety of troglitazone in the treatment of lipodystrophy syndromes. *Ann Intern Med* 2000;**133**:263–74.
106. Owen KR, Donohoe M, Ellard S, Hattersley AT. Response to treatment with rosiglitazone in familial partial lipodystrophy due to a mutation in the LMNA gene. *Diabet Med* 2003;**20**:823–7.
107. Hegele RA. Lessons from human mutations in PPARgamma. *Int J Obes (Lond)* 2005;**29** (Suppl. 1):S31–5.
108. Agostini M, Gurnell M, Savage DB, Wood EM, Smith AG, Rajanayagam O, et al. Tyrosine agonists reverse the molecular defects associated with dominant-negative mutations in human peroxisome proliferator-activated receptor gamma. *Endocrinology* 2004;**145**:1527–38.
109. James OF, Day CP. Non-alcoholic steatohepatitis (NASH): a disease of emerging identity and importance. *J Hepatol* 1998;**29**:495–501.
110. Brennan AM, Mantzoros CS. Drug Insight: the role of leptin in human physiology and pathophysiology—emerging clinical applications. *Nat Clin Pract Endocrinol Metab* 2006;**2**:318–27.
111. Chong AY, Lupsa BC, Cochran EK, Gorden P. Efficacy of leptin therapy in the different forms of human lipodystrophy. *Diabetologia* 2010;**53**:27–35.
112. Petersen KF, Oral EA, Dufour S, Befroy D, Ariyan C, Yu C, et al. Leptin reverses insulin resistance and hepatic steatosis in patients with severe lipodystrophy. *J Clin Invest* 2002;**109**:1345–50.
113. Simha V, Szczepaniak LS, Wagner AJ, DePaoli AM, Garg A. Effect of leptin replacement on intrahepatic and intramyocellular lipid content in patients with generalized lipodystrophy. *Diabetes Care* 2003;**26**:30–5.

114. Mulligan K, Khatami H, Schwarz JM, Sakkas GK, DePaoli AM, Tai VW, et al. The effects of recombinant human leptin on visceral fat, dyslipidemia, and insulin resistance in patients with human immunodeficiency virus-associated lipoatrophy and hypoleptinemia. *J Clin Endocrinol Metab* 2009;**94**:1137–44.
115. Fong LG, Ng JK, Lammerding J, Vickers TA, Meta M, Cote N, et al. Prelamin A and lamin A appear to be dispensable in the nuclear lamina. *J Clin Invest* 2006;**116**:743–52.
116. Fong LG, Ng JK, Meta M, Cote N, Yang SH, Stewart CL, et al. Heterozygosity for Lmna deficiency eliminates the progeria-like phenotypes in Zmpste24-deficient mice. *Proc Natl Acad Sci USA* 2004;**101**:18111–6.
117. Varela I, Pereira S, Ugalde AP, Navarro CL, Suarez MF, Cau P, et al. Combined treatment with statins and aminobisphosphonates extends longevity in a mouse model of human premature aging. *Nat Med* 2008;**14**:767–72.
118. Sagelius H, Rosengardten Y, Schmidt E, Sonnabend C, Rozell B, Eriksson M. Reversible phenotype in a mouse model of Hutchinson–Gilford progeria syndrome. *J Med Genet* 2008;**45**:794–801.
119. Capell BC, Olive M, Erdos MR, Cao K, Faddah DA, Tavarez UL, et al. A farnesyltransferase inhibitor prevents both the onset and late progression of cardiovascular disease in a progeria mouse model. *Proc Natl Acad Sci USA* 2008;**105**:15902–7.
120. Hegele RA, Oshima J. Phenomics and lamins: from disease to therapy. *Exp Cell Res* 2007;**313**:2134–43.
121. Tracy RP. 'Deep phenotyping': characterizing populations in the era of genomics and systems biology. *Curr Opin Lipidol* 2008;**19**:151–7.
122. Hegele RA. Phenomics, lamin A/C, and metabolic disease. *J Clin Endocrinol Metab* 2007;**92**:4566–8.
123. Hegele R. LMNA mutation position predicts organ system involvement in laminopathies. *Clin Genet* 2005;**68**:31–4.
124. Scherer SW, Lee C, Birney E, Altshuler DM, Eichler EE, Carter NP, et al. Challenges and standards in integrating surveys of structural variation. *Nat Genet* 2007;**39**:S7–S15.
125. Conrad DF, Pinto D, Redon R, Feuk L, Gokcumen O, Zhang Y, et al. Origins and functional impact of copy number variation in the human genome. *Nature* 2009;**464**:704–12.
126. Wang J, Ban MR, Hegele RA. Multiplex ligation-dependent probe amplification of LDLR enhances molecular diagnosis of familial hypercholesterolemia. *J Lipid Res* 2005;**46**:366–72.
127. Lanktree M, Hegele RA. Copy number variation in metabolic phenotypes. *Cytogenet Genome Res* 2008;**123**:169–75.
128. Lanktree MB, Fantus IG, Hegele RA. Triple X syndrome in a patient with partial lipodystrophy discovered using a high-density oligonucleotide microarray: a case report. *J Med Case Reports* 2009;**3**:8867.
129. Biesecker LG. Exome sequencing makes medical genomics a reality. *Nat Genet* 2010;**42**:13–4.
130. Yuan G, Hegele RA. Genetic forms of the cardiometabolic syndrome: what can they tell the clinician? *J Cardiometab Syndr* 2007;**2**:45–8.

Cutting the Fat: The Genetic Dissection of Body Weight

Anthony G. Comuzzie,[*,†]
Paul B. Higgins,[†]
Saroja Voruganti,[†] and
Shelley Cole[†]

[*]Group Leader, Chronic Disease and Genomics, Southwest National Primate Research Center, San Antonio, Texas, USA
[†]Department of Genetics, Southwest Foundation for Biomedical Research, San Antonio, Texas, USA

Variation in body weight and composition, as well as adipose tissue function, is regulated by environmental and genetic factors, combined with a variety of interactions, including environment-by-environment, gene-by-gene, and gene-by-environment interactions. Although the recent increase in obesity can in large part be attributed to the increased availability of low-cost but energy-dense foods and an increasingly sedentary lifestyle throughout most of the developed world, the impact of these factors is more pronounced in individuals who are genetically susceptible to these environmental insults. Hence, it is likely to be the response of an individual's genetic background to a given environment that determines susceptibility to obesity. Candidate gene studies, genome-wide linkage studies, and more recently genome-wide association studies (GWASs) have been used to decipher the effect of genetics on obesity. Heritability estimates suggest that 40–70% of the variation in obesity-related phenotypes is attributable to underlying genetic variation. To date, the findings from human cohort studies (both family and case/control studies) summarized in this chapter suggest that there are likely numerous genes impacting the

Progress in Molecular Biology
and Translational Science, Vol. 94
DOI: 10.1016/S1877-1173(10)94007-6

expression of obesity-related phenotypes, with many of these genes seeming to have modest effects. The establishment of many population-based studies that have collected genome-wide data on genetic variation has recently led to the formation of multiple consortia facilitating powerful meta-analyses in order to identify additional genetic variation influencing obesity-related traits through GWAS, as well as to replicate and further characterize previously identified genetic variants and their gene regions. These consortia may have the power to elucidate how genetic and environmental factors act and interact to produce variation in obesity-related phenotypes.

I. Introduction

Body weight and composition are archetypical representations of complex phenotypes. This complexity arises from the contribution of environmental and genetic factors combined with a variety of interaction effects. These interactions include environment-by-environment, gene-by-gene, and gene-by-environment. All of these factors act and interact to produce variation in body weight and fat mass. While clinical obesity, generally defined in Western countries as a body mass index (BMI) ≥ 30 kg/m^2, represents a serious public health issue, the impact of genetics on body weight and composition is not limited to the clinical extremes. Indeed, it can be cogently argued that the clinical manifestation of obesity merely represents one end of the continuum of body weight distribution in a population. While the clinical definition of obesity may be useful for the assessment of health risks and decisions regarding intervention, its placement at any particular body weight is somewhat arbitrary. Therefore, understanding the genetic contribution to variation in body weight and composition, as well as adipose tissue function, is essential for understanding the biology underlying the clinical manifestation known as obesity and its relationship to other chronic diseases such as type 2 diabetes and heart disease.

According to the World Health Organization (WHO), the increasing prevalence of obesity worldwide contributes negatively to the overall public health by increasing an individual's risk for the development of a number of serious comorbidities (e.g., type 2 diabetes, heart disease, and cancer).[1] Current estimates indicate that almost 70% of adults in the United States are overweight. The recent increase in obesity prevalence may be attributed to environmental changes, primarily characterized by increased availability of cheap and energy-dense foods, along with an increasingly sedentary lifestyle. Nevertheless, the effect on obesity is more pronounced in individuals who are genetically susceptible to these environmental insults. Hence, it is likely to be the response of an individual's genetic background to a given environment that determines susceptibility to obesity.

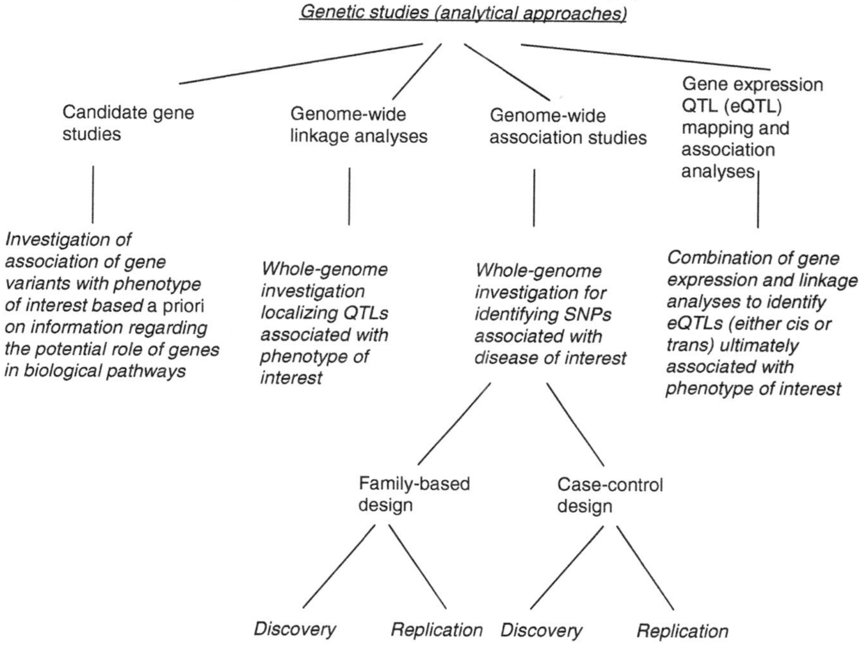

FIG. 1. Genetic study approaches.

Gene mapping studies undertake to localize genes that influence the variation in phenotypes associated with disease risk.[2] Figure 1 shows the types of approaches taken to identify the genetic contribution to prevalent complex metabolic diseases such as obesity. Candidate gene studies, genome-wide linkage studies, and genome-wide association studies (GWASs) have been used to decipher the effect of genetics on obesity. Of late, GWASs have been increasingly used to identify the genetic variation underlying disease phenotypes.[3] This chapter discusses the results of genetic mapping studies of obesity in human populations.

II. Strategies for Finding Genes Contributing to Obesity

A. Candidate Gene Studies

A traditional approach, candidate gene studies test the association between polymorphisms in a specific gene and variation in a given phenotype. This approach requires that the gene has already been identified and characterized before the association analysis.[4] The candidate genes are selected based on their suspected roles in biological pathways or mechanisms related to the phenotype of interest. As a result, candidate gene studies can never lead to the discovery of new or novel genes, but at best can simply verify a gene that is already suspected based on other types of data. While this approach was widely used for much of the 1990s, the results generated from such an approach were fairly disappointing in regards to identifying genes impacting complex biological phenomenon such as obesity, and therefore we will not review that literature here.

However, with the advent of cost-effective, high-throughput genotyping, and more recently, single nucleotide polymorphism (SNP) typing, the field quickly moved away from the use of candidate genes selected *a priori* in favor of genome-wide searches. The key distinction between the genome-wide versus candidate gene studies is that the former make no *a priori* assumptions regarding which genes may be involved and thereby allow for the discovery of potentially novel genes. While the genome-wide searches are largely the approach now used in gene mapping studies, there are two distinctively different methods by which this is done, linkage and association.

B. Genome-Wide Linkage Studies

Historically, penetrance-based linkage analysis has been successfully used to map the genetic variation underlying monogenic diseases. This particular linkage approach is limited when applied to complex phenotypes that are underlain by multiple genes, strong environmental influences, and interaction effects.[5] However, variance-component-based linkage analysis applied in extended family pedigrees is effective in localizing chromosomal regions that cosegregate with complex phenotypes such as those related to obesity.[6] The variance-components approach to linkage can model the effects of multiple genes simultaneously and can be extended to test for gene-by-gene and gene-by-environment interaction effects. Genome-wide linkage has localized several chromosomal regions or quantitative trait loci (QTLs) that contain genetic variation associated with body weight and obesity-related phenotypes.[7]

C. Genome-Wide Association Studies

GWAS investigates the entire genome to identify chromosomal loci associated with disease.[8] GWASs are considered to be powerful in detecting minor genetic effects on the variation in disease phenotypes.[9] These studies involve two phases: a discovery stage and a replication stage. The former stage can now include genotyping over a million SNPs across the genome and analyzing them for association with a disease state or related phenotype. The latter involves attempting to replicate significant associations in other independent studies. In the case–control setting, the frequency of a variant is determined in individuals with (cases) and without (controls) a particular disease. If a particular variant is found to have a higher frequency in individuals with the disease than in those without the disease, then it is inferred that that variant is associated with the particular disease.[10] GWASs have recently identified several novel obesity-related genes.

III. The State of Knowledge in the Genetics of Obesity

A. Findings from Genome-Wide Linkage Studies

QTLs for several anthropometric traits associated with body size and adiposity, as well as adipose tissue function, have been localized in a number of different populations of various ethnic backgrounds. In one of the earliest studies, a region on chromosome 2p22 was linked to fat mass and leptin concentrations,[11] and it was found to be linked to BMI and leptin in two other studies.[12,13] This region contains the proopiomelanocortin (*POMC*) gene, polymorphisms of which have been associated with adiposity traits in other studies.[14,15] Obesity QTLs have also been localized to chromosomes 3,[12,16–18] 4,[19,20] 6,[21,22] and 7.[23,24]

A meta-analysis that combined data from studies conducted in five diverse populations found evidence for linkage of BMI to chromosome 8p12–23.[25] A more recent meta-analysis of genome-wide linkage studies for obesity pooled data from 37 studies in over 10,000 families.[26] Genomic regions with suggestive evidence of linkage to BMI were identified on chromosomes 12q and 13q. Nevertheless, no chromosomal region was found to be significantly linked to BMI at the genome-wide level in this analysis. An extensive summary of the results from human linkage studies and other genetic studies of obesity can be found in the most recent human obesity gene map publication.[7]

B. Findings from GWASs of Obesity

In a quest to identify variants associated with obesity, GWASs have been conducted in several different ethnic populations, with sample sizes ranging from hundreds to several thousands. Several SNP associations have been replicated across studies, while others could not be replicated in follow-up studies.

1. Anthropometrics

Polymorphisms in several genes have been associated with anthropometric traits through GWAS approaches (Table I). Fat mass- and obesity-associated (*FTO*) and melanocortin-4 receptor (*MC4R*) gene regions have been consistently associated with BMI across several studies. The *FTO* gene is known to play a role in nucleic acid methylation in mice; however, the role of *FTO* in humans is not clear yet. In a study conducted as part of the Wellcome Trust Case Control Consortium, a GWAS comparing 490,032 SNPs from 1924 type 2 diabetes patients and 2938 controls found an association between *FTO* SNPs and

TABLE I

Association of SNPs with Obesity-Related Anthropometric Phenotypes: GWAS

Gene (symbol)	Chromosome (SNP)	Trait	Population	References
Catenin, beta-like 1 (CTNNBL1)	(rs 6013029)	BMI/fat mass	U.S. Caucasian	27
Fat mass- and obesity-associated (*FTO*)	16 (rs9930506)	BMI	European Caucasian	28
	16 (rs9930506)	BMI	Hispanic-American	28
	16 (rs9939609)	BMI	European	29
	16 (rs9939609)	BMI	U.S. Caucasian	30
	16 (rs6499640, rs8050136, rs3751812, rs7190492)	BMI	Icelandic/Dutch/ European–American/ African–American	31
	16 (rs1121980)	Early-onset extreme obesity	German	32
	16 (rs1421085)	Early-onset and morbid adult obesity	European	33
Insulin-induced gene 2 (*INSIG2*)	(rs7566605)	BMI	European Caucasian/ U.S. Caucasian/ African–American[a]	34
Mannosyl (alpha-1,3-)-glycoprotein beta-1,2-N-acetylglucosaminyltransferase (*MGAT1*)	(rs12517906)	Body weight	European	35
Melanocortin-4 receptor (*MC4R*)	18 (rs17782313)	BMI/fat mass	European	36
	18 (rs17782313)	BMI	European	37
	18 (rs12970134)	Waist circumference	European/Indian Asian	38
Neurexin 3 (*NRXN3*)	10 (rs10146997)	Waist circumference	Caucasian	39
Niemann-Pick disease, type C1 (*NPC1*)	18 (rs1805081)	BMI	European	33
Phosphotriesterase-related (*PTER*)	10 (rs10508503)	BMI	European	33
SRY-box 6 (*SOX6*)	(rs 297325, rs4756846)	BMI and BMD	Caucasian male	40
V-maf musculoaponeurotic fibrosarcoma oncogene homolog (avian) (*MAF*)	16 (rs1424233)	BMI	European	33

BMI, body mass index; BMD, bone mineral density.

[a]Included cohorts from Iceland, Germany, United States, Costa Rica, and Scandinavia.

increased BMI.[29] Similar observations were reported by Scuteri *et al.*,[28] Dina *et al.*,[41] and Thorleifsson *et al.*,[31] who found a significant association of SNPs in the *FTO* gene region with increased BMI, weight, and hip circumference. Polymorphisms at *FTO* have also been associated with early-onset extreme obesity in young German individuals.[32] Thus, *FTO* seems to be a potential candidate gene regulating BMI or obesity. However, this association was not observed in a Chinese cohort,[42] suggesting that it might be more prevalent in populations of European origin. In addition, there have been gene expression studies suggesting that SNPs in the *FTO* gene region are associated with the expression of nearby genes but not *FTO* itself, suggesting that more studies are needed to characterize the role this region plays in obesity.[43]

Another gene with genetic variation that has been consistently associated with obesity is *MC4R*, which encodes a member of the melanocortin receptor family of 7-transmembrane G-protein-linked receptors. *MC4R* deficiency has been associated with severe childhood obesity.[44] In a GWAS conducted in 16,876 individuals of European descent, Loos *et al.*[36] identified a SNP (rs17782313) in the *MC4R* gene that was significantly associated with BMI and fat mass. This association was replicated by Willer *et al.*[37] in a cohort of European ancestry. The association of *MC4R* SNPs with BMI is not restricted to individuals of European ancestry; it was also observed in a cohort of Asian Indians.[38] However, it should be noted that *MC4R* was also identified through genome-wide linkage analysis in a cohort of hispanic children, as well.[45,46]

In the Framingham Heart Study, 310 families have been followed over a period of 30 years. Association analyses conducted in this population revealed a strong association of BMI and waist circumference with SNPs in genes such as peroxisome proliferator-activated receptor gamma (*PPARG*) and adiponectin, C1Q and collagen domain containing (*ADIPOQ*).[47] Also, researchers were able to replicate an association between a SNP (rs7566605) in the insulin-induced 2 (*INSIG2*) gene and BMI. The INSIG2 protein is affected by insulin and is known to regulate fatty acid synthesis,[48] thus implying a role for *INSIG2* in obesity. The association between a variant in the *INSIG2* gene and BMI was first reported by Herbert *et al.*[49] and then replicated by Lyon *et al.*[34] and Talbert *et al.*[50] However, the association could not be replicated in white, African–American, and Mexican–American individuals.[51] Another novel gene that has been reported to be associated with BMI and fat mass is catenin, beta-like 1 (*CTNNBL1*). This gene is homologous to beta-catenin and may be involved in the induction of apoptosis.[52]

Waist circumference is a surrogate measure of abdominal obesity and correlates well with abdominal fat mass.[53] A GWAS conducted with waist circumference as the focus phenotype found a novel association with SNPs at the neurexin 3 (*NRXN3*) gene locus and replicated previous association results with SNPs in the *FTO* and *MC4R* genes.[39] In Hispanic-Americans, adiposity as measured by computer tomography was investigated in a GWAS, in which

SNPs in the candidate loci regulator of G-protein signaling (*RGS6*) and neuronal guanine nucleotide exchange factor (*NGEF*) were associated with distribution and amount of fat measures including visceral and subcutaneous fat.[54] In a meta-analysis of 16 GWASs, two loci, transcription factor AP2-beta (*TFAP2B*) and peptide methionine sulfoxide reductase A (*MSRA*), were identified as being associated with waist circumference.[55] Another anthropometric measurement that has been utilized in GWAS is height/stature.[56,57]

A combination of linkage and GWAS also has been used to map genes. Linkage is generally considered to be more powerful in detecting rare alleles with large effects, and association is known to identify common variants with small effects. In a study conducted by Johansson *et al.*,[35] a joint linkage and GWAS approach was able to identify a new candidate gene, mannosyl (alpha-1,3-)-glycoprotein beta-1,2-*n*-acetylglucosaminyltransferase (*MGAT1*), for weight in women. This gene is involved in fat absorption in the intestine and is thus an interesting candidate gene for body weight variation.

2. Other Obesity-Related Phenotypes

Obesity is a central and critical risk factor for several metabolic diseases such as type 2 diabetes and cardiovascular disease. Many genes that have been implicated in obesity have also been associated with these metabolic diseases. In a bivariate GWAS of BMI and bone mineral density (BMD), associations with two SNPs in the sex-determining region Y (SRY)-box 6 (*SOX6*) gene were uncovered.[40] This is significant, as SOX6 is known to play a role in cartilage formation as well as obesity-related insulin resistance. The solute carrier family 30, member 8 (*SLC30A8*) is a cofactor in the cellular efflux of zinc and is essential for normal insulin crystallization. *SLC30A8* variants have been known to confer increased risk for type 2 diabetes.[58,59] A GWAS conducted in a French case–control cohort revealed a locus in the *SLC30A8* gene associated with pancreatic beta-cell development and function while confirming an association with a SNP in the transcription factor 7-like 2 (*TCF7L2*) gene.[58] *TCF7L2* has been consistently associated with obesity and type 2 diabetes across multiple populations. In several GWASs, *TCF7L2* variants were linked to increased type 2 diabetes risk.[58,60,61] Most of these GWASs were conducted in European populations, indicating that this gene might have a large effect on type 2 diabetes risk in Europeans.

GWASs have also identified novel genes associated with glucose and type 2 diabetes-related phenotypes. A SNP in the glucokinase (hexokinase 4) regulator (*GCKR*) gene was associated with triglycerides, C-reactive protein, and fasting glucose.[62] SNPs in the gene *MTNR1B* that encodes for melatonin receptor 2 (MT2) were associated with plasma glucose and type 2 diabetes risk.[63] In another meta-analysis of 21 GWASs, several new loci were found to be significantly associated with fasting glucose and type 2 diabetes risk.[64]

IV. Genome-Wide Association Replication Studies

Several replication studies have been undertaken based upon GWAS results, selecting specific polymorphisms identified as being significantly associated with phenotypes of interest. In one respect, this approach differs from candidate gene studies in that the genes selected have been identified statistically and do not, in many cases, have an *a priori* functional basis for association with the phenotype of interest. GWASs have identified several genes and/or SNPs, some novel and some replicated in other GWASs, and some of these genes have been associated with obesity phenotypes across multiple population groups. For example, rs9939609 of the *FTO* gene has been genotyped in multiple populations for the purposes of replication. It was associated with increased BMI in an Australian adult twin cohort study,[65] with BMI and waist circumference in Spanish adults,[66] and with BMI in a Belgian population.[67] Other SNPs in the *FTO* and *MC4R* genes have been replicated in populations such as cohorts of children and youth of European and Hispanic descent[46,68–70]; adults from the Netherlands,[71] France,[72] Sweden,[73] China,[74] Canada,[75] and Japan[76,77]; African–Americans[78]; and Asian Sikhs.[79]

The GWAS conducted by Herbert *et al.* identified the *INSIG2* gene to be associated with BMI. Several studies tested whether the same gene is also implicated in other obesity-related traits. Talbert *et al.*[50] found a significant association of *INSIG2* SNPs with adiposity phenotypes such as visceral and subcutaneous fat depots as measured by computed tomography and glucose homeostasis traits. Likewise, Cauchi *et al.*[80] confirmed the association of *SLC30A8* SNPs with glucose homeostasis traits in a French adult population. The *SLC30A8* variants have also been associated with impaired proinsulin conversion[81] and insulin response to glucose.[82] Among gene polymorphisms that are associated with type 2 diabetes risk, *TCF7L2* has been the most consistent, and the association has been replicated across different ethnicities.[81,83–87] The association of variation at the *GCKR* locus was replicated in a study conducted in 16,853 Danes, in which significant associations were found with triglyceride, fasting glucose, and insulinemia.[88]

V. The "Missing Heritability"

While GWA has repeatedly identified several potential genes influencing various obesity-related traits, the portion of the variance attributable to these variants at the moment appears to be relatively small (Table II). This disparity between the generally reported estimates of heritability for obesity-related traits and the relatively small estimated effects size of these variants raises the issue as to the cause of this "missing heritability".[90,91] There are several

TABLE II
EFFECT SIZES OF OBESITY-RELATED SNPS AS IDENTIFIED BY GWAS

Trait	SNP/gene	Effect size	References
Weight	rs9930506/*FTO*	0.013	28
BMI	rs9939609/*FTO*	0.01	29
	rs9930506/*FTO*	0.013	28
	rs17782313/near *MC4R* (men)	0.058	36
	rs17782313/near *MC4R* (women)	0.039	36
	rs7566605/*INSIG2*	0.002–0.01	34
	rs8050136/*FTO*	0.003	31
	rs9939609/*FTO*	0.003	37
Pediatric BMI	Rs2793821/*NOTCH2*	0.035	89
Waist circumference	Rs10146997/*NRXN3*	0.050	39

possible explanations for this apparent discrepancy, including unaccounted for environmental interactions, epistatic effects, or epigenetic effects. In addition, this situation could, in some cases, be the result of not having yet identified the actual causal variant or variants, but rather ones in linkage disequilibrium with the causal variants. While testing for both gene-by-environment and gene-by-gene interactions will, in most instances, require the collection of larger samples than currently used, the issue of possible epigenetic effects will require looking at other sources of genomic variation such as deletions, duplications, and inversions, as well as imprinting. In the final analysis, the ultimate explanation for this missing heritability is likely to include all of these factors. At the moment, however, the insights which have been and are continuing to be gained from these efforts to identify the genetic contribution to variation in obesity-related traits have helped to identify key, and in some cases novel, metabolic pathways involved, even if they have not fully explained the total heritability of these phenotypes.

VI. Conclusion

Excessive body weight ultimately results from a chronic imbalance between energy intake and energy expenditure, and the genetic variation responsible must influence one or both of these factors. Heritability estimates suggest that 40–70% of body weight variability is attributable to genetic variation.[92] Studies in *Drosophila melanogaster* suggest that several hundred genes may control body weight,[93] and genome-wide searches in mice have identified over 50 chromosomal loci influencing adiposity.[94] The findings from the human studies summarized in this chapter also suggest that the genetic influence on obesity results from a large number of genes with relatively small effects. Environment-by-environment, gene-by-gene, and gene-by-environment interactions and epistasis are also likely to contribute. Until recently, the relatively small sample sizes of most studies did not provide the statistical power to address questions of interaction. The establishment of many population-based studies that have collected genome-wide data on genetic variation has led to the formation of multiple consortia facilitating powerful meta-analyses in order to identify additional genetic variation influencing obesity-related traits through GWAS, as well as to replicate and further characterize previously identified genetic variants and their gene regions. These consortia, such as the Genetic Investigation of Anthropometric Traits (GIANT) Consortium,[37] may have the power to elucidate how genetic and environmental factors act and interact to produce variation in body weight and fat mass.

References

1. Walley AJ, Blakemore AIF, Froguel P. Genetics of obesity and the prediction of risk for health. *Hum Mol Genet* 2006;**15**:R124–30.
2. Althsuler D, Daly MJ, Lander ES. Genetic mapping in human disease. *Science* 2008;**322**:881–8.
3. Amos CI. Successful design and conduct of genome-wide association studies. *Hum Mol Genet* 2007;**16**:R220–5.
4. Walder K, Segal D, Jowett JB, Blangero J, Collier GR. Obesity and diabetes gene discovery approaches. *Curr Pharm Des* 2003;**9**:1357–72.
5. Blangero J, Williams JT, Almasy L. Variance components methods for detecting complex trait loci. *Adv Genet* 2001;**42**:151–81.
6. Almasy L, Blangero J. Human QTL linkage mapping. *Genetica* 2009;**136**:333–40.
7. Rankinen T, Zuberi A, Chagnon YC, Weisnagel SJ, Argyropoulos G, Walts B, et al. The human obesity gene map: the 2005 update. *Obesity* 2006;**14**:529–644.
8. Loos RJ. Recent progress in the genetics of common obesity. *Br J Clin Pharmacol* 2009;**68**:811–29.
9. Kruglyak L. Prospects form whole-genome linkage disequilibrium mapping of common disease genes. *Nat Genet* 1999;**22**:139–44.
10. Lango H, Weedon MN. What will whole genome searches for susceptibility genes for common complex disease offer to clinical practice? *J Intern Med* 2007;**263**:16–27.
11. Comuzzie AG, Hixson JE, Almasy L, Mitchell BD, Mahaney MC, Dyer TD, Stern MP, MacCluer JW, Blangero J. A major quantitative trait locus determining serum leptin levels and fat mass is located on chromosome 2. *Nat Genet* 1997;**15**:273–6.
12. Moslehi R, Goldstein AM, Beerman M, Goldin L, Bergen AW, The Framingham Heart Study. A genome-wide linkage scan for body mass index on Framingham Heart Study families. *BMC Genet* 2003;**4**(Suppl. 1):S97.
13. Rotimi CN, Comuzzie AG, Lowe WL, Luke A, Blangero J, Cooper RS. The quantitative trait locus on chromosome 2 is confirmed in African Americans. *Diabetes* 1999;**48**:643–4.

14. Buono P, Pasanisi F, Nardelli C, Ieno L, Capone S, Liguori R, et al. Six novel mutations in the proopiomelanocortin and melanocortin receptor 4 genes in severely obese adults living in southern Italy. *Clin Chem* 2005;**51**:1358–64.
15. Baker M, Gaukrodger N, Mayosi BM, Imrie H, Farrall M, Watkins H, et al. Association between common polymorphisms of the proopiomelanocortin gene and body fat distribution: a family study. *Diabetes* 2005;**54**:2492–6.
16. Kissebah AH, Sonnenberg GE, Myklebust J, Goldstein M, Broman K, James RG, et al. Quantitative trait loci on chromosomes 3 and 17 influence phenotypes of the metabolic syndrome. *Proc Natl Acad Sci USA* 2000;**97**:14478–83.
17. Wu X, Cooper RS, Borecki I, Hanis C, Bray M, Lewis CE, et al. A combined analysis of genome wide linkage scans for body mass index from the National Heart, Lung, and Blood Institute Family Blood Pressure Program. *Am J Hum Genet* 2002;**70**:1247–56.
18. Luke A, Wu X, Zhu X, Kan D, Su Y, Cooper R. Linkage for BMI at 3q27 region confirmed in an African–American population. *Diabetes* 2003;**52**:1284–7.
19. Stone S, Abkevich V, Hunt SC, Gutin A, Russell DL, Neff CD, et al. A major predisposition locus for severe obesity, at 4p15–p14. *Am J Hum Genet* 2002;**70**:1459–68.
20. Arya R, Duggirala R, Jenkinson CP, Almasy L, Blangero J, O'Connell P, et al. Evidence of a novel quantitative-trait locus for obesity on chromosome 4p in Mexican Americans. *Am J Hum Genet* 2004;**74**:272–8.
21. Geller F, Dempfle A, Görg T, Framingham Heart Study. Genome scan for body mass index and height in the Framingham Heart Study. *BMC Genet* 2003;**31**(Suppl. 1):S91.
22. Meyre D, Lecoeur C, Delplanque J, Francke S, Vatin V, Durand E, et al. A genome-wide scan for childhood obesity-associated traits in French families shows significant linkage on chromosome 6q22.31–q23.2. *Diabetes* 2004;**53**:803–11.
23. Roth H, Hinney A, Ziegler A, Barth N, Gerber G, Stein K, et al. Further support for linkage of extreme obesity to the obese gene in a study group of obese children and adolescents. *Exp Clin Endocrinol Diabetes* 1997;**105**:341–4.
24. Bray MS, Boerwinkle E, Hanis CL. Linkage analysis of candidate obesity genes among the Mexican American population of Starr County, Texas. *Genet Epidemiol* 1999;**16**:397–411.
25. Johnson L, Luke A, Adeyemo A, Deng HW, Mitchell BD, Comuzzie AG, et al. Meta-analysis of five genome-wide linkage studies for body mass index reveals significant evidence for linkage to chromosome 8p. *Int J Obes* 2005;**29**:413–9.
26. Saunders CL, Chiodini BD, Sham P, Lewis CM, Abkevich V, Adeyemo AA, et al. Meta-analysis of genome-wide linkage studies in BMI and obesity. *Obesity* 2007;**15**:2263–75.
27. Liu YJ, Liu XG, Wang L, Dina C, Yan H, Liu JF, et al. Genome-wide association scans identified CTNNBL1 as a novel gene for obesity. *Hum Mol Genet* 2008;**17**:1803–13.
28. Scuteri A, Sanna S, Chen WM, Uda M, Albai G, Strait J, et al. Genome-wide association scan shows genetic variants in the FTO gene are associated with obesity-related traits. *PLoS Genet* 2007;**3**:e115.
29. Frayling TM, Timpson NJ, Weedon MN, Zeggini E, Freathy RM, Lindgren CM, et al. A common variant in the FTO gene is associated with body mass index and predisposes to childhood and adult obesity. *Science* 2007;**316**:889–94.
30. Cotsapas C, Speliotes EK, Hatoum IJ, Greenawalt IJ, Dobrin R, Lum PY, et al. Common body mass index-associated variants confer risk of extreme obesity. *Hum Mol Genet* 2009;**18**:3502–7.
31. Thorleifsson G, Walters GB, Gudbjartsson DF, Steinthorsdottir V, Sulem P, Helgadottir A, et al. Genome-wide association yields new sequence variants at seven loci that associate with measures of obesity. *Nat Genet* 2009;**41**:18–24.
32. Hinney A, Nguyen TT, Schereg A, Friedel S, Bronner G, Muller TD, et al. Genome wide association (GWA) study for early onset extreme obesity supports the role of fat mass and obesity associated gene (FTO) variants. *PLoS ONE* 2007;**26**:e1361.

33. Meyre D, Delplanque J, Chèvre JC, Lecoeur C, Lobbens S, Gallina S, et al. Genome-wide association study for early-onset and morbid adult obesity identifies three new risk loci in European populations. *Nat Genet* 2009;**41**:157–9.
34. Lyon HN, Emilsson V, Hinney A, Heid IM, Lasky-Su J, Zhu X, et al. The association of a SNP upstream of INSIG2 with body mass index is reproduced in several but not all cohorts. *PLoS Genet* 2007;**3**:e61.
35. Johansson A, Marroni F, Hayward C, Franklin CS, Kirichenko AV, Jonasson I, et al. Linkage and genome-wide association analysis of obesity-related phenotypes: association of weight with the MGAT1 gene. *Obesity* 2010;**18**:803–8.
36. Loos RJ, Lindgren CM, Li S, Wheeler E, Zhao JH, Prokopenko I, et al. Common variants near MC4R are associated with fat mass, weight and risk of obesity. *Nat Genet* 2008;**40**:768–75.
37. Willer CJ, Speliotes EK, Loos RJ, Li S, Lindgren CM, Heid IM, et al. Six new loci associated with body mass index highlight a neuronal influence on body weight regulation. *Nat Genet* 2009;**41**:25–34.
38. Chambers JC, Elliott P, Zabaneh D, Zhang W, Li Y, Froguel P, et al. Common genetic variation near MC4R is associated with waist circumference and insulin resistance. *Nat Genet* 2008;**40**:716–8.
39. Heard-Costa NL, Zillikens MC, Monda KL, Johansson A, Harris TB, Fu M, et al. NRXN3 is a novel locus for waist circumference: a genome-wide association study from the CHARGE consortium. *PLoS Genet* 2009;**5**:e1000539.
40. Liu YZ, Pei YF, Liu JF, Yang F, Guo Y, Zhang L, et al. Powerful bivariate genome-wide association analyses suggest the SOX6 gene influencing both obesity and osteoporosis phenotypes in males. *PLoS ONE* 2009;**4**:e6827.
41. Dina C, Meyre D, Gallina S, Durand E, Körner A, Jacobson P, et al. Variation in FTO contributes to childhood obesity and severe adult obesity. *Nat Genet* 2007;**39**:724–6.
42. Li H, Wu Y, Loos RJ, Hu FB, Liu Y, Wang J, et al. Variants in the fat mass- and obesity-associated (FTO) gene are not associated with obesity in a Chinese Han population. *Diabetes* 2008;**57**:264–8.
43. Jowett JB, Curran JE, Johnson MP, Carless MA, Goring HH, Dyer TD, et al. Genetic variation at the FTO locus influences RBL2 gene expression. *Diabetes* 2010;**59**:726–32.
44. Farooqi IS, Keogh JM, Yeo GSH, Lank EJ, Cheetham T, O'Rahilly S. Clinical spectrum of obesity and mutations in the melanocortin 4 receptor gene. *N Engl J Med* 2003;**348**:1085–95.
45. Cai G, Cole SA, Butte N, Bacino C, Diego V, Tan K, et al. A quantitative trait locus on chromosome 18q for physical activity and dietary intake in Hispanic children. *Obesity (Silver Spring)* 2006;**14**:1596–604.
46. Cole SA, Butte NF, Voruganti VS, Cai G, Haack K, Kent Jr. JW, et al. Evidence that multiple genetic variants of MC4R play a functional role in the regulation of energy expenditure and appetite in Hispanic children. *Am J Clin Nutr* 2010;**91**:191–9.
47. Fox CS, Heard-Costa N, Cupples LA, Dupuis J, Vasan RS, Atwood LD. Genome-wide association to body mass index and waist circumference: the Framingham Heart Study 100K project. *BMC Med Genet* 2007;**8**(Suppl. 1):S18.
48. Yabe D, Komuro R, Liang G, Glodstein JL, Brown MS. Liver-specific mRNA for Insig2 downregulated by insulin: implications for fatty acid synthesis. *Proc Natl Acad Sci USA* 2003;**100**:3155–60.
49. Herbert A, Gerry NP, McQueen MB, Heid IM, Pfeufer A, Illig T, et al. A common genetic variant is associated with adult and childhood obesity. *Science* 2006;**312**:279–83.
50. Talbert ME, Langefeld CD, Ziegler JT, Haffner SM, Norris JM, Bowden DW. INSIG2 SNPs associated with obesity and glucose homeostasis traits in Hispanic: the IRAS Family Study. *Obesity* 2009;**17**:1554–62.

51. Bressler J, Kao WH, Pankow JS, Boerwinkle E. Risk of type 2 diabetes and obesity is differentially associated with variation in FTO in Whites and African–Americans in the ARIC study. *PLoS ONE* 2010;**5**:e10521.

52. Jabbour L, Welter JF, Kollar J, Hering TM. Sequence, gene structure and expression pattern of CTNNBL1, a minor class intron-containing gene-evidence for a role in apoptosis. *Genomics* 2003;**81**:292–303.

53. Klein S, Allison DB, Heymsfield SB, Kelley DE, Leibel RL, Nonas C, et al. Waist circumference and cardiometabolic risk. *Am J Clin Nutr* 2007;**85**:1197–202.

54. Norris JM, Langefeld CD, Talbert ME, Wing MR, Haritunians T, Fingerlin TE, et al. Genome-wide association study and follow-up analysis of adiposity traits in Hispanic Americans: the IRAS Family Study. *Obesity* 2009;**17**:1932–41.

55. Lindgren CM, Heid IM, Randall JC, Lamina C, Steinthorsdottir V, Qi L, et al. Genome-wide association meta-analysis identifies three loci influencing adiposity and fat distribution. *PLoS Genet* 2009;**5**:e1000508.

56. Weedon MN, Lettre G, Freathy RM, Lindgren CM, Voight BF, Perry JR, et al. A common variant of HMGA2 is associated with adult and childhood height in the general population. *Nat Genet* 2007;**39**:1245–50.

57. Dempfle A, Wudy SA, Saar K, Hagemann S, Friedel S, Scherag A, et al. Evidence for involvement of the vitamin D receptor gene in idiopathic short stature via a genome-wide linkage study and subsequent association studies. *Hum Mol Genet* 2006;**15**:2772–83.

58. Sladek R, Rocheleau G, Rung J, Dina C, Shen L, Serre D, et al. A genome-wide association study identifies novel risk loci for type 2 diabetes. *Nat Genet* 2007;**445**:881–5.

59. Omori S, Tanaka Y, Takahashi A, Hirose H, Kashiwagi A, Kaku K, et al. Association of CDKAL1, IGF2BP2, CDKN2A/B, HHEX, SLC30A8, and KCNJ11 with susceptibility to type 2 diabetes in a Japanese population. *Diabetes* 2008;**57**:791–5.

60. Zeggini E, Scott LJ, Saxena R, Voight BF, Marchini JL, Hu T, et al. Meta-analysis of genome-wide association data and large-scale replication identifies additional susceptibility loci for type 2 diabetes. *Nat Genet* 2008;**40**:638–45.

61. Scott LJ, Mohlke KL, Bonnycastle LL, Willer CJ, Li Y, Duren WL, et al. A genome-wide association study of type 2 diabetes in Finns detects multiple susceptible variants. *Science* 2007;**316**:1341–5.

62. Orho-Melander M, Melander O, Guiducci C, Perez-Martinez P, Corella D, Roos C, et al. Common missense variant in the glucokinase regulatory protein gene is associated with increased plasma triglycerides and C-reactive protein but lower fasting glucose concentrations. *Diabetes* 2008;**57**:3112–21.

63. Bouatia-Naji N, Bonnefond A, Cavalcanti-Proenca C, Sparso T, Holmkvist J, Marchand M, et al. A variant near MTNR1B is associated with increasing fasting plasma glucose levels and type 2 diabetes risk. *Nat Genet* 2009;**41**:89–94.

64. Dupuis J, Langenberg C, Prokopenko I, Saxena R, Soranzo N, Jackson AU, et al. New genetic loci implicated in fasting glucose homeostasis and their impact on type 2 diabetes risk. *Nat Genet* 2010;**42**:105–16.

65. Cornes BK, Lind PA, Medland SE, Montgomery GW, Nyholt DR, Martin NG. Replication of the association of common rs9939609 variant of FTO with increased BMI in an Australian adult twin population but no evidence for gene by environment (G × E) interaction. *Int J Obes* 2009;**33**:75–9.

66. Gonzalez-Sanchez JL, Zabena C, Marinez-Larrad MT, Martinez-Calatrava MJ, Perez-Barba M, Serrano-Rios M. Variant rs9939609 in the FTO gene is associated with obesity in an adult population from Spain. *Clin Endocrinol* 2009;**70**:390–3.

67. Peeters A, Beckers S, Verrijken A, Roevens P, Peeters P, Van Gaal L, et al. Variants in the FTO gene are associated with common obesity in the Belgian population. *Mol Genet Metab* 2008;**93**:481–4.

68. Zhao J, Bradfield JP, Li M, Wang K, Zhang H, Kim CE, et al. The role of obesity-associated loci identified in genome-wide association studies in the determination of pediatric BMI. *Obesity* 2009;**17**:2254–7.

69. Li S, Zhao JH, Luan J, Luben RN, Rodwell SA, Khaw KT, et al. Cumulative effects and predictive value of common obesity-susceptibility variants identified by genome-wide association studies. *Am J Clin Nutr* 2010;**91**:184–90.

70. Liem ET, Vonk JM, Sauer PJ, van der Steege G, Oosterom E, Stolk RP, et al. Influence of common variants new INSIG2, in FTO and near MC4R genes on overweight and the metabolic profile in adolescence: the TRAILS (Tracking Adolescents' individuals lives survey) Study. *Am J Clin Nutr* 2010;**91**:321–8.

71. Bauer F, Elbers CC, Adan RAH, Loos RJF, Onland-Moret NC, Grobbee DE, et al. Obesity genes identified in genome-wide association studies are associated with adiposity measures and potentially with nutrient-specific food preference. *Am J Clin Nutr* 2009;**90**:951–9.

72. Cauchi S, Stutzmann F, Cavalcanti-Proenca C, Durand E, Pouta A, Hartikainen AL, et al. Combined effects of MC4R and FTP common genetic variants on obesity in European general populations. *J Mol Med* 2009;**87**:537–46.

73. Renström F, Payne F, Nordström A, Brito EC, Rolandsson O, Hallmans G, et al. Replication and extension of genome-wide association study results for obesity in 4923 adults from northern Sweden. *Hum Mol Genet* 2009;**18**:1489–96.

74. Cheung CY, Tso AW, Cheung BM, Xu A, Ong KL, Fong CH, et al. Obesity susceptibility genetic variants identified from recent genome-wide association studies: implications in a Chinese population. *J Clin Endocrinol Metab* 2010;**95**:1395–403.

75. Do R, Bailey SD, Desbiens K, Belisle A, Montpetit A, Bouchard C, et al. Genetic variants of FTO influence adiposity, insulin sensitivity, leptin levels and resting metabolic rate in the Quebec Family study. *Diabetes* 2008;**57**:1147–50.

76. Hotta K, Nakamura M, Nakamura T, Matsuo T, Nakata Y, Kamohara S, et al. Association between obesity and polymorphisms in SEC16B, TMEM18, GNPDA2, BDNF, FAIM2 and MC4R in a Japanese population. *J Hum Genet* 2009;**54**:727–31.

77. Shimaoka I, Kamide K, Ohishi M, Katsuya T, Akasaka H, Saitoh S, et al. Association of gene polymorphisms of the fat-mass amd obesity-associated gene with insulin resistance in Japanese. *Hypertens Res* 2010;**33**:214–8.

78. Liu G, Zhu H, Lagou V, Gutin B, Barbeau P, Treiber FA, et al. Common variants near melanocortin 4 receptor are associated with general and visceral adiposity in European- and African American youth. *J Pediatr* 2010;**156**:598–605.

79. Been LF, Nath SK, Ralhan SK, Wander GS, Mehra NK, Singh J, et al. Replication of association between a common variant near melanocortin-4 receptor gene and obesity-related traits in Asian Sikhs. *Obesity* 2010;**18**:425–9.

80. Cauchi S, Proenca C, Choquet H, Gaqet S, Greave F, Marre M, et al. Analysis of novel risk loci for type 2 diabetes in a general French population: the D.E.S.I.R. study. *J Mol Med* 2008;**86**:341–8.

81. Kirchhoff K, Machicao F, Haupt A, Schafer SA, Tschritter O, Staiger H, et al. Polymorphisms in the TCF7L2, CDKAL1 and SLC30A8 genes are associated with impaired proinsulin conversion. *Diabetologia* 2008;**51**:597–601.

82. Xiang J, Li XY, Xu M, Hong J, Huang Y, Tan JR, et al. Zinc transporter-8 gene (SLC30A8) is associated with type 2 diabetes in Chinese. *J Clin Endocrinol Metab* 2008;**93**:4107–12.

83. Grant SF, Thorleifsson G, Reynisdottir I, Benediktsson R, Manolescu A, Sainz J, et al. Variation of transcription factor 7-like 2 (TCF7L2) gene confers risk of type 2 diabetes. *Nat Genet* 2006;**38**:320–3.
84. Hayashi T, Iwamoti Y, Kaku K, Hirose H, Maeda S. Replication study for the association of TCF7L2 with susceptibility to type 2 diabetes in a Japanese population. *Diabetologia* 2007;**50**:980–4.
85. Mayans S, Lackovic K, Lindgren P, Ruikka K, Agren A, Eliasson M, et al. TCF7L2 polymorphisms are associated with type 2 diabetes in northern Sweden. *Eur J Hum Genet* 2007;**15**:342–6.
86. Ng MCY, Tam CHT, Lam VKL, So WY, Ma RC, Chan JCN. Replication and identification of novel variants at TCF7L2 associated with type 2 diabetes in Hong Kong Chinese. *J Clin Endocrinol Metab* 2007;**92**:3733–7.
87. Miyake K, Horikawa Y, Hara K, Yasuda K, Osawa H, Furuta H, et al. Association of TCF7L2 polymorphisms with susceptibility to type 2 diabetes in 4, 087 Japanese subjects. *J Hum Genet* 2008;**53**:174–80.
88. Sparsø T, Andersen G, Nielsen T, Burgdorf KS, Gjesing AP, Nielsen AL, et al. The GCKR rs780094 polymorphism is associated with elevated fasting serum triacylglycerol, reduced fasting and OGTT-related insulinemia and reduced risk of type 2 diabetes. *Diabetologia* 2008;**51**:70–5.
89. Zhao J, Bradfield JP, Zhang H, Annaiah K, Wang K, Kim CE, et al. Examination of all type 2 diabetes GWAS loci reveals HHEX-IDE as a locus influencing pediatric BMI. *Diabetes* 2010;**59**:751–5.
90. Manolio TA, Collins FS, Cox NJ, Goldstein DB, Hindorff LA, Hunter DJ, et al. Finding the missing heritability of complex diseases. *Nature* 2009;**461**:747–53.
91. Eichler EE, Flint J, Gibson G, Kong A, Leal SM, Moore JH, et al. Missing heritability and strategies for finding the underlying causes of complex disease. *Nat Rev Genet* 2010;**11**:446–50.
92. Comuzzie AG, Allison DB. The search for human obesity genes. *Science* 1998;**280**:1374–7.
93. Dohrmann CE. Target discovery in metabolic disease. *Drug Discov Today* 2004;**9**:785–94.
94. Wuschke S, Dahm S, Schmidt C, Joost HG, Al-Hasani H. A meta-analysis of quantitative trait loci associated with body weight and adiposity in mice. *Int J Obes* 2007;**31**:829–41.

Genetics of Taste and Smell: Poisons and Pleasures

Danielle Renee Reed and
Antti Knaapila

*Monell Chemical Senses Center,
Philadelphia, Pennsylvania, USA*

Eating is dangerous. While food contains nutrients and calories that animals need to produce heat and energy, it may also contain harmful parasites, bacteria, or chemicals. To guide food selection, the senses of taste and smell have evolved to alert us to the bitter taste of poisons and the sour taste and off-putting smell of spoiled foods. These sensory systems help people and animals to eat defensively, and they provide the brake that helps them avoid ingesting foods that are harmful. But choices about which foods to eat are motivated by more than avoiding the bad; they are also motivated by seeking the good, such as fat and sugar. However, just as not everyone is equally capable of sensing toxins in food, not everyone is equally enthusiastic about consuming high-fat, high-sugar foods. Genetic studies in humans and experimental animals strongly suggest that the liking of sugar and fat is influenced by genotype; likewise, the abilities to detect bitterness and the malodors of rotting food are highly variable among individuals. Understanding the exact genes and genetic differences that affect food intake may provide important clues in obesity treatment by allowing caregivers to tailor dietary recommendations to the chemosensory landscape of each person.

Progress in Molecular Biology
and Translational Science, Vol. 94
DOI: 10.1016/S1877-1173(10)94008-8

I. Introduction

Whenever anything is introduced directly into the body, there is a risk that it will be harmful. The air we breathe, the water we drink, and the food we eat give us oxygen, water, and nutrients, but they also have the potential, because of the presence of poisons and pathogens, to make us very ill, and they may even kill us. Our sensory systems are designed to help us detect and avoid these outcomes, through vision (seeing contaminants in our food), touch, taste, smell, and the common chemical sense—a sense that is not well known (with no universally accepted name) and that includes the sting of carbon dioxide and the burn of hot peppers. Together, these senses help protect us from bad food. Yet whenever anything is deliberately eaten or drunk, there is an expectation that it will be beneficial. Beverages, foods, and chemical compounds all are ingested because we are motivated to do so for pleasure or for the relief of unpleasant states such as thirst, hunger, or tiredness (e.g., caffeinated drinks such as coffee). These senses help us to distinguish not only good from bad food but also the good from the great—the sweetest apple, the juiciest meat, and the freshest bread. Eating may be a risk, but it is also a pleasure, and our senses help us find the most desirable food and drink available.

However, what constitutes the best food and drink is often a matter of opinion. When attempting to generalize about what constitutes "good food," counterexamples abound. For example, cheeses such as limburger are desirable to some but repellent to others. Thus, eating is a matter of taste, both in the biological sense and as a matter of individual opinion. Why there should be such diverse views about what constitutes the most desirable food is a worthwhile question and one that can be answered from different perspectives: cultural, social, developmental, and medicinal. People eat what others in their communities and families find desirable;[1] children like different foods than adults do;[2,3] and some people believe diets recommended by their doctors for health reasons are the best food choice.

In this chapter, we provide a genetic and evolutionary perspective on food perception and preference. Humans have changed over time and adapted to specific environments that contain some types of food but not others; this in turn has tailored our sense of taste and, by extension, our genome and individual genes. Nonhuman species provide evidence that the sense of taste has been shaped by evolution; for instance, cats and some other carnivorous species, in addition to chickens,[4] have lost the function of their sweet receptor[5]—they no longer need to taste "sweet" because the foods they eat, the flesh of other animals or starchy grains, contain little sugar. The availability of food may have shaped nearly the entire genome—in yeast, and probably other organisms as well, most genes are involved either directly or indirectly with nutrition and metabolism.[6] Some animals have specialized in eating only particular foods, benefiting from a refined ability to find and ingest those foods and limiting the number of competitors for that food source, whereas other animals, like humans, are generalists that can eat most types of food, which brings other benefits and risks.

At the heart of evolution is individual variation, and perhaps no single human trait has as many person-to-person differences as the abilities to taste and smell.[7,8] However, to what extent these genetic differences in taste and smell affect the food preferences and food intake of contemporary humans is contentious. Information from these senses is one of many influences on decisions about what to eat,[9] and its place on a hierarchy of determinants is unclear. Genes that affect energy metabolism rather than taste and smell might be more powerful determinants of food intake. For instance, the different feelings of satiety and hunger that people experience arise from genetic variation.[10]

Because human food intake will always be determined by many variables, which change in importance with time and circumstances, the controlled settings available with animal models are useful in untangling the relative contributions of taste and smell, metabolism, and experience.[11] In addition, new methods—genome-wide association studies—have recently been developed to survey the contributions of all genes to taste, smell, and food intake. In the sections that follow, we review the genetics of taste and smell, as well as fat and sugar preference, drawing on data collected in humans and other animals, when useful studies are available. We also include results and interpretations of genome-wide studies of taste, smell, and nutrient preference and intake.

II. Psychophysical Measures of Taste and Smell

Before taste and smell can be studied, they must be measured. The field of science devoted to these measurements is called psychophysics. The scientists trained in this field try to understand relationships between physical stimuli (e.g., a tastant or odorant) and the psychological responses they elicit (e.g., taste or odor). Thus, an individual's ability to taste or smell can be determined using psychophysical testing, such as measurement of detection threshold. Detection threshold is the lowest concentration at which a compound can be detected, and subjects often perceive this as only a hint of "something"—just enough to discriminate the stimulus from a blank, but not necessarily enough to recognize its type or quality (e.g., sweet). The lowest concentration at which a stimulus can be named for its quality is called the recognition threshold.

Detection threshold (often referred to as "threshold" and assumed to mean detection threshold) is a frequently used measure in studies on genotype–phenotype associations in taste and smell. The threshold can be measured in several ways. For instance, in odor tests, often the subject's task is to find one bottle with the odorant among a set of three (one bottle with the stimulus and two with blanks), presented in ascending order of concentration. The measurement is regarded as relatively objective because it does not require use of subjective rating scales, in contrast to perceived intensity and pleasantness, which require the subject to offer a judgment. However, measures of perceived intensity are also important because, although biased by a subject's experience,[12] they provide information about the range of concentrations of odorants or tastants most often encountered in everyday life. Likewise, pleasantness is a matter of opinion, but it is also a crucial piece of information because liking is often a prerequisite for consumption.

Identification of stimuli is another common measure of taste and smell. Odor identification tests frequently include between 8 and 40 stimuli, each typically accompanied by four alternative descriptors including one that is meant to be the correct quality of the perceived odor. For instance, amyl acetate (a "fruity" smell) would be offered and the subjects asked to choose among these descriptors: banana, kerosene, burning rubber, and cinnamon. Commercially available odor identification tests, such as the UPSIT,[13] BSIT,[14] and Sniffin' Sticks,[15] are designed for clinical purposes, but because they can be administered quickly, they have also been used in epidemiological studies. Comparable tests for the identification of taste qualities are not widely available.

Some tastants and odorants activate not only the olfactory system, but also the free endings of the trigeminal nerve located in the mouth and nasal cavity, thus contributing to the common chemical sense. This may add complexity to the measurement of odor detection thresholds. For instance, an individual who could not detect an odorant even at a high concentration may still detect the odorant by common chemical sense.[16]

Some people are born with a total absence of the sense of smell (general anosmia, sometimes known only as anosmia) or taste (general ageusia), but these conditions are rare. More common is the inability to detect a specific odorant or tastant, known as a specific anosmia[17] or specific ageusia, the most classic example being the inability to taste the bitterness of sulfur-containing thyroid poisons.[18] However, unlike some people's inability to taste, the inability to smell specific chemicals is more a matter of degree: Specific anosmia often refers not only to total lack of ability to detect a specific odorant but also to reduced sensitivity to the odorant.[19] Subjects are often considered to have a specific anosmia if they have a detection threshold two standard deviations or more above the mean.[20] The most studied specific anosmia is for the odorant androstenone.[21,22] Among its other functions, androstenone is sometimes found in meat from uncastrated male pigs, and it underlies, together with skatole, the characteristic (some say unpleasant) "boar taint" odor.[21]

This section has described ways that taste and smell can be measured (i.e., detection and recognition thresholds, the intensity of the stimulus, its quality, and the criterion for defining specific deficits of taste and smell). When we eat, we attend to the concentration of chemical stimulus in our food (e.g., judging whether our food is too salty or contains a hint of onion). For taste, the qualities of bitter, sweet, sour, etc., determine how much we like our food. Now we consider each quality in turn.

III. Bitter: Poisoned with Pleasure

Bitterness has the simplest relationship with food intake: What is bitter is bad, and what tastes bad is not eaten. Because poisons can kill quickly, their detection in food is paramount. And many poisons are bitter, a taste quality that evokes a classic rejection response.[23,24] This rejection is assumed to be inborn and unlearned because it is apparent in human infants and in nonhuman primates.[25] Furthermore, because it is also present in rodents that have had their neural connection between the brain stem and cortex severed, the rejection of bitterness could even be considered a reflex.[26]

Many people assume that all poisons are bitter, but this viewpoint depends upon the definition of poison. Toxicologists view all chemicals as potentially poisonous—the key issue is determining the relationship between dosage and lethality. Because every chemical is a potential poison but not every chemical is bitter, not all poisons are bitter. From the viewpoint of taste and food intake, a poison is defined as a chemical in a food that is liable to cause illness or death when eaten in sufficient quantity. Even with this narrow and unconventional definition of poison, it is not known how many chemicals are poisons and what proportion of them are bitter. However, when people are offered a range of chemicals to taste, they are overwhelmingly accurate at guessing the toxicity of given compounds using only taste as a guide.[27]

The following are common poisonous plants: castor beans contain ricin, a compound that causes red blood cells to clump together; turnips contain progoitrin, which inhibits thyroid hormones; cassava contains cyanide, which interrupts the ability of cells to make ATP; soybeans contain saponin, which is poorly absorbed into the body but when present in the bloodstream causes red cells to burst. All these chemicals—ricin, progoitrin, cyanide, and saponin—are bitter.[28]

We can see from these examples that many poisonous plant compounds are bitter and that the taste system developed in part to detect and avoid them. However, the relationship between the detection of bitterness of a chemical and its lethality is a puzzle, because some bitter chemicals that are not harmful to humans can nonetheless be perceived at low concentrations.[29] So the ability to sense bitterness may serve other purposes in addition to poison detection. For instance, when proteins are fermented, some of the protein products are bitter,[30] so the bitter taste system may also detect decayed proteins.

Typically, infants and primates immediately and automatically reject bitter stimuli. But for adult humans, the decision about what to do when bitterness is perceived is more complex. Adults sometimes eat foods and drink beverages that are bitter because they contain chemicals that increase feelings of well-being; the most obvious examples are the psychoactive drugs caffeine and alcohol. How much the person likes the effects of the bitter drugs, despite their taste, largely determines whether people ingest them.[31] Even for bitter foods and drinks that offer pharmacological incentives, people often mask the bitterness, for instance adding cream and sugar to coffee. But if adults ingest bitter foods and drinks only when they contain drugs, then we must explain the willingness of people to drink decaffeinated coffee, which is still bitter but contains much less caffeine than regular coffee.[32] It is possible that the overall sensory qualities of coffee become associated with the effect of caffeine[33] and that even during extinction (i.e., when the stimulus is no longer followed by the rewarding response), the association is sufficient to maintain the behavior. Or it could be that the small amounts of caffeine are enough to maintain the consumption of this bitter beverage. But even if bitter substances are willingly ingested for their pharmacological benefit, we still need to explain why some people eat bitter melon (a plant commonly eaten in Asia) or other bitter plants that have no obvious drug-like properties.

This paradox—people eat bitter foods that contain no known psychoactive drug—might be resolved if the bitter compounds make people feel better in other ways. Recent studies suggest that bitter melon may contain secondary chemicals that have favorable metabolic effects, including reducing blood sugar in people with diabetes.[34] Thus, bitter foods might contain healthful compounds that blur the line between nutrient and drug. If bitter-tasting chemicals in plant foods have health benefits, then removing these compounds (through manufacturing of processed foods or selectively breeding plants for low bitterness) may have negative consequences. The harmful effect of increased sugar and fat in the modern human diet has been widely discussed, but the loss of bitter compounds may also contribute to diseases associated with the modern diet, such as obesity and diabetes. Our bitter detection system seems to balance rejection and acceptance for bitterness in order to avoid poisons and to get enough—but not too much—of the bitter substances that make us feel good.

Tests often ask people to sample bitter chemicals dissolved in water, and because these chemicals must be safe to ingest (even so, subjects are usually asked not to swallow the samples), the number of bitter chemicals tested in a laboratory does not reflect the wide range of bitter compounds we could potentially taste. The selection of chemicals tested is further biased toward those that have been previously used for sensory testing, so that data can be compared across studies and because their safe use already has been documented. Some of the frequently tested bitter compounds include quinine (found in the bark of the cinchona tree and used to treat malaria), caffeine (found in coffee beans and widely consumed for its stimulant properties), epicatechin (found in tea), tetralone (found in hops and, by extension, in beer), L-phenylalanine (an amino acid), magnesium sulfate (a mineral found in Epsom salts), urea (a product of nitrogen metabolism), naringin (a compound found in grapefruit), sucrose octaacetate (an acetylated derivative of sucrose), denatonium benzoate (used in consumer products to discourage accidental poisoning), and propylthiouracil (a sulfur-containing drug used to treat hyperthyroid disease). People exhibit marked differences in perceiving these chemicals[18,35,36]: Some find bitter compounds to be very bitter, whereas others experience the same concentration of the same chemical as much less intense.

We know that the origin of these individual differences, at least for most compounds listed above, is partially genetic because people with genetic makeups that are very similar (e.g., identical twins) are more alike in bitter perception than people who differ (e.g., fraternal twins).[37,38] For the least lethal bitter chemicals, which are the most studied in humans, genetic variation is a moderate to strong determinant of how well a person can perceive them. For the most lethal poisons, less individual variation might be expected because people who have lost their ability to taste these chemicals might experience more accidental poisoning, so their genes would be less represented in the population. On the other hand, sensory variation in the worldwide population might be greatest for poisonous chemicals in plants that are found only in some geographic regions. But whether there is greater or lesser individual variation in the perception of lethal bitter chemicals has gone unanswered—ethical concerns obviously prevent testing with these poisons in people. Cell-based assays with one or two human bitter receptors can be used to test the response to a wide range of poisons,[39] but this method provides only a partial answer to the question because artificial systems may not recreate the human taste experience.

In at least one case, a gene's participation in bitter perception is well understood. The inability of some people to taste phenylthiocarbamide (PTC) was discovered in the 1930s by a DuPont chemist named Arthur Fox.[18] It was soon determined that the trait was heritable (i.e., transmitted in families),[40] and 70 years later the responsible gene and allele were identified.[41] The gene, called *TAS2R38*, is a member of the bitter taste receptor family TAS2R. Three alleles in *TAS2R38* account for the bitter-blindness to PTC—they combine to form a haplotype that leads to reduced ability to perceive PTC (and its chemical relative propylthiouracil, one of the commonly studied bitters listed above). The *TAS2R38* haplotype determines most of the variation in people, but alleles in other genes,[42,43] and even age[44] and sex,[45] also contribute to variations in PTC perception. The study of the genetics of this trait is useful because it straddles the divide between the single-gene mode of inheritance found in diseases such as cystic fibrosis and the interactions of many genes found in a complex trait like obesity. Thus, PTC genetics is a useful model for studying genotype/phenotype effects and the influences that modify them.

Genetic differences in bitter taste perception might modify food preferences and intake in a complex manner. Although PTC was first created in a chemistry laboratory and is probably not found in plants, there are many chemical relatives of PTC that stimulate the TAS2R38 bitter taste receptor.[39,46,47] At least one of these compounds is found in plant food (turnips),[48] and less similar but still related compounds are found in other plant species.[49] People with taster and nontaster alleles of *TAS2R38* differ in their perception of vegetables (like watercress) that contain these PTC-like compounds.[50] From here, it is a short step to hypothesize that genetically insensitive people would eat more of these vegetables than would people who find them to be bitter.

If people differ in their intake of some vegetables, bitter perception might ultimately influence body weight, as suggested by some investigators.[51–54] However, a direct relationship between *TAS2R38* genotype, food intake, and body weight has not been detected in epidemiological studies[55,56] or in genome-wide studies of association with body mass index, a measure of obesity.[10,57,58] Thus, if alleles of this bitter receptor gene can directly affect food intake or body weight, they are too weak to be detected in the population as a group. Progress toward understanding genotype/phenotype relationships for PTC taste-blindness and food intake will require narrowing the focus to vegetables that contain these specific compounds. In addition, instead of relying on indirect information about the chemical constitution of vegetables, concentrations of these bitter chemicals should be directly measured in vegetables, because amounts can differ depending on which cultivar is tested or the composition of the soil in which it was grown.

A related point to consider is which aspect of receptor function is most affected by alleles of *TAS2R38*. While it is often referred to as a "bitter *taste* receptor," this receptor and other bitter receptors are also found in the gut[59–61] and in nasal airways, where they detect molecules secreted by bacteria.[62] The expression of the gene *TAS2R38* in the gut is regulated by the amount of cholesterol in the diet, and its expression is highest when cholesterol is low.[63] The interpretation of this observation is that gene expression of bitter receptors is increased when plant foods are consumed, which is logical because bitter compounds are more concentrated in plants than in other foods, like meats. It is therefore reasonable to assume that bitter taste receptors are intimately related to vegetable intake, because vegetables taste bitter and gene expression in the gut is tied to the intake of diets high in plant food. However, the regulation of bitter receptors in the tongue (or the gut) in response to changes in diet had not yet been studied. This is a gap in our knowledge.

A convincing argument can be made that a specific bitter receptor and its alleles might affect food intake, especially of vegetables. But it is important to put these details in context. Although people differ in their ability to taste many bitter chemicals,[36] the complete loss of bitter perception for a particular chemical like PTC is probably rare. (It might be misleading to call this a complete loss because the nontaster form of the receptor might detect different bitter molecules.[64]) Current studies suggest that PTC is unusual because only one of the 25 known bitter receptors is strongly stimulated by it, so the loss of this one receptor (TAS2R38) is consequential.[39] Other bitter molecules stimulate multiple receptors, and the loss of one may decrease but not eliminate the ability to detect that particular bitter molecule.[30,39,47,65–73] The perception of PTC is probably an extreme case of individual variation in bitter perception.

IV. Sourness and Fermentation

Although the relationship between bitter taste and plant poisons is relatively simple (compared to other taste qualities), it is not the only one that signals a warning. Bacteria and fermentation can spoil food, and the end points of these processes are detected by using sour taste as a guide, along with smell, vision, and the common chemical sense. Sometimes bacterial activity in food is wholesome, such as the fermentation of milk, wheat, or grapes to make cheese, bread, or wine. But sometimes it is not, as when meat or vegetables rot. Like bitterness, which can signal either a poison or a beneficial compound, sourness can signal either good or bad bacterial processes. And like bitterness, the preference for sourness is a matter of degree; low concentrations of sourness (and bitterness) must be evaluated in a specific circumstance and a decision made about acceptance or rejection. Context is important: The tartness that is desirable in buttermilk (caused by lactic acid, a by-product of the fermentation process) would be undesirable in ordinary milk. Concentration is also important because there is a continuum, from lower concentrations (which have a pleasant taste) to middle-range concentrations (which may be rejected), to high concentrations (which evoke pain receptors and lead to tissue damage). In other words, we like lemonade, but we don't drink more concentrated acids. From a developmental perspective, bitter and sour differ because sour taste is readily accepted by many children but bitter taste is not.[74] So the sourness of food conveys mixed signals: very bad at very high concentrations, bad in some foods but good in others, and neither universally liked nor rejected by all.

Also like bitter perception, the ability of humans to detect sourness at low concentrations is partially determined by genetics.[75,76] But unlike bitter perception, the genes and their alleles associated with individual differences in sour perception are unknown, because studies to identify them have not been conducted. Furthermore, unlike for bitter taste, studies of sour taste in animals are no more advanced than those in humans. Although genetically distinct strains of inbred mice are known to differ in sour preference,[77] no genes have been identified that account for these differences. One candidate is the polycystic kidney disease 2-like 1 gene (*Pkd2l1*), which is involved in sour taste in mice[78] and possibly in humans.[79] Whether naturally occurring allelic variation in this gene is responsible for differential sour perception (in mice or humans) is not known.

The studies needed are of two types: (1) linkage studies, in which the DNA of family members with a similar trait is searched for shared DNA that would contain the causal genes, and (2) genome-wide association studies, in which subjects are grouped by genotype at many loci throughout the genome and compared for a particular trait, such as intensity of sour perception. Useful future studies would measure individual abilities to: (1) perceive sourness at low concentrations, and (2) judge its intensity at a range of concentrations. It would also be useful to measure how much people like sourness. Genome-wide studies could be performed to find regions of the genome in common among people with similar sour phenotypes. These studies would fill gaps in our current knowledge.

V. Sweet Perception and Liking

One of the pleasures of eating comes from sweet taste, but perception of sweetness and the liking for highly concentrated solutions differ among people (for reviews see Refs. 80–84). Being more or less able to perceive the sweetness of sugar will interact with a person's liking for it. Some people are sweet-likers—no concentration of sweet in food is too much[85,86]—and their enhanced ability to perceive sweetness makes them like the food all the more. Other people have a peak concentration of sweetness they prefer, and as sweetness rises above that point, food becomes too sweet and unpleasant. While this dichotomy (i.e., sweet-liker or disliker) is a useful concept, like most human traits, sweet liking is probably on a continuum and is context-specific[87] (e.g., people who like very sweet ice cream might not prefer very sweet juice). Therefore, because people differ in how much they like sweetness at a given concentration, having a genotype that makes some people more sensitive to sweetness than others will not always result in increased liking. Sweet as a taste quality is complex (from a genetics perspective) because the relationship between perception and liking is complex.

In mice, alleles of the *Tas1r3* gene,[75] which codes for one of the subunits of the sweet receptor (the receptor is a heterodimer, or combination, of Tas1r2 and Tas1r3), determine in part both the sensitivity to and preference for sweet solutions.[88] Alleles in the promoter region of this same gene predict how well people can sort a range of sucrose concentrations into the correct order,[89] but it is not known whether these sensitivity alleles are related to preference. Other genes and their alleles probably also contribute to genetic differences in sweet perception (e.g., second messenger molecules like gustducin). We might expect poor agreement between genes that affect perception and those that predict the preference for and intake of sweet food,[90–92] and in fact, this is the case. Thus far, there is no convergence on particular genomic regions associated with sweet sensory perception and the actual intake of sweet foods.

VI. Umami: Savory or Meaty

Until recently, there was no consensus about whether umami was a true taste quality.[93] The concept of umami, which perhaps translates best into English as "savory" or "meaty," was suggested by Japanese investigators as a unique quality exemplified by monosodium glutamate (MSG). Umami also has a synergistic property: When MSG is combined with ribonucleotides such as inosine monophosphate (compounds often found in meat), the perceived intensity of the mixture is higher than the intensity of either compound alone. Umami was better accepted as a taste quality when its receptor was discovered in taste cells.[94–96] Like the sweet receptor, which is a heterodimer of TAS1R2 and TAS1R3, the umami receptor is a heterodimer of TAS1R1 and TAS1R3. Some people are specifically insensitive to MSG,[97] which is partly caused by alleles of the umami receptor.[98–100] The detection of a genotype/phenotype relationship implies that the trait is at least partially heritable, but we know of no published twin or family studies that estimate the contribution of genes to trait variation.

Since humans differ in their ability to taste MSG, it sparks curiosity about what role this differential response might play in the liking for meat or other foods such as cheese, tomatoes, and mushrooms, which contain glutamate. However, the role that individual differences in umami perception might play in human food intake is unknown and represents another gap in our understanding. That the umami receptor might be a key biological protein in determining meat-eating is speculative, but the idea does have empirical support from comparative studies. The giant panda, which eats only plant food, has lost the umami receptor during evolution.[101] The recent observation that obese women prefer higher concentrations of MSG in soup suggests that this taste quality may be of importance in determining food intake and body weight.[102]

VII. Salt as a Simple Pleasure and Complex Poison

Salt is both a simple pleasure and a complex poison. It is a pleasure in that humans choose to consume more salt than they need and it is added to food to enhance flavor in almost every culture.[103] It is a poison in the sense that it may increase blood pressure and exacerbate other health problems. But whether salt reduction should be a universal mandate is a debated public health position.[104] The Institute of Medicine, a health policy advisory group, recently drafted a report calling for reduced salt consumption.[105]

While the current viewpoint from a biomedical perspective is that nearly everyone overconsumes salt relative to physiological need, there are few studies that concentrate on individual differences in salt perception (e.g., Refs. 7,106,107). Even fewer studies have asked whether there are heritable genetic contributions to variability in salt taste perception[75,108] or preference.[109] Nevertheless, the few results available are consistent: There is no evidence for genetic effects on salt perception or liking. Instead, environment seems to be the major determinant. One's history of sodium exposure can have a substantial impact on preference for, consumption of, and physiological processing of NaCl.[110] Research suggests that time of day[111] or even short-term exposure can have some (temporary) impact on salty taste.[112] Evolutionary forces may have shaped the human ability to recognize salty taste in such a way as to make it very responsive to differences in the environmental mineral and water supply or habitual diet.[113] Therefore, efforts to assess the impact of genetic variation within the salt receptor[114,115] should focus on salt perception of people with similar environmental backgrounds (e.g., early exposure, recent exposure) and be attentive to the current state of the subject (e.g., time of day, thirst).

The molecular aspects of human salt perception are not known, but evidence has accumulated that a sodium channel is important for one component of salt perception in mice.[114] The genes that code for the protein subunits for this channel would be a candidate target for genotype–phenotype studies in humans.

VIII. Calcium as a Basic Taste

One of the most widely debated aspects of human taste is its definition: Most schoolchildren learn that there are four basic tastes, sour, sweet, salty, and bitter. But we now turn to evidence that the list of taste qualities is expanding. One reason for this expansion is that the definition of a basic taste is changing: If there is a working receptor on taste cells, its ligand can be considered to have a "taste." Umami was a controversial taste until its receptor was discovered, and a similar change has occurred with the mineral calcium. A taste receptor sensitive to calcium (Tas1r3) has been identified in mice,[116] and it is possible that the same receptor acts in humans.[117,118] In this regard, it has been suggested that another receptor, CASR, mediates *kokumi* taste,[117] an orosensory quality recognized in Japan but unknown in the West. Although people differ in the perception of calcium solutions,[119] there are no genetic studies to indicate whether this trait is heritable and which genes (including the genes coding the subunits of the receptor) might be involved. It may be that genotype has a potent effect on calcium perception as it does in mice,[120] or it may be that individual differences in calcium perception are tied to the current diet or metabolic need, similar to sodium and salt (as described in the above section), or both could be true. This is an understudied area.

IX. Fat Taste: *Pinguis*

While the controversy of umami as a basic taste quality is largely resolved, the controversy about fat as a basic taste quality is approaching but has not attained resolution. The idea that fat is a basic taste was suggested as early as the 16th century, at which time it was called *pinguis* (Latin for fatty).[121] The evidence has recently been reviewed,[122] as has the heritable aspects of fat perception, liking, and intake.[123] Fat as a taste quality is especially relevant to obesity because of the observation that obese people typically have greater fat preferences than do lean people (reviewed in Ref. 80). Thus far, some genes have been implicated in fat perception, specifically a gene coding a transmembrane protein found in taste cells (Cd36)[124,125] and genes coding several G protein-coupled receptors that respond to fatty acids elsewhere in the body and that are also found in taste cells.[126–128] To date, these studies have been conducted in mice, and it is unclear whether human fat perception occurs through the same mechanism and, if so, whether alleles of these genes might lead some people to be fat-blind in the same way some people are bitter-blind. Thus, as with sour taste, the role of sensory differences in fat perception is another gap in knowledge that can be filled once the influential genes and alleles have been identified.

X. Common Chemical Sense

Perhaps because the common chemical sense has not had a single name in widespread use, it has been poorly integrated into the study of food intake and obesity, but it contributes in several ways to the pain and pleasure of eating: The tingle of carbon dioxide dissolved in soda,[129] the cooling associated with methanol,[130] and the burn of chili peppers[131] all arise from receptors in the mouth (and nose and throat) that convey this information to the brain. These compounds exist primarily in plants and are defenses against insects that would do harm. At low intensity, these defense compounds produce sensations that many people find pleasant; at higher concentrations, the compounds produce sensations that are unpleasant and even painful.

People differ in their perception of these sensations,[132] but little systemic research has focused on whether the differences are heritable in humans. However, one study has reported that the liking for spicy food was highly heritable.[133] From a comparative perspective, birds are indifferent to the main ingredient in hot peppers (capsaicin) that causes the burn, because they lack the receptor TRPV1.[134] Everyday experience suggests some humans, too, may be indifferent to, or even like, the burn of hot peppers. The prevalence of individual differences in the many facets of the common chemical sense (cooling, burning, stinging) in humans, the degree to which genetic variation explains those differences, and their impact on food intake are unknown. But because most of the compounds that stimulate this sense are found in plants,[135] and humans eat plants, sometimes as their only source of food, their effects on human health are probably direct. It is possible that these sensations have effects that are equal to or even greater than those of bitterness in determining individual differences in the liking of vegetables, spices, and condiments such as mustard or chili sauce.

Carbon dioxide also stimulates the common chemical sense and is a constituent of the modern human diet. It is commonly consumed in fizzy soda, but its taste perception may have evolved originally to detect the carbon dioxide produced from rotting food.[136] How carbon dioxide might affect food digestion and metabolism is unknown. For instance, the obesity effects of sugar in soda are often studied (e.g., Ref. 137). Whether these effects are exacerbated or offset by the fizz of carbon dioxide has not yet been examined.

XI. Smell

If taste is the gatekeeper, the sense of smell is the sentinel, evaluating the food for danger before it enters the mouth. When offered an unfamiliar food, we will smell it before we taste it, and smell is one of the key first defenses against spoiled food and an important source of eating enjoyment. Thus, the sense of smell and its loss can have powerful consequences for food intake and quality of life.[138,139] Before addressing genetics of smell and its potential connections to food intake, we introduce the olfactory system for background information.

A. The Olfactory System

When working normally, the sense of smell, or olfaction, enables us to detect a large number of different odorants and to perceive these volatile compounds as odors. Although the stimuli (odorants) are also sometimes called odors, in psychophysics, odor refers to a percept, the result of the process of odor perception, whereas odorant refers to the chemical that elicited the odor.

The airborne molecules from food take one of two paths to sensory cells in the olfactory epithelium: the orthonasal route (through the nostrils, before eating) or the retronasal route (through the nasopharynx, while eating). Both paths are important in food intake, for defense and for pleasure. In the olfactory epithelium, the airborne odorants are detected by olfactory receptors. The receptors lie embedded in the membrane of olfactory sensory neurons, each of which accommodates only one type of receptor.[140] Binding of an odorant molecule by an olfactory receptor initiates a signal transduction cascade, which ultimately leads to the transfer of the olfactory signal to the brain, where the odor percept is generated.

To be a potential odorant, a molecule has to be volatile enough to reach the olfactory epithelium with airflow. Although most volatiles are odorants, some small molecule volatiles, such as carbon monoxide and carbon dioxide, are odorless. In addition, structurally diverse molecules can elicit indistinguishable odors, while similar molecules, such as stereoisomers, can yield distinct odors (e.g., R(−)-carvone smells like spearmint, but S(+)-carvone smells like caraway).[141,142] To date, the type of odor elicited by a volatile compound cannot be reliably predicted by the structure of the molecule.[143]

Humans have about 400 different olfactory receptor types, a number greatly exceeded by the number of potential odorants. Thus, it is unlikely that a particular receptor would bind only one type of odorant or that a certain odorant would attach specifically to only one type of receptor. Instead, the olfactory system is thought to make use of combinatorial receptor coding to gain the capacity to recognize the immense amount of odorants; several types of related receptors bind an odorant with varying affinities, and in turn, multiple related odorants can be detected by the same receptor.[140] The combinatorial coding suggests that most olfactory receptors are selective (broadly tuned) rather than very specific (narrowly tuned). However, the breadth of tuning varies among olfactory receptors.[144,145]

Only a few of the human olfactory receptors have been linked with their odorant ligands (i.e., the molecules that they detect).[146] Development of automated, high-throughput methods for matching the receptors and their ligands in cell-based model systems (or using computational models) will facilitate confirming the functionality of the receptors. These methods, however, cannot replace the measurement of actual human responses in studies of genetic influence on odor percepts. The psychophysical measurement of responses to odorous stimuli remains a time-consuming but essential step when the genetics of the sense of smell and its implications for food intake are studied.

B. Genetics of Olfaction

Humans have nearly 400 potentially functional olfactory receptor genes (OR genes), making this gene family one of the largest in the human genome.[147] In addition to these intact genes, which are thought to produce functional olfactory receptors, humans have at least a similar number of nonfunctional OR genes (pseudogenes) and about 60 genes of which both functional and nonfunctional variants are known to exist (segregating pseudogenes). The exact number of functional genes will be known only after the functionality of the corresponding receptors is demonstrated. However, it is obvious that far more genes encode receptors for smell than for taste. The larger number of olfactory receptors likely reflects the need to detect a wider variety of compounds than is the case for taste. Further, the large number of compounds detected by the sense of smell reflects the wider role of this sense: While the sense of taste serves almost exclusively ingestion, the sense of smell has other functions, too. These include sensing environmental dangers (e.g., smoke) and potential interpersonal chemosignaling (e.g., sexual selection).

The heritability of a trait makes the search for genes influencing that trait reasonable. If little or no heritability is found, the underlying genes, if any, are difficult, if not impossible, to locate in gene-mapping studies. While the ability to smell some odorants is heritable, for other odorants, it is not. For instance, the ability to smell food odors like chocolate or lemon is associated with little or no heritability.[148,149] However, the pleasantness of cinnamon is heritable and has been mapped to chromosome 4 by linkage analysis.[150] If the allelic genes that determine the pleasantness of odors like cinnamon are identified, studies of genotype and food intake might be worthwhile.

Individual variation in perception of some odors has been attributed partly to specific OR genes. The differences among people in the ability to smell androstenone are at least partially determined by genes,[151,152] and an allele of an OR gene, *OR7D4*, contributes to this trait.[153] However, unlike alleles of the taste receptor gene *TAS2R38*, which account for almost 70% of the person-to-person variation in perception of bitter taste from PTC,[41] *OR7D4* alleles account for only a small amount of variance in perception of androstenone.[153] Two other OR genes have been associated with individual variation in the sense of smell: *OR11H7* with isovaleric acid (sweaty odor)[154] and *OR2M7* with the smell of asparagus metabolites in urine.[155] Association between the gene *OR2J3* and detection of *cis*-3-hexen-1-ol (green leaf odor) has also been suggested.[156] Why there is relatively little effect of the alleles of a single olfactory receptor on perception lies in this sense's complex nature: Many olfactory receptors combine to detect a particular odorant,[140] and one odorant may stimulate many receptors, so if one is not working, others may compensate.

Systematic, repeated exposures to individual odorants have been demonstrated to lower detection thresholds (increased sensitivity) to these odorants, suggesting that genes do not entirely determine the perceptions.[157,158] One possibility, yet to be proved, is that there are gene–environment interactions in odor perception (i.e., genes influencing the sense of smell are controlled differently in different environments). Whatever the mechanism, the flexibility of the sense of smell could have been evolutionarily appropriate. When first humans moved to new environments and encountered novel odorants from new threats (e.g., toxins) and opportunities (e.g., food sources), the flexible sense of smell may have helped the population to survive.

C. Implications for Food Intake

Although there may also be some innate preferences, smell is probably more flexible and amenable to learning[159] when compared to taste. This point is particularly relevant when we consider olfaction as a sentry against spoiled food: The products of fermentation can be perceived as wholesome or harmful, depending on context. As an example, isovaleric acid has a pungent odor that people like if they are told it is from cheddar cheese and dislike if they are told it is from body odor.[160] Likewise, people will eat food with a bad smell (e.g., durians or limburger cheese) if they know it is safe and they like the taste. In addition, the pleasant odor of food can stimulate appetite, but the potency of these genetic differences in determining food intake and obesity is unclear.

XII. Alcohol Perception

Ethanol (or in more colloquial terms, alcohol) is a commonly consumed drug that is also a food, and just over 50% of adults living in the United States are regular drinkers.[161] Because of alcohol's popular pharmacological effects, the attractiveness or off-putting taste and smell of alcohol can be overlooked.[162] As a taste, ethanol has a complex quality: Indirect evidence, mostly from the study of mice and rats, suggests that it stimulates the sweet receptor.[163] One explanation for the connection between sweetness and alcohol is that sweet fruits ferment and so this sweetness may help animals gauge the sugar/alcohol ratio in fruit and other fermented products.[164] In addition to sweetness, genotype–phenotype studies in humans suggest that ethanol stimulates at least one bitter receptor.[165] Alcohol may also stimulate receptors for the common chemical sense, at least in rodents.[166] Alcohol also has an odor, and although the exact receptors are not known, based on other typical molecules, it is likely to stimulate several different receptors; the patterns of receptor activation may differ based on concentration.[144]

Individual differences in alcohol intake are studied intensively because of the role of dependence and addiction in human health, yet we are aware of no studies that have examined the heritability of alcohol perception in humans. Although it is reasonable to expect large individual differences that may be due in part to genotype, this is a current gap in scientific understanding.

XIII. Taste, Genetics, and Food Intake

Taste is one reason people report for why they eat the foods they do, but cost, social influences, and food availability all play a role in human food intake.[167–169] What constitutes good food is subtle, but extremes of most taste sensations, including bitterness, sourness, and sting, as well as excessive sweetness, saltiness, and richness from fats, detract from the pleasant experience of food for most people. However, humans can tolerate and even like foods that go too far; for instance, we deliberately add tingling wintergreen oil to candies and drink carbonated sodas that can cause a burning sensation. Or we drink (and even prefer) very bitter coffee. The observation that tingling, burning, and bitterness are so popular deserves more research attention than it receives. The liking for sweetness and fat depends on concentration: For some people, there is no such thing as "too sweet," while others find more than moderate amounts of sweetness to be cloying. And although few people would eat a meal solely of oil or butter, people differ in how much fat is just right or too rich.[123]

Some progress has been made in defining the genes and their alleles associated with the positive and negative aspects of food and flavor. Taking a ham and cheese sandwich as an example (Fig. 1), we might imagine that people with sensitive alleles might differentially detect the mild sweetness of onion (*TAS1R3*),[89] the savory glutamate taste of tomato (*TAS1R3*),[98,100,170] the bitterness of watercress (*TAS2R38*),[50] the smell of cheese (*OR11H7*),[154] or the boar taint odor of ham (*OR7D4*).[153] We envision that a combination of allelic differences might contribute to the range of liking for this sandwich. People who can taste the pleasant components (and not the unpleasant ones) may experience the ham sandwich as more desirable because of its taste.

But how do differences in sensory experience translate to actual food consumption? Whether these individual differences in chemosensory experience affect food selection is the weak link in the chain of causality. People eat what they like, but they also eat for many other reasons. Simple explanations of the links between sensory perception and food intake are misguided: Just as people do not choose art or music based solely on how well they can hear or

Chemical	Gene	Allele
Sucrose	TAS1R3	-1572 C/T
Glutamate	TAS1R3	R757C
Isothiocyanate	TAS2R38	A49P, V262A, I296V
Isovaleric acid	OR11H7P	C/T at nt 679
Androstenone	OR7D4	R88W

Gene	Quality	Genotype	
		Person 1	Person 2
TAS1R3	Sweet in onion	+/+	–/–
TAS1R3	Umami in tomato	+/+	–/–
TAS2R38	Bitter in watercress	–/–	+/+
OR11H7P	Sweat odor in cheese	–/–	+/+
OR7D4	Boar taint in ham	–/–	+/+

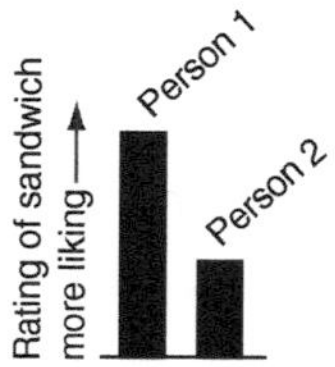

FIG. 1. Example of how taste and smell genotypes may contribute to the perception of common foods. A ham and cheese sandwich contains bread, onion, tomato, watercress, cheese, and ham. The low concentrations of sucrose in the onion will be detected by sweet receptors on the tongue, heterodimers of TAS1R2 and TAS1R3. The glutamate in the tomato, perceived as a savory or umami taste, is sensed by the umami receptor, which is a heterodimer of TAS1R1 and TAS1R3. The bitterness of watercress is due to isothiocyantes (or structurally related compounds) and is detected by one or more bitter receptors (i.e., TAS2R38). Isovaleric acid is a component of cheese and gives it a characteristic odor that some people call "sweaty." This chemical stimulates at least one olfactory receptor, OR11H7. Ham can contain androstenone, which gives the meat an odor called boar taint. Some people perceive this odor as offensive, and the receptor associated with this compound is OR7D4. In this example, people with two positive alleles (+/+) perceive the compound better than people with two negative alleles (–/–). Person 1 can taste the pleasant sweetness of the onion and the umami of the tomato but does not perceive the bitterness of the watercress or the unpleasant odors of the cheese or ham. Thus, Person 1 likes the ham sandwich more than Person 2.

see, we do not choose food based solely on the reactions of the tongue or nose. Although genetic differences determine what we can taste and smell (and at what concentration), our taste is ultimately determined by our experiences, learning, and culture, in an artistic sense, as well as in our likes and dislikes of food and drink. However, perception is the first step toward liking: What cannot be perceived cannot be liked or preferred. Therefore, it is worthwhile to pursue these questions.

This focus on perception and taste is especially important in the realm of human health because most of the chemicals discussed that give rise to bitter taste have metabolic and behavior effects and many are drugs (caffeine, alcohol). People are always urged to eat diets higher in plant foods like vegetables, but these foods are bitter to many. As another example, new medicines that need to be given in liquid forms can taste excessively bitter.[171] And some bitter or stinging compounds are concentrated in plants to help them to fend off insects, but they also tickle our taste buds. Thus, to understand our greater desires for certain types of foods above others, as well as our avoidance of compounds we know we should consume, such as medicines or healthy but bitter vegetables, we must consider our genotype, which dictates our ability to perceive these compounds.

Acknowledgments

Michael G. Tordoff and Gary K. Beauchamp commented on earlier versions of the chapter. Discussions with Julie A. Mennella, Charles J. Wysocki, Johannes Reisert, and Alexander A. Bachmanov improved the quality of this work. The editorial assistance of Patricia J. Watson is gratefully acknowledged. Mary Leonard provided assistance with graphic preparation. This work was funded in part by the National Institutes of Health grant DK56797.

References

1. Counihan C, Van Esterik P. *Food and culture : a reader.* 2nd edit, New York: Routledge; 2008.
2. Mennella JA, Pepino MY, Reed DR. Genetic and environmental determinants of bitter perception and sweet preferences. *Pediatrics* 2005;**115**:e216–22.
3. Pelchat ML, Pliner P. "Try it. You'll like it." Effects of information on willingness to try novel foods. *Appetite* 1995;**24**:153–65.
4. Shi P, Zhang J. Contrasting modes of evolution between vertebrate sweet/umami receptor genes and bitter receptor genes. *Mol Biol Evol* 2006;**23**:292–300.
5. Li X, Li W, Wang H, Cao J, Maehashi K, Huang L, et al. Pseudogenization of a sweet-receptor gene accounts for cats' indifference toward sugar. *PLoS Genet* 2005;**1**:27–35.
6. Hillenmeyer ME, Fung E, Wildenhain J, Pierce SE, Hoon S, Lee W, et al. The chemical genomic portrait of yeast: uncovering a phenotype for all genes. *Science* 2008;**320**:362–5.
7. Blakeslee AF, Salmon TN. Genetics of Sensory Thresholds: Individual Taste Reactions for Different Substances. *Proc Natl Acad Sci USA* 1935;**21**:84–90.
8. Menashe I, Man O, Lancet D, Gilad Y. Different noses for different people. *Nat Genet* 2003;**34**:143–4.
9. Jacobs HL, Sharma KN. Taste versus calories: sensory and metabolic signals in the control of food intake. *Annals of the New York Academy of Sciences* 1969;**157**:1084–125.
10. Bauer F, Elbers CC, Adan RA, Loos RJ, Onland-Moret NC, Grobbee DE, et al. Obesity genes identified in genome-wide association studies are associated with adiposity measures and potentially with nutrient-specific food preference. *Am J Clin Nutr* 2009;**90**:951–9.
11. Reed DR. Animal models of gene-nutrient interactions. *Obesity (Silver Spring)* 2008;**16** (**Suppl 3**):S23–7.
12. Bartoshuk LM, Duffy VB, Green BG, Hoffman HJ, Ko CW, Lucchina LA, et al. Valid across-group comparisons with labeled scales: the gLMS versus magnitude matching. *Physiol Behav* 2004;**82**:109–14.
13. Doty RL, Marcus A, Lee WW. Development of the 12-item Cross-Cultural Smell Identification Test (CC-SIT). *Laryngoscope* 1996;**106**:353–6.

14. Doty RL, Shaman P, Dann M. Development of the University of Pennsylvania Smell Identification Test: a standardized microencapsulated test of olfactory function. *Physiol Behav* 1984;**32**:489–502.
15. Kobal G, Hummel T, Sekinger B, Barz S, Roscher S, Wolf S. "Sniffin' sticks": screening of olfactory performance. *Rhinology* 1996;**34**:222–6.
16. Boyle JA, Lundstrom JN, Knecht M, Jones-Gotman M, Schaal B, Hummel T. On the trigeminal percept of androstenone and its implications on the rate of specific anosmia. *J Neurobiol* 2006;**66**:1501–10.
17. Amoore JE. Stereochemical theory of olfaction. *Nature* 1963;**199**:912–3.
18. Fox AL. The relationship between chemical constitution and taste. *Proc Natl Acad Sci USA* 1932;**18**:115–20.
19. Labows JN, CJ W. Individual differences in odor perception. *Perfumer Flavorist* 1984;**9**:24–6.
20. Amoore JE, Venstrom D, Davis AR. Measurement of specific anosmia. *Percept Mot Skills* 1968;**26**:143–64.
21. Griffiths NM, Patterson RL. Human olfactory responses to 5-alpha-androst-16-en-3-one–principal component of boar taint. *J Sci Food Agric* 1970;**21**:4–6.
22. Amoore JE, Pelosi P, Forrester LJ. Specific anosmias to 5α-androst-16-en-3-one and ω-pentadecalactone: the urinous and musky primary odors. *Chemical Senses and Flavour* 1977;**2**:401–25.
23. Steiner JE. The gusto-facial response: Observation on normal and anencephalic newborn infants. In: Bosmas JF, editor. *4th Symposium on Oral sensation and Perception*. Washington, DC: US Government Printing Office; 1973.
24. Steiner J. Facial expressions of the neonate infant indicating the hedonics of food-related chemical stimuli. In: *Taste and development: the genesis of sweet preference*. Washington, DC: U.S. Government Printing Office; 1977. p. 173–88.
25. Steiner JE, Glaser D, Hawilo ME, Berridge KC. Comparative expression of hedonic impact: affective reactions to taste by human infants and other primates. *Neurosci Biobehav Rev* 2001;**25**:53–74.
26. Grill HJ, Norgren R. The taste reactivity test. II. Mimetic responses to gustatory stimuli in chronic thalamic and chronic decerebrate rats. *Brain Res* 1978;**143**:281–97.
27. Schiffman SS, Erickson RP. A psychophysical model for gustatory quality. *Physiology and Behavior* 1971;**7**:617–33.
28. Liener IE. *Toxic constituents of plant foodstuffs*. 2d edit, New York: Academic Press; 1980.
29. Glendinning JI. Is the bitter rejection response always adaptive? *Physiol Behav* 1994;**56**:1217–27.
30. Maehashi K, Matano M, Wang H, Vo LA, Yamamoto Y, Huang L. Bitter peptides activate hTAS2Rs, the human bitter receptors. *Biochem Biophys Res Commun* 2008;**365**:851–5.
31. Mattes RD. Influences on acceptance of bitter foods and beverages. *Physiol Behav* 1994;**56**:1229–36.
32. Fujioka K, Shibamoto T. Chlorogenic acid and caffeine contents. *Food Chemistry* 2008;**106**:217–21.
33. Yeomans MR, Durlach PJ, Tinley EM. Flavour liking and preference conditioned by caffeine in humans. *Q J Exp Psychol B* 2005;**58**:47–58.
34. Leung L, Birtwhistle R, Kotecha J, Hannah S, Cuthbertson S. Anti-diabetic and hypoglycaemic effects of Momordica charantia (bitter melon): a mini review. *Br J Nutr* 2009;**102**:1703–8.
35. Kameswaran L, Gopalakrishnan S, Sukumar M. Phenylthiocarbamide and naringin taste threshold in South Indian medical students. *Ind J Pharmac* 1974;**6**:134–40.
36. Delwiche JF, Buletic Z, Breslin PA. Covariation in individuals' sensitivities to bitter compounds: evidence supporting multiple receptor/transduction mechanisms. *Percept Psychophys* 2001;**63**:761–76.

37. Hansen JL, Reed DR, Wright MJ, Martin NG, Breslin PA. Heritability and genetic covariation of sensitivity to PROP, SOA, quinine HCl, and caffeine. *Chem Senses* 2006;**31**:403–13.
38. Smith SE, Davies PD. Quinine taste thresholds: a family study and a twin study. *Ann Hum Genet* 1973;**37**:227–32.
39. Meyerhof W, Batram C, Kuhn C, Brockhoff A, Chudoba E, Bufe B, et al. The molecular receptive ranges of human TAS2R bitter taste receptors. *Chem Senses* 2009;**35**:157–70.
40. Snyder LH. Inherited taste deficiency. *Science* 1931;**74**:151–2.
41. Kim UK, Jorgenson E, Coon H, Leppert M, Risch N, Drayna D. Positional cloning of the human quantitative trait locus underlying taste sensitivity to phenylthiocarbamide. *Science* 2003;**299**:1221–5.
42. Reed DR, Bartoshuk LM, Duffy V, Marino S, Price RA. PROP tasting: determination of underlying thresholds distributions using maximum likelihood. *Chemical Senses* 1995;**20**:529–33.
43. Drayna D, Coon H, Kim UK, Elsner T, Cromer K, Otterud B, et al. Genetic analysis of a complex trait in the Utah Genetic Reference Project: a major locus for PTC taste ability on chromosome 7q and a secondary locus on chromosome 16p. *Hum Genet* 2003;**112**:567–72.
44. Mennella JA, Duke F, Pepino MY, Reed DR. Age modifies the genotype-phenotype relationship for the bitter receptor TAS2R38. *BMC Genetics* 2010;**11**:60.
45. Guo SW, Reed DR. The genetics of phenylthiocarbamide perception. *Ann Hum Biol* 2001;**28**:111–42.
46. Barnicot N, Harris H, Kalmus H. Taste thresholds of further eighteen compounds and their correlation with PTC thresholds. *Annals of Eugenics* 1951;**16**:119–28.
47. Bufe B, Breslin PA, Kuhn C, Reed DR, Tharp CD, Slack JP, et al. The molecular basis of individual differences in phenylthiocarbamide and propylthiouracil bitterness perception. *Curr Biol* 2005;**15**:322–7.
48. Boyd WC. Taste reactions to antithyroid substances. *Science* 1950;**112**:153.
49. VanEtten C. Goitrogens. In: Liener IE, editor. *Toxic constituents of plant foodstuffs. Food science and technology.* New York: Academic Press; 1969. p. 103–42.
50. Sandell MA, Breslin PA. Variability in a taste-receptor gene determines whether we taste toxins in food. *Curr Biol* 2006;**16**:R792–4.
51. Lumeng JC, Cardinal TM, Sitto JR, Kannan S. Ability to taste 6-n-propylthiouracil and BMI in low-income preschool-aged children. *Obesity (Silver Spring)* 2008;**16**:1522–8.
52. Tepper BJ, Koelliker Y, Zhao L, Ullrich NV, Lanzara C, d'Adamo P, et al. Variation in the bitter-taste receptor gene TAS2R38, and adiposity in a genetically isolated population in Southern Italy. *Obesity (Silver Spring)* 2008;**16**:2289–95.
53. Keller KL, Tepper BJ. Inherited taste sensitivity to 6-n-propylthiouracil in diet and body weight in children. *Obes Res* 2004;**12**:904–12.
54. Keller KL, Reid A, Macdougall MC, Cassano H, Lee Song J, Deng L, et al. Sex differences in the effects of inherited bitter thiourea sensitivity on body weight in 4-6-year-old children. *Obesity (Silver Spring)* 2009;**18**:1194–200.
55. Drewnowski A, Henderson SA, Cockroft JE. Genetic sensitivity to 6-n-propylthiouracil has no influence on dietary patterns, body mass indexes, or plasma lipid profiles of women. *J Am Diet Assoc* 2007;**107**:1340–8.
56. Timpson NJ, Christensen M, Lawlor DA, Gaunt TR, Day IN, Ebrahim S, et al. TAS2R38 (phenylthiocarbamide) haplotypes, coronary heart disease traits, and eating behavior in the British Women's Heart and Health Study. *Am J Clin Nutr* 2005;**81**:1005–11.
57. Frayling TM, Timpson NJ, Weedon MN, Zeggini E, Freathy RM, Lindgren CM, et al. A common variant in the FTO gene is associated with body mass index and predisposes to childhood and adult obesity. *Science* 2007;**316**:889–94.

58. Willer CJ, Speliotes EK, Loos RJ, Li S, Lindgren CM, Heid IM, et al. Six new loci associated with body mass index highlight a neuronal influence on body weight regulation. *Nat Genet* 2009;**41**:25–34.
59. Dyer J, Salmon KS, Zibrik L, Shirazi-Beechey SP. Expression of sweet taste receptors of the T1R family in the intestinal tract and enteroendocrine cells. *Biochem Soc Trans* 2005;**33**:302–5.
60. Chen MC, Wu SV, Reeve Jr. JR, Rozengurt E. Bitter stimuli induce Ca2+ signaling and CCK release in enteroendocrine STC-1 cells: role of L-type voltage-sensitive Ca2+ channels. *Am J Physiol Cell Physiol* 2006;**291**:C726–39.
61. Rozengurt E. Taste receptors in the gastrointestinal tract. I. Bitter taste receptors and alpha-gustducin in the mammalian gut. *Am J Physiol Gastrointest Liver Physiol* 2006;**291**:G171–7.
62. Shah AS, Ben-Shahar Y, Moninger TO, Kline JN, Welsh MJ. Motile cilia of human airway epithelia are chemosensory. *Science* 2009;**325**:1131–4.
63. Jeon TI, Zhu B, Larson JL, Osborne TF. SREBP-2 regulates gut peptide secretion through intestinal bitter taste receptor signaling in mice. *J Clin Invest* 2008;**118**:3693–700.
64. Henkin RI, Gillis WT. Divergent taste responsiveness to fruit of the tree Antidesma bunius. *Nature* 1977;**265**:536–7.
65. Chandrashekar J, Mueller KL, Hoon MA, Adler E, Feng L, Guo W, et al. T2Rs function as bitter taste receptors. *Cell* 2000;**100**:703–11.
66. Behrens M, Brockhoff A, Kuhn C, Bufe B, Winnig M, Meyerhof W. The human taste receptor hTAS2R14 responds to a variety of different bitter compounds. *Biochem Biophys Res Commun* 2004;**319**:479–85.
67. Bufe B, Hofmann T, Krautwurst D, Raguse JD, Meyerhof W. The human TAS2R16 receptor mediates bitter taste in response to beta-glucopyranosides. *Nat Genet* 2002;**32**:397–401.
68. Conte C, Guarin E, Marcuz A, Andres-Barquin PJ. Functional expression of mammalian bitter taste receptors in Caenorhabditis elegans. *Biochimie* 2006;**88**:801–6.
69. Kuhn C, Bufe B, Winnig M, Hofmann T, Frank O, Behrens M, et al. Bitter taste receptors for saccharin and acesulfame K. *J Neurosci* 2004;**24**:10260–5.
70. Behrens M, Brockhoff A, Batram C, Kuhn C, Appendino G, Meyerhof W. The human bitter taste receptor hTAS2R50 is activated by the two natural bitter terpenoids andrographolide and amarogentin. *J Agric Food Chem* 2009;**57**:9860–6.
71. Sainz E, Cavenagh MM, Gutierrez J, Battey JF, Northup JK, Sullivan SL. Functional characterization of human bitter taste receptors. *Biochem J* 2007;**403**:537–43.
72. Pronin AN, Xu H, Tang H, Zhang L, Li Q, Li X. Specific alleles of bitter receptor genes influence human sensitivity to the bitterness of aloin and saccharin. *Curr Biol* 2007;**17**:1403–8.
73. Pronin AN, Tang H, Connor J, Keung W. Identification of ligands for two human bitter T2R receptors. *Chem Senses* 2004;**29**:583–93.
74. Liem DG, Mennella JA. Heightened sour preferences during childhood. *Chem Senses* 2003;**28**:173–80.
75. Wise PM, Hansen JL, Reed DR, Breslin PA. Twin study of the heritability of recognition thresholds for sour and salty taste. *Chem Senses* 2007;**32**:749–54.
76. Kaplan AR, Fischer R, Karras A, Griffin F, Powell W, Marsters RW, et al. Taste thresholds in twins and siblings. *Acta Genet Med Gemellol (Roma)* 1967;**16**:229–43.
77. Bachmanov AA, Reed DR, Tordoff MG, Price RA, Beauchamp GK. Intake of ethanol, sodium chloride, sucrose, citric acid, and quinine hydrochloride solutions by mice: a genetic analysis. *Behavior Genetics* 1996;**26**:563–73.
78. Huang AL, Chen X, Hoon MA, Chandrashekar J, Guo W, Trankner D, et al. The cells and logic for mammalian sour taste detection. *Nature* 2006;**442**:934–8.

79. Huque T, Cowart BJ, Dankulich-Nagrudny L, Pribitkin EA, Bayley DL, Spielman AI, et al. Sour ageusia in two individuals implicates ion channels of the ASIC and PKD families in human sour taste perception at the anterior tongue. *PLoS One* 2009;**4**:e7347.
80. Reed DR, Bachmanov AA, Beauchamp GK, Tordoff MG, Price RA. Heritable variation in food preferences and their contribution to obesity. *Behav Genet* 1997;**27**:373–87.
81. Reed DR, Li X, Bachmanov AA, Mascioli K, Beauchamp GK. The molecular basis of the mammalian sweet tooth. In: Medeiros-Neto G, Halpern A, Bouchard C, editors. *Progress in Obesity Research. Vol.***9**:London; 2003. p. 304–6.
82. McDaniel AH, Reed DR. The human sweet tooth and its relationship to obesity. In: Berndanier CD, Moustaid-Moussa N, editors. *Genomics and Proteomics in Nutrition.* New York: Marcel Dekker, Inc; 2004.
83. Reed DR, McDaniel AH. The human sweet tooth. *BMC Oral Health* 2006;**6**(**Suppl 1**):S17.
84. Reed DR, Tanaka T, McDaniel AH. Diverse tastes: Genetics of sweet and bitter perception. *Physiol Behav* 2006;**88**:215–26.
85. Thompson DA, Moskowitz HR, Campbell R. Effects of body weight and food intake on pleasantness ratings for a sweet stimulus. *Journal of Applied Psychology* 1976;**41**:77–83.
86. Thompson DA, Moskowitz HR, Campbell RG. Taste and olfaction in human obesity. *Physiol Behav* 1977;**19**:335–7.
87. Conner MT, Booth DA. Preferred sweetness of a lime drink and preference for sweet over non-sweet foods, related to sex and reported age and body weight. *Appetite* 1988;**10**:25–35.
88. Reed DR, Li S, Li X, Huang L, Tordoff MG, Starling-Roney R, et al. Polymorphisms in the taste receptor gene (*Tas1r3*) region are associated with saccharin preference in 30 mouse strains. *J Neurosci* 2004;**24**:938–46.
89. Fushan AA, Simons CT, Slack JP, Manichaikul A, Drayna D. Allelic polymorphism within the TAS1R3 promoter is associated with human taste sensitivity to sucrose. *Curr Biol* 2009;**19**:1288–93.
90. Collaku A, Rankinen T, Rice T, Leon AS, Rao DC, Skinner JS, et al. A genome-wide linkage scan for dietary energy and nutrient intakes: the Health, Risk Factors, Exercise Training, and Genetics (HERITAGE) Family Study. *Am J Clin Nutr* 2004;**79**:881–6.
91. Cai G, Cole SA, Bastarrachea RA, Maccluer JW, Blangero J, Comuzzie AG. Quantitative trait locus determining dietary macronutrient intakes is located on human chromosome 2p22. *Am J Clin Nutr* 2004;**80**:1410–1.
92. Keskitalo K, Knaapila A, Kallela M, Palotie A, Wessman M, Sammalisto S, et al. Sweet taste preferences are partly genetically determined: identification of a trait locus on chromosome 16. *Am J Clin Nutr* 2007;**86**:55–63.
93. Bachmanov AA, Beauchamp GK. Taste receptor genes. *Annu Rev Nutr* 2007;**27**:389–414.
94. Nelson G, Chandrashekar J, Hoon MA, Feng L, Zhao G, Ryba NJ, et al. An amino-acid taste receptor. *Nature* 2002;**416**:199–202.
95. Zhao GQ, Zhang Y, Hoon MA, Chandrashekar J, Erlenbach I, Ryba NJ, et al. The receptors for mammalian sweet and umami taste. *Cell* 2003;**115**:255–66.
96. Zhang Y, Hoon MA, Chandrashekar J, Mueller KL, Cook B, Wu D, et al. Coding of sweet, bitter, and umami tastes: different receptor cells sharing similar signalling pathways. *Cell* 2003;**112**:293–301.
97. Lugaz O, Pillias AM, Faurion A. A new specific ageusia: some humans cannot taste L-glutamate. *Chem Senses* 2002;**27**:105–15.
98. Chen QY, Alarcon S, Tharp A, Ahmed OM, Estrella NL, Greene TA, et al. Perceptual variation in umami taste and polymorphisms in TAS1R taste receptor genes. *Am J Clin Nutr* 2009;**90**:770S–779S.
99. Shigemura N, Shirosaki S, Ohkuri T, Sanematsu K, Islam AS, Ogiwara Y, et al. Variation in umami perception and in candidate genes for the umami receptor in mice and humans. *Am J Clin Nutr* 2009;**90**:764S–769S.

100. Shigemura N, Shirosaki S, Sanematsu K, Yoshida R, Ninomiya Y. Genetic and molecular basis of individual differences in human umami taste perception. *PLoS One* 2009;**4**:e6717.
101. Li R, Fan W, Tian G, Zhu H, He L, Cai J, et al. The sequence and de novo assembly of the giant panda genome. *Nature* 2009;**463**:311–7.
102. Pepino MY, Finkbeiner S, Beauchamp GK, Mennella JA. Obese women have lower monosodium glutamate taste sensitivity and prefer higher concentrations than do normal-weight women. *Obesity (Silver Spring)* 2010;**18**:959–65.
103. Denton D. *The hunger for salt: An anthropological, physiological and medical analysis*. Berlin: Springer-Verlag; 1984.
104. Appel LJ, Anderson CA. Compelling evidence for public health action to reduce salt intake. *N Engl J Med* 2010;**362**:650–2.
105. Anonymous. In:Medicine Io, editor. *Strategies to reduce salt intake*. Consensus study; 2010.
106. Odeigah PG, Obieze AC. Differences in sodium chloride taste sensitivity in a rural and an urban population in Nigeria: implications for the incidence of hypertension. *East Afr Med J* 1986;**63**:236–43.
107. Hladik CM. Primate models for taste and food preferences. In:Macbeth H, editor. *Food preferences and taste. Community and change*. Berghahn Books, Providence; 1997.
108. Beauchamp GK, Bertino M, Engelman K. Sensory basis for human salt consumption. In: Horan MJ, Blaustein MP, Dunbar JB, Kachadorian W, Kaplan NM, Simopoulos AP, editors. *NIH Workshop on Nutrition and Hypertension. Proceedings from a Symposium*. New York: Biomedical Information Corporation; 1985.
109. Greene LS, Desor JA, Maller O. Heredity and experience: their relative importance in the development of taste preference in man. *Journal of Comparative and Physiological Psychology* 1975;**89**:279–84.
110. Beauchamp GK. The human preference for excess salt. *American Scientist* 1987;27–33.
111. Irvin DL, Goetzl FR. Diurnal variations in acuity of sense of taste for sodium chloride. *Proc Soc Exp Biol Med* 1952;**79**:115–8.
112. Ayya N, Beauchamp GK. Short-term effects of diet on salt taste preferences. *Appetite* 1992;**18**:77–82.
113. Jackson FL. An evolutionary perspective on salt, hypertension, and human genetic variability. *Hypertension* 1991;**17**:I129–32.
114. Chandrashekar J, Kuhn C, Oka Y, Yarmolinsky DA, Hummler E, Ryba NJ, et al. The cells and peripheral representation of sodium taste in mice. *Nature* 2010;**464**:297–301.
115. Stahler F, Riedel K, Demgensky S, Neumann K, Dunkel A, Taubert A, et al. A role of the epithelial sodium channel in human salt taste transduction? *Chemosensory Perception* 2008;**1**:78–90.
116. Tordoff MG, Shao H, Alarcon LK, Margolskee RF, Mosinger B, Bachmanov AA, et al. Involvement of T1R3 in calcium-magnesium taste. *Physiol Genomics* 2008;**34**:338–48.
117. Ohsu T, Amino Y, Nagasaki H, Yamanaka T, Takeshita S, Hatanaka T, et al. Involvement of the calcium-sensing receptor in human taste perception. *J Biol Chem* 2010;**285**:1016–22.
118. San Gabriel A, Uneyama H, Maekawa T, Torii K. The calcium-sensing receptor in taste tissue. *Biochem Biophys Res Commun* 2009;**378**:414–8.
119. Tordoff MG. Some basic psychophysics of calcium salt solutions. *Chemical Senses* 1996;**21**:417–24.
120. Tordoff MG, Reed DR, Shao H. Calcium taste preferences: genetic analysis and genome screen of C57BL/6J x PWK/PhJ hybrid mice. *Genes Brain Behav* 2008;**7**:618–28.
121. Fernelius I. *Therapeutices universalis seu medendi rationis, libri septem*. Vol.IV, Frankfurt: Andream Wechelum; 1581.
122. Montmayeur J-P, Le Coutre J. Fat detection : taste, texture, and post ingestive effects, CRC Press/Taylor & Francis, Boca Raton; 2009.

123. Reed DR. Heritable variation in fat preference. In:Montmayeur JP, de Coutre J, editors. *Fat detection: taste, texture, and post-investive effects*. Boca Raton: Taylor & Francis; 2009.
124. Laugerette F, Passilly-Degrace P, Patris B, Niot I, Febbraio M, Montmayeur JP, et al. CD36 involvement in orosensory detection of dietary lipids, spontaneous fat preference, and digestive secretions. *J Clin Invest* 2005;**115**:3177–84.
125. Sclafani A, Ackroff K, Abumrad NA. CD36 gene deletion reduces fat preference and intake but not post-oral fat conditioning in mice. *Am J Physiol Regul Integr Comp Physiol* 2007;**293**: R1823–32.
126. Tsuzuki S. *5th International Symposium on Molecular and Neural Mechanisms of Taste and Olfactory Perception, Fukuoka* 2007.
127. El-Yassimi A, Hichami A, Besnard P, Khan NA. Linoleic acid induces calcium signaling, Src kinase phosphorylation, and neurotransmitter release in mouse CD36-positive gustatory cells. *J Biol Chem* 2008;**283**:12949–59.
128. Matsumura S, Eguchi A, Mizushige T, Kitabayashi N, Tsuzuki S, Inoue K, et al. Colocalization of GPR120 with phospholipase-Cbeta2 and alpha-gustducin in the taste bud cells in mice. *Neurosci Lett* 2009;**450**:186–90.
129. Chandrashekar J, Yarmolinsky D, von Buchholtz L, Oka Y, Sly W, Ryba NJ, et al. The taste of carbonation. *Science* 2009;**326**:443–5.
130. Bautista DM, Siemens J, Glazer JM, Tsuruda PR, Basbaum AI, Stucky CL, et al. The menthol receptor TRPM8 is the principal detector of environmental cold. *Nature* 2007;**448**:204–8.
131. Caterina MJ, Schumacher MA, Tominaga M, Rosen TA, Levine JD, Julius D. The capsaicin receptor: a heat-activated ion channel in the pain pathway. *Nature* 1997;**389**:816–24.
132. Wysocki CJ, Wise PM. Methods, approaches, and caveats for functionally evaluating olfaction and chemesthesis. In:Deibler K, Delwiche JF, editors. *Handbook of Flavor Characterization: Sensory Chemical and Physiological*. New York: Marcel-Dekker; 2003.
133. Faust J. A twin study of personal preferences. *J Biosoc Sci* 1974;**6**:1975–91.
134. Jordt SE, Julius D. Molecular basis for species-specific sensitivity to "hot" chili peppers. *Cell* 2002;**108**:421–30.
135. Jordt SE, McKemy DD, Julius D. Lessons from peppers and peppermint: the molecular logic of thermosensation. *Curr Opin Neurobiol* 2003;**13**:487–92.
136. Frommer WB. Biochemistry. CO2mmon sense. *Science* 2010;**327**:275–6.
137. Tordoff MG, Alleva AM. Effect of drinking soda sweetened with aspartame or high-fructose corn syrup on food intake and body weight. *American Journal of Clinical Nutrition* 1990;**51**:963–9.
138. Cowart BJ, Young IM, Feldman RS, Lowry LD. Clinical disorders of smell and taste. *Occup Med* 1997;**12**:465–83.
139. Hummel T, Nordin S. Olfactory disorders and their consequences for quality of life. *Acta Otolaryngol* 2005;**125**:116–21.
140. Malnic B, Hirono J, Sato T, Buck LB. Combinatorial receptor codes for odors. *Cell* 1999;**96**:713–23.
141. Friedman L, Miller JG. Odor incongruity and chirality. *Science* 1971;**172**:1044–6.
142. Russell GF, Hills JI. Odor differences between enantiomeric isomers. *Science* 1971;**172**:1043–4.
143. Triller A, Boulden EA, Churchill A, Hatt H, Englund J, Spehr M, et al. Odorant-receptor interactions and odor percept: a chemical perspective. *Chem Biodivers* 2008;**5**:862–86.
144. Saito H, Chi Q, Zhuang H, Matsunami H, Mainland JD. Odor coding by a Mammalian receptor repertoire. *Sci Signal* 2009;**2**:ra9.
145. Reisert J, Restrepo D. Molecular tuning of odorant receptors and its implication for odor signal processing. *Chem Senses* 2009;**34**:535–45.

146. Krautwurst D. Human olfactory receptor families and their odorants. *Chem Biodivers* 2008;**5**:842–52.
147. Niimura Y, Nei M. Evolution of olfactory receptor genes in the human genome. *Proc Natl Acad Sci USA* 2003;**100**:12235–40.
148. Hubert HB, Fabsitz RR, Feinleib M, Brown KS. Olfactory sensitivity in humans: genetic versus environmental control. *Science* 1980;**208**:607–9.
149. Knaapila A, Tuorila H, Silventoinen K, Wright MJ, Kyvik KO, Keskitalo K, et al. Environmental effects exceed genetic effects on perceived intensity and pleasantness of several odors: a three-population twin study. *Behav Genet* 2008;**38**:484–92.
150. Knaapila A, Keskitalo K, Kallela M, Wessman M, Sammalisto S, Hiekkalinna T, et al. Genetic component of identification, intensity and pleasantness of odours: a Finnish family study. *Eur J Hum Genet* 2007;**15**:596–602.
151. Wysocki CJ, Beauchamp GK. Ability to smell androstenone is genetically determined. *Proc Natl Acad Sci USA* 1984;**81**:4899–902.
152. Knaapila A, Tuorila H, Silventoinen K, Wright MJ, Kyvik KO, Cherkas LF, et al. Genetic and environmental contributions to perceived intensity and pleasantness of androstenone odor: an international twin study. *Chemosensory Perception* 2008;**1**:34–42.
153. Keller A, Zhuang H, Chi Q, Vosshall LB, Matsunami H. Genetic variation in a human odorant receptor alters odour perception. *Nature* 2007;**449**:468–72.
154. Menashe I, Abaffy T, Hasin Y, Goshen S, Yahalom V, Luetje CW, et al. Genetic elucidation of human hyperosmia to isovaleric acid. *PLoS Biol* 2007;**5**:e284.
155. Eriksson N, Macpherson JM, Tung J, Hon L, Naughton B, Saxonov S, et al. Web-based, participant-driven studies yield novel genetic asociations for common traits. *PLoS Genet* 2010;**6**:e1000993.
156. Jaeger SR, McRae JF, Salzman Y, Williams L, Newcomb RD. A preliminary investigation into a genetic basis for cis-3-hexen-1-ol odour perception: A genome-wide association approach. *Food Quality and Preference* 2010;**21**:121–31.
157. Wysocki CJ, Dorries KM, Beauchamp GK. Ability to perceive androstenone can be acquired by ostensibly anosmic people. *Proc Natl Acad Sci USA* 1989;**86**:7976–8.
158. Dalton P, Doolittle N, Breslin PA. Gender-specific induction of enhanced sensitivity to odors. *Nat Neurosci* 2002;**5**:199–200.
159. Schmidt HJ, Beauchamp GK. Adult-like odor preferences and aversions in three-year-old children. *Child Dev* 1988;**59**:1136–43.
160. de Araujo IE, Rolls ET, Velazco MI, Margot C, Cayeux I. Cognitive modulation of olfactory processing. *Neuron* 2005;**46**:671–9.
161. Pleis J, Lucas J, Ward B. In:Stat VH, editor. *Summary health statistics for U.S. adults: National Health Interview Survey: 2008*. Hyattsville: 10, National Center for Health Statistics; 2009.
162. Bachmanov AA, Kiefer SW, Molina JC, Tordoff MG, Duffy VB, Bartoshuk , et al. Chemosensory factors influencing alcohol perception, preferences, and consumption. *Alcohol Clin Exp Res* 2003;**27**:220–31.
163. Brasser SM, Norman MB, Lemon CH. T1r3 taste receptor involvement in gustatory neural responses to ethanol and oral ethanol preference. Physiol Genomics **Epub ahead of print**.
164. Dudley R. Evolutionary origins of human alcoholism in primate frugivory. *Q Rev Biol* 2000;**75**:3–15.
165. Duffy VB, Davidson AC, Kidd JR, Kidd KK, Speed WC, Pakstis AJ, et al. Bitter receptor gene (TAS2R38), 6-n-propylthiouracil (PROP) bitterness and alcohol intake. *Alcohol Clin Exp Res* 2004;**28**:1629–37.
166. Blednov YA, Harris RA. Deletion of vanilloid receptor (TRPV1) in mice alters behavioral effects of ethanol. *Neuropharmacology* 2009;**56**:814–20.

167. Monsivais P, Drewnowski A. Lower-energy-density diets are associated with higher monetary costs per kilocalorie and are consumed by women of higher socioeconomic status. *J Am Diet Assoc* 2009;**109**:814–22.
168. Salvy SJ, Vartanian LR, Coelho JS, Jarrin D, Pliner PP. The role of familiarity on modeling of eating and food consumption in children. *Appetite* 2008;**50**:514–8.
169. Wansink B, Painter JE, Lee YK. The office candy dish: proximity's influence on estimated and actual consumption. *Int J Obes (Lond)* 2006;**30**:871–5.
170. Raliou M, Wiencis A, Pillias AM, Planchais A, Eloit C, Boucher Y, et al. Nonsynonymous single nucleotide polymorphisms in human tas1r1, tas1r3, and mGluR1 and individual taste sensitivity to glutamate. *Am J Clin Nutr* 2009;**90**:789S–799S.
171. Mennella JA, Beauchamp GK. Optimizing oral medications for children. *Clin Ther* 2008;**30**:2120–32.

Genetic Findings in Anorexia and Bulimia Nervosa

Anke Hinney, Susann Scherag, and Johannes Hebebrand

Department of Child and Adolescent Psychiatry and Psychotherapy, University of Duisburg-Essen, Essen, Germany

Anorexia nervosa (AN) and bulimia nervosa (BN) are complex disorders associated with disordered eating behavior. Heritability estimates derived from twin and family studies are high, so that substantial genetic influences on the etiology can be assumed for both. As the monoaminergic neurotransmitter systems are involved in eating disorders (EDs), candidate gene studies have centered on related genes; additionally, genes relevant for body weight regulation have been considered as candidates. Unfortunately, this approach has yielded very few positive results; confirmed associations or findings substantiated in meta-analyses are scant. None of these associations can be considered unequivocally validated. Systematic genome-wide approaches have been performed to identify genes with no *a priori* evidence for their relevance in EDs. Family-based scans revealed linkage peaks in single chromosomal regions for AN and BN. Analyses of candidate genes in one of these regions led to the identification of genetic variants associated with AN. Currently, an international consortium is conducting a genome-wide association study for AN, which will hopefully lead to the identification of the first genome-wide significant markers.

Progress in Molecular Biology
and Translational Science, Vol. 94
DOI: 10.1016/S1877-1173(10)94009-X

I. Introduction

The Diagnostic and Statistical Manual of Mental Disorders (DSM-IV)[1] and the International Statistical Classification of Diseases and Related Health Problems (ICD-10)[2] are commonly used to define the eating disorders (EDs) anorexia nervosa (AN; Fig. 1), bulimia nervosa (BN; Fig. 2), and EDs not otherwise specified (EDNOS; Fig. 3). The preliminary criteria of the DSM-V working group for EDs have recently been published (www.dsm5.org). Because work on these criteria is still in progress, we refer to the DSM-IV criteria throughout this article. Controversial diagnostic issues have been summarized recently (e.g., Refs. 3–6). Binge-eating disorder (BED), for which DSM-IV provided research criteria, is to become a distinct ED in DSM-V. Currently, this disorder is diagnosed as an EDNOS.

Our review focuses on AN and BN. Both clinical and epidemiological studies demonstrate frequent crossover from AN to BN (8–54%) and from BN to AN (4–27%), typically within the first 5 years of illness.[7–19] Whereas these findings suggest an overlapping of the etiological predisposition, AN and BN in genetic terms have mostly been treated as distinct disorders; both twin and other family studies based on index patients with one or the other disorder have frequently based their heritability estimates on narrowly and broadly defined phenotypes. Broad phenotypes typically include AN, BN, and EDNOS.

A complex etiology encompassing environmental and genetic factors is common to EDs. Indeed, heritability estimates suggest a substantial genetic contribution to the etiology of both AN and BN.[20–29] Candidate gene studies and genome-wide approaches have been performed; however, up to now, none of these analyses has led to unequivocal evidence for the underlying biological mechanisms.

II. Twin and Other Family Studies

Family studies are used to analyze the role of environmental and genetic contributions to a phenotype. Twin studies compare the phenotypic similarity of monozygotic or identical twins, who share nearly 100% of their genetic makeup, to that of dizygotic or fraternal twins, who on average share 50% of their genetic variants. The analysis of twin pairs allows evaluation of the genetic (or heritable) part and the shared and unique (nonshared) environmental contributions to the variance of the analyzed phenotype or disease.[20] Heritability estimates are useful in quantifying the genetic component of a complex disease. Their usefulness depends on the validity of the assumptions on which they are based.[20,24,25,30] Caveats are mainly statistical power and ascertainment bias.[20] If a clinical sample is used for the twin study, the severity of the disorder

A Refusal to maintain body weight at or above a minimally normal weight for age and height (e.g., weight loss leading to maintenance of body weight less than 85% of that expected; or failure to make expected weight gain during period of growth, leading to body weight less than 85% of that expected).

B Intense fear of gaining weight or becoming fat, even though underweight.

C Disturbance in the way one's body weight or shape is experienced, undue influence of body weight or shape on self-evaluation, or denial of the seriousness of the current low body weight.

D In postmenarcheal females, amenorrhea, i.e., the absence of at least three consecutive menstrual cycles. (A woman is considered to have amenorrhea if her periods occur only following hormone, e.g., estrogen, administration).

Specify type

Restricting Type: during the current episode of Anorexia Nervosa, the person has not regularly engaged in binge-eating or purging behavior (i.e., self-induced vomiting or the misuse of laxatives, diuretics, or enemas).

Binge Eating/Purging Type: during the current episode of Anorexia Nervosa, the person has regularly engaged in binge-eating or purging behavior (i.e., self-induced vomiting or the misuse of laxatives, diuretics, or enemas).

FIG. 1. DSM-IV[1] criteria for anorexia nervosa.

is usually greater. Therefore, the genetic loading might be different from that of patients ascertained from the general population, and the heritability might be overestimated.[27]

Twin studies based on patients with a specific ED or on epidemiological samples[28] showed heritability estimates between 48% and 88% for AN and 28% and 83% for BN.[20,23–26,31–35] There is little evidence for a strong role of shared environmental factors so that the remaining liability stems from the individual-specific (nonshared) environment. Epidemiological twin studies are less prone to ascertainment bias. However, Walters and Kendler[36] were not able to detect a genetic basis of AN in their population-based female twin sample ($n = 2163$); only a small number of twins had retrospectively diagnosed AN, thus revealing a major drawback of epidemiological studies. Concordance rates were higher for dizygotic than monozygotic twins. The larger Swedish Twin Study of Adults ($n = 7000$) provided evidence for the contribution of additive genetic factors to AN and BN. A moderate overlap of both genetic and unique environmental factors that influence the two conditions was demonstrated. Common concurrent and sequential comorbidity of AN and BN can in part be accounted for by shared genetic and environmental influences on liability, although independent factors are also operative.[37]

A Recurrent episodes of binge eating. An episode is characterized by both of the following:

(1) eating, in a discrete period of time (e.g., within any 2-hour period), an amount of food that is definitely larger than most people would eat during a similar period of time and under similar circumstances.

(2) a sense of lack of control over eating during the episode (e.g., a feeling that one cannot stop eating or control what or how much one is eating).

B Recurrent inappropriate compensatory behavior in order to prevent weight gain, such as self-induced vomiting; misuse of laxatives, diuretics, enemas, or other medications; fasting; or excessive exercise.

C The binge eating and inappropriate compensatory behavior both occur, on average, at least twice a week for 3 months.

D Self evaluation is unduly influenced by body shape and weight.

E The disturbance does not occur exclusively during episodes of Anorexia Nervosa.

Specify type

Purging Type: during the current episode of Bulimia Nervosa, the person has regularly engaged in self-induced vomiting or the misuse of laxatives, diuretics, or enemas.

Nonpurging Type: during the current episode of Bulimia Nervosa, the person has used other inappropriate compensatory behavior, such as fasting or excessive exercise, but has not regularly engaged in self-induced vomiting or the misuse of laxatives, diuretics, or enemas.

FIG. 2. DSM-IV[1] criteria for bulimia nervosa.

Family studies were performed on patients with an ED and their family members. Some of these studies had a case–control design (controlled family study); hence the "case families" were recruited via a patient with AN or BN, and the "control families" were ascertained via a proband without an ED.[38]

A large controlled family study (1831 relatives of 504 probands: 152 AN, 171 BN, 181 controls) showed that first-degree female relatives of patients with AN or BN have at least a 10-fold increased risk (relative risks of 11.3 and 12.3, respectively) of developing the full syndrome of AN compared to individuals from the general population.[38] The risk of developing BN was approximately fourfold higher for first-degree female relatives of patients with AN or BN (relative risks of 4.2 and 4.4, respectively).[38] Full and partial syndromes of AN and BN aggregated in female relatives of either AN or BN patients. Hence, it was suggested that specific genes can predispose to both EDs.[38]

1 For females, all diagnostic criteria for Anorexia Nervosa are met except that the individual has regular menses.

2 All of the criteria for Anorexia Nervosa are met except that, despite significant weight loss, the individual's current weight is in the normal range.

3 All of the criteria for Bulimia Nervosa are met except that the binge eating and inappropriate compensatory mechanisms occur at a frequency of less than twice a week or for a duration of less than 3 months.

4 The regular use of inappropriate compensatory behavior by an individual of normal body weight after eating small amounts of food (e.g., self-indiced vomiting after consumption of two cookies).

5 Repeatedly chewing and spitting out, but not swallowing, large amounts of food.

6 Binge-eating disorder: recurrent episodes of binge eating in the absence of the regular use of inappropriate compensatory behaviors characteristic of Bulimia Nervosa.

FIG. 3. DSM-IV[1] criteria for eating disorder not otherwise specified (EDNOS).

III. Molecular Genetic Approaches

Basically, two approaches are used for the molecular genetic elucidation of a complex phenotype: (i) candidate gene studies based on *a priori* available physiological, biochemical, pharmacological, or genetic evidence (see Table I) and (ii) genome-wide approaches without *a priori* hypotheses. The latter include family-based linkage studies and more recently genome-wide association studies (GWAS).

A. Statistical Considerations: Association Studies

Association studies measure the variation in allele or genotype frequencies between patients (cases) and controls. Cases and controls have to be matched carefully, ethnicity being particularly important. If the analyzed phenotype is not the only difference between the groups, a positive association could be due to a confounder (false-positive association). For both AN and BN, the full phenotype, subtype(s), endophenotypes, age at onset, minimal body mass index (BMI), and severity of the disease have been analyzed. However, if too many parameters are used to subdifferentiate EDs, the number of individuals in each group is reduced substantially, entailing a decreased power. Furthermore, multiple comparisons are required; thus appropriate statistical correction needs to be performed, again implying a reduction in power.[39]

TABLE I

ASSOCIATION STUDIES OF CANDIDATE GENES (ALPHABETICAL ORDER) FOR ANOREXIA NERVOSA AND BULIMIA NERVOSA

			Genetic variant		Evidence for association to	
Area[a]	Gene	Chromosome	rs-Number	SNP or variation[b] (amino acid change)	AN	BN
iii	β3-adrenoceptor (*ADRB3*)	8p11.23	rs4994	190T/C (Trp64Arg)	No[157]	n.d.
	Agouti-related protein (*AGRP*)	16q22.1	n.a.	Epistasis between 4-bp ins/del in AAGG4 repeat island and 5-HTTLPR	No[158]	n.d.
			n.a.	526G/A in linkage disequilibrium with Ala67Thr	Yes[76,105]	n.d.
			n.a.	605C/T	No[104]	n.d.
	Brain-derived neurotrophic factor (*BDNF*)	11p14.1	rs7934165	−270C/T	Yes[126,127,131,133] No[105,130]	Yes[133] No[130]
			rs6265	196G/A (Val66Met)	Yes[126,127,131,133,159] No[105,130]	Yes[160] No[130]
v	Calcium-activated potassium channel (*KCNN3*)	1q21.3	n.a.	Coding CAG repeat (Gln insertion)	Yes[161]	n.d.
ii, iii	Cannabinoid receptor 1 (*CNR1*)	6q15	rs1049353	1359G/A	Yes[92,c] No[91]	Yes[92]
			rs2180619	126G/A	No[91]	n.d.
			rs806379	615A/T		
			rs1535255	1080T/G		
			rs2023239	−3163A/G		
			n.a.	AAT trinucleotide repeat	Yes[162]	n.d.

i	Catechol-*O*-methyltransferase (*COMT*)	22q11.21	rs4633	186C/T (His102His)	Yes[163]	Yes[164]
			rs4680	472G/A (Val158Met)	Yes[23,66,80,165]	Yes[66]
					No[166]	No[80]
	Dopamine D2 receptor (*DRD2*)	11q23.2	rs1800497	2137G/A TaqA1	Yes[83]	Yes[83]
			n.a.	− 141 Indel	Yes[140]	n.d.
	Dopamine D3 receptor (*DRD3*)	3q13.31	n.a.	25A/G (Ser9Gly)	No[70]	n.d.
	Dopamine D4 receptor (*DRD4*)	11p15.5	n.a.	13 bp deletion	No[85]	n.d.
iv	Estrogen receptor β (*ESR2*)	14q23.2	n.a.	1082G/A	Yes[165,167]	Yes[165,167]
			n.a.	1730A/G	Yes[165,167]	Yes[165,167]
ii, iii	Fatty acid amide hydrolase (*FAAH*)	1p33	rs324420	385C/A (Pro129Thr)	Yes[92,c]	Yes[92]
					No[91]	
			rs932816	4811A/G	No[91]	n.d.
			rs324419	897C/T		
			rs873978	18965C/T		
			rs2295632	24624T/G		
iii	Leptin (*LEP*)	7q32.1	n.a.	− 1387G/A	No[102]	No[168]
iii	Melanocortin-4 receptor (*MC4R*)	18q21.32	rs2229616	307G/A (Val103Ile)	No[86]	No[86]
			rs52820871	751A/C (Ile251Thr)	No[86]	No[86]
ii, iii	Monoglyceride lipase (*MGLL*)	3q21.3	rs893294	1661/A	No[91]	n.d.
	N-Acylethanolamine-hydrolyzing acid amidase (*NAAA*)	4q21.1	rs2292534	349G/A	No[91]	n.d.
			rs4859567	727A/T		
			rs10518142	478C/A		
			rs6819442	884G/A		
v	Norepinephrine transporter (*NET, SLC6A2*)	16q12.2	n.a.	4-bp ins/del in AAGG4 repeat island	Yes[169]	n.d.
	Nerve growth factor (*NGF*)	1p13.2	rs17033706	8234C/T	Yes[132,c]	Yes[132,a]

(*Continues*)

TABLE I (*Continued*)

			Genetic variant		Evidence for association to	
Area[a]	Gene	Chromosome	rs-Number	SNP or variation[b] (amino acid change)	AN	BN
iii	Neuropeptide Y receptor Y1 (*NPY1R*)	4q32.2	n.a.	*Pst I*-polymorphism in first intron	No[170]	n.d.
	Neuropeptide Y receptor Y5 (*NPY5R*)	4q32.2	n.a.	1333G/A	No[170]	n.d.
	Neurotrophic tyrosine kinase receptor type 3 (*NTRK3*)	15q25.3	rs7180942	1696A/G	Yes[132,171,c]	Yes[132]
	Proopiomelanocortin (*POMC*)	2p23.3	n.a.	9 bp insertion between codons 73 and 74	No[172]	n.d.
i	Serotonin 5-HT$_{1B/1D\beta}$ receptor (*HTR1B*)	6q14.1	rs6296	861G/C	Yes[87]	n.d.
			rs130060	371T/C (Phe124Cys)	No[173]	n.d.
	Serotonin 5-HT$_{2A}$ receptor (*HTR2A*)	13q14.2	n.a.	− 1438 G/A	Yes[29,65–69,174–177]	Yes[73,176,177]
			rs1805055	74C/A (Thr25Asn)	No[23,72–75,178,179]	No[66,67,75,174]
			rs6305	102T/C	No[67,77]	No[67]
			n.a.	516C/T	No[67,77]	n.d.
			rs6314	1354C/T (His452Tyr)	No[67,77]	No[67]
	Serotonin 5-HT$_{2C}$ receptor (*HTR2C*)	Xq23	rs6318	68G/C (Cys23Ser)	No[180]	No[67,180]
	Serotonin 5-HTR$_{3B}$ receptor (*HTR3B*)	11q23.2	n.a.	Haplotype exon III VNTR	Yes[74] No[85,86]	Yes[88]
			n.a.	120 bp repeat	Yes[74]	n.d.
			n.a.	− 521C/T	Yes[74]	n.d.
			rs936461	− 809A/G	Yes[74]	n.d.
			rs1176744	386A/C (Tyr129Ser)	Yes[181]	No[181]
	Serotonin 5-HT receptor 7 (*HTR7*)	10q21	n.a.	836C/T (Pro279Leu)	No[85]	n.d.

i	Serotonin 5-HT transporter (*SERT, 5-HTT, SCL6A4*)	17q11.2	n.a.	5-HTTLPR; intron 2 VNTR	Yes[29,78,90,177,182] No[71,77,79,158,175,176]	Yes[71,90,177] No[80]
iii	Tumor necrosis factor alpha (*TNF*)	6p21.3	n.a.	308G/A	Yes[160]	n.d.
			n.a.	− 857C/T	No[71]	n.d.
			n.a.	− 863C/A		
			n.a.	− 1031T/C		
i	Tryptophan hydroxylase 1 (*TPH1*)	22q22.11	rs1800532	218A/C	No[183]	No[183]
			rs56151798	1095T/C	No[168]	n.d.
iii	Uncoupling protein 2 (*UCP2*)	11p13.4	n.a.	Flanking microsatellite markers	Yes[71] No[184]	n.d.
	Uncoupling protein 3 (*UCP3*)	11p13.4	n.a.	Flanking microsatellite markers	Yes[71] No[184]	n.d.

SNP = single nucleotide polymorphism, bp = base pair, AN = anorexia nervosa, BN = bulimia nervosa, n.d. = no data, n.a. = not available, HTTLPR = serotonin-transporter-linked polymorphic region, HT = hydroxytryptamine, VTNR = variable number of tandem repeats.

[a]Five areas of candidate genes: (i) monoaminergic pathways, (ii) endogenous reward system, (iii) weight regulation, (iv) female puberty, and (v) comorbid psychiatric disorders.

[b]The position of the variation is given in relation to the first nucleotide from the mRNA found in dbSNP (http://www.ncbi.nlm.nih.gov/snp/) or in the Ensembl (http://www.ensembl.org/index.html) database.

[c]Synergistic effect.

The identification of an optimal control group for association studies in EDs is difficult. It appears a reasonable requirement to exclude individuals who report a lifetime occurrence of an ED. Ideally, females aged ≥ 25 years who have passed the critical manifestation age should be used as controls. An even more scrutinized approach would also exclude individuals with subclinical EDs, high anxiousness, obsessive–compulsive features, and high depression scores. Such an approach is cumbersome. Because of the low prevalence of these disorders in the general population, the use of unscreened controls has also been proposed.

It is uncertain whether controls should be selected for body weight. If (single) gene variants that predispose to obesity (or leanness) are also involved in EDs, the results of an association study will depend on the composition of the control group with respect to weight status. In contrast, if a genetic overlap is nonexistent or minimal, selection for weight status of the controls will have no or little impact on the strength of association signals. As premorbid body weights of AN patients have been shown to cover the whole weight range,[40] BMI of controls might not be important. In contrast, in BN, both premorbid and familial obesity occur more frequently than can be expected by chance,[41] suggesting a potential genetic overlap of BN with obesity.

Stratification problems can be circumvented by use of the transmission disequilibrium test (TDT[42]), which tests for both association and linkage. The TDT uses family trios consisting of an index patient and both parents. The nontransmitted parental alleles are used as internal controls. The frequency of the parental transmission of a given allele to the diseased child is compared to the frequency of its nontransmission. If the analyzed allele is transmitted more often than expected by chance, this allele is thought to predispose to the disease; if the transmission is less frequent than expected, the allele seems to be protective.[42] The TDT is simple and powerful and takes genetic heterogeneity into account.[43]

The identification of a candidate gene can also be based on its location in a linkage region (see below) or a region identified via GWAS. Up to 2 million genetic variants can currently be analyzed simultaneously for association with a given phenotype in GWAS. Large study groups need to be screened, as the effect sizes are usually small.[44]

B. Statistical Considerations: Linkage Studies

For linkage analyses, specific markers are analyzed in both affected and unaffected members of one or more pedigrees. In simplistic terms, the test statistic is based on the cosegregation of the genetic marker with the disease phenotype in individual pedigrees. The further apart the marker is from the disease locus, the higher the frequency of recombination events. Within each pedigree, the analysis is based on the frequency of sharing (nonsharing) of the specific marker allele by affected (unaffected) family members. Commonly, microsatellite markers that comprise simple repeat units (e.g., CA–CA–CA– . . .) have been used for these studies. The standard scale for linkage studies is the LOD score, where LOD is defined as the logarithm (log 10) of the likelihood ratio for linkage versus nonlinkage. In complex disorders, an LOD score > 3.3 has been proposed as the threshold for a positive linkage result.[45] If a linkage between a marker and disease is identified, the surrounding chromosomal region (peak region), which as a rule encompasses several megabases, is thoroughly screened for candidate genes.[39] Fine mapping and resequencing are required to pinpoint the genetic variant(s) underlying the linkage peak.

For complex diseases that involve many genes, association and linkage studies have limited power.[43] Hence, if a variant confers only a small effect on the predisposition to the disease, individuals carrying this allele will have only a slightly elevated risk of developing the disorder. The detection of underlying alleles in complex disorders thus requires (very) large sample sizes. Similar to other complex disorders, it is assumed that several gene variants, each of small effect size, constitute the molecular basis of EDs.[44]

IV. Candidate Gene Studies

The selection of candidate genes in molecular genetic studies of EDs has been based on several major hypotheses (see Table I for a synopsis of candidate gene studies):

1. Because prevalence rates of both AN and BN are considerably higher in females (ratio 9:1) and both disorders typically manifest during puberty and late adolescence, genes relevant to female puberty have been screened.
2. The serotonergic and dopaminergic systems are presumed to play an important role in the etiology of EDs; neuroendocrine studies have revealed deviations both in the acute stage and after recovery.[45,46–52] Mainly serotonin and dopamine receptor genes have been analyzed.
3. According to an alternative hypothesis, the endogenous reward system figures prominently in the etiology of EDs.[48,53–57] Both dieting and binge eating can be viewed as behaviors that originate or are maintained within this system. Apart from the aforementioned candidate gene studies of the dopaminergic system, more recent studies have focused on genes of the endogenous cannabinoid system.

4. Because of the occurrence of underweight in AN and the elevated rates of both premorbid and parental obesity in BN patients,[41] genes found to be involved in obesity subsequently have been assessed in candidate gene studies of AN and BN.[58]
5. Finally, because of both the high comorbidity and lifetime occurrence of mood and anxiety disorders in ED patients, gene variants predisposing to these disorders could potentially also be associated with AN and/or BN.

A. Serotonergic and Dopaminergic System

Serotonin (5-hydroxytryptamine; 5-HT) is involved in a broad range of behavioral, biological, and physiological functions.[59,60] In particular, this neurotransmitter system is implicated in body weight regulation, eating behavior[59–61], and EDs.[62,63] Studies in long-term weight-restored patients with AN or BN revealed elevated mean 5-hydroxyindolacetic acid (5-HIAA) levels in cerebrospinal fluid in comparison with healthy controls. Hence, hyperserotonergic function might be regarded as a trait marker in ED patients.[50] This elevated tone could account for characteristic psychopathological features frequently associated with AN, such as rigidity, perfectionism, and obsessiveness.[45,46] However, most of the molecular genetic studies pertaining to the serotonergic system were negative or equivocal.[64]

Two promising analyses pertain to the genes coding for the 5-HT_{2A} receptor and the serotonin transporter:

1. Association with AN was shown for the A-allele of the promoter polymorphism (− 1438G>A) of the 5-HT_{2A} receptor gene (*HTR2A*).[65] Whereas some of the replication studies confirmed this result,[66–69] others did not.[70–74] Additionally, an early meta-analysis showed a lack of association.[75] However, a subsequent meta-analysis comprising a larger number of studies revealed evidence for an association of the − 1438A allele with AN. In addition, large heterogeneity between the different samples was detected.[76] Hence, tentatively, *HTR2A* might be considered as a relevant gene for AN.
2. The functional insertion/deletion polymorphism in the promoter region of the serotonin-transporter gene (*SLC6A4*) harbors two frequent alleles (short: *S and long: *L). The frequency of the *S allele was assessed in AN[77–79] and BN,[80] with equivocal results. Meta-analyses suggest that the *S allele might be implicated in both EDs, the evidence being stronger for an involvement in AN.[29,81]

Major symptoms related to AN such as weight loss, repulsion to food, hyperactivity, body image distortion, and obsessive–compulsive behavior have all been related in part to dopamine activity.[47] Trait-related disturbances of dopamine metabolism may contribute to the vulnerability to EDs.[45,51,52]

Numerous candidate gene studies were conducted, none of them with solidly confirmed results. Initially, promising analyses pertained to the genes of the dopamine D2 and D4 receptors (*DRD2*; *DRD4*). Association of two of seven polymorphisms in *DRD2* with AN[82] was not confirmed in an independent study.[83] Polymorphisms in *DRD4*, mainly the 7-repeat allele of the exon 3 repeat, yielded positive[84] and negative[85,86] associations for AN. The 7-repeat allele is possibly relevant for both BN and BED, as an elevated rate of binge eaters was detected among carriers of the 7-repeat allele.[87] The same allele contributed to weight gain in women with BN.[88] Neither of these two results has been confirmed independently.

Replication in large independent samples or meta-analyses is warranted for most of the studies on genes of the monoaminergic systems in the etiology of AN and BN (see Table I[64]) as the reported studies had individually insufficient statistical power.

B. Endogenous Reward System: Body Weight Regulation

The endogenous reward system might well be disturbed in patients with EDs (e.g., Refs. 48,53–57a). For example, in recovered patients with AN[53] or BN,[54] altered striatal activity was observed, implicating a disturbed reward system as a trait marker of both AN and BN.

Because of its therapeutic potential in the treatment of obesity,[89] the endocannabinoid system, which is part of the endogenous reward system, was analyzed in EDs. Cannabinoids, both exogenous and endogenous, stimulate food intake by activation of the cannabinoid receptor 1 (CNR1). The central CNR1 as well as the major endocannabinoid-degrading enzymes are implicated in the orexigenic effects of cannabinoids. Analysis of an (AAT)*n* repeat in the downstream flanking region of *CNR1* indicated that a specific allele (14-repeat allele) was preferentially transmitted to AN patients with binge eating/purging type whereas the 13-repeat allele was transmitted slightly more often to patients with the restricting type.[90] This finding should be regarded with great caution as the whole study comprised only 52 families. Thus, it is not surprising that a larger, independent study (91 families) did not find evidence for association of the (AAT)*n* repeat or any of 15 analyzed SNPs with AN or subtypes of the disorder.[91]

Recently, one SNP in *CNR1* and one in fatty acid amide hydrolase (*FAAH*) were investigated in patients with AN and BN and in normal-weight, healthy controls. Genotypes of these SNPs were associated with both disorders. Additionally, a synergistic effect of the two SNPs was shown for AN.[92] However, several statistical tests were performed without correction for this multiple testing. Again, the size of the study (approximately 150 individuals per group) precludes a definitive evaluation of the results. As a previous study (Ref. 91; see above) did not find evidence for the association of SNPs in *CNR1* or *FAAH* to AN, confirmation is warranted in independent and sufficiently sized samples.

C. Body Weight Regulation: Leptin-Mediated Signaling Pathways

The past 15 years have witnessed a tremendous increase in the number of molecular genetic studies related to body weight regulation (see also Ref. 93) The discovery of leptin (LEP) in 1994 led to a focus on LEP-mediated signaling pathways. Apart from being a key hormone for the regulation of energy balance, LEP is also a pleiotropic hormone involved in various neuroendocrine and behavioral processes associated with profound changes in energy storage. LEP plays a major role in the adaptation of the organism to semistarvation.[94,95] In acute AN, hypoleptinemia is a cardinal feature; LEP levels in patients are typically below those of healthy gender- and age-matched controls. Circulating LEP levels in AN patients are highly correlated with percentage body fat[96–98] and, to a lesser extent, with BMI on admission.[95,96,98] Single studies suggest that AN patients who have recovered from EDs nevertheless have reduced serum LEP levels, adjusted for BMI and/or fat mass.[99,100] This suggests that relative hypoleptinemia might be a trait marker in EDs; alternatively, this finding could be due to the frequent persistence of minor eating abnormalities in such patients. However, evidence for such a persistent relative hypoleptinemia is not unequivocal.[101] Molecular genetic studies of the *LEP* and leptin receptor (*LEPR*) genes yielded no evidence for their involvement in EDs in other studies.[102,103]

Downstream of LEP, the agouti-related peptide (AGRP) acts as an important orexigenic regulator of energy balance. Endogenous AGRP is an inverse agonist at melanocortin receptors, which are involved in body weight regulation (see below). Mutations leading to a loss of AGRP function could be envisaged as leading to reduced starvation signaling in the brain. One (Ala76Thr) of the three polymorphisms in *AGRP* has been found to be associated with AN.[104] This result was not confirmed in an independent study,[105] according to which the other allele of the polymorphism was more frequently transmitted to patients with AN (114 AN-trios).

The melanocortin-4 receptor (MC4R) is a downstream effector of the leptin signal. Alpha-melanocyte-stimulating hormone (α-MSH) is an agonist at MC4R, whereas AGRP is an inverse agonist (see above). It is well known that mutations in *MC4R* that lead to reduced receptor function confer a codominant form of obesity (e.g., Refs. 106–112). As obesity has been identified as a risk factor for the development of BN,[28] genotypes predisposing to obesity might be detected more commonly in patients with BN than in normal-weight controls. In fact, a mutation screen of *MC4R* revealed a single extremely obese patient with BN harboring a haplo-insufficiency mutation in *MC4R* (total sample size: 87 BN patients). Hence, a validated genotype predisposing to obesity was for the first time detected in a BN patient.[113]

Two studies reported that all analyzed obese carriers of *MC4R* variants fulfilled criteria for BED.[114,115] The initial study[114] described, among other variants, the Ile103 variant of the Val103Ile polymorphism as an obesity risk variant; however, this variant was convincingly found to be associated with a slightly reduced BMI[116,117] and is thus protective against obesity. Furthermore, subsequent studies in independent samples did not detect any *MC4R* mutations among extremely obese adults with BED.[118] Finally, in obesity families with a total of 43 *MC4R* mutation carriers, evidence for elevated rates of binge-eating episodes was not detected in the carriers of *MC4R* mutations.[119] This lack of confirmation led to the conclusion that it is unlikely that BED is strongly influenced by *MC4R* mutations.

The brain-derived neurotrophic factor (BDNF) acts downstream of MC4R.[120] It regulates synaptic efficiency through the modulation of relevant neurotransmitter systems that have been shown to be involved in EDs. The expression of *BDNF* is found in hypothalamic nuclei that are associated with weight regulation and feeding control. *Bdnf*$^{+/-}$ mice show altered eating behavior, increased body weight, and hypertrophic adipocytes. Decreased serum BDNF levels were shown in acute patients with AN or BN.[121–124] In patients with AN, serum BDNF levels depend on the state of illness and the degree of hypoleptinemia. Upregulation of BDNF in weight-recovered (minimum 3 months) patients with AN has been described.[122,125]

The role of the nonsynonymous polymorphism Val66Met in *BDNF* was intensively studied in EDs. The initial study showed lack of association with AN in general for the 66Met variant; the same allele was, however, associated with the restricting type of AN.[126] A larger, European-wide, family-based study subsequently showed association of the 66Met variant with AN, AN restricting type, AN binge eating/purging type, and BN[127,128]. These findings were later confirmed by some[129] but not all studies.[105,130,131] Analysis of the BDNF receptors neurotrophic tyrosine kinase receptor types 2 and 3 (NTRK2 and NTRK3) also revealed associations with these disorders.[132,133] These data indicate that the neurotrophin family is a key regulator of eating behavior and might be a susceptibility factor for EDs.

D. Genes with Effect on Body Weight Derived from Genome-Wide Screens

The first GWAS for body weight and obesity marked the beginning of a new era to unravel (neuro)biological mechanisms involved in body weight regulation (e.g., Refs. 134–136). The identified genes and the respective pathways might also be of importance for EDs (see above). A recent study has for the first time focused on GWAS-based obesity gene variants in AN. However, associations between individual SNPs and AN were not detected.[137]

V. Genome-Wide Approaches

Despite considerable efforts (see Table I), none of the candidate gene analyses has yielded unequivocal and clearly confirmed evidence for the involvement of specific alleles in the etiology of EDs. As there is no clear-cut evidence implicating genes of a specific regulatory system in these disorders, the candidate gene approach is difficult. Hence, a systematic genome-wide approach that does not rely on *a priori* hypotheses pertaining to the underlying genes or pathological mechanisms offers a useful alternative to detect genes involved in the etiology of EDs.

An international, multisite, collaborative group (the Price Foundation Collaborative Group[138]) collected a group of 196 AN patients and 237 relatives affected with AN, BN, or EDNOS. Microsatellite markers ($n = 386$) were used for a genome-wide linkage study.[139] A single, nonsignificant, multipoint nonparametric linkage (NPL) score above 1.5 was detected on chromosome 4; NPL scores above 1 were observed for markers on three additional chromosomes.[139] In a subset ($n = 37$) of the families based on index patients with restricting AN, the highest multipoint NPL score observed was 3.03 on chromosome 1p.

Candidate genes in the linkage region on chromosome 1p[140,141] have been investigated by further linkage and association analyses. Polymorphisms in the serotonin 1D receptor gene (*HTR1D*) and the opioid delta receptor gene (*OPRD1*), together with 33 microsatellite markers, substantially increased the evidence for linkage of this region to restricting AN (37 families as mentioned above). Genotypic, allelic, and haplotypic associations to AN were observed at both genes.[140] In an independent replication study, one of four genotyped SNPs in *HTR1D* showed association to AN, whereas in *OPRD1*, three out of six investigated SNPs were nominally associated.[141] Nonetheless, further analyses are required to validate these findings. Currently, there is no evidence to suggest that any of the polymorphisms examined has a functional consequence on the biological activity of the two gene products.[141]

To analyze specific covariates, a multipoint affected sibling pair (ASP) linkage analysis was performed on the same data set. By exploring seven attributes specific for individuals with EDs, two variables (drive for thinness and obsessionality) were identified for which clusters of ASPs were highly concordant. When these attributes were incorporated into the ASP linkage analysis, both jointly and separately, three regions of suggestive linkage were identified.[142] For these, candidate gene analyses have not yet been reported.

For BN, a linkage analysis based on 308 multiplex families that were identified through a patient with BN yielded a peak, with the highest nonparametric, multipoint maximum LOD score on chromosome 10.[143] Linkage analysis was performed in a subset of 133 families in which at least two affected relatives reported self-induced vomiting; again, the highest linkage peak was on chromosome 10p.[143] This region, interestingly, already showed up in full and partial genome scans for obesity[144–147]; however, attempts to identify the underlying gene variant(s) have not led to replicable results.[148,149] In addition, GWAS performed for the phenotype obesity have not identified genes in this region.[135,136,150] Despite these negative findings, a gene on chromosome 10p might be involved in both BN and obesity.

VI. Genome-Wide Association Studies

During the past 4 years, the number of studies using large chip-based sets of genetic markers (up to 2 million) to search for genetic variation underlying common diseases has increased dramatically. GWAS rely on the assumption that linkage disequilibrium (LD) enables one SNP to act as a surrogate marker for association to other sequence variants in the same region.[151] By analyzing a large number of SNPs, there is a reasonable chance that a SNP(s) will be in LD with common functional mutation(s)/polymorphism(s) relevant for the investigated phenotype. GWAS represent a major step forward in the study of common genetic variation in complex diseases like EDs. Several GWAS have revealed previously unknown gene–disease associations. The first GWAS for AN were currently being conducted. Within the Wellcome Trust Case Control Consortium (WTCCC3), 4000 patients with AN, who were recruited worldwide, will be screened by GWAS chips. The identification of new genes/pathways for EDs is anticipated in the very near future. Because the effect sizes of gene variants identified with GWAS have proven to be mostly small, it is unclear whether the initial sample size is sufficiently powered. Substantially larger samples were required to obtain genome-wide significant association signals for other psychiatric disorders such as depression and schizophrenia (see Ref. 44).

VII. Gene–Environment Interactions

Both genetic and environmental factors contribute to the development of EDs. If an environmental factor is necessary for a genetic factor to lead to an increased risk for a phenotype (or vice versa), a gene–environment (G×E) interaction applies (e.g., Ref. 152). Small to moderate effects are typical for these interactions. For example, an individual with a genetic vulnerability to

AN could become ill upon exposure to the socioculturally driven and media-promoted thinness ideal for females. While the current knowledge on both genetic and environmental risk factors has been analyzed in depth previously,[153] the knowledge of G×E interaction is limited. Distinct findings for G×E interactions in EDs have not been published. This might in part be attributed to design challenges of G×E studies.[154] Investigation of G×E interactions may be more rewarding for the detailed characterization of identified disease genes (i.e., at more advanced stages of genetic research).[154]

Currently, variations in or near the *HTR2A* and *BDNF* genes represent the most consistently supported genetic findings for EDs, and they are eligible candidates for investigation of interactions with environmental factors. The first large-scale GWAS for EDs will presumably increase this list. Ideally, genetic research will help to improve the understanding of environmental risk factors for these disorders[153] and, in particular, the mechanisms by which they induce their effects.

VIII. Conclusions and Future Perspectives

Neither candidate gene studies nor genome-wide linkage studies have led to the identification of truly validated genes for EDs. Whereas GWAS have for the first time allowed large-scale identification of genes involved in the etiology of complex disorders, effect sizes have proven to be small. The results of the first GWAS in EDs will become available shortly for AN. It remains to be seen whether the 4000 patients are sufficient to provide a genome-wide significant finding. Substantial efforts will be required to ascertain thousands of additional patients to allow for further GWAS and meta-analyses. Such studies will also reveal the extent of overlapping findings with other psychiatric disorders. If overlapping with "obesity genes" is detected, further research will be required to clarify whether this overlap is due to an above-average premorbid BMI of ED patients. Such analyses will be of particular interest for both BN and BED because of their known (premorbid) association with obesity.[28]

As heritability estimates based on family and twin studies for both AN and BN have mostly been shown to be high, the *a priori* expectations are high for gene identification via GWAS. However, similar to other complex disorders for which high heritability estimates also have been determined, the initial GWAS results will at best lead to the identification of one (or a few) genome-wide significant SNPs. Maher[155] has termed this phenomenon, observed for all complex disorders, as the case of the missing heritability.

Highly dense SNP chips cover a major part but not all variation in the human genome. Furthermore, haplotypes instead of mere SNPs might prove more relevant in the etiology of complex disorders such as EDs. Future research will also be based on novel technologies including exome and whole-genome sequencing. In addition, copy number variants need to be analyzed for their potential role in the etiology of EDs.

The quality of the formal genetic studies of AN and BN needs to be evaluated critically. Most twin and family studies are hampered by small sample sizes; in addition, the ascertainment procedures are frequently not suited to derive unbiased results. Potentially, the heritability has been overestimated. Similarly, the previous molecular genetic studies have been characterized by inadequate power to detect associations with small effect sizes. Additionally, multiple *post hoc* comparisons have been performed, which increase the likelihood of false-positive findings. To address these issues, more stringent significance levels, large-scale meta-analyses, or studies with sufficient sample sizes are needed. The recent success of genetic analyses in other psychiatric disorders[44,156] has convincingly demonstrated that large sample sizes, including several thousands of cases and controls, are required to identify and confirm molecular genetic findings. Because some of the candidate gene studies in EDs have been suggested in independent study groups, these genes can be further analyzed within the context of the upcoming GWAS.

In sum, for the identification of genes involved in various complex disorders and phenotypes, GWAS has been a very successful tool (e.g., reviewed in Ref. 44). The results of the first GWAS for EDs will hopefully pave the way to a deeper understanding of as yet unknown pathological mechanisms. It is at the same time of crucial importance to recruit further patients with these disorders, so that confirmation of positive GWAS signals becomes possible. Depending on our progress, G×E interactions can be subsequently investigated. The insight into the genetic mechanisms will prove valuable for understanding the etiologies of EDs. In the longer run, it is hoped that the identification of ED genes and the respective pathways will lead to the development of new pharmacological treatments.

Acknowledgments

This work was supported by grants from the German Federal Ministry of Education and Research (NGFNPlus: 01GS0820 and EDNET: 01GV0905). We thank Dr. Carla I.G. Vogel and Anna-Lena Volckmar, MSc, for assistance with the table. We thank Prof. Robert D. Oades for carefully reading and improving the manuscript.

References

1. American Psychiatric Association (APA). *Diagnostic and statistical manual of mental disorders*. 4th ed. Washington: American Psychiatric Association; 1994 Text Revision.
2. World Health Organization. *ICD-10. International classification of mental and behavioural disorders. Clinical description and diagnostic guidelines*. Geneva: WHO; 1992.
3. Becker AE, Eddy KT, Perloe A. Clarifying criteria for cognitive signs and symptoms for eating disorders in DSM-V. *Int J Eat Disord* 2009;**42**:611–9.
4. Hebebrand J, Casper R, Treasure J, Schweiger U. The need to revise the diagnostic criteria for anorexia nervosa. *J Neural Transm* 2004;**111**:827–40.
5. van Hoeken D, Veling W, Sinke S, Mitchell JE, Hoek HW. The validity and utility of subtyping bulimia nervosa. *Int J Eat Disord* 2009;**42**:595–602.
6. Wilson GT, Sysko R. Frequency of binge eating episodes in bulimia nervosa and binge eating disorder: diagnostic considerations. *Int J Eat Disord* 2009;**42**:603–10.
7. Anderluh M, Tchanturia K, Rabe-Hesketh S, Collier D, Treasure J. Lifetime course of eating disorders: design and validity testing of a new strategy to define the eating disorders phenotype. *Psychol Med* 2009;**1**:1–10.
8. Bulik C, Sullivan PF, Fear J, Pickering A. Predictors of the development of bulimia nervosa in women with anorexia nervosa. *J Nerv Ment Dis* 1997;**185**:704–7.
9. Eckert ED, Halmi KA, Marchi P, Grove W, Crosby R. Ten-year follow-up of anorexia nervosa: clinical course and outcome. *Psychol Med* 1995;**25**:143–56.
10. Eddy KT, Dorer DJ, Franko DL, Tahilani K, Thompson-Brenner H, Herzog DB. Diagnostic crossover in anorexia nervosa and bulimia nervosa: implications for DSM-V. *Am J Psychiatry* 2008;**165**:245–50.
11. Fichter MM, Quadflieg N, Hedlund S. Twelve-year course and outcome predictors of anorexia nervosa. *Int J Eat Disord* 2006;**39**:87–100.
12. Fichter MM, Quadflieg N. Long-term stability of eating disorder diagnoses. *Int J Eat Disord* 2007;**40**:S61–6.
13. Gillberg IC, Rastam M, Gillberg C. Anorexia nervosa outcome: six-year controlled longitudinal study of 51 cases including a population cohort. *J Am Acad Child Adol Psychiatry* 1994;**33**:729–39.
14. Keel P, Dorer D, Franko D, Jackson S, Herzog D. Postremission predictors of relapse in women with eating disorders. *Am J Psychiatry* 2005;**162**:2263–8.
15. Milos G, Spindler A, Schnyder U, Fairburn CG. Instability of eating disorder diagnoses: prospective study. *Br J Psychiatry* 2005;**187**:573–8.
16. Nishimura H, Komaki G, Ando T, Nakahara T, Oka T, Kawai K, et al. Psychological and weight-related characteristics of patients with anorexia nervosa-restricting type who later develop bulimia nervosa. *Biopsychosoc Med* 2008;**2**:5.
17. Strober M, Freeman R, Morrell W. The long-term course of severe anorexia nervosa in adolescents: survival analysis of recovery, relapse, and outcome predictors over 10–15 years in a prospective study. *Int J Eat Disord* 1997;**22**:339–60.
18. Tozzi F, Thornton L, Klump K, Bulik C, Fichter M, Halmi K, et al. Symptom fluctuation in eating disorders: correlates of diagnostic crossover. *Am J Psychiatry* 2005;**162**:732–40.
19. van der Ham T, van Strien D, van Engeland H. A four-year prospective follow-up study of 49 eating-disordered adolescents: differences in course of illness. *Acta Psychiatr Scand* 1994;**90**:229–35.
20. Bulik CM, Sullivan PF, Wade TD, Kendler KS. Twin studies of eating disorders: a review. *Int J Eat Disord* 2000;**27**:1–20.

21. Holland AJ, Hall A, Murray R, Russell GF, Crisp AH. Anorexia nervosa: a study of 34 twin pairs and one set of triplets. *Br J Psychiatry* 1984;**145**:414–9.
22. Holland AJ, Sicotte N, Treasure J. Anorexia nervosa: evidence for a genetic basis. *J Psychosom Res* 1988;**32**:561–71.
23. Kipman A, Gorwood P, Mouren-Siméoni MC, Adès J. Genetic factors in anorexia nervosa. *Eur Psychiatry* 1999;**14**:189–98.
24. Klump KL, Kaye WH, Strober M. The evolving genetic foundations of eating disorders. *Psychiatr Clin North Am* 2001;**24**:215–25.
25. Klump KL, Miller KB, Keel PK, McGue M, Iacono WG. Genetic and environmental influences on anorexia nervosa syndromes in a population-based twin sample. *Psychol Med* 2001;**31**:737–40.
26. Klump KL, Wonderlich S, Lehoux P, Lilenfeld LR, Bulik CM. Does environment matter? A review of nonshared environment and eating disorders. *Int J Eat Disord* 2002;**31**:118–35.
27. Bulik CM, Slof-Op't Landt MC, van Furth EF, Sullivan PF. The genetics of anorexia nervosa. *Annu Rev Nutr* 2007;**27**:263–75.
28. Fairburn CG, Harrison PJ. Eating disorders. *Lancet* 2003;**361**:407–16.
29. Gorwood P. Eating disorders, serotonin transporter polymorphisms and potential treatment response. *Am J Pharmacogenomics* 2004;**4**:9–17.
30. Klump KL, Holly A, Iaconom WG, McGue M, Willson LE. Physical similarity and twin resemblance for eating attitudes and behaviors: a test of the equal environments assumption. *Behav Genet* 2000;**30**:51–8.
31. Bulik CM, Sullivan PF, Kendler KS. Heritability of binge-eating and broadly defined bulimia nervosa. *Biol Psychiatry* 1998;**44**:1210–8.
32. Kendler KS, MacLean C, Neale MC, Kessler RC, Heath AC, Eaves LJ. The genetic epidemiology of bulimia nervosa. *Am J Psychiatry* 1991;**148**:1627–37.
33. Kendler KS, Walters EE, Neale MC, Kessler RC, Heath AC, Eaves LJ. The structure of the genetic and environmental risk factors for six major psychiatric disorders in women. Phobia, generalized anxiety disorder, panic disorder, bulimia, major depression, and alcoholism. *Arch Gen Psychiatry* 1995;**52**:374–83.
34. Wade TD, Bulik CM, Neale M, Kendler KS. Anorexia nervosa and major depression: shared genetic and environmental risk factors. *Am J Psychiatry* 2000;**157**:469–71.
35. Kortegaard LS, Hoerder K, Joergensen J, Gillberg C, Kyvik KO. A preliminary population-based twin study of self-reported eating disorder. *Psychol Med* 2001;**31**:361–5.
36. Walters EE, Kendler KS. Anorexia nervosa and anorexic-like syndromes in a population-based female twin sample. *Am J Psychiatry* 1995;**152**:64–71.
37. Bulik CM, Thornton LM, Root TL, Pisetsky EM, Lichtenstein P, Pedersen NL. Understanding the relation between anorexia nervosa and bulimia nervosa in a Swedish national twin sample. *Biol Psychiatry* 2010;**67**:71–7.
38. Strober M, Freeman R, Lampert C, Diamond J, Kaye W. Controlled family study of anorexia nervosa and bulimia nervosa: evidence of shared liability and transmission of partial syndromes. *Am J Psychiatry* 2000;**157**:393–401.
39. Hinney A, Remschmidt H, Hebebrand J. Candidate gene polymorphisms in eating disorders. *Eur J Pharmacol* 2000;**410**:147–59.
40. Coners H, Remschmidt H, Hebebrand J. The relationship between premorbid body weight, weight loss, and weight at referral in adolescent patients with anorexia nervosa. *Int J Eat Disord* 1999;**26**:171–8.
41. Fairburn CG, Cooper Z, Doll HA, Welch SL. Risk factors for anorexia nervosa: three integrated case–control comparisons. *Arch Gen Psychiatry* 1999;**56**:468–76.

42. Spielman RS, McGinnis RE, Ewens WJ. Transmission test for linkage disequilibrium: the insulin gene region and insulin-dependent diabetes mellitus (IDDM). *Am J Hum Genet* 1993;**52**:506–16.
43. Risch N, Merikangas K. The future of genetic studies of complex human diseases. *Science* 1996;**273**:1516–7.
44. Hebebrand J, Scherag A, Schimmelmann BG, Hinney A. Child and adolescent psychiatric genetics. *Eur Child Adolesc Psychiatry* 2010;**19**:259–79.
45. Lander E, Kruglyak L. Genetic dissection of complex traits: guidelines for interpreting and reporting linkage results. *Nat Genet* 1995;**11**:241–7.
46. Kaye WH, Frank GK, McConaha C. Altered dopamine activity after recovery from restricting-type anorexia nervosa. *Neuropsychopharmacology* 1999;**21**:503–6.
47. Kaye WH, Klump KL, Frank GK, Strober M. Anorexia and bulimia nervosa. *Annu Rev Med* 2000;**51**:299–313.
48. Kaye WH, Bulik CM, Thornton L, Barbarich N, Masters K. Comorbidity of anxiety disorders with anorexia and bulimia nervosa. *Am J Psychiatry* 2004;**161**:2215–21.
49. Kaye WH, Fudge JL, Paulus M. New insights into symptoms and neurocircuit function of anorexia nervosa. *Nat Rev Neurosci* 2009;**10**:573–84.
50. Kaye WH, Greeno CG, Moss H, Fernstrom J, Fernstrom M, Lilenfeld LR, et al. Alterations in serotonin activity and psychiatric symptoms after recovery from bulimia nervosa. *Arch Gen Psychiatry* 1998;**55**:927–35.
51. Kaye WH, Gwirtsman HE, George DT, Ebert MH. Altered serotonin activity in anorexia nervosa after long-term weight restoration. Does elevated cerebrospinal fluid 5-hydroxyindoleacetic acid level correlate with rigid and obsessive behavior? *Arch Gen Psychiatry* 1991;**48**:556–62.
52. Barry VC, Klawans HL. On the role of dopamine in the pathophysiology of anorexia nervosa. *J Neural Transm* 1976;**38**:107–22.
53. Golden NH, Shenker IR. Amenorrhea in anorexia nervosa. Neuroendocrine control of hypothalamic dysfunction. *Int J Eat Disord* 1994;**16**:53–60.
54. Wagner A, Aizenstein H, Venkatraman VK, Fudge J, May JC, Mazurkewicz L, et al. Altered reward processing in women recovered from anorexia nervosa. *Am J Psychiatry* 2007;**164**:1842–9.
55. Wagner A, Aizenstein H, Venkatraman VK, Bischoff-Grethe A, Fudge J, May JC, et al. Altered striatal response to reward in bulimia nervosa after recovery. *Int J Eat Disord* 2010;**43**:289–94.
56. Keating C. Theoretical perspective on anorexia nervosa: the conflict of reward. *Neurosci Biobehav Rev* 2010;**34**:73–9.
57. Vaz-Leal FJ, Santos LR, García-Herráiz MA, Monge-Bautista M, López-Vinuesa B. Bulimia nervosa with history of anorexia nervosa: could the clinical subtype of anorexia have implications for clinical status and treatment response? *Int J Eat Disord* 2010; [Epub ahead of print].
58. Hebebrand J, Remschmidt H. Anorexia nervosa viewed as an extreme weight condition: genetic implications. *Hum Genet* 1995;**95**:1–11.
59. Blundell JE, Lawton CL, Halford JC. Serotonin, eating behavior, and fat intake. *Obes Res* 1995;**3**:471S–476S.
60. Halford JC, Blundell JE. Separate systems for serotonin and leptin in appetite control. *Ann Med* 2000;**32**:222–32.
61. Wurtman RJ, Wurtman JJ. Brain serotonin, carbohydrate-craving, obesity and depression. *Adv Exp Med Biol* 1996;**398**:35–41.
62. Brewerton TD, Jimerson DC. Studies of serotonin function in anorexia nervosa. *Psychiatry Res* 1996;**62**:31–42.

63. Jimerson DC, Lesem MD, Kaye WH, Brewerton TD. Low serotonin and dopamine metabolite concentrations in cerebrospinal fluid from bulimic patients with frequent binge episodes. *Arch Gen Psychiatry* 1992;**49**:132–8.
64. Scherag S, Hebebrand J, Hinney A. Eating disorders: the current status of molecular genetic research. *Eur Child Adolesc Psychiatry* 2010;**19**:211–26.
65. Collier DA, Arranz MJ, Li T, Mupita D, Brown N, Treasure J. Association between 5-HT2A gene promoter polymorphism and anorexia nervosa. *Lancet* 1997;**350**:412.
66. Enoch MA, Kaye WH, Rotondo A, Greenberg BD, Murphy DL, Goldman D. 5-HT2A promoter polymorphism—1438G/A, anorexia nervosa, and obsessive–compulsive disorder. *Lancet* 1998;**351**:1785–6.
67. Nacmias B, Ricca V, Tedde A, Mezzani B, Rotella CM, Sorbi S. 5-HT2A receptor gene polymorphisms in anorexia nervosa and bulimia nervosa. *Neurosci Lett* 1999;**277**:134–6.
68. Rybakowski F, Slopien A, Dmitrzak-Weglarz M, Czerski P, Rajewski A, Hauser J. The 5-HT2A – 1438 A/G and 5-HTTLPR polymorphisms and personality dimensions in adolescent anorexia nervosa: association study. *Neuropsychobiology* 2006;**53**:33–9.
69. Sorbi S, Nacmias B, Tedde A, Ricca V, Mezzani B, Rotella CM. 5-HT2A promoter polymorphism in anorexia nervosa. *Lancet* 1998;**351**:1785.
70. Ando T, Komaki G, Karibe M, Kawamura N, Hara S, Takii M, et al. 5-HT2A promoter polymorphism is not associated with anorexia nervosa in Japanese patients. *Psychiatr Genet* 2001;**11**:157–60.
71. Campbell DA, Sundaramurthy D, Markham AF, Pieri LF. Lack of association between 5-HT2A gene promoter polymorphism and susceptibility to anorexia nervosa. *Lancet* 1998;**351**:499.
72. Gorwood P, Adès J, Bellodi L, Cellini E, Collier DA, Di Bella D, et al. The 5-HT(2A) – 1438G/A polymorphism in anorexia nervosa: a combined analysis of 316 trios from six European centres. *Mol Psychiatry* 2002;**7**:90–4.
73. Nishiguchi N, Matsushita S, Suzuki K, Murayama M, Shirakawa O, Higuchi S. Association between 5HT2A receptor gene promoter region polymorphism and eating disorders in Japanese patients. *Biol Psychiatry* 2001;**50**:123–8.
74. Hinney A, Ziegler A, Nöthen MM, Remschmidt H, Hebebrand J. 5-HT2A receptor gene polymorphisms, anorexia nervosa, and obesity. *Lancet* 1997;**350**:1324–5.
75. Ziegler A, Hebebrand J, Görg T, Rosenkranz K, Fichter M, Herpertz-Dahlmann B, et al. Further lack of association between the 5-HT2A gene promoter polymorphism and susceptibility to eating disorders and a meta-analysis pertaining to anorexia nervosa. *Mol Psychiatry* 1999;**4**:410–2.
76. Gorwood P, Kipman A, Foulon C. The human genetics of anorexia nervosa. *Eur J Pharmacol* 2003;**480**:163–70.
77. Hinney A, Barth N, Ziegler A, von Prittwitz S, Hamann A, Hennighausen K, et al. Serotonin transporter gene-linked polymorphic region: allele distributions in relationship to body weight and in anorexia nervosa. *Life Sci* 1997;**61**:295–303.
78. Matsushita S, Suzuki K, Murayama M, Nishiguchi N, Hishimoto A, Takeda A, et al. Serotonin transporter regulatory region polymorphism is associated with anorexia nervosa. *Am J Med Genet B Neuropsychiatr Genet* 2004;**128**:114–7.
79. Sundaramurthy D, Pieri LF, Gape H, Markham AF, Campbell DA. Analysis of the serotonin transporter gene linked polymorphism (5-HTTLPR) in anorexia nervosa. *Am J Med Genet* 2000;**96**:53–5.
80. Lauzurica N, Hurtado A, Escartí A, Delgado M, Barrios V, Morandé G, et al. Polymorphisms within the promoter and the intron 2 of the serotonin transporter gene in a population of bulimic patients. *Neurosci Lett* 2003;**352**:226–30.

81. Calati R, De Ronchi D, Bellini M, Serretti A. The 5-HTTLPR polymorphism and eating disorders: a meta-analysis. *Int J Eat Disord* 2010; [Epub ahead of print].
82. Bergen AW, Yeager M, Welch RA, Haque K, Ganjei JK, van den Bree MB, et al. Association of multiple DRD2 polymorphisms with anorexia nervosa. *Neuropsychopharmacology* 2005;**30**:1703–10.
83. Nisoli E, Brunani A, Borgomainerio E, Tonello C, Dioni L, Briscini L, et al. D2 dopamine receptor (DRD2) gene Taq1A polymorphism and the eating-related psychological traits in eating disorders (anorexia nervosa and bulimia) and obesity. *Eat Weight Disord* 2007;**12**:91–6.
84. Bachner-Melman R, Lerer E, Zohar AH, Kremer I, Elizur Y, Nemanov L, et al. Anorexia nervosa, perfectionism, and dopamine D4 receptor (DRD4). *Am J Med Genet B Neuropsychiatr Genet* 2007;**144**:748–56.
85. Hinney A, Schneider J, Ziegler A, Lehmkuhl G, Poustka F, Schmidt MH, et al. No evidence for involvement of polymorphisms of the dopamine D4 receptor gene in anorexia nervosa, underweight, and obesity. *Am J Med Genet* 1999;**88**:594–7.
86. Karwautz A, Rabe-Hesketh S, Hu X, Zhao J, Sham P, Collier DA, et al. Individual-specific risk factors for anorexia nervosa: a pilot study using a discordant sister-pair design. *Psychol Med* 2001;**31**:317–29.
87. Levitan RD, Masellis M, Basile VS, Lam RW, Kaplan AS, Davis C, et al. The dopamine-4 receptor gene associated with binge eating and weight gain in women with seasonal affective disorder: an evolutionary perspective. *Biol Psychiatry* 2004;**56**:665–9.
88. Kaplan AS, Levitan RD, Yilmaz Z, Davis C, Tharmalingam S, Kennedy JL. A DRD4/BDNF gene–gene interaction associated with maximum BMI in women with bulimia nervosa. *Int J Eat Disord* 2008;**41**:22–8.
89. Di Marzo V, Bifulco M, De Petrocellis L. The endocannabinoid system and its therapeutic exploitation. *Nat Rev Drug Discov* 2004;**3**:771–84.
90. Siegfried Z, Kanyas K, Latzer Y, Karni O, Bloch M, Lerer B, et al. Association study of cannabinoid receptor gene (CNR1) alleles and anorexia nervosa: differences between restricting and binging/purging subtypes. *Am J Med Genet B Neuropsychiatr Genet* 2004;**125**:126–30.
91. Müller TD, Reichwald K, Brönner G, Kirschner J, Nguyen TT, Scherag A, et al. Lack of association of genetic variants in genes of the endocannabinoid system with anorexia nervosa. *Child Adolesc Psychiatry Ment Health* 2008;**2**:33.
92. Monteleone P, Bifulco M, Di Filippo C, Gazzerro P, Canestrelli B, Monteleone F, et al. Association of CNR1 and FAAH endocannabinoid gene polymorphisms with anorexia nervosa and bulimia nervosa: evidence for synergistic effects. *Genes Brain Behav* 2009;**8**:728–32.
93. Hinney A, Vogel CI, Hebebrand J. From monogenic to polygenic obesity: recent advances. *Eur Child Adolesc Psychiatry* 2010;**19**:297–310.
94. Ahima RS, Prabakaran D, Mantzoros C, Qu D, Lowell B, Maratos-Flier E, et al. Role of leptin in the neuroendocrine response to fasting. *Nature* 1996;**382**:250–2.
95. Hebebrand J, Blum WF, Barth N, Coners H, Englaro P, Juul A, et al. Leptin levels in patients with anorexia nervosa are reduced in the acute stage and elevated upon short-term weight restoration. *Mol Psychiatry* 1997;**2**:330–4. (a)Hebebrand J, Müller TD, Holtkamp K, Herpertz-Dahlmann B. The role of leptin in anorexia nervosa: clinical implications. *Mol Psychiatry* 2007;**12**:23–35.
96. Grinspoon S, Gulick T, Askari H, Landt M, Lee K, Anderson E, et al. Serum leptin levels in women with anorexia nervosa. *J Clin Endocrinol Metab* 1996;**81**:3861–3.
97. Haas V, Onur S, Paul T, Nutzinger DO, Bosy-Westphal A, Hauer M, et al. Leptin and body weight regulation in patients with anorexia nervosa before and during weight recovery. *Am J Clin Nutr* 2005;**81**:889–96aHebebrand J, van der Heyden J, Devos R, Köpp W, Herpertz S,

Remschmidt H, et al. Plasma concentrations of obese protein in anorexia nervosa. *Lancet* 1995;**346**:1624–5.
98. Holtkamp K, Mika C, Grzella I, Heer M, Pak H, Hebebrand J, et al. Reproductive function during weight gain in anorexia nervosa. Leptin represents a metabolic gate to gonadotropin secretion. *J Neural Transm* 2003;**110**:427–35.
99. Frey J, Hebebrand J, Müller B, Ziegler A, Blum WF, Remschmidt H, et al. Reduced body fat in long-term followed-up female patients with anorexia nervosa. *J Psychiatr Res* 2000;**34**:83–8.
100. Haluzík M, Papezová M, Nedvídková J, Kábrt J. Serum leptin levels in patients with anorexia nervosa before and after partial refeeding, relationships to serum lipids and biochemical nutritional parameters. *Physiol Res* 1999;**48**:197–202.
101. Gendall KA, Kaye WH, Altemus M, McConaha CW, La Via MC. Leptin, neuropeptide Y, and peptide YY in long-term recovered eating disorder patients. *Biol Psychiatry* 1999;**46**:292–9.
102. Hinney A, Bornscheuer A, Depenbusch M, Mierke B, Tölle A, Middeke K, et al. No evidence for involvement of the leptin gene in anorexia nervosa, bulimia nervosa, underweight or early onset extreme obesity: identification of two novel mutations in the coding sequence and a novel polymorphism in the leptin gene linked upstream region. *Mol Psychiatry* 1998;**3**:539–43.
103. Quinton ND, Meechan DW, Brown K, Eastwood H, Blakemore AI. Single nucleotide polymorphisms in the leptin receptor gene: studies in anorexia nervosa. *Psychiatr Genet* 2004;**14**:191–4.
104. Vink T, Hinney A, van Elburg AA, van Goozen SH, Sandkuijl LA, Sinke RJ, et al. Association between an agouti-related protein gene polymorphism and anorexia nervosa. *Mol Psychiatry* 2001;**6**:325–58.
105. Dardennes RM, Zizzari P, Tolle V, Foulon C, Kipman A, Romo L, et al. Family trios analysis of common polymorphisms in the obestatin/ghrelin, BDNF and AGRP genes in patients with Anorexia nervosa: association with subtype, body-mass index, severity and age of onset. *Psychoneuroendocrinology* 2007;**32**:106–13.
106. Farooqi IS, Yeo GS, Keogh JM, Aminian S, Jebb SA, Butler G, et al. Dominant and recessive inheritance of morbid obesity associated with melanocortin 4 receptor deficiency. *J Clin Invest* 2000;**106**:271–9.
107. Hinney A, Bettecken T, Tarnow P, Brumm H, Reichwald K, Lichtner P, et al. Prevalence, spectrum, and functional characterization of melanocortin-4 receptor gene mutations in a representative population-based sample and obese adults from Germany. *J Clin Endocrinol Metab* 2006;**91**:1761–9.
108. Hinney A, Hohmann S, Geller F, Vogel C, Hess C, Wermter AK, et al. Melanocortin-4 receptor gene: case–control study and transmission disequilibrium test confirm that functionally relevant mutations are compatible with a major gene effect for extreme obesity. *J Clin Endocrinol Metab* 2003;**88**:4258–67.
109. Hinney A, Schmidt A, Nottebom K, Heibült O, Becker I, Ziegler A, et al. Several mutations in the melanocortin-4 receptor gene including a nonsense and a frameshift mutation associated with dominantly inherited obesity in humans. *J Clin Endocrinol Metab* 1999;**84**:1483–6.
110. Vaisse C, Clement K, Durand E, Hercberg S, Guy-Grand B, Froguel P. Melanocortin-4 receptor mutations are a frequent and heterogeneous cause of morbid obesity. *J Clin Invest* 2000;**106**:253–62.
111. Vaisse C, Clement K, Guy-Grand B, Froguel P. A frameshift mutation in human MC4R is associated with a dominant form of obesity. *Nat Genet* 1998;**20**:113–4.
112. Yeo GS, Farooqi IS, Aminian S, Halsall DJ, Stanhope RG, O'Rahilly S. A frameshift mutation in MC4R associated with dominantly inherited human obesity. *Nat Genet* 1998;**20**:111–2.

113. Hebebrand J, Fichter M, Gerber G, Görg T, Hermann H, Geller F, et al. Genetic predisposition to obesity in bulimia nervosa: a mutation screen of the melanocortin-4 receptor gene. *Mol Psychiatry* 2002;**7**:647–51.
114. Branson R, Potoczna N, Kral JG, Lentes KU, Hoehe MR, Horber FF. Binge eating as a major phenotype of melanocortin 4 receptor gene mutations. *N Engl J Med* 2003;**348**:1096–103.
115. Potoczna N, Branson R, Kral JG, Piec G, Steffen R, Ricklin T, Hoehe MR, Lentes KU, Horber FF. Gene variants and binge eating as predictors of comorbidity and outcome of treatment in severe obesity. *J Gastrointest Surg* 2004;**8**:971–81 [discussion 981–982].
116. Geller F, Reichwald K, Dempfle A, Illig T, Vollmert C, Herpertz S, et al. Melanocortin-4 receptor gene variant I103 is negatively associated with obesity. *Am J Hum Genet* 2004;**74**:572–81.
117. Wang D, Ma J, Zhang S, Hinney A, Hebebrand J, Wang Y, et al. Association of the MC4R V103I polymorphism with obesity: a Chinese case–control study and meta-analysis in 55, 195 individuals. *Obesity (Silver Spring)* 2010;**18**:573–9.
118. Herpertz S, Siffert W, Hebebrand J. Binge eating as a phenotype of melanocortin 4 receptor gene mutations. *N Engl J Med* 2003;**349**:606–9 [author reply 606–609].
119. Hebebrand J, Geller F, Dempfle A, Heinzel-Gutenbrunner M, Raab M, Gerber G, et al. Binge-eating episodes are not characteristic of carriers of melanocortin-4 receptor gene mutations. *Mol Psychiatry* 2004;**9**:796–800.
120. Xu B, Goulding EH, Zang K, Cepoi D, Cone RD, Jones KR, et al. Brain-derived neurotrophic factor regulates energy balance downstream of melanocortin-4 receptor. *Nat Neurosci* 2003;**6**:736–42.
121. Nakazato M, Hashimoto K, Shimizu E, Kumakiri C, Koizumi H, Okamura N, et al. Decreased levels of serum brain-derived neurotrophic factor in female patients with eating disorders. *Biol Psychiatry* 2003;**54**:485–90.
122. Nakazato M, Hashimoto K, Yoshimura K, Hashimoto T, Shimizu E, Iyo M. No change between the serum brain-derived neurotrophic factor in female patients with anorexia nervosa before and after partial weight recovery. *Prog Neuropsychopharmacol Biol Psychiatry* 2006;**30**:1117–21.
123. Monteleone P, Fabrazzo M, Martiadis V, Serritella C, Pannuto M, Maj M. Circulating brain-derived neurotrophic factor is decreased in women with anorexia and bulimia nervosa but not in women with binge-eating disorder: relationships to co-morbid depression, psychopathology and hormonal variables. *Psychol Med* 2005;**35**:897–905.
124. Monteleone P, Tortorella A, Martiadis V, Serritella C, Fuschino A, Maj M. Opposite changes in the serum brain-derived neurotrophic factor in anorexia nervosa and obesity. *Psychosom Med* 2004;**66**:744–8.
125. Ehrlich S, Salbach-Andrae H, Eckart S, Merle JV, Burghardt R, Pfeiffer E, et al. Serum brain-derived neurotrophic factor and peripheral indicators of the serotonin system in underweight and weight-recovered adolescent girls and women with anorexia nervosa. *J Psychiatry Neurosci* 2009;**34**:323–9.
126. Ribasés M, Gratacòs M, Armengol L, de Cid R, Badía A, Jiménez L, et al. Met66 in the brain-derived neurotrophic factor (BDNF) precursor is associated with anorexia nervosa restrictive type. *Mol Psychiatry* 2003;**8**:745–51.
127. Ribasés M, Gratacòs M, Fernández-Aranda F, Bellodi L, Boni C, Anderluh M, et al. Association of BDNF with anorexia, bulimia and age of onset of weight loss in six European populations. *Hum Mol Genet* 2004;**13**:1205–12.
128. Ribasés M, Gratacòs M, Fernández-Aranda F, Bellodi L, Boni C, Anderluh M, et al. Association of BDNF with restricting anorexia nervosa and minimum body mass index: a family-based association study of eight European populations. *Eur J Hum Genet* 2005;**13**:428–34.

129. Dmitrzak-Weglarz M, Skibinska M, Slopien A, Szczepankiewicz A, Rybakowski F, Kramer L, et al. BDNF Met66 allele is associated with anorexia nervosa in the Polish population. *Psychiatr Genet* 2007;**17**:245–6.
130. Friedel S, Horro FF, Wermter AK, Geller F, Dempfle A, Reichwald K, et al. Mutation screen of the brain derived neurotrophic factor gene (BDNF): identification of several genetic variants and association studies in patients with obesity, eating disorders, and attention-deficit/hyperactivity disorder. *Am J Med Genet B Neuropsychiatr Genet* 2005;**132**:96–9.
131. Rybakowski F, Dmitrzak-Weglarz M, Szczepankiewicz A, Skibinska M, Slopien A, Rajewski A, et al. Brain derived neurotrophic factor gene Val66Met and – 270C/T polymorphisms and personality traits predisposing to anorexia nervosa. *Neuro Endocrinol Lett* 2007;**28**:153–8.
132. Mercader JM, Saus E, Agüera Z, Bayés M, Boni C, Carreras A, et al. Association of NTRK3 and its interaction with NGF suggest an altered cross-regulation of the neurotrophin signaling pathway in eating disorders. *Hum Mol Genet* 2008;**17**:1234–44.
133. Ribases M, Gratacos M, Badia A, Jimenez L, Solano R, Vallejo J, et al. Contribution of NTRK2 to the genetic susceptibility to anorexia nervosa, harm avoidance and minimum body mass index. *Mol Psychiatry* 2005;**10**:851–60.
134. Frayling TM, Timpson NJ, Weedon MN, Zeggini E, Freathy RM, Lindgren CM, et al. A common variant in the FTO gene is associated with body mass index and predisposes to childhood and adult obesity. *Science* 2007;**316**:889–94.
135. Hinney A, Nguyen TT, Scherag A, Friedel S, Brönner G, Müller TD, et al. Genome wide association (GWA) study for early onset extreme obesity supports the role of fat mass and obesity associated gene (FTO) variants. *PLoS One* 2007;**2**:1361.
136. Willer CJ, Speliotes EK, Loos RJ, Li S, Lindgren CM, Heid IM, et al. Six new loci associated with body mass index highlight a neuronal influence on body weight regulation. *Nat Genet* 2009;**41**:25–34.
137. Brandys MK, van Elburg AA, Loos RJ, Bauer F, Hendriks J, van der Schouw YT, et al. Are recently identified genetic variants regulating BMI in the general population associated with anorexia nervosa? *Am J Med Genet B Neuropsychiatr Genet* 2010;**153**:695–9.
138. Kaye WH, Lilenfeld LR, Berrettini WH, Strober M, Devlin B, Klump KL, et al. A search for susceptibility loci for anorexia nervosa: methods and sample description. *Biol Psychiatry* 2000;**47**:794–803.
139. Grice DE, Halmi KA, Fichter MM, Strober M, Woodside DB, Treasure JT, et al. Evidence for a susceptibility gene for anorexia nervosa on chromosome 1. *Am J Hum Genet* 2002;**70**:787–92.
140. Bergen AW, van den Bree MB, Yeager M, Welch R, Ganjei JK, Haque K, et al. Candidate genes for anorexia nervosa in the 1p33-36 linkage region: serotonin 1D and delta opioid receptor loci exhibit significant association to anorexia nervosa. *Mol Psychiatry* 2003;**8**:397–406.
141. Brown KM, Bujac SR, Mann ET, Campbell DA, Stubbins MJ, Blundell JE. Further evidence of association of OPRD1 & HTR1D polymorphisms with susceptibility to anorexia nervosa. *Biol Psychiatry* 2007;**61**:367–73.
142. Devlin B, Bacanu SA, Klump KL, Bulik CM, Fichter MM, Halmi KA, et al. Linkage analysis of anorexia nervosa incorporating behavioral covariates. *Hum Mol Genet* 2002;**11**:689–96.
143. Bulik CM, Devlin B, Bacanu SA, Thornton L, Klump KL, Fichter MM, et al. Significant linkage on chromosome 10p in families with bulimia nervosa. *Am J Hum Genet* 2003;**72**:200–7.
144. Dong C, Wang S, Li WD, Li D, Zhao H, Price RA. Interacting genetic loci on chromosomes 20 and 10 influence extreme human obesity. *Am J Hum Genet* 2003;**72**:115–24.

145. Hager J, Dina C, Francke S, Dubois S, Houari M, Vatin V, et al. A genome-wide scan for human obesity genes reveals a major susceptibility locus on chromosome 10. *Nat Genet* 1998;**20**:304–8.
146. Hinney A, Ziegler A, Oeffner F, Wedewardt C, Vogel M, Wulftange H, et al. Independent confirmation of a major locus for obesity on chromosome 10. *J Clin Endocrinol Metab* 2000;**85**:2962–5.
147. Price RA, Li WD, Bernstein A, Crystal A, Golding EM, Weisberg SJ, et al. A locus affecting obesity in human chromosome region 10p12. *Diabetologia* 2001;**44**:363–6.
148. Boutin P, Dina C, Vasseur F, Dubois S, Corset L, Séron K, et al. GAD2 on chromosome 10p12 is a candidate gene for human obesity. *PLoS Biol* 2003;**1**:68.
149. Swarbrick MM, Waldenmaier B, Pennacchio LA, Lind DL, Cavazos MM, Geller F, et al. Lack of support for the association between GAD2 polymorphisms and severe human obesity. *PLoS Biol* 2005;**3**:315.
150. Thorleifsson G, Walters GB, Gudbjartsson DF, Steinthorsdottir V, Sulem P, Helgadottir A, et al. Genome-wide association yields new sequence variants at seven loci that associate with measures of obesity. *Nat Genet* 2009;**41**:18–24.
151. Freimer NB, Sabatti C. Human genetics: variants in common diseases. *Nature* 2007;**445**:828–30.
152. Wermter AK, Laucht M, Schimmelmann BG, Banaschweski T, Sonuga-Barke EJ, Rietschel M, et al. From nature versus nurture, via nature and nurture, to gene x environment interaction in mental disorders. *Eur Child Adolesc Psychiatry* 2010;**19**:199–210.
153. Mazzeo SE, Bulik CM. Environmental and genetic risk factors for eating disorders: what the clinician needs to know. *Child Adolesc Psychiatr Clin N Am* 2009;**18**:67–82.
154. Dempfle A, Scherag A, Hein R, Beckmann L, Chang-Claude J, Schäfer H. Gene–environment interactions for complex traits: definitions, methodological requirements and challenges. *Eur J Hum Genet* 2008;**16**:1164–72.
155. Maher B. Personal genomes: the case of the missing heritability. *Nature* 2008;**456**:18–21.
156. Psychiatric GWAS Consortium Coordinating Committee. Cichon S, Craddock N, Daly M, Faraone SV, Gejman PV, et al. Genomewide association studies: history, rationale, and prospects for psychiatric disorders. *Am J Psychiatry* 2009;**166**:540–56.
157. Hinney A, Herrmann H, Lohr T, Rosenkranz K, Ziegler A, Lehmkuhl G, et al. No evidence for an involvement of alleles of polymorphisms in the serotonin (1D beta and 7) receptor genes in obesity, underweight or anorexia nervosa. *Int J Obes* 1999;**23**:760–3.
158. Urwin RE, Bennetts BH, Wilcken B, Beumont PJV, Russell JD, Nunn KP. Investigation of epistasis between the serotonin transporter and norepinephrine transporter genes in anorexia nervosa. *Neuropsychopharmacology* 2003;**28**:1351–5.
159. Dmitrzak-Weglarz M, Skibinska M, Slopien A, Szczepankiewicz A, Rybakowski F, Kramer L, et al. BDNF Met66 allele is associated with anorexia nervosa in the Polish population. *Psychiatr Genet* 2007;**17**:245–6.
160. Hu X, Murphy F, Karwautz A, Li T, Freeman B, Franklin D, et al. Analysis of microsatellite markers at the UCP2/UCP3 locus on chromosome 11q13 in anorexia nervosa. *Mol Psychiatry* 2002;**7**:276–7.
161. Koronyo-Hamaoui M, Danziger Y, Frisch A, Stein D, Leor S, Laufer N, et al. Association between anorexia nervosa and the hsKCa3 gene: a family-based and case control study. *Mol Psychiatry* 2002;**7**:82–5.
162. Rosenkranz K, Hinney A, Ziegler A, Hermann H, Fichter M, Mayer H, et al. Systematic mutation screening of the estrogen receptor beta gene in probands of different weight extremes: identification of several genetic variants. *J Clin Endocrinol Metab* 1998;**83**:4524–7.

163. Mikolajczyk E, Grzywacz A, Samochowiec J. The association of catechol-*O*-methyltransferase genotype with the phenotype of women with eating disorders. *Brain Res* 2010;**1307**:142–8.
164. Michaelovsky E, Frisch A, Leor S, Stein D, Danziger Y, Carel C, et al. Haplotype analysis of the COMT–ARVCF gene region in Israeli anorexia nervosa family trios. *Am J Med Genet B Neuropsychiatr Genet* 2005;**139**:45–50.
165. Di Bella D, Catalano M, Cavallini MC, Riboldi C, Bellodi L. Serotonin transporter linked polymorphic region in anorexia nervosa and bulimia nervosa. *Mol Psychiatry* 2000;**5**:233–4.
166. Frieling H, Gozner A, Römer KD, Lenz B, Bönsch D, Wilhelm J, et al. Global DNA hypomethylation and DNA hypermethylation of the alpha synuclein promoter in females with anorexia nervosa. *Mol Psychiatry* 2007;**12**:229–30.
167. Rosenkranz K, Hinney A, Ziegler A, von Prittwitz S, Barth N, Roth H, et al. Screening for mutations in the neuropeptide Y Y5 receptor gene in cohorts belonging to different weight extremes. *Int J Obes Relat Metab Disord* 1998;**22**:157–63.
168. Gabrovsek M, Brecelj-Anderluh M, Bellodi L, Cellini E, Di Bella D, Estivill X, et al. Combined family trio and case–control analysis of the COMT Val158Met polymorphism in European patients with anorexia nervosa. *Am J Med Genet B Neuropsychiatr Genet* 2004;**124**:68–72.
169. Steiger H, Richardson J, Schmitz N, Joober R, Israel M, Bruce KR, et al. Association of trait-defined, eating-disorder sub-phenotypes with (biallelic and triallelic) 5HTTLPR variations. *J Psychiatr Res* 2009;**43**:1086–94.
170. Ricca V, Nacmias B, Cellini E, Di Bernardo M, Rotella CM, Sorbi S. 5-HT2A receptor gene polymorphism and eating disorders. *Neurosci Lett* 2002;**323**:105–8.
171. Martaskova D, Slachtova L, Kemlink D, Zahorakova D, Papezova H. Polymorphisms in serotonin-related genes in anorexia nervosa. The first study in Czech population and meta-analyses with previously performed studies. *Folia Biol* 2009;**55**:192–7.
172. Hammer C, Kapeller J, Endele M, Fischer C, Hebebrand J, Hinney A, et al. Functional variants of the serotonin receptor type 3A and B gene are associated with eating disorders. *Pharmacogenet Genomics* 2009;**19**:790–9.
173. Hinney A, Becker I, Heibült O, Nottebom K, Schmidt A, Ziegler A, et al. Systematic mutation screening of the pro-opiomelanocortin gene: identification of several genetic variants including three different insertions, one nonsense and two missense point mutations in probands of different weight extremes. *J Clin Endocrinol Metab* 1998;**83**:3737–41.
174. Kim YR, Woo JM, Heo SY, Kim JH, Lim SJ, Yu BH. An association study of the A218C polymorphism of the tryptophan hydroxylase 1 gene with eating disorders in a Korean population: a pilot study. *Psychiatry Investig* 2009;**6**:44–9.
175. Lee Y, Lin PY. Association between serotonin transporter gene polymorphism and eating disorders: a meta-analytic study. *Int J Eat Disord* 2009; [Epub ahead of print].
176. Mikolajczyk E, Smiarowska M, Grzywacz A, Samochowiec J. Association of eating disorders with catechol-*O*-methyltransferase gene functional polymorphism. *Neuropsychobiology* 2006;**54**:82–6.
177. Steiger H, Joober R, Gauvin L, Bruce KR, Richardson J, Israel M, et al. Serotonin-system polymorphisms (5-HTTLPR and -1438G/A) and responses of patients with bulimic syndromes to multimodal treatments. *J Clin Psychiatry* 2008;**69**:1565–71.
178. Burnet PW, Smith KA, Cowen PJ, Fairburn CG, Harrison PJ. Allelic variation of the 5-HT2C receptor (HTR2C) in bulimia nervosa and binge eating disorder. *Psychiatr Genet* 1999;**9**:101–4.
179. Fumeron F, Betoulle D, Aubert R, Herbeth B, Siest G, Rigaud D. Association of a functional 5-HT transporter gene polymorphism with anorexia nervosa and food intake. *Mol Psychiatry* 2001;**6**:9–10.

180. Bruins-Slot L, Gorwood P, Bouvard M, Blot P, Ades J, Feingold J, et al. Lack of association between anorexia nervosa and D3 dopamine receptor gene. *Biol Psychiatry* 1998;**43**:76–8.
181. Han L, Nielsen DA, Rosenthal NE, Jefferson K, Kaye W, Murphy D, et al. No coding variant of the tryptophan hydroxylase gene detected in seasonal affective disorder, obsessive–compulsive disorder, anorexia nervosa, and alcoholism. *Biol Psychiatry* 1999;**45**:615–9.
182. Frisch A, Laufer N, Danziger Y, Michaelovsky E, Leor S, Carel C, et al. Association of anorexia nervosa with the high activity allele of the COMT gene: a family-based study in Israeli patients. *Mol Psychiatry* 2001;**6**:243–5.
183. Kanbur N, Mesci L, Derman O, Turul T, Cuhadaroğlu F, Kutluk T, et al. Tumor necrosis factor alpha-308 gene polymorphism in patients with anorexia nervosa. *Turk J Pediatr* 2008;**50**:219–22.
184. Hinney A, Lentes KU, Rosenkranz K, Barth N, Roth H, Ziegler A, et al. Beta 3-adrenergic-receptor allele distributions in children, adolescents and young adults with obesity, underweight or anorexia nervosa. *Int J Obes Relat Metab Disord* 1997;**21**:224–30.

Driven to Be Inactive?—The Genetics of Physical Activity

Trudy Moore-Harrison* and
J. Timothy Lightfoot*,†

*Department of Kinesiology, University of North Carolina at Charlotte, Charlotte, North Carolina, USA

†Huffines Institute, Department of Health and Kinesiology, Texas A&M University, College Station, TX, USA

The health implications of physical inactivity, including its integral role in promoting obesity, are well known and have been well documented. Physical activity is a multifactorial behavior with various factors playing a role in determining individual physical activity levels. Research using both human and animal models in the past several years has clearly indicated that genetics is associated with physical activity. Furthermore, researchers have identified several significant and suggestive genomic quantitative trait loci associated with physical activity. To date, the identities of the causal genes underlying physical activity regulation are unclear, with few strong candidate genes. The current research provides a foundation from which future confirmatory research can be launched as well as determination of the mechanisms through which the genetic factors act. The application of this knowledge could significantly augment the information available for physical activity behavior change interventions resulting in more efficient programs for those predisposed to be inactive.

I. Introduction

Research has convincingly demonstrated the benefits of physical activity on health and disease.[1] Despite this evidence, the prevalence of physical activity is continuing to decline with directly measured data suggesting that less than 5%

Progress in Molecular Biology
and Translational Science, Vol. 94
DOI: 10.1016/S1877-1173(10)94010-6

1877-1173/10 $35.00

of adults engage in moderate activity on a regular basis.[2] Researchers recognize that physical inactivity is a risk factor for many health outcomes such as cardiovascular disease, diabetes, some forms of cancer, and obesity.[3] For example, the population attributable risk increases significantly for several chronic diseases such as stroke (24.3%), hypertension (13.8%), and Type 2 diabetes (21.1%), due to physical inactivity and the risk for hypertension (34%), and Type 2 diabetes (28.6%) increases even more due to obesity.[4] Consequently, physical inactivity has been held responsible for approximately two million deaths per year worldwide[5] and is defined as an actual cause of chronic diseases by the Centers for Disease Control and Prevention.[6] The percentage of the total population of the United States that is physically inactive[2] far exceeds the percentage of those who smoke (19.7%), are hypertensive (27.5%), or have elevated total cholesterol (37.5%).[7] Therefore, the investigation of the causes and mechanisms of physical inactivity should be considered an area of critical importance for our national health policy[6] especially in regards to the prevention and treatment of obesity.

Physical activity can be defined as any daily movement of the body and exists on a continuum ranging from frank inactivity to constant movement. Physical activity can consist of numerous types of activity such as household chores, occupational tasks, leisure activity, sports, and care-giving duties. While the preponderance of research has focused on determining the demographic, social, psychological, and environmental factors affecting physical activity, a growing body of literature using both human and animal models has demonstrated significant genetic influence on physical activity.[8–13] Given that genetic factors are often an important determinant in limiting the response and rate of response to a physiological stressor, it would be advantageous to determine the genetic influences on physical activity so that this knowledge can be used to limit or augment the role of genetic factors on activity.

II. The Heritability of Physical Activity

Heritability is the influence of genetic factors on the variance of any phenotype between individuals;[14] in our case, we are interested in the mechanisms of how genetic factors influence physical activity. However, first, it is important to determine whether genetics actually play a role. Estimates of the magnitude of the heritable influence on any phenotype is expressed as values that range from zero to one (or 0–100%) with a value of zero indicating no influence of heredity with mainly environmental effects influencing that trait[15] and a value of one indicating that all individual variance in the characteristic of interest arose from genetic factors.[16] These heritability estimates are usually derived from either broad-sense and/or narrow-sense heritability equations. Narrow-sense heritability estimates the amount of phenotypic variation in physical activity that is transmitted from parent to child and since just parental transmission is involved, the estimates of narrow-sense heritability are thought to include only additive genetic effects and are lower (i.e., more conservative) than other measures of heritability.[17] Narrow-sense heritabilities are often calculated using offspring–parent regression. Conversely, the contribution of all genetic factors to the phenotypic variation in physical activity is considered broad-sense heritability. Because all genetic factors are considered, including additive and dominant effects, this estimate is higher than narrow-sense heritabilities and thus considered a more liberal estimation of heritability.[17–19]

Research designs that have investigated physical activity heritability have fallen into two general camps: those using human subjects and those using animal models. Within the human subjects' side of the literature, both family resemblance models and twin studies have been used. Determining familial resemblance in physical activity is an approach that primarily uses human families and examines the variance in activity amongst parents and children. One of the earliest family studies of physical activity was the Quebec Family Study in 300 families.[20] In this study, the investigators determined that genetic factors explained 20–29% of the familial resemblance in habitual physical activity as measured with 3-day activity survey data. In a second phase of this study, physical activity was assessed using 3-day activity diaries and a 1-year recall questionnaire in 200 families.[21] In this phase, heritability estimates were lower (16–25%) and were explained more completely by a combination of common environment and genetic factors. Interestingly, the inactivity phenotype, based on a lack of activity reported in the subjects' activity diaries, had a slightly higher heritability level than past and current physical activity levels. A further extension of the family resemblance design was a study by Sallis and coworkers that examined physical activity in 95 Anglo families and 111 Mexican-American families from the San Diego Family Health Project using a 7-day recall questionnaire. They found moderate correlations (0.25–0.55) of familial aggregation for kilocalories expended per day and 0.20–0.35 correlations for vigorous leisure physical activity.[22]

Similar to the values reported by Simonen *et al.*,[21] heritability estimates from the Southwest Ohio Family Study[23] were 0.17–0.26 for sport and leisure physical activity across 521 relatives. Activity in this study was measured by the Baecke Questionnaire of Habitual Physical Activity. Supporting the general trend toward more modest heritability estimates, the Canada Fitness Survey observed low familial correlations (0.08–0.38) for physical activity level in 13,804 individuals.[24] Exhibiting an even lower heritability of activity, a study of 1364 Mexican-Americans family members did not demonstrate a significant familial effect (9%) on physical activity levels as determined by a 7-day recall questionnaire.[25] Interestingly, while all of the family studies that have used indirect means to estimate activity have reported lower estimations of activity heritability, the Viva la Familia Study that used accelerometers to measure activity in 631 Hispanic parents and 1030 of their children found generally higher ranges of heritability (0.32–0.60) for physical activity.[26] Similarly, in the Framingham Children's Study, Moore *et al.*[27] found that when using the Caltrac accelerometer to measure physical activity in 100 children and parents, the children were 5.8 times likely to be active if both parents were active. Thus, the majority of the family resemblance studies have shown activity heritability estimates in the range of 0.25. Future family studies of activity heritability should follow the lead of Butte *et al.*[26] and Moore *et al.*[27] and employ methods that give a more direct estimation of daily activity to eliminate this potential source of deflation in the heritability estimates.

Monozygous twins are genetically identical and therefore phenotypic differences between pairs are assumed to be due to environmental factors and/or measurement error.[28] Dizygous pairs are approximately 50% genetically identical and thus, in conjunction with monozygous twins, can further illuminate the contribution of common versus unique environmental factors on physical activity.

One of the first twin studies to consider physical activity using a twin design was a Finnish study that measured physical activity from recall.[29] One thousand five hundred and thirty-seven monozygotic (MZ) and 3507 dizygotic (DZ) male twin pairs were asked the amount, intensity, and duration of current physical activity and the number of years of activity in adult life. Intraclass correlations were 0.57 for MZ twins and 0.26 for DZ twins, with an overall age adjusted of 0.62 and a common environmental effect of zero.[29] While not stated, these results suggested a 38% contribution of unique environmental effects and/or measurement error. These heritability estimates are naturally higher than the majority of those seen in the family studies due to the use of the more liberal broad-sense heritability statistics. However, several years later, as a continuation from the Finnish study using a cohort of the male MZ twins, familial aggregation still accounted for a large portion (43%) of exercise participation in adulthood suggesting that childhood influences might have affected these subjects' physical activity through their lifespan.[30] Other extensive twin studies since Kaprio *et al.*'s[29] study have also reported higher levels of heritability than seen in the family studies. For example, a twin study of over 3000 male MZ and DZ pairs from the Vietnam Era Twin Registry assessed physical activity levels using questionnaires and determined that there was genetic influence in physical activity levels since the MZ pairs had higher correlations than the DZ pairs.[11] Specifically, the activity questionnaires regarding intense physical activity resulted in heritability estimates of 39–58%, with the activity index of "running at least 10 miles per week" having a significant genetic component (53%). The heritability estimates resulting from the moderate exercise portion of the questionnaire were generally lower, but all were significant, with a 38% heritability estimate for the overall index of moderate activity.

The largest twin study to date examined exercise participation survey responses from 13,676 MZ and 23,375 DZ pairs 19–40 years old, from seven different countries (total $n = 85{,}198$).[31] This study found that heritability for leisure-time exercise participation of at least 60 min per week ranged from 27% to 71% depending on the nationality and sex of the subject.[31] Males in Norway had the lowest activity heritability (27%) whereas females in the United Kingdom (UK) had the highest heritability (71%). The median exercise participation heritability for all countries adjusted for sex was 62%. However, a study of 1003 MZ and 386 DZ twins comparing a higher threshold (150 min per week) to a lower threshold (60 min per week) of exercise participation found that unique environmental and genetic factors accounted for 55% and 45%, respectively, of the variance for the lower threshold.[32] The higher threshold (150 min) had 72% and 28% of the variance accounted for by unique and common environmental factors, respectively, suggesting that genetic influence on exercise might be dependent on duration. Therefore, while both genetics and unique environmental factors appear to contribute significantly to exercise participation, common environmental factors may have a stronger influence on higher duration of activity.

There has been ambiguity in the literature regarding whether there is differential heritability of activity in males and females. With a sizable cohort (over 5000 MZ and 8000 DZ male and female twins), the Swedish Twin Study reported significant genetic heritability of physical activity levels using a question about leisure-time activity over the past year. Heritability of physical activity levels between the sexes was similar, being reported as 57% for males and 50% for females, while the unique environmental factors accounted for 40% in males and 44% in females and influence of common environmental factors ranged between 3% and 6%.[33] In contrast, in a study of over 400 Portuguese twin pairs, 12–25 years old, researchers demonstrated that males had a higher genetic influence for leisure-time physical activity assessed by a questionnaire than females (63% vs. 32%).[34] The male pairs had no significant common environmental influence and 37% arose from unique environmental factors. However, females, while exhibiting similar unique environmental influences as males (30%), showed a significant common environmental influence on activity (38%). The finding of differential heritability of activity was originally suggested by Boomsma *et al.*[35] when they examined sports participation by a single question in a group 44 MZ and 46 DZ Dutch twins aged 14–20 years.[35] They found that heritability was 77% for the males and 35% for females, which is consistent with Maia and colleagues' estimates.[34] Supporting the lack of common environment influence in male twins were findings from the Leuven Longitudinal Study on over 180 Flemish 15-year-old male and female twins.[36] Using a single question survey, variation in sports participation in males was explained 83% by genetic factors and 17% by common environmental factors. However, in females, genetic factors explained 44% and environmental factors explained 54% of the variance for sport participation. Only a small percentage (2%) of the variation was explained by unique environmental factors. Thus, in a large number of twin studies, there are reported higher levels of activity heritability with a lack of consensus regarding differential heritability in males versus females.

Interestingly, there is early evidence that the potential difference in male and female heritability of activity may be more complex and involve environmental factors. In a three-generation study, physical activity patterns showed no association between twins, their parents, and their grandparents, as assessed by questionnaires.[37] While there were no generational associations in activity level, there were strong correlations in MZ boys (0.72) and MZ girls (0.64) in activity levels. Thus, the authors suggested that the lack of association between intergenerational physical activities might be due to social changes over time and not necessarily due to the transmission of genetic factors. Furthermore, a twin study of 62 MZ and 38 DZ pairs aged 4–10 years found that additive genetic factors did not explain any variance in physical activity level or physical activity energy expenditure after adjustment for body weight, indicating that body weight was actually the factor genetically controlled and not activity.[38] Both Aarnio *et al.*[37] and Franks *et al.*[38] results may be partially explained by Stubbe *et al.*'s[13] findings that before the age of 18, genetics played virtually no part in determining physical activity, but rather activity levels were due primarily to common environmental influence.

Much like the family resemblance studies, the majority of the twin studies have used indirect estimations of physical activity, usually very-short activity recall surveys. Recall surveys of physical activity are well known to grossly overestimate actual physical activity levels,[39] and thus, because the phenotype is inaccurate, it is probable that the heritability estimates may also be inaccurate. Therefore, direct measures of activity are important to integrate in twin studies. However, the use of direct measures necessarily decreases the number of human subjects that can be tested. Thus, researchers have to determine whether the error controlled with directly measuring physical activity outweighs the lower statistical power associated with the necessarily more limited sample sizes that arise from directly measuring activity. An elegant example of these issues and the resulting effect on activity heritability estimates is the study by Joosen *et al.*,[28] where physical activity was measured by accelerometry and energy expenditure was measured using doubly labeled water in both a respiration chamber and free-living conditions. In this study, genetic factors accounted for 72% and 78% of the variance in activity-induced energy expenditure and physical activity, respectively, during free-living conditions in 12 MZ and 8 DZ twins. Broad-sense estimates of heritability of activity in the free-living conditions suggested that 92% of the individual variance in activity was explained by genetics. While these estimates were generally higher than those seen using survey methods, the very low number of subjects (20 twin pairs) used for these heritability estimates makes them suspect. Further twin studies where activity is measured using direct methods are required to resolve the wide range of heritability estimates present in the literature.

It is clear that human studies using family and twin models have confirmed that there is a genetic component to physical activity. However, there are large inconsistencies in the reported magnitude of the association between heritability and activity. While it is possible that the range of genetic heritability of activity will actually vary significantly due to individual differential genetic regulation, we also cannot dismiss the possibility that the large range is due to inconsistencies and measurement error associated with the various phenotypes and measurements of physical activity. Additionally, the requirement for literally thousands of subjects to provide sufficient statistical power is also responsible for the confusion found in the published literature on this topic. Thus, to control for some of these limitations, other researchers have turned to animal models to consider the heritability of activity.

Animal models can be a practical resource to measure heritability of voluntary physical activity, since most environmental conditions can be standardized. Additionally, the use of inbred and selectively bred mice provides the ability to measure genetic variability in large cohorts that have homozygous genomes and using breeding schemes, researchers can introduce "controlled heterozygosity" of the genome with the ability to determine from which parent arose particular genomic regions and genes. Additionally, with the shorter lifespans of many animals, and in particular rodents, extensive breeding designs and lifespan-length measures can be used. Furthermore, measures of physical activity are generally easy to conduct and robust; wheel running activity, which has been claimed as the best analogue of human voluntary activity,[40] has been used since the early 1920s providing a deep and rich literature base.

One of the earliest studies of the heritability of activity in mice found narrow-sense heritability to be 0.20 in two inbred lines of house mice.[41] A few years later, a larger study found broad-sense heritability for 24- to 48-h activity measures ranging from 0.26 to 0.29 for 26 inbred strains.[42] Similarly, Lightfoot *et al.* found overall broad-sense heritabilities for 21 days of wheel running activity of 0.25, 0.18, and 0.14 for duration, distance, and average velocity, respectively, in 14 strains of inbred mice.[43] In addition, another study measured wheel running in seven strains of male inbred mice and demonstrated slightly higher broad-sense heritability estimates of 0.42, 0.39, and 0.24 for duration, distance, and average speed, respectively.[44] Thus, in these mouse strain screen designs, it was shown that genetics played a significant role in determining physical activity levels. However, these estimates were generally broad-sense in nature and did not consider the amount of transmissibility from parents to the offspring.

Using a research design that selectively bred mice for high wheel running activity, Swallow and coworkers[45] observed lower heritabilities of activity after 10 generations of selective breeding, demonstrating a mean narrow-sense heritability of 0.19 and an adjusted value for within-family selection of 0.28. When Swallow and colleagues[45] estimated heritability using broad-sense methods, they found values of 0.46 for males and 0.53 for females with an average of 0.49 for full sibs in generation 0. These values were similar to those derived from a larger study using 310 F_2 mice produced from high active C57L/J and low active C3H/HeJ inbred progenitor mice. In this study, reported broad-sense heritability estimates of 0.59, 0.50, and 0.47 for duration, distance, and average wheel running speed, respectively, were reported.[46]

Similar to human work by Boomsma *et al.*[35] and Maia *et al.*,[34] differential heritabilities of activity by sex were also observed by Lightfoot *et al.*[43] These investigators showed 12% genetic influence of activity for females and 31% for males in the amount of distance run daily. Interestingly, there has also been noted an influence of sex on activity with female mice running longer, further, and faster in several studies.[43,47,48] Whether this sex effect on activity is caused by genetic regulation is not clear.[46]

Human and animal research has made it apparent that physical activity is partly heritable, but the variations in study design make it difficult to estimate the exact magnitude of heritability on physical activity. Human studies have mainly used subjective measures that tend to overestimate physical activity whereas wheel running as a measure of physical activity in mice is consistent and repeatable.[43,49–51] Since the genetic homology between humans and mice is significant,[52] the heritability of physical activity found in mice may have implications for understanding the role of genetic variation in spontaneous activity in humans.

III. Heritability of Physical Activity across the Lifespan

Given the difficulty and expense of tracking people for long periods of time, the majority of activity heritability studies that have employed human designs have not considered possible changes in the heritability of activity with age. Since the data are clear that activity decreases with age in both humans and animals,[53–56] it is interesting to consider whether genetic influence on activity is altered across the lifespan and as a result, may be a factor in the age-related decrease in activity. Thus, while difficult, a few studies have attempted to address this question.

Using a simple binary question ("Do you participate in sports regularly?"), a study using the Netherlands Twin Registry found in over 2600 Dutch male and female twin pairs, aged 13–20 years, that sport participation decreased with age.[13] Common environmental factors were found to significantly influence sport participation from ages 13 to 16 years, whereas this pattern was reversed with sports participation almost entirely associated with heritability after the age of 18. In addition, a study that examined lifetime exercise participation in 147 MZ and 153 DZ male twins aged 35–70 years from the Finnish Twin Cohort showed heritability increased from 17% in adolescence to 51% in adulthood.[57]

The finding of an increase in genetic influence on activity with aging has been supported by the one existing animal study on the topic that was released almost simultaneously with the study by Stubbe *et al.*[13] Turner *et al.*[58] examined physical activity through the first 9 months of the lifespan (approximately 35% of the lifespan) in female mice from 10 inbred strains. In this model, overall physical activity measured by wheel running distance decreased throughout a 26-week time period. However, average broad-sense heritabilities for wheel running distance increased from 41% at 12 weeks of age to 76% at 30 weeks of age, after which the heritabilities dropped to the mid-60% range for the last 6 weeks of the experiment. These findings, in conjunction with those from the two human studies, suggest that the genetic influence on physical activity tends to increase with age.

IV. Genomic Locations Associated with Physical Activity

While it is interesting to debate the level of heritability associated with activity as discussed above, differences in activity quantification, the model used, and the contributions of age, disagreements about the precise magnitude of activity heritability are likely to persist. However, it is safe to say that genetic factors do influence physical activity. This brings us to the genetic mechanisms that are involved with the regulation of physical activity. Given the size of both the human and mouse genomes, an intermediary step has been to identify the genomic regions associated with activity. The determination of these genomic locations—called quantitative trait loci (QTL)—provides a foundation from which gene identification efforts and mechanistic studies can be launched.

Researchers have found significant and suggestive genomic locations for physical activity using both rodent and human models. In rodents, the best analogue for human voluntary activity has been considered to be wheel running.[39] Using a cohort of 310 F_2 mice derived from high active C57L/J and low active C3H/HeJ mice, Lightfoot *et al.*[46] identified four significant QTL, three of which colocalized on chromosome 13 (*DUR13.1*, *DIST13.1*, and *SPD13.1*) and one on chromosome 9 (*SPD9.1*) associated with speed of activity. The chromosome 13 QTL accounted for approximately 6% of the variability and the *SPD9.1* QTL accounting for approximately 11% of the variability in the speed of activity. However, the four significant QTL and 14 suggestive QTL in this study only explained 11–34% of the phenotypic variance (depending on the activity index used), indicating there were other QTL or genetic factors that explained additional variance in physical activity levels. Subsequently, Leamy *et al.*,[59] using the same F_2 database,[46] uncovered a significant number of epistatic QTL indicating that any consideration of activity QTL would need to account for potential interaction between genes. Interestingly, Leamy and coworkers noted that the inclusion of these epistatic QTL, none of which was significant by itself, explained between 18% and 36% of the variance in physical activity. Thus, the combination of the direct effect QTL[46] with the epistatic QTL,[59] explained most of the genetic variance in physical activity in this F_2 model. Interestingly, there were minimal relationships between any of the activity indices and weight of the animals.

However, Leamy proposed that perhaps there were various pleiotropic models that described the relationship between weight and physical activity indices arising even in the controlled heterozygosity of the F_2 generation. Leamy and coworkers[60] discovered 19 relationship QTL (relQTL) associated with pleiotropic relationship of the physical activity traits (distance, duration, and speed) and body weight. Seventy-nine percent of these relQTL influenced the relationship between one of the physical traits and body weight. In addition, the relQTL had 40 significant interactions with 31 of the epistatic QTL that had been discovered earlier.[59] Therefore, Leamy and coworkers observed that even in a population where controlled genomic heterozygosity was introduced using breeding schemes, there were at least three different models describing the genetic pleiotropic relationship between activity indices (i.e., distance, duration, and speed) and weight, leading to a rather complex picture of the genetic relationship between activity and weight.

Using mice from the unique selective breeding model of Garland and Kelly,[61] both Nehrenberg and coworkers[62] and Hartmann and colleagues[63] have identified additional QTL associated with physical activity. First, Hartmann and colleagues[63] identified a QTL on chromosome 11 that is strongly associated with the recessive mini-muscle phenotype that Garland's lab group has discovered in their high active animals. When controlling for the influence of this QTL, Nehrenberg and colleagues[60] also discovered QTL associated with high activity on chromosomes 5, 6, and 7. While several of these QTL overlap with those previously discovered by Lightfoot and colleagues,[46] several are unique and provide additional regions for candidate gene exploration.

While more difficult and necessarily requiring large cohorts of subjects, there have been efforts to identify genomic QTL associated with activity in humans. In an earlier, smaller genome-wide scans in humans, Cai and coworkers[64] identified one highly significant QTL on chromosome 18q associated with sedentary activity in Hispanic American children. Colocalizing within this QTL, the authors suggested that the melanocortin-4 receptor (*MC4R*) gene was a potential gene linked with physical activity in their cohort. In a larger genome-wide association study (GWAS) that was part of the Quebec Family Study, Simonen *et al.* observed one significant and several suggestive linkages for physical inactivity and activity.[65] In Simonen *et al.*'s study,[65] the significant QTL was on chromosome 2 (2p22–p16) and suggestive linkages were found on chromosomes 7p11.2 and 20q13.1 for physical inactivity, chromosomes 11p15 and 15q13.3 for time spent in physical activity, chromosomes 13q22–q31 for total daily physical activity, and chromosomes 4q28.2, 7p11.2, 9q31.1, and 13q22–q31 for moderately strenuous activity.

Recently, the most extensive of the human GWAS was published. De Moor *et al.*[66] studied 2622 adults of European ancestry using an exercise participation survey and a moderately dense single nucleotide polymorphism (SNP) map of 1,607,535 measured or imputed common SNPs. This GWAS revealed 37 novel SNPs for exercise participation clustered in three distinct genomic regions. The QTL containing the majority of the associated SNPs colocated with the *PAPSS2* gene on chromosome 10q23.2, while the other two QTL fell in intergenic areas on chromosomes 2 (2q33.1) and 18 (18p11.32). Beside the potential for novel QTL discovery, the strength of this type of study was the ability to cross-reference previously suggested potential candidate genes to determine if these genes colocalized within the genomic region of the QTL. In this study, De Moor and colleagues found that while not significant, suggestive associations were found with the location of the leptin receptor (*LEPR*) gene and the *GABRG3* gene on chromosome 15 (15q12–13). However, none of the other potential candidate genes that had been suggested earlier angiotensin-converting enzyme (*ACE*), calcium sensing receptor (*CASR*), *CYP19A1*, dopamine receptor 2 (*DRD2*), and *MC4R* [67,66,69,70–72] localized within any of the three significant QTL found in this study.

Surprisingly, when considered in total, there are few (> 10) significant QTL that have been associated with physical activity. While it is possible that there are actually just a few genes that influence physical activity level, the likelihood is that the existing models have had limited power to detect numerous less-powerful QTL that may play a role in regulating physical activity; this is a point that has been made by both De Moor *et al.*[66] and Lightfoot *et al.*[46] and a point that needs to be addressed by using a GWAS approach and much larger datasets. However, the available data do provide identification of genomic regions that presumably contain candidate genes associated with physical activity level.

V. Potential Candidate Genes for Physical Activity

Once potential QTL are found, it is tempting to declare candidate genes based solely on apparent functional relevance and localization within these QTL. While these criteria are important, they have led to a less than stellar track record of identifying causal genes from QTL studies.[73] Therefore, additional criteria should be used to sort through potential candidate genes before a gene's candidacy for involvement in a complex trait such as physical activity is made.[74] Additional candidate criteria can take several forms, ranging from demonstration of haplotype differences in gene structure to actual phenotypical changes due to gene manipulation. As such, multiple genes have been speculated to be involved in physical activity; however, few at this time present multiple lines of evidence to confirm their candidacy as being part of the regulatory mechanisms of physical activity.

Several studies have suggested that dopaminergic function is part of the biological regulation of physical activity in both animal and human studies.[67,75,76] Pharmacologically based studies have demonstrated that altered dopaminergic function in selectively bred high active mice was at least partially responsible for the high activity of these mice.[75,76] Injection with dopamine reuptake blockers cocaine and GBR 12909[75] and subsequently ritalin and apomorphine[76] resulted in altered wheel running activity in the high active mice leading to identification of dopamine receptor 1 (*Drd1*) as a potential physical activity-regulating gene. Supporting this hypothesis, Knab *et al.*[51] found significant differences in expression level between the high (C57L/J) and low (C3H/HeJ) active mice for both *Drd1* and tyrosine hydroxylase (*TH*) dopaminergic genes in the nucleus accumbens independent of wheel running exposure, but no difference for dopamine receptors 2–5 genes and the dopamine transporter gene (*Dat*). Supporting the suggestion of *Drd1* as a gene regulating physical activity, regional haplotype analysis conducted by Ceaser *et al.*[77] found differences in *Drd1* haplotype distributions between high active C57L/J mice and low active C3H/HeJ mice. These three independent lines of research, in conjunction with the well-known functional relevance of *Drd1* in locomotor disorders (e.g., Parkinson's) and the location of the *Drd1* gene in one of the QTL identified by Lightfoot *et al.*,[46] suggest that *Drd1* is an appropriate candidate gene for further investigation regarding its role in physical activity regulation.

In addition to *Drd1*, there is some evidence that *DRD2* is also a potential candidate gene for activity regulation. In fact, *Drd2* localizes to a genomic area that was identified as an epistatic QTL[59] and a *DRD2* gene polymorphism was found to be associated with past physical activity in white women in the Quebec Family and HERITAGE Family studies.[67] Subsequently, in a cohort of selectively bred high active mice, the *Drd2* gene was expressed at a 20% higher rate than in a control group[78] with an additional study showing that *Drd2* gene knockout mice exhibited reduced locomotor activity compared to wild-type mice.[79] However, there have been no differences reported in either *Drd2* expression or haplotype structure between high and low active mouse strains.[51,77] Furthermore, *DRD2* was also not replicated by the only GWAS using over one million SNP in Dutch and American adults.[66] Thus, while certainly a compelling candidate gene, further evidence is needed to strengthen the case of *DRD2* as a gene regulating activity level.

Another potential candidate gene regulating physical activity is the *Nhlh2* gene.[80] While relatively new, the *Nhlh2* gene has been known to be associated with leptin, and its encoded protein is a precursor to the formation of endorphins, both potentially functionally relevant to activity levels.[80] It has been shown that running wheel activity in *Nhlh2* knockout mice (N2KO) was reduced by approximately 50% as compared to wild-type mice;[80] further, *Nhlh2* is located on chromosome 3 and colocalizes with one of the epistatic physical activity QTL identified by Leamy *et al.*[59] Thus, *Nhlh2* has functional relevance to physical activity, localizes within one of the identified QTL, exhibits partial haplotype differences between high and low active mice,[77] and, when manipulated, directly affects physical activity. Thus, *Nhlh2* is a good candidate for further investigation of its activity-regulating effects.

Several other genes have been suggested as potential candidate genes regulating physical activity. While these genes may in fact be involved in activity regulation, evidence at this time is lacking to confidently declare them candidate genes. For example, Stefan *et al.*[68] found in Pima Indians that the Arg223-encoding allele of the *LEPR* gene predisposed to lower energy expenditure and physical activity levels compared to individuals with the Gln223-encoding allele. Another study showed that *LEPR* was related to physical activity energy expenditure in young boys.[81] However, *LEPR* does not localize into any of the significant QTL in either human[66] or mouse[46,59] and does not show haplotype differences between high and low active animals.[77]

Another potential candidate gene regulating physical activity is the *MC4R* gene.[69] Using a cohort from the Quebec Family Study, the *MC4R* gene on chromosome 18 was determined to be associated with past and current physical activities measured using a questionnaire and 3-day activity diary,[69] and individuals with the homozygous (T/T) or heterozygous (C/T) variation of *MC4R* were significantly more inactive than individuals with the homozygous (C/C) allele. Furthermore, a study of Hispanic children that measured physical activity using accelerometry suggested that a mutation of *MC4R*, which colocalized to a QTL for chromosome 18q, was associated with activity levels.[64] However, *MC4R* did not colocalize to any of the QTL identified in the larger human GWAS[66] or in any of the identified animal QTL. Thus, while *MC4R* shows promise as a candidate gene, further evidence is needed.

To this point, the majority of suggested candidate genes are postulated to work in a central manner, usually affecting or altering the "motivation" of exercise through the reward system. However, when the gene for glucose transporter 4 (*Slc2a4*, also known as *Glut-4*) was overexpressed in fast-twitch skeletal muscle of mice, these mice ran four times farther than control mice.[82] The authors speculated that the increase in muscle glucose availability secondary to the increased transport might be the reason the mice were able to sustain higher wheel running. Supporting this hypothesis, a study using highly active selectively bred mice reported that *Slc2a4* expression was 2.4-fold higher in the gastrocnemius[83] as compared to the control mice. Additionally, *Slc2a4* colocalizes with one of the QTL (chromosome 11, 40 cM) associated with physical activity.[62] Like *MC4R*, *LEPR*, and *DRD2*, *SLC2A4* continues to be a viable potential candidate gene and awaits further confirmatory evidence.

Other potential candidate genes have been identified as possibly associated with physical activity. In a bone mineral density study of 97 Caucasian girls with a mean age of 16.9 ± 1.2 years, the girls with the S allele for *CASR* were found to be less physically active.[70] In a study of 355 mild hypertensive men and women, the *ACE* gene was linked to leisure physical activity.[71] Individuals with DD polymorphism were more inactive and individuals with the II polymorphism for *ACE* engaged more frequently in sport activities.[71] Additionally, an aromatase (*CYP19*) gene polymorphism was associated with physical activity at baseline in a study of 331 early postmenopausal women.[72] Even though these possible associations were significant, none of these genes localize to any of the association sites discovered by De Moor *et al.*,[66] and only the mouse variants of *Cyp19* and *Ace* map to QTL discovered in mouse models.[46,59] Thus, with available evidence ambiguous at best, additional work is needed to further understand the relationship between *CASR*, *ACE*, and *CYP19* and physical activity.

Given the suggestion by Leamy *et al.*[60] of potential pleiotropic relationships between weight and activity, a few of the potential candidate genes for physical activity have been linked with obesity. A study of *Nhlh2* transcription factor knockout mice (N2KO) showed that a reduction in voluntary wheel activity led to weight gain in male N2KO mice after 12 weeks.[80,84] Other authors have shown that access to a running wheel and genetic differences in activity level are not necessarily related to the caloric intake of an animal, nor protective of body mass or body fat.[85] However, *Nhlh2* is known to have transcriptional control of *MC4R*,[80] of which polymorphisms in the *MC4R* gene have been associated with obesity and reduced physical activity in several studies.[64,69,86] An example of this association was shown in the Quebec Family Study where the inactive offspring in the cohort that had the *MC4R-C-2745T* variant had a lower body mass index (BMI); however, there was no association between inactivity and BMI in the parents.[69] While associated with *MC4R*, the *Nhlh2* gene is also known to be associated with leptin, a key regulator of food intake and energy expenditure.[87] In Pima Indians, individuals homozygotic for the leptin (*LEPR*) receptor Arg223-encoding allele had lower physical activity levels along with larger fat cell size.[68] Taken together, these findings, in conjunction with Leamy and coworkers' recent suggestion of the pleiotropic nature of the genes involved in weight maintenance and activity,[60] suggest that there is a complex relationship between physical activity and obesity that may involve genetic determinants. However, the lack of localization of many of the attractive candidate genes (e.g., *MC4R, LEP, LEPR*, etc.) within any published physical activity QTL greatly increases the complexity of the interpretation of these relationships. Future research will need to not only confirm these relationships, but further delineate potential mechanistic pathways involved.

VI. Summary

There are still considerable limitations and differences in study design, methodology, culture, and cohort composition that make determining an exact magnitude of the heritability on activity level and subsequent effects on obesity, difficult. However, in general, it appears that genetics has a moderate influence on physical activity level. Additionally, there is early evidence that age and sex may both be regulators of the heritability of activity. The initial genomic maps that exist associating specific genomic locations with activity, while limited at this early stage, provide a foundation from which further efforts to identify genes and mechanisms of regulation can be launched. Furthermore, research suggests that the full range of physical activity levels (inactivity to vigorous) may be regulated by different genomic regions. In particular, while the genetic structures responsible for physical activity regulation remain unknown, new genetic and molecular methods are making identification of specific physical activity genes an attainable goal. *DRD1* and *NHLH2* have the most evidence linking them with activity, while *SLC2A4 and PAPSS2* appear to be promising candidate genes with several other genes, while attractive as potential candidate genes, still lacking clear evidence of their involvement with activity. In addition, growing evidence supports a complex and pleiotrophic genetic association between physical activity and body weight regulation; however, the amount, nature, and mechanism(s) of these genetic associations are still undefined.

When the genetic regulators of physical activity are identified, this knowledge could significantly impact health promotion strategies focused on increasing physical activity levels and decreasing obesity. Knowledge of the genetic mechanisms associated with spontaneous physical activity level could lead to individualized programs and behavior change strategies tailored for those predisposed to be inactive. Besides the positive influence on the currently estimated cost of physical inactivity to our health care system ($507 billion per year)[88], an increase in physical activity levels would provide increases in both quality and quantity of life for all citizens.

Acknowledgments

A generous grant (R01AR050085) from the NIH National Institute of Arthritis and Musculoskeletal Diseases (NIAMS) supported the writing of this chapter and some of the data collected within.

References

1. Roberts CK, Barnard RJ. Effects of exercise and diet on chronic disease. *J Appl Physiol* 2005;**98**:3–30.
2. Troiano RP, Berrigan D, Dodd KW, Masse LC, Tilert T, McDowell M. Physical activity in the United States measured by accelerometer. *Med Sci Sports Exerc* 2008;**40**:181–8.
3. Rankinen T, Bouchard C. Invited commentary: physical activity, mortality, and genetics. *Am J Epidemiol* 2007;**166**:260–2.
4. Katzmarzyk PT, Janssen I. The economic costs associated with physical inactivity and obesity in Canada: an update. *Can J Appl Physiol* 2004;**29**:90–115.
5. World Health Organization. *The World Health Report 2002: reducing risks to health, promoting healthy life*. Geneva: World Health Organization; 2002.
6. Booth FW, Lees SJ. Fundamental questions about genes, inactivity, and chronic diseases. *Physiol Genomics* 2006;**28**:146–57.
7. Centers for Disease Control and Prevention (CDC). *Behavioral risk factor surveillance system survey data*. Atlanta, Georgia: U.S. Department of Health and Human Services, Centers for Disease Control and Prevention; 2007.
8. Stephens T, Jacobs Jr. DR, White CC. A descriptive epidemiology of leisure-time physical activity. *Public Health Rep* 1985;**100**:147–58.
9. King AC, Blair SN, Bild DE, Dishman RK, Dubbert PM, Marcus BH, et al. Determinants of physical activity and interventions in adults. *Med Sci Sports Exerc* 1992;**24**:S221–36.
10. Sallis JF, Simons-Morton BG, Stone EJ, Corbin CB, Epstein LH, Faucette N, et al. Determinants of physical activity and interventions in youth. *Med Sci Sports Exerc* 1992;**24**:S248–57.
11. Lauderdale DS, Fabsitz R, Meyer JM, Sholinsky P, Ramakrishnan V, Goldberg J. Familial determinants of moderate and intense physical activity: a twin study. *Med Sci Sports Exerc* 1997;**29**:1062–8.
12. Varo JJ, Martinez-Gonzalez MA, De Irala-Estevez J, Kearney J, Gibney M, Martinez JA. Distribution and determinants of sedentary lifestyles in the European Union. *Int J Epidemiol* 2003;**32**:138–46.
13. Stubbe JH, Boomsma DI, De Geus EJ. Sports participation during adolescence: a shift from environmental to genetic factors. *Med Sci Sports Exerc* 2005;**37**:563–70.
14. Falconer DS, Mackay TFC. *Introduction to quantitative genetics*. 4th ed. Essex, England: Longman; 1996.
15. Carlier M, Spitz E, Vacher-Lavenu MC, Villeger P, Martin B, Michel F. Manual performance and laterality in twins of known chorion type. *Behav Genet* 1996;**26**:409–17.

16. Silventoinen K, Kaprio J, Lahelma F. Genetic and environmental contributions to the association between body height and educational attainment: a study of adult Finnish twins. *Behav Genet* 2000;**30**:477–85.
17. Sesardic N. *Making sense of heritability*. Cambridge University Press: Cambridge, England; 2005.
18. Festing MFW. *Strategy in the use of inbred strains. Inbred strains in biomedical research*. New York, NY: Oxford University Press; 1979, pp. 122–131.
19. Festing MFW. *Notes on genetic analysis. Inbred strains in biomedical research*. New York, NY: Oxford University Press; 1979, pp. 80–98.
20. Perusse L, Tremblay A, LeBlanc C, Bouchard C. Genetic and environmental influences on level of habitual physical activity and exercise participation. *Am J Epidemiol* 1989;**129**:1012–22.
21. Simonen RL, Perusse L, Rankinen T, Rice T, Rao DC, Bouchard C. Familial aggregation of physical activity levels in the Quebec family study. *Med Sci Sports Exerc* 2002;**34**:1137–42.
22. Sallis JF, Patterson TL, Buono MJ, Atkins CJ, Nader PR. Aggregation of physical activity habits in Mexican-American and Anglo families. *J Behav Med* 1988;**11**:31–41.
23. Choh AC, Demerath EW, Lee M, Williams KD, Towne B, Siervogel RM, et al. Genetic analysis of self-reported physical activity and adiposity: the Southwest Ohio Family Study. *Public Health Nutr* 2009;**12**:1052–60.
24. Perusse L, LeBlanc C, Bouchard C. Inter-generation transmission of physical fitness in the Canadian population. *Can J Sport Sci* 1988;**13**:8–14.
25. Mitchell BD, Rainwater DL, Hsueh WC, Kennedy AJ, Stern MP, Maccluer JW. Familial aggregation of nutrient intake and physical activity: results from the San Antonio Family Heart Study. *Ann Epidemiol* 2003;**13**:128–35.
26. Butte NF, Cai G, Cole SA, Comuzzie AG. Viva la familia study: genetic and environmental contributions to childhood obesity and its comorbidities in the Hispanic population. *Am J Clin Nutr* 2006;**84**:646–54.
27. Moore LL, Lombardi DA, White MJ, Campbell JL, Oliveria SA, Ellison RC. Influence of parents' physical activity levels on activity levels of young children. *J Pediatr* 1991;**118**:215–9.
28. Joosen AM, Gielen M, Vlietnick R, Westerterp KP. Genetic analysis of physical activity in twins. *Am J Clin Nutr* 2005;**82**:1253–9.
29. Kaprio JM, Koskenvuo M, Sarna S. Cigarette smoking, use of alcohol and leisure-time activity among same-sexed adult male twins. In: Gedda L, Parisi P, Nance WE, editors. *Progress in clinical and biological research. Twin research 3: epidemiological and clinical studies*, vol. 45. New York: Alan R Liss; 1981, p. 37–46.
30. Simonen RL, Videman T, Kaprio J, Levalahti E, Battie MC. Factors associated with exercise lifestyle—a study of monozygotic twins. *Int J Sports Med* 2003;**24**:499–505.
31. Stubbe JH, Boomsma DI, Vink JM, Cornes BK, Martin NG, Martin NG, et al. Genetic influences on exercise participation in 37051 twin pairs from seven countries. *PLoS ONE* 2006;**1**:e22.
32. Duncan GE, Goldberg J, Noonan C, Moudon AV, Hurvitz P, Buchwald D. Unique environmental effects on physical activity participation: a twin study. *PLoS ONE* 2008;**3**:e2019.
33. Carlsson S, Andersson T, Lichtenstein P, Michaelsson K, Ahlbom A. Genetic effects on physical activity: results from the Swedish Twin Registry. *Med Sci Sports Exerc* 2006;**38**:1396–401.
34. Maia JA, Thomis M, Beunen G. Genetic factors in physical activity levels. A twin study. *Am J Prev Med* 2002;**23**:87–91.
35. Boomsma DI, van den Bree MBM, Orlebeke JF, Molenaar PCM. Resemblances of parents and twins in sports participation and heart rate. *Behav Genet* 1989;**19**:123–41.
36. Beunen G, Thomis M. Genetic determinants of sports participation and daily physical activity. *Int J Obes* 1999;**23**:S55–63.

37. Aarnio M, Winter T, Kujala UM, Kaprio J. Familial aggregation of leisure-time physical activity—a three generation study. *Int J Sports Med* 1997;**18**:549–56.
38. Franks PW, Ravussin E, Hanson RL, Harper IT, Allison DB, Knowler WC, et al. Habitual physical activity in children: the role of genes and the environment. *Am J Clin Nutr* 2005;**82**:901–8.
39. Shephard RJ. Limits to the measurement of habitual physical activity by questionnaires. *Br J Sports Med* 2003;**37**:197–206.
40. Eikelboom R. Human parallel to voluntary wheel running: exercise. *Anim Behav* 1999;**57**: F11–2.
41. Oliverio A, Castellano C, Messeri P. Genetic analysis of avoidance, maze, and wheel-running behaviors in the mouse. *J Comp Physiol Psychol* 1972;**79**:459–73.
42. Festing MFW. Wheel activity in 26 strains of mouse. *Lab Anim* 1977;**11**:257–8.
43. Lightfoot JT, Turner MJ, Daves M, Vordermark A, Kleeberger SR. Genetic influence on daily wheel running activity level. *Physiol Genomics* 2004;**19**:270–6.
44. Lerman I, Harrison BC, Freeman K, Hewett TE, Allen DL, Robbins J, et al. Genetic variability in forced and voluntary endurance exercise performance in seven inbred mouse strains. *J Appl Physiol* 2002;**92**:2245–55.
45. Swallow JG, Carter PA, Garland Jr. T. Artificial selection for increased wheel-running behavior in house mice. *Behav Genet* 1998;**28**:227–37.
46. Lightfoot JT, Turner MJ, Pomp D, Kleeberger SR, Leamy LJ. Quantitative trait loci (QTL) for physical activity traits in mice. *Physiol Genomics* 2008;**32**:401–8.
47. Koteja P, Swallow JG, Carter PA, Garland Jr. T. Energy cost of wheel running in house mice: implications for coadaptation of locomotion and energy budgets. *Physiol Biochem Zool* 1999;**72**:238–49.
48. Koteja Jr. P, Swallow JG, Carter PA, Garland Jr. T. Different effects of intensity and duration of locomotor activity on circadian period. *J Biol Rhythms* 2003;**18**:491–501.
49. Swallow JG, Carter PA, Zhan WZ, Sieck GC. Effects of voluntary activity and genetic selection on aerobic capacity in house mice. *J Appl Physiol* 1998;**84**:69–76.
50. Houle-Leroy P, Garland Jr. T, Swallow JG, Guderley H. Effects of voluntary activity and genetic selection on muscle metabolic capacities in house mice *Mus domesticus*. *J Appl Physiol* 2000;**89**:1608–16.
51. Knab A, Bowen RS, Hamilton AT, Gulledge AA, Lightfoot JT. Altered dopaminergic profiles: implications for the regulation of voluntary physical activity. *Behav Brain Res* 2009;**204**:147–52.
52. Paigen K. One hundred years of mouse genetics: an intellectual history. II. The molecular revolution (1981–2002). *Genetics* 2003;**163**:1227–35.
53. Lhotellier L, Cohen-Salmon C. Genetics and senescence: age-related changes in activity and exploration in three inbred strains of mice. *Physiol Behav* 1989;**45**:491–3.
54. Sallis JF. Epidemiology of physical activity and fitness in children and adolescents. *Crit Rev Food Sci Nutr* 1993;**33**:403–8.
55. Caspersen CJ, Merritt RK. Physical activity trends among 26 states: 1986–1990. *Med Sci Sports Exerc* 1995;**27**:713–20.
56. Crespo CJ, Keteyian SJ, Heath GW, Sempos CT. Leisure-time physical activity among US adults. Results from the Third National Health and Nutrition Examination Survey. *Arch Intern Med* 1996;**8**:93–8.
57. Simonen R, Levalahti E, Kaprio J, Videman T, Battie MC. Multivariate genetic analysis of lifetime exercise and environmental factors. *Med Sci Sports Exerc* 2004;**36**:1559–66.
58. Turner MJ, Kleeberger SR, Lightfoot JT. Influence of genetic background on daily running-wheel activity differs with aging. *Physiol Genomics* 2005;**22**:76–85.
59. Leamy LJ, Pomp D, Lightfoot JT. An epistatic genetic basis for physical activity traits in mice. *J Hered* 2008;**99**:639–46.

60. Leamy LJ, Pomp D, Lightfoot JT. Genetic variation in the pleitropic association between physical activity and body weight in mice. *Genet Sel Evol* 2009;**41**:41.
61. Garland Jr. T, Kelly SA. Phenotypic plasticity and experimental evolution. *J Exp Biol* 2006;**209**:2344–61.
62. Nehrenberg DL, Wang S, Hannon RM, Garland Jr. T, Pomp D. QTL underlying voluntary exercise in mice: interactions with the "mini muscle" locus and sex. *J Hered* 2010;**101**:42–53.
63. Hartmann J, Garland Jr. T, Hannon RM, Kelly SA, Munoz G, Pomp D. Fine mapping of "mini muscle", a recessive mutation causing reduced hindlimb muscle mass in mice. *J Hered* 2008;**99**:679–87.
64. Cai G, Cole SA, Butte N, Bacino O, Diego V, Tan K, et al. A quantitative trait locus on chromosome 18q for physical activity and dietary intake in Hispanic children. *Obesity* 2006;**14**:1596–604.
65. Simonen RL, Rankinen T, Perusse L, Rice T, Rao DC, Chagnon YC, et al. Genome-wide linkage scan for physical activity levels in the Quebec Family Study. *Med Sci Sports Exerc* 2003;**35**:1355–9.
66. De Moor MHM, Liu Y, Boomsma D, Li J, Hamilton JJ, Hottenga JJ, et al. Genome-wide association study of exercise behavior in Dutch and American adults. *Med Sci Sports Exerc* 2009;**41**:1887–95.
67. Simonen RL, Rankinen T, Perusse L, Leon AS, Skinner JS, Wilmore JH, et al. A dopamine D2 receptor gene polymorphism and physical activity in two family studies. *Physiol Behav* 2003;**78**:751–7.
68. Stefan N, Vozarova B, Del Parigi A, Ossowski V, Thompson DB, Hanson RL, et al. The Gln223Arg polymorphism of the leptin receptor in Pima Indians: influence on energy expenditure, physical activity, and lipid metabolism. *Int J Obes Relat Metab Disord* 2002;**26**:1629–32.
69. Loos RJ, Rankinen T, Tremblay A, Perusse L, Chagnon Y, Bouchard C. Melanocortin-4 receptor gene and physical activity in the Quebec Family Study. *Int J Obes (Lond)* 2005;**29**:420–8.
70. Lorentzon M, Lorenzton R, Lerner UH, Nordstrom P. Calcium sensing receptor gene polymorphism, circulating calcium concentrations, and bone mineral density in healthy adolescent girls. *Eur J Endocrinol* 2001;**144**:257–67.
71. Winnicki M, Accurso V, Hoffmann M, Pawlowski R, Dorigatti F, Santonastaso M, et al. Physical activity and angiotensin-converting enzyme gene polymorphism in mild hypertensives. *Am J Med Genet* 2004;**125A**:38–44.
72. Salmen T, Heikkinen AM, Mahonen A, Kroger H, Komulainen M, Pallonen H, et al. Relation of aromatase gene polymorphism and hormone replacement therapy to serum estradiol levels, bone mineral density, and fracture risk in early postmenopausal women. *Ann Med* 2003;**35**:282–8.
73. Flint J, Valdar W, Shifman S, Mott R. Strategies for mapping and cloning quantitative trait genes in rodents. *Nat Rev Genet* 2005;**6**:271–86.
74. Paigen K, Szatkiewicz JP, Sawyer K, Leahy N, Parvanov ED, Ng SHS, et al. The recombinational anatomy of a mouse chromosome. *PLoS Genet* 2008;**4**:e1000119.
75. Rhodes JS, Hosack GR, Girard I, Kelley AE, Mitchell GS, Garland Jr. T. Differential sensitivity to administration of cocaine, GBR 12909, and fluoxetine in mice selectively bred for hyperactive wheel-running behavior. *Psychopharmacology* 2001;**158**:120–31.
76. Rhodes JS, Garland Jr. T. Differential sensitivity to administration of Ritalin, apormorphine, SCH 23390, but not raclopride in mice selectively bred for hyperactive wheel-running behavior. *Psychopharmacology* 2003;**167**:242–50.
77. Ceaser T. Interval specific haplotype analysis between high and low active mice within identified quantitative trait loci for physical activity. Theses, UNC Charlotte, 2009.

78. Bronikowski AM, Rhodes JS, Garland Jr. T, Prolla TA, Awad TA, Gammie SC. The evolution of gene expression in mouse hippocampus in response to selective breeding for increased locomotor activity. *Evolution* 2004;**58**:2079–86.
79. Kelly MA, Rubinstein M, Phillips TJ, Lessov CN, Burkhart-Kasch S, Zhang G, et al. Locomotor activity in D2 dopamine receptor-deficient mice is determined by gene dosage, genetic background, and developmental adaptations. *J Neurosci* 1998;**18**:3470–9.
80. Good DJ, Coyle CA, Fox DL. Nhlh2: a basic helix–loop–helix transcription factor controlling physical activity. *Exerc Sport Sci Rev* 2008;**36**:187–92.
81. Richert L, Chevalley T, Manen D, Bonjour JP, Rizzoli R, Ferrari S. Bone mass in prepubertal boys is associated with a Gln223Arg amino acid substitution in the leptin receptor. *J Clin Endocrinol Metab* 2007;**92**:4380–6.
82. Tsao TS, Li J, Change KS, Stenbit AE, Galuska D, Anderson JE, et al. Metabolic adaptations in skeletal muscle overexpressing GLUT4: effects on muscle and physical activity. *FASEB J* 2001;**15**:958–69.
83. Gomes FR, Rezende EL, Malisch JL, Lee SK, Rivas DA, Kelly SA, et al. Glycogen storage and muscle glucose transporters (GLUT-4) of mice selectively bred for high voluntary wheel running. *J Exp Biol* 2009;**212**:238–48.
84. Coyle CA, Jing E, Hosmer T, Powers JB, Wade G, Good D. Reduced voluntary activity precedes adult-onset obesity in Nhlh2 knockout mice. *Physiol Behav* 2002;**77**:287–302.
85. Jung AP, Luthin DR. Wheel access does not attenuate weight gain in mice fed high-fat or high-CHO diets. *Med Sci Sports Exerc* 2010;**42**:355–60.
86. Farooqi S, O'Rahilly S. Genetics of obesity in humans. *Endocr Rev* 2006;**27**:710–8.
87. Vella KR, Burnside AS, Brennan KM, Good DJ. Expression of the hypothalamic transcription factor Nhlh2 is dependent on energy availability. *Neuroendocrinology* 2007;**19**:499–510.
88. Chenoweth D, Leutzinger J. The economic cost of physical inactivity and excess weight in American adults. *J Phys Act Health* 2006;**3**:148–63.

Epigenetics and Obesity

Javier Campión, Fermin Milagro, and J. Alfredo Martínez

Department of Food Sciences, Physiology and Toxicology, University of Navarra, Pamplona, Spain

The etiology of obesity is multifactorial, involving complex interactions among the genetic makeup, neuroendocrine status, fetal programming, and different unhealthy environmental factors, such as sedentarism or inadequate dietary habits. Among the different mechanisms causing obesity, epigenetics, defined as the study of heritable changes in gene expression that occur without a change in the DNA sequence, has emerged as a very important determinant. Experimental evidence concerning dietary factors influencing obesity development through epigenetic mechanisms has been described. Thus, identification of those individuals who present with changes in DNA methylation profiles, certain histone modifications, or other epigenetically related processes could help to predict their susceptibility to gain or lose weight. Indeed, research concerning epigenetic mechanisms affecting weight homeostasis may play a role in the prevention of excessive fat deposition, the prediction of the most appropriate weight reduction plan, and the implementation of newer therapeutic approaches.

Progress in Molecular Biology and Translational Science, Vol. 94
DOI: 10.1016/S1877-1173(10)94011-8

1877-1173/10 $35.00

I. Breakthroughs and Perspectives in Obesity and Epigenetics

A. Genes, the Environment, and Their Roles and Interactions in Obesity

Excessive fat accumulation and obesity conditions depend not only on the genetic makeup controlling energy balance, but also on lifestyle and perinatally programmed behavior. In this context, the impact of environmental agents (poor dietary and activity patterns), fetal programming, and even assortative mating could also result in a genetic predisposition and likely play a part in the obesity epidemic. Indeed, it has been hypothesized that the maintenance of body weight and composition depends upon an axis with three interrelated and self-controlled components: (a) food intake; (b) nutrient utilization and thermogenesis; and (c) body fat deposition. All three elements underlie complex interrelated feedback mechanisms and are affected by the individual's genetic background.[1]

Also, from an evolutionary point of view, the survival of individuals with a genome resistant to malnutrition could partly explain the susceptibility of certain population segments most at risk of becoming obese.[2] The interactions between genes and the environment, in particular unhealthy dietary patterns and sedentarism, undoubtedly obscure the specific roles of each determinant.[3]

B. Early Research in the Field

The genetics of obesity has achieved universal recognition only in the last decades.[1] However, the first attempts to link genetic inheritance to excess body mass were reported in the early 1920s by Davenport, in "Body build and its inheritance," which were extended by Vague and coworkers in the 1940s when they reported the occurrence of obesity subtypes associated with sexual dimorphisms. Later, the "thrifty gene" theory rose in a seminal paper by Neel in 1962, outlining the concept that genes predisposing to diabetes would confer relative advantages in populations that often experienced starvation in the past, thus contributing to hypothesize a relationship between genetics, diabetes, caloric intake, and body weight control.[2] The classic books published in the 1970s about "Energy balance and obesity in man" by Garrow (1974) and "The obese patient" by Bray (1976) made minor references to monogenic forms of obesity.[4,5] In 1977, the Nation, Heart, Lung, and Blood Twins Study first indicated the possibility that the observed aggregation for obesity was due to genetic factors rather than environment.[6] Later, at the International Congress of Obesity held at Jerusalem in 1986, the pioneering studies by Bouchard and coworkers concerning the role of genetic predisposition on obesity onset were debated. These studies were further supported by the classic experiment on weight gain in overfed twins.[7]

Subsequently, classic textbooks about obesity by Bjorntorp (2001)[8] as well as by Bouchard *et al.* (1998 and later editions)[9] reinforced the fundamentals of inheritance contributions and gene–lifestyle interactions on obesity. More recently, the impact of biomics, particularly genomics and transcriptomics, on obesity research is leading to new breakthroughs by using specific tools and methodologies such as microarrays.[10] By 2006, 12 editions of the human obesity map had been published showing the continuous advances in the field.[11]

C. Further Research in the Field

Today's view is that about 40–70% of body weight variability is under genetic control and that the growing prevalence of obesity is the result of interactions between a genetic susceptibility with inadequate lifestyles (increase in consumption of high-energy foods enriched with sugar and fats combined with reductions of physical activity levels, etc.). It is also assumed that minor participation of several genes with a larger cumulative effect is more likely to occur than the individual involvement of a major locus. In any case, more than 600 chromosomal regions have been identified up to now as putatively involved in body weight regulation and about 20–30 genes are supported by more than five independent investigations for their role in body thermodynamics.[11]

In this context, the word *epigenetics* was launched as a conceptual model for explaining putative unrevealed interactions between genes and environmental surroundings to produce a phenotype.[12] Thus, an early definition for epigenetics involved "the study of the mechanisms of temporal and spatial control of gene activity describing pathways different from those directly attributable to the underlying DNA sequence and with an influence on the adaptive response of an organism."[12] The prefix *epi-* in *epigenetics* implies features that are "on top of" or "in addition to" genetics, adding to the conventional inheritance paradigm assigned to DNA.

The "epigenetic code," encompassing the chromatin information mainly encrypted by histone signatures and DNA methylation profiles, represents "the sum of the alterations to the chromatin template that collectively establish, modulate, and propagate different patterns of gene expression and/or silencing from the same genome."[13] The nucleosome is the basic repeating unit of chromatin and is constituted of a protein octamer containing two molecules of each core histone (H2A, H2B, H3, and H4) around which is wrapped 147 bp of DNA. Each successive nucleosome core is separated by a DNA linker of approximately 60 bp associated with a single molecule of histone H1.[14]

A number of epigenetic changes appear to occur within the course of each individual organism's lifetime, but some epigenetic information may be passed from one generation to the next.[15] Even though the mammalian genome is wiped clean of most genetic modifications during early embryogenesis, these changes are progressively reestablished during embryonic development, since some evidence suggests that the modifications can be transmitted past the immediately succeeding generation through female and perhaps male lines.[16]

Specific epigenetic processes include bookmarking, imprinting, gene silencing, X chromosome inactivation, position effects, reprogramming, maternal effects, DNA methylation, histone modifications, or chromatin folding and, in general, all those phenomena eventually affect gene expression patterns without changes in the nucleotide sequence (iRNAs, DNA repair, transposons, chaperones, copy number variations, etc.).[12] Two characteristic and outstanding features of epigenetic processes are cellular memory transmission or transgenerational heritability and the involvement in spatial and temporal cell differentiation from totipotent cells.[17]

Therefore, epigenetics can provide some insights not only into unsolved mysteries such as cellular identity, stem cell plasticity, tissue regeneration, tumorigenesis and aging, but also into understanding genetic fetal programming, monozygotic twin differences and chronic disease onset in the adult. The latter requires development of newer and more appropriate experimental models.[18] Epigenetic research helps to explain the way that cells carrying identical DNA can differentiate into distinct cell types by mechanisms such as DNA methylation, small and noncoding RNAs, and chromatin architecture changes.[12] These mechanisms, together with other transcriptional regulatory events, ultimately regulate gene activity and expression during development and differentiation or in response to environmental stimuli.[13]

Furthermore, DNA methylation is commonly linked to transcriptional silencing occurring preferentially in CpG areas or islands that have been found in the promoter region of about half of human genes. The addition of a methyl group to create a 5-methylcytosine is catalyzed by DNA methyltransferases (DNMTs), including DNMT1 for methyl group maintenance or DNMT3a and DNMT3b for *de novo* methylation.[19] Distinctive distribution patterns of CpG methylation are believed critical for the control of gene silencing and chromosomal stability. Also, acetylation of histones H3 and H4 at lysine residues at the N-terminal of histone tails by histone acetyltransferase enzymes (HATs) or deacetylation by histone deacetylases (HDACs) influence the chromatin structure and the accessibility to transcriptional factors, given that acetylation is assumed to open the condensed chromatin structure and allow the transcriptional machinery an easier access to promoter regions. Some inhibitors of HDACs have been hypothesized as potential therapeutic drugs to combat obesity.[13] Histone kinases and phosphatases also may be involved in these regulatory processes.

Another major mechanism by which transitions in the chromatin template are induced is the recruitment of chromatin remodeling complexes that use energy (ATP hydrolysis) to change chromatin and nucleosome composition in a noncovalent manner.[13] While hyperacetylation is often positively correlated with a more active transcription, histone methylation is associated with transcriptional activation, inactivation, or silent genomic regions depending on the methylated region and the degree of methylation.[12] Additional covalent histone modifications that have been identified and characterized as affecting gene expression include histone phosphorylation, ubiquitination, sumoylation, ADP-ribosylation, biotinylation, proline isomerization, and likely others that await description.[20] The discovery and characterization of a vast array of small noncoding RNAs have been associated with interfering properties (siRNAs) or the formation of silent heterochromatin (RNAi), which may produce an epigenetic modulation of targeted promoters. Indeed, RNAi screens have been able to discover new potential therapeutic targets to fight obesity.[17]

The occurrence of transposons or parasitic and repetitive mobile elements that are dispersed through the genome (representing about 45% of the genetic mass) have also been considered another potential epigenetic-related mechanism, which may underlie a substantial amount of phenotypical variability. In this context, metastable epialleles, which may maintain their epigenetic state transgenerationally, have been associated with a transposon insertion and may be influenced by early nutrition.[21]

So far, the better known epigenetic processes affecting phenotype and the gene expression (transcription and silencing) patterns[12] in the same genome involve

- DNA methylation
- Histone variants with their own pattern of modifications
- Copy number variation
- Chaperones, that is, proteins that assist the noncovalent folding/unfolding of macromolecular structures that can remodel chromatin affecting the binding of transcriptional factors to promoter sequences in DNA
- Polycomb group/trithorax proteins, establishing a silenced chromatin state or propagating gene activity as, for example, in the inactivation of the X chromosome in female mammals

Currently, it is assumed that the combination or cumulative effects of several signals over an extended chromatin region can stabilize and propagate epigenetic states. In this context, one can hypothesize that epigenomic markers in specific gene promoters may enable the identification of individuals with an increased susceptibility to chronic diseases in the adulthood and eventually may help in preventing or treating such individuals, as well as predicting therapeutic outcomes. Furthermore, hitherto unidentified imprinted genes and mosaicism are two emerging challenges that can be explained by epigenetic processes involved in the struggle for resources (the imprinting kinship theory) or as phenomena accounting for differences among people in body weight gain and other physiologically regulated processes.[22] Epigenetics is also potentially involved in the differences between homozygous twin brothers or sisters with aging that relate to differences in environmental influences.

Furthermore, genes together with epigenetic mechanisms may determine afferent and efferent signals as well as central mechanisms involved in body weight regulation.[2] Thus, the genetic and epigenetic information involved in short- and long-term body weight regulation and body composition maintenance is acting via (a) different peptides and monoamines involved in the regulation of the appetite, (b) variations in energy and nutrient utilization, resting metabolic rate or energy expenditure of physical activity, and (c) individual differences in adipocyte metabolism.[23] Possible mechanisms mediated by the genetic and epigenetic factors include reduced rates of basal metabolism and macronutrients oxidation, alterations of adipogenesis, and quantitative and qualitative deviations of food intake or activity energy expenditure.[1]

The functional links between the RNAi machinery, covalent histone modifications and DNA methylation contribute to the complex control required for cell-fate determination during development, including the role of enzymatic activities and unveiled chromatin alterations.[13] The major issues in epigenetic research are not only uncovering the epigenetic code and characterizing the epigenetic inheritance patterns as well as their impact on cellular identity and memory, germline imprint,[18] regeneration and aging,[24] but also finding the roles of noncoding RNAs, modeling enzymes, and histone variants on epigenetic marks and dysfunctions.

II. Methods for Epigenome Study of Obesity

The arsenal of genomic assays available and the rapid technological developments concerning next-generation molecular techniques have made possible multiple approaches to study epigenetics and obesity. These technologies range from classical molecular biology (Western blot, RT-PCR) to the most recent genome wide association screens. Thus, depending on the epigenetic question to be investigated (DNA methylation, covalent histone modifications, enzymes involved in modifications, polycombs, microRNA—miRNA, etc.) and the issues of interest (global, locus specific, or massive), a large number of techniques are available for research (Table I).

Many methods have been used to assess DNA methylation. For example, high-pressure liquid chromatography (HPLC) is a simple method to determine the overall content of 5-methyl cytosine in genomic DNA. Other techniques, such as LUMA (LUminometric Methylation Assay),[25] identifier repetitive elements,[26] or high-performance capillary electrophoresis (HPCE)[27] are also

TABLE I
AVAILABLE TECHNIQUES FOR RESEARCH IN EPIGENETICS

	DNA methylation	Histone modifications and nonhistone proteins	Enzymes involved	miRNA
Global	LUMA, IRE, HPLC	HPLC, Western Blot, PAGE, HPCE, RP-HPLC, and hydrophilic interaction LC	Western blot, enzymatic activity, RT-PCR	
Locus-specific	BSP, MM, Pirosequencing, MS-SnuPE, HMR, COBRA, MSP	ChIP assay	ChIP assay	RT-PCR
Genome wide analysis	Human methylation 27 BeadChips, MeDIP	ChIP-on-chip Chip-Seq	ChIP-on-chip, microarrays, proteomics	Microarrays

BSP, bisulfite-sequencing PCR; ChIP, chromatin immunoprecipitation; ChIP-on-chip, ChIP with microarray technology; COBRA, combined bisulfite restriction analysis; HPCE, high-performance capillary electrophoresis; HPLC, high-pressure liquid chromatography; LC, liquid chromatography; LUMA, LUminometric methylation assay; MeDIP, methylated DNA immunoprecipitation; MSP, methylation-specific PCR; MS-SnuPE, methylation sensitive single nucleotide primer extension; PAGE, polyacrylamide gel electrophoresis; RP-HPLC, reversed phase-high-pressure liquid chromatography.

good tools for tissue-specific analysis of the global genomic DNA methylation. These methods give global methylation information, but are not sensitive enough to discriminate pattern differences among genomic loci. Thus, in the search of a set of CpG sites from a specific DNA region, the gold standard method is bisulfite sequencing, based on the treatment of DNA with sodium bisulfite, which converts cytosine residues to uracil, but leaves 5-methyl cytosine residues unaltered.[28] The sequencing of the subsequently generated PCR products enables a reliable analysis of the percentage of methylated CpG sites and the total DNA region. This method is highly time-consuming and expensive, and the quantification depends on the analyzed clones (from $n = 8$ to $n = 20$ is the normal rule).

Other methods to optimize the detection of specific DNA methylation changes are being developed. The most common assay is the methylation-specific PCR (MSP).[29] Two primer sets, one specific for the C-containing sequence and the other binding to T-containing DNA, discriminate between the methylated and unmethylated status of genomic DNA. Useful alternatives to existing high-throughput methylation assays are quantitative methylation-specific PCR (QMSP),[30] MethylLight,[31] and methylation pyrosequencing[32] combining simple CpG analysis with accurate quantitation. Another approach is combined bisulfite restriction analysis (COBRA),[33] which uses bisulfite treatment followed by PCR-amplification of the relevant regions, enzymatic restriction, digestion, and quantification of the resulting fragments. The methylation sensitive single nucleotide primer extension (or MS-SnuPE), like MSP, is a reliable method with several advantages over existing methods because it allows rapid quantitation and requires small amounts of DNA.[34] Finally, matrix-assisted laser desorption/ionization-time of flight (MALDI-TOF) mass spectrometry analysis of a specific PCR product after a base-specific cleavage process results in a distinct signal pattern from the methylated and nonmethylated template DNA.[35]

Two major techniques have later appeared as an alternative to bisulfite conversion immunoprecipitation with 5-methyl cytosine (methylated DNA immunoprecipitation—MeDIP),[36] and methyl-dependent enzyme digestion with the endonuclease McrBC.[37] Both methods have been successfully used with quantitative PCR and even with microarrays, establishing high-throughput methods that allow a detailed analysis of the DNA methylation pattern of individual CpG islands. Indeed, oligonucleotide microarrays have been used to investigate DNA methylation in different tumors,[38,39] and could be an excellent way to process large numbers of obese patients and genetic regions. In this regard, human methylation 27 BeadChips from Illumina[40] is nowadays an excellent approach for genome methylation wide analysis for human samples.

Histone modifications play a critical role in epigenetic control of gene expression. Global analysis in different tissues could indicate the metabolic state of a patient using classical methods such as Western blot or more specialized techniques such as HPLC. The multitude of these posttranslational modifications and the existence of numerous histone variants require particular separative strategies for their analysis, a prerequisite for studying biological processes. The most widely utilized techniques for the separation of histones are polyacrylamide gel electrophoresis, HPCE, reversed-phase-HPLC and hydrophilic interaction liquid chromatography.[41] However, locus specific analyses are necessary also for the histone modifications, for both a deeper understanding of the involved mechanisms and description of the role of histones in gene regulation together with the specific DNA methylation pattern. Chromatin immunoprecipitation assay (ChIP assay), the gold standard method for these locus-specific studies, allows determination of the location of DNA binding sites on the genome for a particular protein of interest. This technique provides a picture of protein–DNA interactions that occur inside the nucleus of living cells or tissues. Of course, ChIP assays also offer the possibility of analyzing the role of transcription factors and nonhistone proteins that influence chromatin structure by interacting with histones and DNA in a number of ways, such as methyl-CpG binding proteins or polycomb group complex.[42]

ChIP-on-chip and Chip-Seq methods are used primarily to determine the way that transcription factors and other chromatin-associated proteins influence phenotype-affecting mechanisms.[42,43] Determining those proteins that interact with DNA to regulate gene expression is essential for the understanding of many biological processes and disease states. This epigenetic information is complementary to genotype and expression analysis. The use of modification-specific antibodies in ChIPs coupled to gene array technology (ChIP-on-chip) has revolutionized the ability to monitor the global incidence of histone modifications. Such combined global and locus specific analysis has been performed only on a subset of modifications (mainly acetylation and lysine methylation), but the results clearly show that modifications are not uniformly distributed.[44] As with regular ChIP, ChIP-on-chip is used to investigate *in vivo* interactions between proteins and DNA. Specifically, it allows the identification of binding sites of DNA-binding proteins on a genome-wide basis.[45] Whole-genome analysis can be performed to determine the locations of binding sites for almost any protein of interest.[45] The distribution of modifications and their localizations may offer new insights into the mechanisms of regulation of cellular processes.

ChIP-Seq technology is also seen as an alternative to ChIP-on-chip. The latter requires a hybridization array and necessarily introduces some bias, as an array is restricted to a fixed number of probes. Sequencing, by contrast, is thought in principle to be less biased, although the sequencing bias of the current technologies is not yet fully understood. Specific DNA sites in direct physical interaction with transcription factors and other proteins can be isolated by ChIP that produces a library of target DNA sites, which bound to a target *in vivo*. Massively parallel sequence analyses are used in conjunction with whole-genome sequence databases to analyze the interaction pattern of any protein with DNA,[46] or the pattern of any epigenetic chromatin modifications. This strategy can be applied to the set of ChIP-able proteins and modifications, such as transcription factors, polymerases and transcriptional machinery, structural proteins, protein modifications, and DNA modifications.[43]

The biological mechanisms underlining the changes in DNA methylation and covalently modified histones involve enzymes that remodel chromatin, including the DNMT family (for DNA), HATs, histone methyltransferases (HMTs), protein kinases, and ubiquitin- and SUMO-protein ligases. The combination of modifications ("marks") produced by these enzymes acts as a code that regulates downstream processes such as transcription, DNA repair, and apoptosis.[47] Hence, the techniques for investigating changes in mRNA (RT-PCR), protein levels (Western blot, immunohistochemistry), and specific enzymatic activity facilitate the study of epigenomics.[47–50] In addition, it is possible to similarly analyze histone modifications in the binding of these enzymes to specific promoters by ChIP assays and ChIP-on-chip and ChIP-

Seq, in order to investigate relationships between changes in DNA methylation, patterns of histone modification, and the presence of a putative enzyme on a particular gene.[51]

Epigenetic research requires elegant models and a wide array of molecular techniques to investigate DNA-mediated epigenetic phenomena from bisulfite sequencing to ChIP (together with its large-scale variants ChIP-on-chip and ChIP-Seq). Bioinformatics plays an increasing role (computational epigenetics)[42] and is essential to analyze the massive amount of data generated by these new techniques to cross-link them to the global epigenetic field (DNA methylation or histone modifications vs. gene expression, histone modifications vs. DNA methylation, enzymatic mechanisms vs. gene expression, DNA methylation or histone modifications vs. enzymatic mechanisms, etc.). Thus, computational epigenetics[42] uses bioinformatic methods to complement experimental research in epigenetics. Due to the recent surge in epigenome datasets, computational methods play an increasing role in all areas of epigenetic research.[42]

In this context, online resources are useful tools for data analysis and computational studies. There are several consortia and networks with news and information regarding epigenomic phenomena (Table II); for example, the Human Epigenome Consortium or Epigenome Network of Excellence. These Web sites and others can enhance computational epigenetics, from extraction of raw information (promoter sequence) to statistical analysis of the results. To start with a reliable DNA region (generally a promoter sequence in charge of controlling the expression of a gene), the UCSC Genome Bioinformatics Site is widely used, although others, such as PROMOSER and Transcriptional Regulatory Element Database (TRED), aim specifically at the extraction of a large number of promoter sequences from mammalian genome databases.[52,53]

Several Web-based programs can detect changes in the methylation patterns after bisulfite conversion of a DNA region of interest (Table II). Thus, after bisulfite conversion of genomic DNA, introducing a sequence of interest in MethPrimer or Bisearch serves to design bisulfite-sequencing PCR (BSP) or MSP.[54] Useful tools for bisulfite sequencing analysis of CpG methylation are BiQ Analyzer,[55] Methtools and QUMA,[56] software designed for DNA methylation profile analyses, including easy visualization and quality control of DNA methylation data. For planning DNA methylation assays beside BSP, Methmaker from Max-Planck-Institut für Informatik (Germany) facilitates the design for COBRA, bisulfite SNuPE, bisulfite pyrosequencing, MethyLight, and MSP. This excellent online resource (http://methmarker.mpi-inf.mpg.de/) also implements a systematic workflow for design, optimization, and computational validation of DNA methylation biomarkers.[57] Finally, another interesting resource for COBRA analysis is Snake Charmer.

TABLE II
SOME EXAMPLES OF WEB-BASED RESOURCES FOR EPIGENETICS RESEARCH

Name	Webpage
Computational epigenetics	http://en.wikipedia.org/wiki/Computational_epigenetics
Epigenetics society	http://es.landesbioscience.com/index.php
Human epiogenome Project	http://www.epigenome.org/index.php
Epigenome-NOE	http://www.epigenome-noe.net/
Epigenetics research	http://www.epidna.com/
Chromatin structure and function	http://www.chromatin.us/chrom.html
Epigenie	http://www.epigenie.com/index.html
Extraction of a large number of promoter sequences from mammalian genome databases	
UCSC Genome Bioinformatics Site	http://genome.ucsc.edu/
PROMOSER	http://biowulf.bu.edu/zlab/PromoSer/
TRED	http://rulai.cshl.edu/TRED/
Design of primers for DNA methylation assays after bisulfite conversion	
Methmaker	http://methmarker.mpi-inf.mpg.de/
Methprimer	http://www.urogene.org/methprimer/index1.html
BiSearch	http://bisearch.enzim.hu/
MethprimerDB	http://medgen.ugent.be/methprimerdb/index.php
Tools for bisulfite sequencing analysis of CpG methylation	
BIQ Analyzer	http://biq-analyzer.bioinf.mpi-inf.mpg.de
QUMA	http://quma.cdb.riken.jp/
MethTools 2.0	http://194.167.139.26/methtools/MethTools2_submit.html
Tools for the selection of restriction enzymes for COBRA	
Snake charmer	http://insilico.ehu.es/restriction/two_seq/snake_charmer.html
Matrix-based searches of factor binding sites in regulatory DNA sequences	
MatInspector	http://www.genomatix.de
Match profile	http://www.gene-regulation.com
Regulatory sequence analysis tools (RSAT)	http://rsat.ulb.ac.be/rsat/index.html
Vista	http://genome.lbl.gov/vista/index.shtml
INCLUSive	http://homes.esat.kuleuven.be/~dna/BioI/Software.html
DNA methylation database	
MethDB	http://www.methdb.de/
Histone modifications databases	
Human histone modification database	http://www.hhmd.org/

It may be interesting for modeling and monitoring therapies, to identify those transcription factors whose binding sites could be affected by the methylation patterns at a specific CpG site. Actually, MatInspector,[58] RSAT, Vista, and Match Profile[59] are some Web-based resources that utilize large libraries of matrix descriptions for transcription factor binding sites to locate matches in DNA sequences. INCLUSive Suite, involving motifsampler and motifscanner software, can find specific sequences in a genome-wide analysis. Finally, databases on DNA methylation (MethD) and histone posttranscriptional modifications (HHMD) focus on integrating useful DNA methylation and histone modification information from experimental data.

III. Fetal Programming and Epigenetics

The fetal or developmentally programmed origins of adult disease hypothesis states that environmental factors and maternal lifestyles, particularly nutrition, act in early life to drive the risks for the onset of metabolic diseases including cardiovascular events, insulin resistance, and excessive weight gain in later life stages.[60] These relationships between adult diseases and perinatal nutritional status were initially identified through epidemiological strategies, but further confirmed and characterized in animal experiments and may be partly explained by epigenetic mechanisms.[16] Indeed, maternal nutrition can epigenetically program the embryo with gene expression patterns that persist into adulthood and may contribute to the appearance of typical metabolic syndrome features such as hypertension, type 2 diabetes, hyperlipemia, and excess weight or obesity.[61]

The thrifty phenotype hypothesis proposed that, when the fetal environment is nutritionally unbalanced by excess or deficiency, an adaptive response prioritizes the functions of key organs at the expense of less metabolically relevant organs.[62] This hypothesis is consistent with the putative existence of temporal fetal and embryonic "windows" to a suboptimal intrauterine environment believed to predispose to future adverse consequences in specific metabolic pathways. In this context, the term metabolic plasticity describes the interactions between changing nutritional *in utero* conditions and subsequent health outcomes that may persist for several generations. These effects are at least in part mediated through epigenetic processes that may alter the risk of obesity later in life.[60]

The mechanisms by which specific tissues could be affected permanently as a result of nutritional impairments include epigenetic changes in gene regulation, variations in organ structure and cell number or size, apoptosis processes, clonal selection of specific cell lineages, hormonal imprinting, and metabolic differentiation.[63]

In this context, the relation between birthweight and adiposity, measured in childhood and adulthood is generally positive, but some investigations have reported a J-shaped or even a U-shaped association between birthweight and adult obesity, with a higher tendency to central adiposity at low birthweights, particularly when followed by nutritional abundance in the postnatal environment.[64]

Given that early nutrition can influence covalent modifications of DNA and core histones, current research is focused on epigenetic processes that are sensitive to nutritional regulation in early life. Thus, there is convincing evidence from *in vitro* embryo cultures that the methylation status of genomically imprinted genes, including insulin-like growth factor 2 (*Igf2*), histocompatibility 1 (*H1*), and insulin-like growth factor 2 receptor (*Igf2r*), is affected by the nutritional incubation conditions and that epigenetic lability may extend to the early postnatal period in rodents.[63] Dietary methyl donor supplementation may also prevent transgenerational amplification of adiposity in obesity-prone mice. Pioneering experiments have demonstrated that maternal diets enriched in methyl donors such as choline, folate, or methionine may result in epigenetic changes in gene expression in the fetus with permanent impact on appetite control and energy homeostasis that can be transmitted from one generation to the next.[65]

Parental conditions and lifestyles, especially maternal obesity, famine at perinatal periods, the use of nutritional supplements or alcohol, or drug abuse, as well as the administration of therapeutic agents in this critical period, may alter epigenetic processes with an impact on embryonic, placental, and fetal growth, organogenesis, or regulatory set points for system functions affecting adiposity.[66] All these environmental–epigenetic interactions have potential consequences on the health and body composition of the next and subsequent generations by affecting epigenetically mediated gene regulation.

While the mechanisms underlying these associations are still poorly characterized, the two primary targets for the perinatal programming of obesity appear to be the adipocyte and the neurocrine networks that control food intake and fuel homeostasis in adult life. It has been reported that fetal nutrition influences the activation of some adipogenic/lipogenic genes, such as peroxisome proliferator-activated receptor gamma (*PPARG*)[67] or lipoprotein lipase (*LPL*),[68] which affect subcutaneous adipocytes development predominantly around the time of birth. Another possibility is that maternal nutrition participates in the generation of metabolic or endocrine signals acting on genes encoding neuropeptides, such as cocaine- and amphetamine-regulated transcript (*CART*), neuropeptide Y (*NPY*), agouti-related protein homolog (mouse) (*AGRP*), or proopiomelanocortin (*POMC*).[69] Furthermore, some nutritional regimens used to investigate early life developmental programming of obesity have shown that poor maternal nutrition during pregnancy involves glucocorticoid metabolism in the offspring by altering nuclear receptor subfamily 3, group C, member 1 (*NR3C1*, also known as *GR*), hydroxysteroid (11-beta) dehydrogenase 1 (*HSD11B1*), and *HSD11B2* gene expression, all of which may be modulated by leptin.[70] The contribution of other hormones or factors, such as ghrelin and uncoupling proteins, on the long term nutritional programming of the energy balance is still a matter of debate, although insulin and glycemia appear as consistent mediators of some of the putatively involved epigenetic processes and predictive adaptive responses to unbalanced nutrition in early life and particularly during fetal development.[71]

In this context, researchers have found that infants of women with gestational diabetes mellitus had significantly greater skinfold measurements and adiposity and that children exposed to acute malnutrition in early pregnancy may be more likely to be obese in later life.[72] Indeed, a series of experiments has suggested the possibility of epigenetic heredity induced by a high-fat diet, since the offspring of mothers fed a hyperlipidic regimen had hyperinsulemia and hyperglycemia as well as increased *LPL* and leptin gene expressions.[73] These findings support the idea that epigenetic marks are not necessarily erased between generations and that epigenetic reprogramming could occur during gametogenesis and early embryogenesis affecting gene expression, especially when influenced by environmental and nutritional factors (Fig. 1). Thus, it has been suggested that some methylated sites survive and then are replicated by DNMTs when the cell divides, and this epigenetic code is passed along with associated histones.[74]

Furthermore, it has been hypothesized that genomic imprinting or inheritance depending on parental-specific epigenetic information may constitute a way of competing for resources during pregnancy as stated by the kinship theory.[75] For instance, the Prader-Willi syndrome, frequently associated with hyperphagia resulting in morbid obesity, is caused by the underexpression of genes at 15q11-q13 that are normally expressed when paternally derived. The expression of these imprinted genes is regulated by the methylation of CpG islands in their promoters, whose markings are stably replicated during cell division, but are reversed when inherited through an individual of the opposite sex.[76]

IV. Dietary Factors and Epigenetic Regulation

A. DNA Methylation

Epigenetic mechanisms can be controlled by dietary compounds, metabolic state, endocrine impairments, and diseases, all of them keys in the development, comorbidities, and treatment of obesity.[66] Early studies of DNA methylation postulated possible effects of one-carbon compounds on

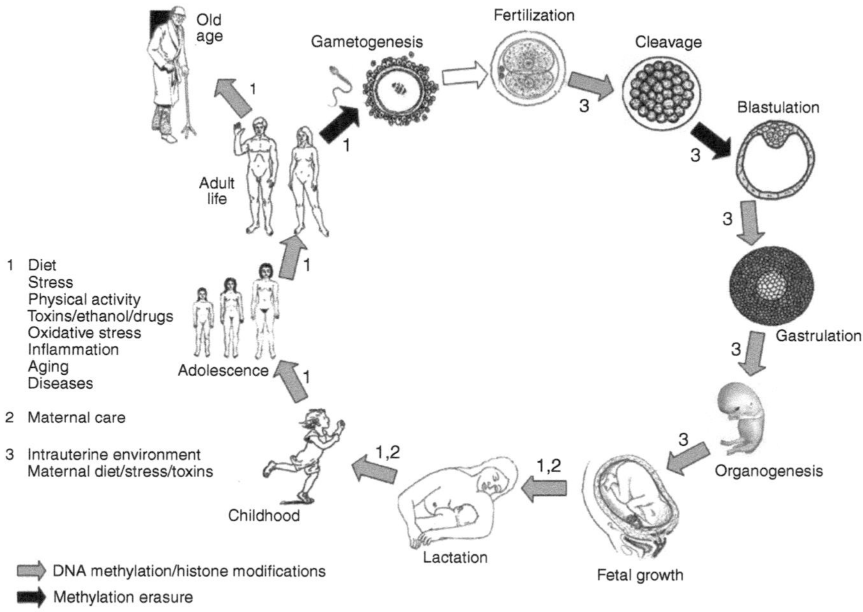

FIG. 1. Periods of life in which DNA methylation processes have been reported. Erasure of methylation imprints is almost exclusively seen at two stages: primordial germ cells and blastocyst development. Maternal care, aging, and environmental factors, such as dietary components, toxins and drugs, inflammation and, maybe, physical activity, could regulate the methylation processes at various periods of life.

methyltransferases action. DNA and HMTs all use *S*-adenosylmethionine (SAM) as a methyl donor,[47] its bioavailability being directly influenced by the diet. SAM is formed from methyl groups derived from choline, methionine, or methyl-tetrahydrofolate. Betaine, choline, methionine, zinc, and folate are necessary for homocysteine conversion into methionine,[76] so their effects on epigenetic marks are interrelated to alter epigenetic DNA methylation processes. In fact, the perturbation of the metabolism of one of the methyl donors results in compensatory changes in the other methyl donors because of the intermingling of these metabolic pathways.[47]

The underlying concept is that if a methyl donor such as folate affecting epigenetic status is interrupted, the subsequent loss of epigenetic control will alter gene expression. Other disruptions include mutations in genes that metabolize folate before the synthesis of SAM.[77] Limiting folate intake decreases the SAM/*S*-adenosylhomocysteine (SAH) ratio to inhibit DNA methylation by affecting the activity of DNMTs.[68] Experimental *in vivo* studies modifying one-carbon nutrients (Table III) show that dietary restriction of methyl donors

TABLE III

Examples of Dynamic Changes in DNA Methylation as Affected by Dietary Methyl Donors Supply and Reported Relationships with Obesity

Species	Tissue	Nutritional condition	Effect	Gene	Relation to obesity	Ref.
Human	Erythrocytes	High serum folate	No changes in methylation status	ec-SOD gene		85
Human	Whole blood samples	Periconceptional folic acid use	Higher methylation	IGF2 DMR	Modulation of growth processes	84
Mouse	Liver, kidney, and brain	Extra folic acid, vitamin B_{12}, choline, and betaine during development	DNA hypermethylation	A(vy) allele	Maternal obesity induces transgenerational amplification of obesity	65
Mouse	Colon	Folate supplementation	Increased levels of methylation	Global DNA and p16 promoter		82
Mouse apoe-deficient	Aorta	Betaine supplementation	No effects	TNF-alpha	Betaine may exert its antiatherogenic effect by inhibiting inflammation	86
Rat	Aorta	Hyperhomocysteinemia induced by high methionine diet	Hypomethylation	B1 repetitive elements		88
Rat	Liver	Diet deficient in methionine and choline	Hypomethylation	c-Ha-ras, c-Ki-ras, and c-fos proto-oncogenes		80
Rat	Liver	Postweaning dietary folate deficiency	Increased methylation	Genomic DNA		89
Rat	Liver	Folate supplementation	Increased levels of methylation	Global DNA		81

Rat	Mammary tumors	Maternal choline nutrition	Highly methylated	Stratifin gene		83
Rat	Fetal liver and brain	Maternal choline deficient	Hypermethylation	G9a and Suv39h1 genes		88
Rat	Liver	B_{12} deficient	Hypomethylation	Cystathionine beta-synthase promoter		79
Rat	Liver	Choline-devoid diet	DNA undermethylation	Global DNA		78
Sheep	Fetal liver	Restricting the supply of specific vitamins and methionine	Multiple changes	Altered methylation status of 4% of 1400 CpG islands examined	Adult offspring were both heavier and fatter	90

induces, in general, DNA hypomethylation,[78–80] and supplementation produces DNA hypermethylation.[81–84] It is possible also to find studies showing lack of effects[85,86] or opposite results,[87–89] due in some cases to the time window of the dietary treatment (maternal, postweaning, or nonmaternal nutrition) or more probably mediated by locus and sex-specific changes.[90]

The worldwide consensus for women planning pregnancy is to consume a daily intake of 400 μg of synthetic folic acid in the periconceptional period to prevent birth defects in children. Periconceptional folic acid administration is associated with epigenetic changes in *IGF2* in the embryo that may affect intrauterine programming of growth and development with consequences for health and disease throughout life.[84] However, relevant reductions in specific dietary inputs to the methionine/folate cycles during the periconceptional period can lead to widespread epigenetic alterations to DNA methylation in offspring, and modify adult health-related phenotypes.[90] Also, early dietary interventions led to adult offspring who were both heavier and fatter, had altered immune responses to antigenic challenge, elicited insulin-resistance, and had elevated blood pressure-effects that were more obvious in males. Also, feeding mice a lipogenic methyl-deficient diet caused liver injury similar to human nonalcoholic steatohepatitis and alterations in the expression and levels of the maintenance DNMT1 and *de novo* DNMT3A proteins in the liver.[91] Finally, the elegant model of Waterland *et al.*,[65] using agouti viable yellow (A(vy)) mice and methyl donors in the diet, shows that in a population with a genetic tendency for obesity, the effects of maternal obesity accumulate over successive generations and suggest that DNA methylation is involved in this process.

When the dietary model was designed to reduce the dietary protein content during a rodent pregnancy, instead of modifying the one-carbon nutrient supply (Table IV), induced changes were seen in DNMTs mRNA. Several obesity-related genes appeared to be hypomethylated, including acetylcholinesterase (*Ache,* also known as *Ace-1*), *Ppara*, phosphoenolpyruvate carboxykinase 1, cytosolic (*Pepck*), and *Nr3c1*(or *Gr*).[92–94] The protein source was observed to be a determinant in epigenetic changes when human vegetarians were compared to omnivores,[95] although some discrepancies were found probably due to the DNA region and CpGs sites analyzed.[96] In this context, Waterland *et al.*[97] found differential metabolic patterns of the imprinted maternally expressed transcript (nonprotein coding) (*H19*) gene using different protein sources, although the cause of this effect could also be a dietary nucleotide deficiency.

Interventional hypocaloric regimens are widely used for treating overweight in humans,[98] but few reports have dealt with their potential effects on DNA methylation patterns. Individuals who were prenatally exposed to famine during the Dutch Hunger Winter in 1944–1945 had, six decades later, less

TABLE IV

Examples of Dynamic Changes in DNA Methylation as Induced by Different Macronutrients or Calorie Intake and Reported Relationships with Obesity

Species	Tissue	Nutritional condition	Effect	Gene	Relation to obesity	Ref.
Fetal baboons	Brain, kidney, liver, and heart	Nutrient-restricted mothers fed 70% of controls	Decreased methylation in 0.5G kidney and increased methylation in 0.9G kidney and frontal cortex	Global DNA		101
Human	Whole blood	Famine during periconceptional exposure	Hypomethylation	IGF2 gene		99
Human	PBMC cells	Calorie restriction of obese men	Successful weight loss in hypomethylated promoter	TNF	DNA methylation could be a good inflammation marker predicting the hypocaloric diet-induced weight loss	103
Human	Buccal mucosa	Vegetarian versus omnivores	Decreased CpG methylation in the vegetarian group	MnSOD gene		95
Human	Colon	Vegetable intake	No changes in DNA methylation	PKCb1, ornithine decarboxylase 1, c-fos and MTHFR		96
Human	White adipose	Caloric restriction	Hypo- and hypermethylated	Several loci involved in body weight loss	DNA methylation is responsive to caloric restriction	102
Mouse	Kidney	Protein origin/dietary nucleotide deficiency	Hypermethylation	H19		97

(*Continues*)

TABLE IV (*Continued*)

Species	Tissue	Nutritional condition	Effect	Gene	Relation to obesity	Ref.
Mouse	Fetal offspring	Antenatal maternal low-protein diet	Hypomethylation	ACE-1		93
Mouse	Visceral adipose tissue	Diet-induced obesity	Hypermethylation	PPARgamma	High-fat diet and diabetes increase promoter methylation	104
Mouse	Hypothalamus	Maternal high-fat diet	Hypomethylation	GH secretagogue receptor	Heritability of body length and glucose homeostasis are modulated by maternal diet	73
Primiparous sows	Embryo	Feed restriction of primiparous sows during the last week of lactation	Hypomethylation	Global DNA		100
Rat	Liver	Dietary protein restriction of pregnant rats	Hypomethylation	GR and PPARalpha		94
Rat	Liver	Dietary protein restriction of pregnant rats	Hypomethylation in F1	PPARalpha and GR promoter		92
Rat	Hypothalamus	Neonatal overfeeding by rearing Wistar rats in small litters (SL)	Hypermethylation	*POMC*	Overfeeding is an epigenetic risk factor of obesity	106
Rat	Adipose tissue	Diet-induced obesity	Hypermethylation	Leptin	High-fat diet increases promoter methylation	105

DNA methylation of the imprinted *IGF2* gene compared with their unexposed same-sex siblings.[99] The association was specific for periconceptional exposure, reinforcing the notion that very early mammalian development is a crucial period for establishing and maintaining epigenetic marks. Studies using nutrient restriction during pregnancy in primiparous sows[100] and in baboons[101] showed changes in global DNA methylation from brain, kidney, liver, and heart demonstrating organ specificity, gestational age specificity, and maternal nutritional status implications.

In the adipose tissue from humans, it has been recently demonstrated that DNA methylation and gene expression are responsive to caloric restriction.[102] This finding provides new insights about the epigenetic pathways involved in body weight loss as well as methylation regulation during adulthood. In this regard, our group described a successful prediction of weight loss based on several CpGs sites from the tumor necrosis factor (TNF) promoter of blood mononuclear cells.[103] Finally, we have also recently found that several CpGs sites of aquaporine 9 (*AQP9*) gene from blood cells were hypomethylated following calorie restriction in humans (unpublished data).

Examples of obesity-related modifications in DNA methylation have been reported in relation to changes in dietary macronutrients. Thus, *in vivo* studies using rats fed a high-fat diet showed hypermethylation of several CpG sites at the *Pparg*[104] and leptin[105] genes from white adipose tissue and transgenerational hypomethylation of the growth hormone receptor (*Ghr*) gene[73] in the hypothalamus, all of them suggesting a role of high-fat diet and DNA methylation in the pathogenesis of obesity and metabolic syndrome. Moreover, it has been described that early overfeeding in rats alters DNA methylation patterns of promoter regions of hypothalamic genes critically involved in the lifelong regulation of food intake and body weight.[106] All these studies demonstrate that a nutritionally driven alteration of the methylation pattern plays a critical role in body weight homeostasis.

Dietary micronutrient supplementation and plant compounds have been proposed for the improvement of different disease states. Arsenite, selenite, or iron levels influence CpG methylation.[107–110] Soy isoflavones of different origins partially reversed the hypermethylation status of tumors in mice and humans[111–113] or induced *per se* hypermethylation at some genes.[112,114]

DNA methylation levels within the cells can be altered by dietary nutrients. Hence, cellular metabolic compounds and hormones affected indirectly by diet and nutrients, and by endocrine imbalance and inflammatory disease, can have a role in epigenetic mechanisms.[66,115,116] The chronic low-grade inflammatory state that accompanies obesity and other related disorders, such as insulin resistance, cardiovascular diseases, and metabolic syndrome, may also be under epigenetic regulation. Evidence suggests an epigenetic control of inflammation by several genes, as suggested, for example, by the hypermethylation of the interleukin 8 (*IL8*) gene in the oral mucosa[117] or a direct effect of TNF-alpha treatment on non-CpG methylation of peroxisome proliferative activated receptor, gamma, coactivator 1 (*PGC-1*) alpha.[118] Moreover, inflammatory diseases such as ischemia[119] or inflammatory bowel[120] produce changes in the DNA methylation pattern of ER alpha (now ESR1) gene or polycomb target genes.

In addition, changes in the DNA methylation level by specific hormones could contribute to the control of transcription, although these processes are not yet well understood.[121] Insulin induces global hypomethylation in human hepatic HepG2 cells,[122] and the exposure of preimplantation embryos to insulin alters expression of imprinted genes, such as *H19* and *IGF2*. Recently, it has been reported that DNA methylation/demethylation is endocrinally switched to control transcription of the cytochrome p450 27B1 (*CYP27B1*) gene[121] by vitamin D and parathyroid hormone, some estrogens hypomethylated genes (fragile X mental retardation syndrome 1 homolog (*Fmr-1*) and *O*-6-methylguanine-DNA methyltransferase—*MGMT*),[123,124] and miRNA-9-3 promoter region.[125] These findings link miRNAs with hormonal control of epigenetics and DNA methylation.

Finally, some metabolic-related diseases (e.g., coronary artery disease,[126] hypertension,[127] diabetes,[86,118,122,128,129] anorexia nervosa,[130] and bulimia) have been related to aberrant methylation of certain obesity genes including *H19*, *IGF2*, peroxisome proliferative activated receptor, gamma, coactivator 1 alpha (*PPARGC1A*), *HSD11B2*, and atrial natriuretic peptide (*NPPA*), suggesting that obese unhealthy patients could present a different methylation pattern in comparison to healthy normal-weight subjects. Indeed, epigenetics markers could act as prognosis biomarkers of obesity-related diseases.

B. Histone Modifications

Histones, especially the accessible residues of the amino termini of H3 and H4 and the amino and carboxyl termini of H2A, H2B, and H1, are susceptible to a variety of posttranslational modifications, including phosphorylation of serine and threonine, lysine acetylation, methylation of lysine and arginine, lysine ubiquitination, lysine sumoylation, lysine biotinylation, deimination, ADP ribosylation, glycosylation, and carbonylation.[131] Such modifications of histones can create or stabilize binding sites for regulatory proteins, such as transcription factors, proteins involved in chromatin condensation or DNA repair, but may also have the opposite effect, disrupting or occluding chromatin-binding sites.[12] Some of these modifications are compatible with other reactions, facilitating other modifications, or be present simultaneously whereas still other modifications are incompatible, and negatively affect other histone changes.

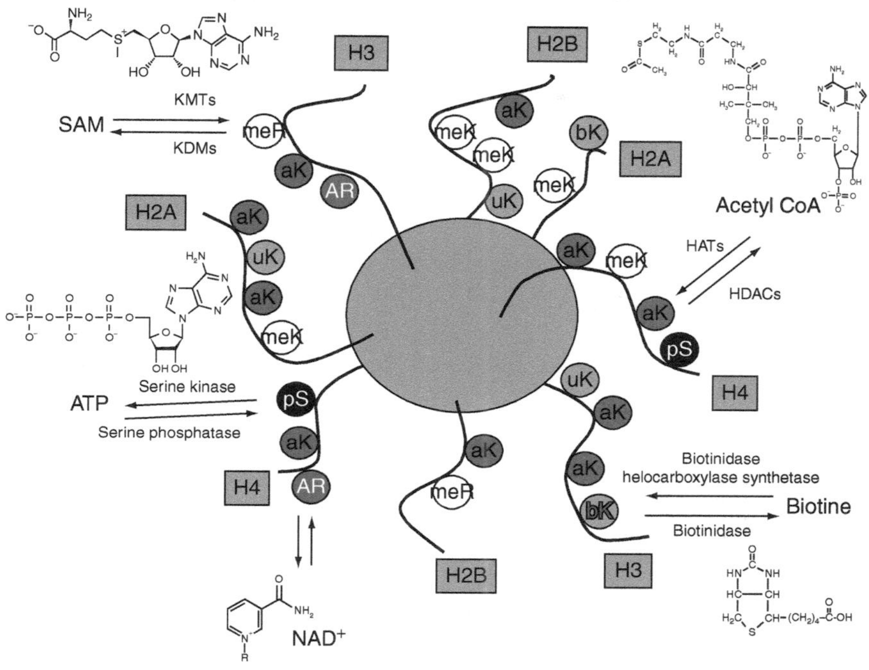

FIG. 2. Main histone modifications, enzymes involved, and donors. aK, acetyl lysine; meR, methyl arginine; meK, methyl lysine; uK-ubiquinated lysine; pS, phosphorylated serine; bK, biotinylated lysine; AR, ADP-ribosylation.

Different enzymes are involved in posttranslational histone modifications. A number of HDACs and HATs, HMTs, or more recently histone demethylases (HDMs) have been described in humans.[12,13] Other enzymes that mediate histone modifications are serine and threonine kinases and phosphatases, ubiquitin ligases, biotinidases, sumoylases and desumoylases, proline isomerases, ADP-ribosylases, or deiminases.[132] Functions of these enzymes depend on cofactors, such as acetyl CoA, biotin, NAD, and SAM. Intranuclear levels of these proteins are crucial in the function of epigenetic enzymatic modifications, but they depend not only on the dietary intake of related nutrients but also on the metabolic, inflammatory, or redox status of the cell (Fig. 2).

As chromatin modifications correlate and perhaps directly regulate gene repression and induction, these modifications have been broadly classified into repressing and activating mechanisms. In general, histone acetylation and phosphorylation are considered activators of gene expression, whereas histone

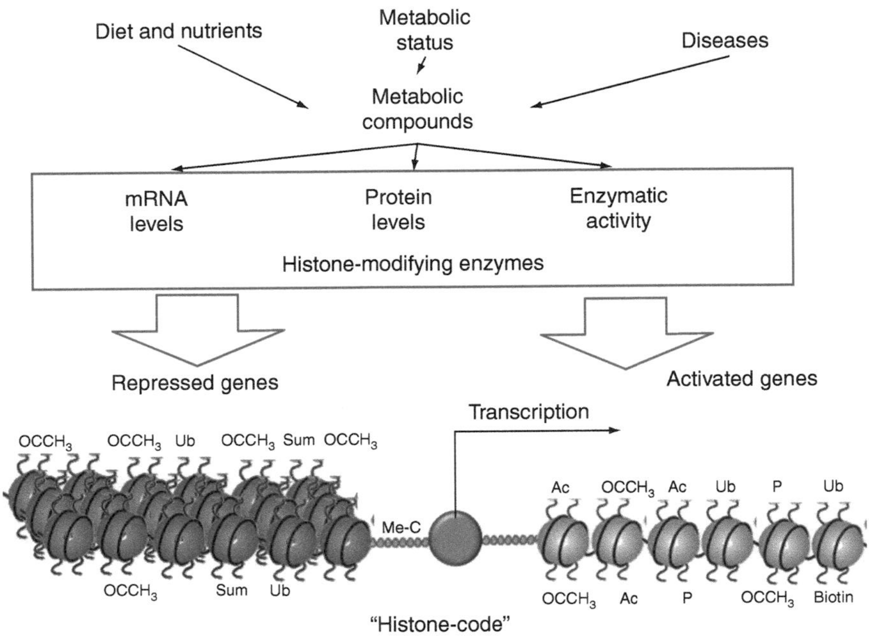

FIG. 3. Influence of metabolic compounds and diseases on the activity of histone-modifying enzymes and gene transcription regulation.

deacetylation, sumoylation, and biotinylation inhibit gene expression (Fig. 3). Finally, methylation and ubiquitination can act as repressors or activators depending on the affected histone residue.[133]

Acetylation of histones: Histone lysine acetylation is a reversible posttranslational process governed by the opposing actions of acetyltransferases and deacetylases.[12,13] Histone acetylation is performed by HATs, which transfer an acetyl group from acetyl CoA to lysine. This process neutralizes the positively charged histones; disrupting histone binding to negatively charged DNA and thus promoting transcription factor binding. Deacetylation of histones by HDACs results in a decrease in the space between the nucleosome and the DNA, diminishing the accessibility for transcription factors and modifying the chromatin in the vicinity of nearby genes, silencing heterochromatin structure (Fig. 4).

In relation to the increasing prevalence of obesity, one factor that stimulates histone acetylation is chronic hypoxia, which is exacerbated in obese adipose tissue, triggering macrophage infiltration, and the secretion of proinflammatory cytokines.[134] In prostate cells, it has been demonstrated that hypoxia increases global DNA methylation and H3K9 histone acetylation.[135]

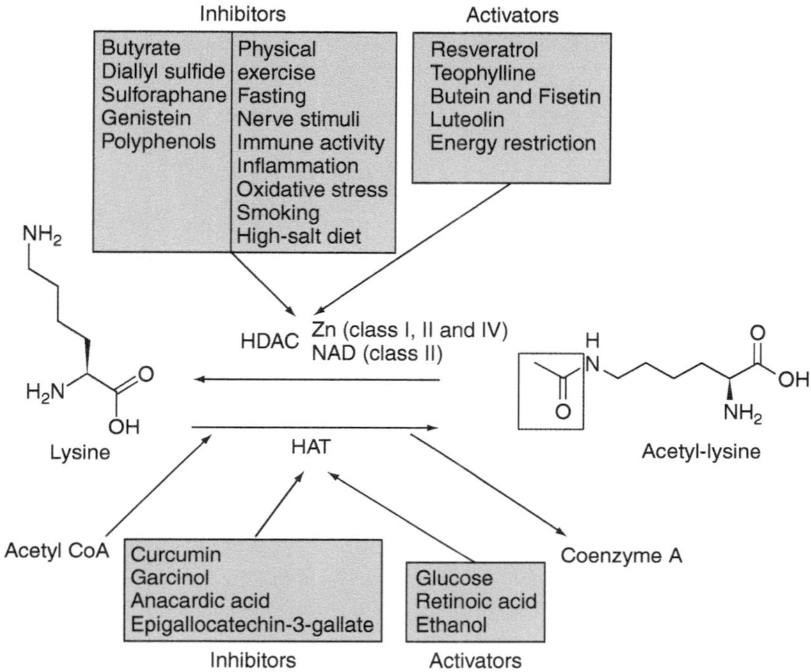

FIG. 4. Different dietary and metabolic factors influencing histone acetylation.

Among other obesity-related stimuli, reactive oxygen species (ROS) and TNF-alpha act as HDAC inhibitors, indirectly increasing HAT activity and DNA unwinding, facilitating the binding of transcription factors (i.e., AP-1 and NF-kB), and enhancing proinflammatory gene expression.[136]

Several dietary factors have been linked to increased histone acetylation (Table V), including biotin deficiency,[137] quercetin,[138] or caffeine.[139] Also, glucose is able to regulate insulin gene transcription by hyperacetylation of histone H4,[140] which depends on the recruitment of the HAT p300 to the insulin gene promoter.[141]

Among the natural HDAC inhibitors that can be provided by the diet, the most relevant are short-chain fatty acids produced by microbial fermentation of dietary fiber in the large intestine, such as butyrate and (more weakly) propionate, which inhibit type I, II, and IV HDACs.[142] Other inhibitors of classical HDACs are diallyl disulfide, present in garlic;[143] sulforaphane from cruciferous vegetables;[144] and genistein, a soy isoflavone with antiestrogenic activities.[145]

TABLE V

EXAMPLES OF DIETARY AND METABOLIC FACTORS AFFECTING HISTONE ACETYLATION

Category	Effect	Refs.
HAT inhibitors		
Garcinol	Inhibits histone acetyltransferases p300 and PCAF	156
Anacardic acid	Inhibits p300 and p300/CBP-associated factor histone acetyltranferase	157
Epigallocatechin-3-gallate	Abrogates p300-induced p65 and IL6 acetylation and increases HDAC3 recruitment	158
Curcumin	Inhibits the p300/CREB-binding protein	155
HAT activators		
Glucose	Acetylation of H3 K9, 18, 27 and H4 K5, 8, 12 by stimulating Gcn5 and Esa1	160
Glucose	Hyperacetylation of histone H4 by recruiting histone acetyl-transferase p300	140,141
Retinoic acid	Acetylates H3K9	50
Ethanol	Acetylates H3K9	161
High carbohydrate/fat ratio	Induces H3 and H4 acetylation in SGLT1 and sucrose-isomaltase genes	162,163
HDAC activators		
Resveratrol	Activation of SIRT1, an HDAC	48
Teophylline	Enhances HDAC activity	151
Caloric restriction	Induction of the SIRT1 and SIRT2 deacetylases	153,154
HDAC inhibitors		
Glucose	Reduction of the class III HDACs	150
Butyrate	Inhibition of type I, II, and IV HDACs	142
Diallyl disulfide	Increases histone H4 and/or H3 acetylation by inhibiting HDACs	143
Sulforaphane	HDAC inhibition	144
Genistein	Inhibition of HDAC6	145
Black tea polyphenols	Inhibits the expression of histone deacetylase HDAC-1	146
High salt	Blockade of the functioning of class IIa HDACs	147
Zinc	Inhibition of type I, II, and IV HDACs	148
High glucose	Reduction of SIRT1 expression levels and enzyme activity	150
Low levels of NAD+	Inhibition of sirtuin activity	149
Others		
Biotin deficiency	Alteration of the synthesis of the enzymes involved in histone acetylation	137
Quercetin	Induction of histone hyperacetylation	138
Caffeine	Increases GLUT4 expression by acetylating the MEF2 site	139

Other polyphenols have been reported as weak inhibitors of HDACs in the colon, such as coumaric acid, 3-(4-OH-phenyl)-propionate and caffeic acid,[142] as well as a mix of black tea polyphenols.[146]

High sodium induces the expression of a kinase that can block the functioning of class IIa HDACs in rats.[147] Finally, as all the classical HDACs are zinc-dependent hydrolases, zinc levels in the diet could affect their activity.[148]

Regarding sirtuin inhibitors, it has been reported that high nicotinamide levels[149] and high glucose levels[150] can reduce the class III HDAC activity. As sirtuins need to bind to NAD^+, low levels of NAD^+ inhibit their activity.[149] Other natural compounds, however, can act as activators of HDACs. The most interesting one is resveratrol, a plant antioxidant found in nuts, grapes, and red wine that acts as a sirtuin activator.[48] Another interesting anti-inflammatory plant compound, teophylline, found in small amounts in green tea and cocoa beans, shows activating properties on HDACs.[151] The plant polyphenols butein and fisetin[48] and the flavonoid luteolin[152] also act as SIRT1-activating compounds. *In vivo*, activation of sirtuins mimics caloric restriction, which is linked to positive effects on longevity biomarkers.[153,154]

Although less abundant than HDAC inhibitors, several natural products that inhibit HAT activity have been described. The most potent of them is curcumin, the main curcuminoid of curry,[155] that specifically inhibits the p300/CREB-binding protein through the involvement of ROS. Other HAT natural inhibitors are garcinol, a benzophenone isolated from *Garcinia indica* fruit,[156] anacardic acid, isolated from cashew nut shell liquid,[157] and epigallocatechin-3-gallate, a catechin abundant in green tea.[158] Finally, also copper induces histone hypoacetylation by inhibiting HAT activity *in vitro*.[159]

Regarding HAT activators, it has been reported that glucose induces overall acetylation of H3K9, K18, K27 and H4K5, K8, K12 in yeasts,[160] retinoic acid (an oxidized form of vitamin A) acetylates H3K9,[50] and ethanol acetylates H3K9 and H3K18 in rat hepatocytes and increases p300 protein levels,[161] all of which probably contribute to alcohol-induced hepatotoxicity.

Finally, a diet with a high carbohydrate/fat ratio induces H3K9 acetylation on the murine sodium-glucose cotransporter (*Sglt1*) gene and its expression in jejunum,[162] as well as H3 and H4 acetylation on the sucrase-isomaltase (*Sis*) gene.[163] These results suggest an adaptive response to dietary exposure, possibly leading to epigenetic changes in the genes responsible for carbohydrate digestion and absorption.

Among the intrinsic factors able to activate HDACs, we find the glucocorticoids,[164] which are widely used as anti-inflammatory drugs. They decrease inflammatory gene transcription by recruiting HDAC-2 to the transcription activation complex by the glucocorticoid receptor, resulting in deacetylation of histones.[165]

Some key genes in body weight regulation, such as adiponectin (*ADIPOQ*),[166] are suppressed by HDAC inhibitors (i.e., valproic acid). However, other important metabolic genes, such as thyroid hormones[167] or leptin (*LEP*),[168] increase histone acetylation and transcriptional activation. Other factors potentially involved in obesity susceptibility that have been reported to alter histone methylation patterns are maternal behavior in the lactation period,[169] physical exercise, and fasting.[170]

Methylation of histones: Histone methylation has strong combinatorial potential with respect to other modifications, given that a lysine residue can house either mono-, di-, or trimethyl moieties on the amine group, whereas an arginine residue can carry mono- or dimethyl groups on the guanidinyl group.[12] The dimethyl arginine state is further defined by whether the modification exists in a symmetric or an asymmetric configuration.[49,171] Both, lysine and arginine methylations, can act either as activators or repressors of gene transcription, that is, methylation of H3K4, H3K36, and H3K79 correlates with gene activation, whereas methylation of H3K9 and H3K27 correlates with repression.[172,173]

All lysine methyltransferases transfer a methyl group from SAM to the amino group of a lysine residue, leaving the cofactor by-product SAH.[174] SAM is formed from methyl groups derived from nutrients such as choline, methionine, or methyl-tetrahydrofolate; meanwhile betaine, choline, methionine, zinc, and folate are necessary for homocysteine conversion into methionine. Thus, the effects of these nutrients on epigenetic marks are interrelated, which influences epigenetic DNA and histone methylation processes.[66]

Regarding lysine demethylases, two mechanisms have been demonstrated: Amine oxidation by LSD1 requires flavin adenine dinucleotide (FAD) and produces H2O2, which functions as a signaling molecule in a variety of physiological processes, including insulin signaling, transcriptional regulation, and different pathological conditions.[175] However, hydroxylation by JmjC-domain containing proteins requires α-ketoglutarate, Fe2+, and oxygen.[176]

Hypoxia, which is usually exacerbated in obese adipose tissue, induces a pool of histone modifications typically associated with transcriptional activation or repression, including H3K4, H3K9, H3K27, H3K36, and H3K79.[177,178] Moreover, most of the JmjC domain-containing proteins are upregulated by hypoxia.[179]

An inflammatory condition, related to different metabolic diseases such as obesity and cardiovascular diseases, is also linked to an increase in histone methylation. Thus, LPS, a proinflammatory macrophage-inducing compound, upregulates KDM6B/JMJD3, an H3K27me3 demethylase,[180] and increases H3K4me3 and H3K4 levels in relation to iNOS and TNFA genes and SET1MLL methyltransferase.[181] TNF-alpha induces the recruitment of p65, p300, and SET7/9 (H3K4 methyltransferase) at the MCP-1 and TNFA promoters, accompanied by an increase in H3K4me3 levels.[182] Finally, the chronic upregulation of the v-rel reticuloendotheliosis viral oncogene homolog A (avian) (NF-kB-p65 or RELA) gene as a result of ambient or prior hyperglycemia has been associated with increased H3K4m1 and reduced H3K9m2 and H3K9m3 methylation on the p65 promoter.[183] However, senescence increases expression of the HDM Jmjd3 and decreases the level of histone H3K27 trimethylation (H3K27me3).[184]

A number of nutrients and dietary compounds have been related to changes in histone methylation (Table VI). Thus, high doses of the phytoestrogen genistein demethylates H3K9 at the PTEN and CYLD promoters,[185] whereas high levels of different minerals, such as chromium, nickel, and arsenite, induce changes in histone methylation in different residues (H3K4, H3K9, and H3K27).[186–188]

As for DNA methylation, methyl donors are crucial for the correct process of protein methylation, and a deficiency hampers methylation in the cell. Thus, the cellular ratio SAM/SAH, especially due to the inhibitory capacity of SAH, inhibits methyltransferase activities.[189] Additionally, a methyl-deficient diet reduces protein and mRNA levels of some lysine and arginine methyltransferases in mice.[190]

However, different studies have demonstrated varied findings on tissue-, residue-, and dose-specific patterns of histone methylation.[12,91,190,191] Thus, the consumption of a methyl-deficient diet (low methionine, choline, and folate free), which is used to develop hepatic steatosis in mice, resulted in a loss of histone H4K20me3 in liver but a compensating increase in H3K9me3,[91] indicating that histone methylation is subject to complex regulation. Another study showed that methyl-deficient diet increased the levels of histone macroH2A and H3K9me3 in mice, but reduced those of H3K27me3, H3K4me2 and H4K20me3,[191] H3K9me2, and H3K9me3.[190]

Choline supplementation during gestation[192] or *in vitro* experiments[181] increased lysine methyltransferase mRNA levels. It also enhanced the levels of the repressing marks H3K9Me2 and H3K27Me3, whereas those of H3K4Me2, associated with active promoters, were higher in choline-deficient rats.[192] However, general food restriction[193] and protein deprivation[194] increased H3K27me3, H3K9me3, and H3K9me1 and reduced H3K4me3, with changes in glucocorticoid receptor (NR3C1) and IGf2 mRNA gene expression.

Furthermore, ethanol has been reported to induce deep changes in histone methylation in rat hepatocytes. Thus, alcohol increases H3K27me3, which silences gene expression, and H3K4me2, which activates gene expression,[195] decreases H3K9me2 methylation and increases H3K4me2[196] and has been related to an increase in p300 protein levels.[197]

Circulating levels of glucose are increased in many obese patients, which have been associated to an enhanced methylation of H3K4 and H3K9 on the coding region of the L-PK gene. cAMP produces opposite effects,[198] and

TABLE VI

Effects of Different Metabolic and Dietary Experimental Models on Specific Histone Methylation

Region	Methyl deficiency	Methyl deficiency	Low-protein diet	Hypocaloric diet	Gestational choline	Chromium	Nickel	Transient hyperglycemia	Glucose	Insulin	Glucose reduction	Type 1 diabetes	TNF-alpha	LPS *in vitro* (inflammation)	Hypoxia
H3K4	–			–	+	+		+	+	+		+	+	+	+
H3K9	+	+	+/–	+	+	+	+		+	+	+	+/–			+
H3K27	–	–		+	+	–									+/–
H3K36															+
H3K79															+
H4K20	–	–		–											
H3R2						–									
H3R17										–					
H4R3															+
Refs.	191	91,190	194	193	192	187	186	199	198	200	204	202,203	182	181	135,178

+, Hypermethylation; –, hypomethylation.

transient hyperglycemia increases H3K4me1 monomethylation and mobilizes histone methyltransferase Set7.[199] Insulin reduces the methylation level at H3R17 and downregulates PEPCK and G6Pase.[200] Finally, the diabetic state, frequently accompanied by excessive fat deposition, induces a decrease of H3K9me3 in vascular smooth muscle cells from db/db mice,[201] an increase of histone H3K9m2 in the lymphocytes of type 1 diabetic patients,[202] and a decrease of H3K9me2 and an increase of H3K4me2 in human peripheral blood cells of the same patients.[203]

Some of these physiological factors regulate histone methylation via redox status. Thus, glucose deprivation induces a higher NAD+/NAD ratio which is related to a redox-induced increase in the HMT Suv39H1 and in H3K9me2,[204] and insulin-induced increases in H3K4me1 and H3K9me1 via ROS.[205]

Other histone modifications: Histone phosphorylation, although less studied than histone methylation and acetylation, is believed to play a direct role in mitosis, cell death, repair, replication, and recombination, being usually related to gene transcription activation.[206] Different dietary factors (including genistein, a methylating agent mentioned earlier) facilitate this modification, found to be associated to H1 and H3 phosphorylation in cancer cells,[207] sulforaphane,[208] the vinca alkaloid vinblastine,[209] or the mycotoxin zearalenone.[210] Histone H2AX, which mediates DNA repair, is rapidly phosphorylated when DNA is damaged by hypoxia.[211]

Finally, histone biotinylation depends on biotin availability, with biotin deficiency producing important effects on chromatin structure,[212] whereas a nickel overdose is deleterious for histone ubiquitination of histones H2A and H2B.[213]

V. Fetal and Early Postnatal Life Influence on Future Obesity Risk

It is becoming clear that epigenetics is a significant contributor to obesity incidence and other related diseases.[22,66] Recent studies have highlighted the importance of adverse perinatal environments, during pregnancy or lactation, in the future development of obesity, suggesting that mother's nutrition or lifestyle choices could alter the developmental programming of the fetus.[214] The role of nutrition in adulthood in modifying the epigenetic patterns of DNA and their possible transmission through the gametes is a matter of serious debate.[215] Changes in DNA methylation patterns could be the result of complex interplay between various dietary factors, such as imbalances in the supply of methyl donors, particularly folate, and exposure to DNMT inhibitors such as polyphenols and possibly isothiocyanates from plants.[216]

In comparative biological sciences, the concept of "maternal effects," whereby maternal environmental influences affecting one generation can have significant impact on the next generation and potentially subsequent generations, is well recognized.[72] In addition, maternal stress and dietary intake during gestation influence later life behavior and physiology.[60] These observations have led to the definition of "metabolic imprinting," which refers to the long-term effects that the prenatal and perinatal environment have on the developing fetus and the neonate.[217] This concept asserts that the control of energy homeostasis is not only genetically determined, but that it is also highly plastic and readily modified by environmental inputs during developmental stages, especially the perinatal period.[63]

Thus, endocrine exchanges between fetus and mother, intrauterine positions, pre- or perinatal nutrition, and maternal attention may contribute to specific epigenetic marks.[218] Indeed, maternal undernutrition or stress, sometimes associated with low birthweight, increases the risk of metabolic and cardiovascular illness in the offspring, which has been partially explained by alterations in glucocorticoid metabolism.[219,220] Some of these events are due to physiological, endocrine or nutritional imbalances, but could be also related to epigenetic mechanisms.

Among the recognized environmental factors, a sedentary lifestyle, the meal size, or dietary fat and carbohydrate intake in the servings may contribute to obesity development.[221] For instance, postnatal dietary intake is associated with insulin resistance in fetally malnourished, low-birthweight rats,[222] intrauterine ethanol exposure with abnormal glucose homeostasis in adult female rat offspring,[223] and maternal dietary fat and alcohol consumption with insulin resistance in adult rat offspring.[224] However, fish oil supplementation in pregnant rats reduces adult disease risks in their offspring[225] and improves cognitive performance.[226] Perinatal temperature also may play a role in this process, as demonstrated in Osborne-Mendel rats. Those reared at 18 °C gained more weight during adult life than those reared at 30 °C, irrespective of their diet.[227]

Perinatal environment is particularly relevant during developmental stages because of its impact on the DNA methylation profile of different cell types.[228] This notion has been linked to the concept of developmental plasticity, a mechanism by which a growing organism, especially in the perinatal period, can adapt to its environment to maximize its chances of survival.[229] However, such impairments could result in permanent changes in specific physiological axes and subsequent predisposition to diverse pathological conditions in the adulthood, as has been reported for relevant hormonal pathways, such as those involved in glucocorticoid metabolism.[230]

There are several examples about the role of nutritional interventions in pregnancy, such as protein restriction, caloric deprivation, micronutrient deficiency or supplementation, and excessive fat feeding, which determine a cluster of disorders in the offspring. Two of the most interesting reports come from Ravelli *et al.*,[231] who showed that most children born after the Dutch Famine of 1944 developed obesity in adult life, and from Eriksson *et al.*,[232] who concluded that fetal growth retardation is a risk factor for obesity. In fact, reduced fetal growth has been repeatedly associated with adiposity and excess weight in adult life, as well as increased morbidity and mortality from cardiovascular disease, type 2 diabetes and metabolic syndrome complications.[63,233] Prenatal undernutrition decreases levels of spontaneous physical activity in adult rats,[234] and has also been associated to elevated blood pressure,[235] glucose intolerance and insulin resistance,[236] and central adiposity.[237] The underlying mechanism is unknown but may result from epigenetically and permanently modified gene expression in susceptible cells.[238]

The early postnatal diet is also relevant concerning the predisposition to later obesity. Many studies suggest that breastfed infants are less likely to become obese than formula-fed children.[239] Neonatal overfeeding in rodents, caused by small litter size or by insulin injections, lead to increased rates of obesity, diabetes, and altered sympathetic activity in adult animals, similar to early exposure to cool temperatures or to a maternal diet containing refined carbohydrates.[240] The early nutritional environment can induce a hyperphagic state by modifying brain structure and vasculature.[241] A good example is the study by Vickers *et al.*[242] showing that growth-retarded pups became hyperphagic and, when later fed on a high-caloric diet, developed central adiposity. Regarding specific diets, maternal iron deficiency,[243] and high-saturated-fat diet[244] increase blood pressure in the offspring. Protein restriction during gestation and lactation causes adverse transgenerational effects on glucose metabolism in rat progeny,[245] and impairs appetite via preferential fat intake over carbohydrates, at least in female rats.[246]

Some studies suggest that maternal obesity itself is associated with a higher risk of obesity in offspring, probably by altered brain monoamine function in obesity-prone offspring.[247] Also, diabetes (types 1 and 2) during pregnancy increases the risk of offspring to develop hypertension,[248] obesity, and type 2 diabetes.[249] Finally, human studies have shown that maternal smoking could lead to adult obesity in the offspring,[250,251] but these findings are still a matter of debate.

Parental attention has been reported to be important factor in some metabolic and mental diseases.[220,252] Thus, in rats, it has been reported that maternal care, high-fat diet, and different forms of malnutrition can influence the maturation of neural systems controlling activity, spatial orientation, and scanning time, and those affecting behavior.[253] Even in adult life, social structures lead to different patterns of dominance status and can modify behavior in test paradigms focused on aggressiveness, anxiety, and locomotion.[218] These effects could be related to epigenetic mechanisms, because it has been shown that stressing rat mothers during lactation can alter DNA methylation pattern in the offspring.[254]

Thus, the ratio of leptin to fat mass was significantly greater in the children who had received a nutrient-enriched preterm formula than in those who received a banked breast milk. This finding suggests that programming of relative leptin concentrations by early diet may be a mechanism linking perinatal nutrition with later obesity through epigenetic marks affecting central nervous networks and specific neuropeptides (NPY, Agro, POMC, CART) controlling appetite and the satiety set point.[63] One example is that the hypothalamic POMC promoter becomes hypermethylated by early overfeeding, an epigenetic risk for obesity programming.[106]

Furthermore, the epigenetic control of gene expression through DNA methylation and histone modification is influenced by free radicals and oxygen availability, which are nutrition-dependent and would be mediated by DNMTs and histone modifying enzymes.[47]

Methyl donor supplementation during pregnancy may prevent transgenerational amplification of obesity in an obesity-prone mouse strain, which is believed to be mediated by epigenetic mechanisms.[65] Studies in sheep have shown that limiting folate and vitamin B_{12} supply in the immediate periconceptual period can result in fatter adult offspring.[255] Pioneering empirical data concerning prenatal exposure to famine in humans have revealed a lower DNA methylation of the imprinted *IGF2* gene in siblings from deprived mothers as compared to controls, in agreement with the developmental onset of health and disease hypothesis. Interesting features of the epigenetic programming are its potential reversibility, which has been reported after methyl supplementation,[256] and the changes in DNA methylation over time.[257]

Furthermore, dietary protein restriction of pregnant rats induces, while folic acid supplementation prevents, epigenetic modifications as evidenced by the expression of genes such as *Gr* (*Nr3c1*), *Ppar*, and acyl-Coenzyme A oxidase 1, palmitoyl (*Acox1 or Aox*) in the liver of offspring,[94] whose epigenotype and lipid profile phenotype can be modified by extra folic acid administration during the juvenile-pubertal period.[258]

Interestingly, multivitamin supplementation of Wistar rats during pregnancy increases food intake and accelerates the development of obesity in offspring fed an obesogenic diet, which has been hypothesized as mediated by the epigenetic machinery, although the specific vitamins involved remain to be identified.[259]

Recent evidence has demonstrated that some chromatin epigenetic remodeling events can orchestrate gametogenesis in a sex-dependent way.[260] In humans, maternal levels of serum folate during pregnancy positively correlate with the offspring adiposity and insulin resistance at 6 years of age.[261] Lower H3 methylation and higher expression of proinflammatory genes have been reported in diabetic animals.[201]

Thus, the challenges concerning the role of epigenetics in obesity include the description of nutritional factors influencing epigenomic marking,[66] the characterization of the epigenetically vulnerable periods of life, the identification of putatively involved gene promoter regions, and the application of epigenomics in designing reliable obesity biomarkers as has been found for cancer and other chronic diseases.[57,262]

VI. Genetics and Epigenetics of Obesity: Epiobesogenes

Continuous scientific advances promote understanding of the implication of genetic and epigenetic mechanisms in the development and therapy of several important diseases, such as cancer, type 2 diabetes, and obesity.[263–265] The heritability estimates for obesity are high compared with other complex diseases.[266] Thus, multigenic interactions are more relevant to human obesity, for which the genetic contribution has been estimated to account for 25–80% of the variability in weight gain.[267] There is general agreement that most susceptibility genes for obesity do not have a major etiological role, but are interlinked with environmental factors such as diet or physical activity.[268] There are at least 250 genes that, when mutated or expressed as transgenes in the mouse, can affect adiposity and body weight, and more than 150 candidate genes for obesity have been associated with overweight-related phenotypes.[11]

Evidence of the polygenic nature of obesity is indirectly demonstrated by the description of only 11 single-gene mutations, affecting less than 200 patients with an obesity phenotype[11]: corticotropin-releasing hormone receptor 1 and 2 (*CRHR1* and *CRHR2*), G-protein-coupled receptor 24 (*MCHR1*, also known as *GPR24*), *LEP* and its receptor (*LEPR*), melanocortin receptors 3 and 4 (*MC3R* and *MC4R*), neurotrophic tyrosine kinase receptor type 2 (*NTRK2*), *POMC*, proprotein convertase subtilisin/kexin type 1 (*PCSK1*), and single-minded homolog 1 (*SIM1*).

Moreover, the number of publications describing genes regulated by epigenetic mechanisms is rising every month, in relation to cancer and metabolic diseases such as hypertension, atherosclerosis, diabetes, and obesity (Table VII). In this context, the search for gene promoters both susceptible to epigenetic regulation and associated with the development of obesity is of great interest. Thus, from approximately 800 human genes under putative epigenetic regulation (http://www.ncbi.nlm.nih.gov/sites/entrez), approximately 13% of them could be associated to obesity and are defined as epiobesogenes (Table VII).

TABLE VII

Examples of Several Human Genes Recently Described as Regulated by Epigenetic Mechanisms and Involved in Obesity

Obesity-related condition	Gene symbol	Name	Epigenetic evidence	Refs.
Adipogenesis	*PPARGC1A*	Peroxisome proliferator activated receptor gamma coactivator-1 alpha	Important in human islet insulin secretion	118,128
Adipogenesis	*ESR1*	Estrogen receptor alpha	Prognostic value of ER hypermethylation	277
Adipogenesis	*NR3C1*	Glucocorticoid receptor	Methylation status is sensitive to prenatal maternal mood	252
Adipogenesis and cell cycle	*CDKN1A*	Cyclin-dependent kinase inhibitor 1A (p21, Cip1)	Aberrant promoter methylations related to cancer	113
Adipogenesis and inflammation	*LEP*	Leptin	Postzygotic development, adipocyte maturation, and cellular aging	284,285,313
Energy metabolism	*COX7A1*	Cytochrome *c* oxidase subunit VIIa polypeptide 1 (muscle)	Age influences DNA methylation	287
Inflammation	*SOD3*	Extracellular superoxide dismutase	Development of foam cells	283
Inflammation	*SOCS1*	Suppressor of cytokine signaling 1	Severity of liver fibrosis and hepatocarcinoma	280
Inflammation	*SOCS3*	Suppressor of cytokine signaling 3	Role in cellular growth and migration and melanomas	281
Inflammation and apoptosis	*CASP8*	Caspase 8	Hypermethylation in neuroblastomas and medulloblastomas	278
Inflammation and insulin resistance	*IL8*	Interleukin 8	Inflammation in the oral mucosa by periodontitis induce hypomethylation	117

Inflammation and insulin resistance	*TNF*	Tumor necrosis factor alpha	Epigenetic silencing during endotoxin tolerance and myeloid differentiation	270
Inflammation and insulin resistance	*SIRT1*	Class III histone deacetylase	CpG island methylator phenotype in cancer	279
Insulin resistance	*IGF2*	Insulin-like growth factor 2	Biomarker associated with pathway-specific predisposition to cancer	282
Insulin resistance	*SSTR2*	Somatostatin receptor 2	Tissue-specific related	286
Insulin resistance	*IGFBP3*	Insulin-like growth factor binding protein 3	Hypermethylation is associated with nonsmall cell lung cancer	274
Insulin resistance and adiposity	*HSD11B2*	11 beta-hydroxysteroid dehydrogenase 2	*In vivo* epigenetic repression and relation to hypertension	127
Insulin signaling	*CAV1*	Caveolin 1	Aberrant methylation is associated with hepatocellular carcinoma	275
Lipid metabolism	*LPL*	Lipoprotein lipase	Commonly methylated in prostate tumors	276
Lipid metabolism	*FABP5*	Fatty acid binding protein 5	Aberrant methylation in esophageal squamous cell carcinoma	273
Lipid metabolism	*NNPA*	Atrial natriuretic peptide	Epigenetic downregulation in females with eating disorders	130

An understanding of variation in their methylation modification patterns due to dietary factors, age, inflammation, or other obesity-related physiological agents is crucial for the investigation of the roles of these mechanisms in the prevention, development, and therapy of obesity.[98]

Increasing our knowledge about epigenetic patterns, their role in development, evolution, and adaptation, and small molecules (nutrients, drugs, etc.) that can reverse epigenetic effects would be useful to unlocking silenced genes or enhancing genes for therapeutic purposes, including those targeting obesity.[116] In addition, certain genetic variants of methyl metabolism enzymes such as the 5,10-methylenetetrahydrofolate reductase (NADPH) (*MTHFR*), 5-methyltetrahydrofolate-homocysteine methyltransferase (*MTR*), and 5-methyltetrahydrofolate-homocysteine methyltransferase reductase (*MTRR*) genes could be risk factors predisposing to obesity, and, by affecting methylation reactions, could be viewed as being indirect epigenetic mechanisms.[269]

A bioinformatic analysis of promoter regions in the search of epigenetic biomarkers of obesity identified affected methylation patterns on several obesity-related genes, such as fibroblast growth factor 2 (basic) (*FGF2*), phosphatase and tensin homolog (*PTEN*), cyclin-dependent kinase inhibitor 1A (p21, Cip1) (*CDKN1A*), and estrogen receptor 1 (*ESR1*), implicated in adipogenesis; suppressor of cytokine signaling 1 (*SOCS1*)/*SOCS3*, in inflammation; and cytochrome *c* oxidase subunit VIIa polypeptide 1 (muscle) (*COX7A1*), *LPL*, caveolin 1, caveolae protein, 22 kDa (*CAV1*), and insulin-like growth factor binding protein 3 (*IGFBP3*), in intermediate metabolism and insulin signaling.[66] The characterization of those individuals who at an early age undergo changes in the methylation profiles of specific genes could help to predict their susceptibility to later develop obesity, helping prevention, and aiding in following up its progress, as well as in researching and developing better therapeutic approaches. Although promoters of other putative candidate genes do not contain CpG-rich sequences, they could also be implicated in epiobesogenic mechanisms regulating body weight.[270] For instance, *NR3C1* and *TNF* genes do not have CpG islands but are under epigenetic control and play an important role in obesity development.[194,270–272]

Certain human genes recently reported as having an epigenetic role in a disease or a physiological trait (Table VII) are related to enhanced susceptibility to several kinds of tumors, such as *ESR1*, *CDKN1A*, *SOCS3*, caspase 8, apoptosis-related cysteine peptidase (*CASP8*), sirtuin (silent mating type information regulation 2 homolog) 1 (*Saccharomyces cerevisiae*) (*SIRT1*), *IGF2*, *IGFBP3*, *CAV1*, *LPL*, and fatty acid binding protein 5 (psoriasis-associated) (*FABP5*).[113,273–282] Moreover, changes in the methylation patterns of several genes have been shown to be involved in hypertension (*HSD11B2*),[127] anorexia/bulimia (*NPPA*),[130] diabetes (*PPARGC1*),[118,128] periodontitis

(*IL8*),[117] tissue and cellular differentiation (*LEP*, superoxide dismutase 3, extracellular (*SOD3*), *TNF*, somatostatin receptor 2—*SSTR2*),[270,283–286] aging (*COX7A1*),[287] and depression (*NR3C1*).[252]

These genes also play a role in obesity development and related processes, especially adipogenesis, inflammation, lipid metabolism disturbances, or insulin signaling.[288–291] The control of adipose tissue differentiation and proliferation is crucial to understand the complexity of overweight.[292] Genes reported to be involved in adipogenesis and susceptible to epigenetic control are *PPARGC1A*,[288] *CDKN1A*,[289] *ESR1*,[291] and *NR3C1*.[290] The process involves the continued control of chromatin structure, histone modifications, and the regulation by miRNAs, which coordinately modulate the expression of sets of genes that maintain pluripotency, direct cells toward a particular lineage, or contribute to maintain the differentiated state of mature cells.[293] Histone acetylation increases at the promoter of adipogenic marker genes during adipocyte differentiation, whereas a decrease in HDAC expression takes place. Unexpectedly, a negative effect of HDAC inhibitors on adipogenesis has also been described.[294] In this sense, a master epiobesogene is *SIRT1*, a class II HDAC involved in obesity, insulin resistance, inflammation, and lipid metabolism.[295–297]

Also, a key adipokine involved in body weight and food intake regulation with CpG islands in the promoter region that could be affected by dynamic methylation is leptin. In obese rats fed a high-fat diet, a CpG in the leptin promoter was found to be hypermethylated[105] and associated with lower circulating leptin levels, suggesting that this position could be important in the regulation of leptin gene expression. This finding indicates that leptin is a target sequence of different transcription factors, with a promoter pattern influenced by diet-induced obesity, and suggests that epigenetic mechanisms could be involved in the recent obesity pandemic by regulating the expression of important epiobesogenic genes.[66]

In addition to adipogenesis, inflammation is now recognized as a process involved in obesity and associated comorbidities.[3,298] Inflammation could induce hypermethylation and DNA damage in several promoter genes,[299] providing a potential link between excess weight and epigenetics. Moreover, increased expression of *Dnmt3a* in the adipose tissue of transgenic mice may contribute to obesity-related inflammation.[300] Thus, the methylation patterns of promoters of several inflammatory genes in different cell populations, such as blood cells, hepatocytes, or adipocytes, could be useful markers of obesity development or of the response to diet-induced weight loss.[98] Thus, TNF-alpha is a proinflammatory cytokine, commonly elevated in obese subjects and with a promoter susceptible to regulation by cytosine methylation. Epigenetic regulation of human *TNFA* promoter by cytosine methylation can be involved in the predisposition to lose body weight after following a balanced hypocaloric diet.[103] Thus, *TNFA* promoter methylation together with other inflammatory-related genes, such as *SOD3*,[301] *SOCS1* and *SOCS3*,[302] *IL8*,[303] and *CASP8*,[304] could be inflammation markers predicting hypocaloric diet-induced weight loss, and could represent a first step toward personalized nutrition based on epigenetic criteria.

Finally, lipid metabolism and insulin resistance-related genes are excellent targets for the study of epigenetic regulation of obesity and related comorbidities, including those which encode proteins implicated in insulin-signaling pathways such as *IGF2*,[305] *SSTR2*,[306] *IGFBP3*,[307] and *CAV1*.[308] Moreover, epigenetic-related genes responsible for lipid metabolism and obesity development include *LPL*,[291] *FABP5*,[309] and *NPPA*,[310], which relate to metabolic deterioration through actions in both macrophages and adipocytes.

In summary, epiobesogenes are a set of genes involved in obesity and tend to be regulated by epigenetic mechanisms.[66] In this sense, one can speculate that it may be feasible to develop algorithms to score accurately the power in body weight management and energy metabolism control of every methylation CpG of the epiobesogene to create an epigenetic score.[311] An epigenotype score may serve as a diagnostic/prescription/prognostic tool to study body weight response to hypocaloric diet and risk factors for the development of obesity and other related diseases, including metabolic syndrome.[312–314]

VII. Concluding Remarks

Obesity is a multifactorial disorder characterized by a disproportionately high adipose tissue mass, associated with a disequilibrium in the energy balance equation.[9] Today's view is that a large fraction of body weight variability is under genetic control and that the growing prevalence of obesity is the result of the interaction of a genetic susceptibility with inadequate lifestyles (increases in consumption of foods rich in carbohydrates and fats and reduction in physical activity level, etc.). It is also commonly assumed that cumulative small effects of several genes are more likely than the major involvement of a single locus.

Additionally, there is growing evidence suggesting that interindividual differences in obesity susceptibility depend not only on the DNA sequence (genetics), but also on epigenetic factors affecting gene expression such as DNA methylation, covalent histone modifications, chromatin folding, and, more recently described, the regulatory action of miRNAs and polycomb group complexes. Moreover, different environmental agents and metabolic compounds may regulate these epigenetics mechanisms, such as DNA and histone modifying enzymes. In this context, epigenetics can provide some insights not only on cellular identity, stem cell plasticity, tissue regeneration, tumorigenesis, and aging, but also in understanding genetic fetal programming, monozygotic twin differences, and chronic disease onset, such as obesity in the adult.

Epigenetic research is critical to overcoming some hidden phenomena that have not been explained by the completion of the Human Genome Project. A future focus should be on quantifying the importance of epigenetic regulation in the etiology and development of obesity with adequate support from epigenetic research techniques to report and characterize new epiobesogenic genes potentially involved in these processes.

Two main objectives are currently pursued in the field of obesity and epigenetics. First, the identification at an early age of individuals who could present specific methylation profiles and other epigenetic marks on specific genes suggesting susceptibility to different metabolic diseases, including excess body weight gain and type 2 diabetes. Second, the use of dietary supplementation to counteract adverse epigenomic profiles in an individualized manner, similar to the administration of inhibitors of DNMTs and HDAC inhibitors in cancer therapy.

Some of the issues that must be investigated in the coming years include (a) the regulation of epigenetic processes by endocrinological, nutritional, and lifestyle factors and in what doses, (b) the periods of the life course in which the genome is especially vulnerable to altered epigenetic markings by dietary factors, as well as the transmission of these marks between generations, (c) the importance of maternal and nutritional care of the infant and the young child with respect to epigenetic events, (d) the thorough understanding of epigenetic events in obesity for developing novel and early diet-related biomarkers of disease risk, and finally, (e) new applications of drugs targeting HDACs, HATs, or DNMTs. Experimental advances concerning the impact of nutritional status and dietary exposures on the epigenome and their implications for health are anxiously awaited, as they will likely result in newer technologies and improved risk factor profile and more healthy longevity.

References

1. Marti A, Moreno-Aliaga MJ, Hebebrand J, Martinez JA. Genes, lifestyles and obesity. *Int J Obes Relat Metab Disord* 2004;**3**(28 Suppl.):S29–36.
2. Bell CG, Walley AJ, Froguel P. The genetics of human obesity. *Nat Rev Genet* 2005;**6**:221–34.
3. Moreno-Aliaga MJ, Santos JL, Marti A, Martinez JA. Does weight loss prognosis depend on genetic make-up? *Obes Rev* 2005;**6**:155–68.
4. Bray GA. *The obese patient.* Philadelphia: Saunders; 1976.
5. Garrow J. *Energy balance and obesity in man.* Amsterdam: Elsevier Science Ltd; 1974.
6. Feinleib M, Garrison RJ, Fabsitz R, Christian JC, Hrubec Z, Borhani NO, et al. The NHLBI twin study of cardiovascular disease risk factors: methodology and summary of results. *Am J Epidemiol* 1977;**106**:284–5.

7. Bouchard C. Genetics of human obesity: recent results from linkage studies. *J Nutr* 1997;**127**:1887S–90S.
8. Bjorntorp P. *International textbook of obesity.* Wiley:West Sussex, UK; 2001.
9. Bray GA, Bouchard C, James WP. *Handbook of obesity.* New York: Marcel Dekker; 1998.
10. Corthesy-Theulaz I, den Dunnen JT, Ferre P, Geurts JM, Muller M, van Belzen N, et al. Nutrigenomics: the impact of biomics technology on nutrition research. *Ann Nutr Metab* 2005;**49**:355–65.
11. Rankinen T, Zuberi A, Chagnon YC, Weisnagel SJ, Argyropoulos G, Walts B, et al. The human obesity gene map: the 2005 update. *Obesity (Silver Spring)* 2006;**14**:529–644.
12. Allis CD, Jenuwein T, Reinberg D. *Epigenetics.* New York: Cold Spring Harbor Laboratory Press; 2007.
13. Milagro FI, Campion J, Martinez JA. Metabolism and chromatin dynamics/remodelling. In: Tollefsbol Trygu, editor. *Handbook of epigenetics: the new molecular and medical genetics.* Oxford, UK: Grupo Elsevier; 2010 (in press).
14. Luger K. Structure and dynamic behavior of nucleosomes. *Curr Opin Genet Dev* 2003;**13**:127–35.
15. Morgan DK, Whitelaw E. The case for transgenerational epigenetic inheritance in humans. *Mamm Genome* 2008;**19**:394–7.
16. Godfrey KM, Lillycrop KA, Burdge GC, Gluckman PD, Hanson MA. Epigenetic mechanisms and the mismatch concept of the developmental origins of health and disease. *Pediatr Res* 2007;**61**:5R–10R.
17. Campion J, Milagro FI, Martinez JA. Genetic manipulation in nutrition, metabolism, and obesity research. *Nutr Rev* 2004;**62**:321–30.
18. Mathers JC. Early nutrition: impact on epigenetics. *Forum Nutr* 2007;**60**:42–8.
19. Turek-Plewa J, Jagodzinski PP. The role of mammalian DNA methyltransferases in the regulation of gene expression. *Cell Mol Biol Lett* 2005;**10**:631–47.
20. Berger SL. The complex language of chromatin regulation during transcription. *Nature* 2007;**447**:407–12.
21. Dolinoy DC, Das R, Weidman JR, Jirtle RL. Metastable epialleles, imprinting, and the fetal origins of adult diseases. *Pediatr Res* 2007;**61**:30R–37R.
22. Stoger R. Epigenetics and obesity. *Pharmacogenomics* 2008;**9**:1851–60.
23. Palou A, Pico C, Bonet ML. The molecular basis of body weight control. *Forum Nutr* 2003;**56**:164–8.
24. Calvanese V, Lara E, Kahn A, Fraga MF. The role of epigenetics in aging and age-related diseases. *Ageing Res Rev* 2009;**8**:268–76.
25. Karimi M, Johansson S, Ekstrom TJ. Using LUMA: a luminometric-based assay for global DNA-methylation. *Epigenetics* 2006;**1**:45–8.
26. Kim HH, Park JH, Jeong KS, Lee S. Determining the global DNA methylation status of rat according to the identifier repetitive elements. *Electrophoresis* 2007;**28**:3854–61.
27. Fraga MF, Rodriguez R, Canal MJ. Rapid quantification of DNA methylation by high performance capillary electrophoresis. *Electrophoresis* 2000;**21**:2990–4.
28. Grunau C, Clark SJ, Rosenthal A. Bisulfite genomic sequencing: systematic investigation of critical experimental parameters. *Nucleic Acids Res* 2001;**29**:E65.
29. Herman JG, Graff JR, Myohanen S, Nelkin BD, Baylin SB. Methylation-specific PCR: a novel PCR assay for methylation status of CpG islands. *Proc Natl Acad Sci USA* 1996;**93**:9821–6.
30. Harden SV, Sanderson H, Goodman SN, Partin AA, Walsh PC, Epstein JI, et al. Quantitative GSTP1 methylation and the detection of prostate adenocarcinoma in sextant biopsies. *J Natl Cancer Inst* 2003;**95**:1634–7.
31. Eads CA, Danenberg KD, Kawakami K, Saltz LB, Blake C, Shibata D, et al. MethyLight: a high-throughput assay to measure DNA methylation. *Nucleic Acids Res* 2000;**28**:E32.

32. Uhlmann K, Brinckmann A, Toliat MR, Ritter H, Nurnberg P. Evaluation of a potential epigenetic biomarker by quantitative methyl-single nucleotide polymorphism analysis. *Electrophoresis* 2002;**23**:4072–9.
33. Xiong Z, Laird PW. COBRA: a sensitive and quantitative DNA methylation assay. *Nucleic Acids Res* 1997;**25**:2532–4.
34. Gonzalgo ML, Liang G. Methylation-sensitive single-nucleotide primer extension (Ms-SNuPE) for quantitative measurement of DNA methylation. *Nat Protoc* 2007;**2**:1931–6.
35. Docherty SJ, Davis OS, Haworth CM, Plomin R, Mill J. Bisulfite-based epityping on pooled genomic DNA provides an accurate estimate of average group DNA methylation. *Epigenetics Chromatin* 2009;**2**:3.
36. Weber M, Davies JJ, Wittig D, Oakeley EJ, Haase M, Lam WL, et al. Chromosome-wide and promoter-specific analyses identify sites of differential DNA methylation in normal and transformed human cells. *Nat Genet* 2005;**37**:853–62.
37. Ibrahim AE, Thorne NP, Baird K, Barbosa-Morais NL, Tavare S, Collins VP, et al. MMASS: an optimized array-based method for assessing CpG island methylation. *Nucleic Acids Res* 2006;**34**:e136.
38. Mund C, Beier V, Bewerunge P, Dahms M, Lyko F, Hoheisel JD. Array-based analysis of genomic DNA methylation patterns of the tumour suppressor gene p16INK4A promoter in colon carcinoma cell lines. *Nucleic Acids Res* 2005;**33**:e73.
39. Shi H, Maier S, Nimmrich I, Yan PS, Caldwell CW, Olek A, et al. Oligonucleotide-based microarray for DNA methylation analysis: principles and applications. *J Cell Biochem* 2003;**88**:138–43.
40. Brunner AL, Johnson DS, Kim SW, Valouev A, Reddy TE, Neff NF, et al. Distinct DNA methylation patterns characterize differentiated human embryonic stem cells and developing human fetal liver. *Genome Res* 2009;**19**:1044–56.
41. Lindner H, Sarg B, Helliger W. Application of hydrophilic-interaction liquid chromatography to the separation of phosphorylated H1 histones. *J Chromatogr A* 1997;**782**:55–62.
42. Bock C, Lengauer T. Computational epigenetics. *Bioinformatics* 2008;**24**:1–10.
43. Won KJ, Ren B, Wang W. Genome-wide prediction of transcription factor binding sites using an integrated model. *Genome Biol* 2010;**11**:R7.
44. Kouzarides T. Chromatin modifications and their function. *Cell* 2007;**128**:693–705.
45. Aparicio O, Geisberg JV, Struhl K. Chromatin immunoprecipitation for determining the association of proteins with specific genomic sequences in vivo. *Curr Protoc Cell Biol* 2004;**17**(Unit 17):17.
46. Johnson DS, Mortazavi A, Myers RM, Wold B. Genome-wide mapping of in vivo protein–DNA interactions. *Science* 2007;**316**:1497–502.
47. Hitchler MJ, Domann FE. Metabolic defects provide a spark for the epigenetic switch in cancer. *Free Radic Biol Med* 2009;**47**:115–27.
48. Yang J, Kong X, Martins-Santos ME, Aleman G, Chaco E, Liu GE, et al. Activation of SIRT1 by resveratrol represses transcription of the gene for the cytosolic form of phosphoenolpyruvate carboxykinase (GTP) by deacetylating hepatic nuclear factor 4alpha. *J Biol Chem* 2009;**284**:27042–53.
49. Nottke A, Colaiacovo MP, Shi Y. Developmental roles of the histone lysine demethylases. *Development* 2009;**136**:879–89.
50. Manzo F, Tambaro FP, Mai A, Altucci L. Histone acetyltransferase inhibitors and preclinical studies. *Expert Opin Ther Pat* 2009;**19**:761–74.
51. Evertts AG, Zee BM, Garcia BA. Modern approaches for investigating epigenetic signaling pathways. *J Appl Physiol* 2009;**109**(3):927–33.
52. Halees AS, Leyfer D, Weng Z. PromoSer: a large-scale mammalian promoter and transcription start site identification service. *Nucleic Acids Res* 2003;**31**:3554–9.

53. Jiang C, Xuan Z, Zhao F, Zhang MQ. TRED: a transcriptional regulatory element database, new entries and other development. *Nucleic Acids Res* 2007;**35**:D137–40.
54. Li LC, Dahiya R. MethPrimer: designing primers for methylation PCRs. *Bioinformatics* 2002;**18**:1427–31.
55. Bock C, Reither S, Mikeska T, Paulsen M, Walter J, Lengauer T. BiQ Analyzer: visualization and quality control for DNA methylation data from bisulfite sequencing. *Bioinformatics* 2005;**21**:4067–8.
56. Kumaki Y, Oda M, Okano M. QUMA: quantification tool for methylation analysis. *Nucleic Acids Res* 2008;**36**:W170–5.
57. Schuffler P, Mikeska T, Waha A, Lengauer T, Bock C. MethMarker: user-friendly design and optimization of gene-specific DNA methylation assays. *Genome Biol* 2009;**10**:R105.
58. Cartharius K, Frech K, Grote K, Klocke B, Haltmeier M, Klingenhoff A, et al. MatInspector and beyond: promoter analysis based on transcription factor binding sites. *Bioinformatics* 2005;**21**:2933–42.
59. Matys V, Fricke E, Geffers R, Gossling E, Haubrock M, Hehl R, et al. TRANSFAC: transcriptional regulation, from patterns to profiles. *Nucleic Acids Res* 2003;**31**:374–8.
60. Gluckman PD, Hanson MA, Cooper C, Thornburg KL. Effect of in utero and early-life conditions on adult health and disease. *N Engl J Med* 2008;**359**:61–73.
61. Fernandez-Twinn DS, Ozanne SE. Mechanisms by which poor early growth programs type-2 diabetes, obesity and the metabolic syndrome. *Physiol Behav* 2006;**88**:234–43.
62. Hales CN. Fetal and infant growth and impaired glucose tolerance in adulthood: the "thrifty phenotype" hypothesis revisited. *Acta Paediatr Suppl* 1997;**422**:73–7.
63. McMillen IC, Robinson JS. Developmental origins of the metabolic syndrome: prediction, plasticity, and programming. *Physiol Rev* 2005;**85**:571–633.
64. Greenberg JA. Biases in the mortality risk versus body mass index relationship in the NHANES-1 epidemiologic follow-up study. *Int J Obes Relat Metab Disord* 2001;**25**:1071–8.
65. Waterland RA, Travisano M, Tahiliani KG, Rached MT, Mirza S. Methyl donor supplementation prevents transgenerational amplification of obesity. *Int J Obes (Lond)* 2008;**32**:1373–9.
66. Campion J, Milagro FI, Martinez JA. Individuality and epigenetics in obesity. *Obes Rev* 2009;**10**:383–92.
67. Muhlhausler BS, Duffield JA, McMillen IC. Increased maternal nutrition stimulates peroxisome proliferator activated receptor-gamma, adiponectin, and leptin messenger ribonucleic acid expression in adipose tissue before birth. *Endocrinology* 2007;**148**:878–85.
68. Boualga A, Bouchenak M, Belleville J. Low-protein diet prevents tissue lipoprotein lipase activity increase in growing rats. *Br J Nutr* 2000;**84**:663–71.
69. Terroni PL, Anthony FW, Hanson MA, Cagampang FR. Expression of agouti-related peptide, neuropeptide Y, pro-opiomelanocortin and the leptin receptor isoforms in fetal mouse brain from pregnant dams on a protein-restricted diet. *Brain Res Mol Brain Res* 2005;**140**:111–5.
70. Bertram C, Trowern AR, Copin N, Jackson AA, Whorwood CB. The maternal diet during pregnancy programs altered expression of the glucocorticoid receptor and type 2 11beta-hydroxysteroid dehydrogenase: potential molecular mechanisms underlying the programming of hypertension in utero. *Endocrinology* 2001;**142**:2841–53.
71. Metges CC. Early nutrition and later obesity: animal models provide insights into mechanisms. *Adv Exp Med Biol* 2009;**646**:105–12.
72. Gluckman PD, Hanson MA, Beedle AS. Non-genomic transgenerational inheritance of disease risk. *Bioessays* 2007;**29**:145–54.
73. Dunn GA, Bale TL. Maternal high-fat diet promotes body length increases and insulin insensitivity in second-generation mice. *Endocrinology* 2009;**150**:4999–5009.
74. Rakyan VK, Preis J, Morgan HD, Whitelaw E. The marks, mechanisms and memory of epigenetic states in mammals. *Biochem J* 2001;**356**:1–10.

75. Haig D, Wharton R. Prader-Willi syndrome and the evolution of human childhood. *Am J Hum Biol* 2003;**15**:320–9.
76. Zeisel SH. Importance of methyl donors during reproduction. *Am J Clin Nutr* 2009;**89**:673S–7S.
77. Friso S, Choi SW, Girelli D, Mason JB, Dolnikowski GG, Bagley PJ, et al. A common mutation in the 5, 10-methylenetetrahydrofolate reductase gene affects genomic DNA methylation through an interaction with folate status. *Proc Natl Acad Sci USA* 2002;**99**:5606–11.
78. Locker J, Reddy TV, Lombardi B. DNA methylation and hepatocarcinogenesis in rats fed a choline-devoid diet. *Carcinogenesis* 1986;**7**:1309–12.
79. Uekawa A, Katsushima K, Ogata A, Kawata T, Maeda N, Kobayashi K, et al. Change of epigenetic control of cystathionine beta-synthase gene expression through dietary vitamin B12 is not recovered by methionine supplementation. *J Nutrigenet Nutrigenomics* 2009;**2**:29–36.
80. Zapisek WF, Cronin GM, Lyn-Cook BD, Poirier LA. The onset of oncogene hypomethylation in the livers of rats fed methyl-deficient, amino acid-defined diets. *Carcinogenesis* 1992;**13**:1869–72.
81. Choi SW, Friso S, Keyes MK, Mason JB. Folate supplementation increases genomic DNA methylation in the liver of elder rats. *Br J Nutr* 2005;**93**:31–5.
82. Keyes MK, Jang H, Mason JB, Liu Z, Crott JW, Smith DE, et al. Older age and dietary folate are determinants of genomic and p16-specific DNA methylation in mouse colon. *J Nutr* 2007;**137**:1713–7.
83. Kovacheva VP, Davison JM, Mellott TJ, Rogers AE, Yang S, O'Brien MJ, et al. Raising gestational choline intake alters gene expression in DMBA-evoked mammary tumors and prolongs survival. *FASEB J* 2009;**23**:1054–63.
84. Steegers-Theunissen RP, Obermann-Borst SA, Kremer D, Lindemans J, Siebel C, Steegers EA, et al. Periconceptional maternal folic acid use of 400 microg per day is related to increased methylation of the IGF2 gene in the very young child. *PLoS ONE* 2009;**4**:e7845.
85. Hirsch S, Ronco AM, Guerrero-Bosagna C, de la Maza MP, Leiva L, Barrera G, et al. Methylation status in healthy subjects with normal and high serum folate concentration. *Nutrition* 2008;**24**:1103–9.
86. Lv S, Fan R, Du Y, Hou M, Tang Z, Ling W, et al. Betaine supplementation attenuates atherosclerotic lesion in apolipoprotein E-deficient mice. *Eur J Nutr* 2009;**48**:205–12.
87. Davison JM, Mellott TJ, Kovacheva VP, Blusztajn JK. Gestational choline supply regulates methylation of histone H3, expression of histone methyltransferases G9a (Kmt1c) and Suv39h1 (Kmt1a), and DNA methylation of their genes in rat fetal liver and brain. *J Biol Chem* 2009;**284**:1982–9.
88. Jiang Y, Sun T, Xiong J, Cao J, Li G, Wang S. Hyperhomocysteinemia-mediated DNA hypomethylation and its potential epigenetic role in rats. *Acta Biochim Biophys Sin (Shanghai)* 2007;**39**:657–67.
89. Kotsopoulos J, Sohn KJ, Kim YI. Postweaning dietary folate deficiency provided through childhood to puberty permanently increases genomic DNA methylation in adult rat liver. *J Nutr* 2008;**138**:703–9.
90. Sinclair KD, Allegrucci C, Singh R, Gardner DS, Sebastian S, Bispham J, et al. DNA methylation, insulin resistance, and blood pressure in offspring determined by maternal periconceptional B vitamin and methionine status. *Proc Natl Acad Sci USA* 2007;**104**:19351–6.
91. Pogribny IP, Tryndyak VP, Bagnyukova TV, Melnyk S, Montgomery B, Ross SA, et al. Hepatic epigenetic phenotype predetermines individual susceptibility to hepatic steatosis in mice fed a lipogenic methyl-deficient diet. *J Hepatol* 2009;**51**:176–86.
92. Burdge GC, Slater-Jefferies J, Torrens C, Phillips ES, Hanson MA, Lillycrop KA. Dietary protein restriction of pregnant rats in the F0 generation induces altered methylation of

hepatic gene promoters in the adult male offspring in the F1 and F2 generations. *Br J Nutr* 2007;**97**:435–9.
93. Goyal R, Galffy A, Field SA, Gheorghe CP, Mittal A, Longo LD. Maternal protein deprivation: changes in systemic renin–angiotensin system of the mouse fetus. *Reprod Sci* 2009;**16**:894–904.
94. Lillycrop KA, Phillips ES, Jackson AA, Hanson MA, Burdge GC. Dietary protein restriction of pregnant rats induces and folic acid supplementation prevents epigenetic modification of hepatic gene expression in the offspring. *J Nutr* 2005;**135**:1382–6.
95. Thaler R, Karlic H, Rust P, Haslberger AG. Epigenetic regulation of human buccal mucosa mitochondrial superoxide dismutase gene expression by diet. *Br J Nutr* 2009;**101**:743–9.
96. van Breda SG, van Delft JH, Engels LG, Kleinjans JC, Mathers JC. Methylation status of CpG islands in the promoter region of genes differentially expressed in colonic mucosa from adenoma patients and controls in response to altered vegetable intake. *Br J Nutr* 2009;**101**:1295–9.
97. Waterland RA, Lin JR, Smith CA, Jirtle RL. Post-weaning diet affects genomic imprinting at the insulin-like growth factor 2 (Igf2) locus. *Hum Mol Genet* 2006;**15**:705–16.
98. Martinez JA, Parra MD, Santos JL, Moreno-Aliaga MJ, Marti A, Martinez-Gonzalez MA. Genotype-dependent response to energy-restricted diets in obese subjects: towards personalized nutrition. *Asia Pac J Clin Nutr* 2008;**17**(Suppl. 1):119–22.
99. Heijmans BT, Tobi EW, Stein AD, Putter H, Blauw GJ, Susser ES, et al. Persistent epigenetic differences associated with prenatal exposure to famine in humans. *Proc Natl Acad Sci USA* 2008;**105**:17046–9.
100. Vinsky MD, Murdoch GK, Dixon WT, Dyck MK, Foxcroft GR. Altered epigenetic variance in surviving litters from nutritionally restricted lactating primiparous sows. *Reprod Fertil Dev* 2007;**19**:430–5.
101. Unterberger A, Szyf M, Nathanielsz PW, Cox LA. Organ and gestational age effects of maternal nutrient restriction on global methylation in fetal baboons. *J Med Primatol* 2009;**38**:219–27.
102. Bouchard L, Rabasa-Lhoret R, Faraj M, Lavoie ME, Mill J, Perusse L, et al. Differential epigenomic and transcriptomic responses in subcutaneous adipose tissue between low and high responders to caloric restriction. *Am J Clin Nutr* 2010;**91**:309–20.
103. Campion J, Milagro FI, Goyenechea E, Martinez JA. TNF-alpha promoter methylation as a predictive biomarker for weight-loss response. *Obesity (Silver Spring)* 2009;**17**:1293–7.
104. Fujiki K, Kano F, Shiota K, Murata M. Expression of the peroxisome proliferator activated receptor gamma gene is repressed by DNA methylation in visceral adipose tissue of mouse models of diabetes. *BMC Biol* 2009;**7**:38.
105. Milagro FI, Campion J, Garcia-Diaz DF, Goyenechea E, Paternain L, Martinez JA. High fat diet-induced obesity modifies the methylation pattern of leptin promoter in rats. *J Physiol Biochem* 2009;**65**:1–9.
106. Plagemann A, Harder T, Brunn M, Harder A, Roepke K, Wittrock-Staar M, et al. Hypothalamic proopiomelanocortin promoter methylation becomes altered by early overfeeding: an epigenetic model of obesity and the metabolic syndrome. *J Physiol* 2009;**587**:4963–76.
107. Davis CD, Uthus EO, Finley JW. Dietary selenium and arsenic affect DNA methylation in vitro in Caco-2 cells and in vivo in rat liver and colon. *J Nutr* 2000;**130**:2903–9.
108. Huang YC, Hung WC, Chen WT, Yu HS, Chai CY. Sodium arsenite-induced DAPK promoter hypermethylation and autophagy via ERK1/2 phosphorylation in human uroepithelial cells. *Chem Biol Interact* 2009;**181**:254–62.
109. Mas S, Lafuente MJ, Crescenti A, Trias M, Ballesta A, Molina R, et al. Lower specific micronutrient intake in colorectal cancer patients with tumors presenting promoter hypermethylation in p16(INK4a), p4(ARF) and hMLH1. *Anticancer Res* 2007;**27**:1151–6.

110. Uthus EO, Ross SA, Davis CD. Differential effects of dietary selenium (se) and folate on methyl metabolism in liver and colon of rats. *Biol Trace Elem Res* 2006;**109**:201–14.
111. Fang MZ, Chen D, Sun Y, Jin Z, Christman JK, Yang CS. Reversal of hypermethylation and reactivation of p16INK4a, RARbeta, and MGMT genes by genistein and other isoflavones from soy. *Clin Cancer Res* 2005;**11**:7033–41.
112. Kato K, Long NK, Makita H, Toida M, Yamashita T, Hatakeyama D, et al. Effects of green tea polyphenol on methylation status of RECK gene and cancer cell invasion in oral squamous cell carcinoma cells. *Br J Cancer* 2008;**99**:647–54.
113. Majid S, Kikuno N, Nelles J, Noonan E, Tanaka Y, Kawamoto K, et al. Genistein induces the p21WAF1/CIP1 and p16INK4a tumor suppressor genes in prostate cancer cells by epigenetic mechanisms involving active chromatin modification. *Cancer Res* 2008;**68**:2736–44.
114. Guerrero-Bosagna CM, Sabat P, Valdovinos FS, Valladares LE, Clark SJ. Epigenetic and phenotypic changes result from a continuous pre and post natal dietary exposure to phytoestrogens in an experimental population of mice. *BMC Physiol* 2008;**8**:17.
115. Hitchler MJ, Domann FE. An epigenetic perspective on the free radical theory of development. *Free Radic Biol Med* 2007;**43**:1023–36.
116. Junien C, Gallou-Kabani C, Vige A, Gross MS. Nutritional epigenomics of metabolic syndrome. *Med Sci (Paris)* 2005;**21**:396–404.
117. Oliveira NF, Damm GR, Andia DC, Salmon C, Nociti Jr. FH, Line SR, et al. DNA methylation status of the IL8 gene promoter in oral cells of smokers and non-smokers with chronic periodontitis. *J Clin Periodontol* 2009;**36**:719–25.
118. Barres R, Osler ME, Yan J, Rune A, Fritz T, Caidahl K, et al. Non-CpG methylation of the PGC-1alpha promoter through DNMT3B controls mitochondrial density. *Cell Metab* 2009;**10**:189–98.
119. Westberry JM, Prewitt AK, Wilson ME. Epigenetic regulation of the estrogen receptor alpha promoter in the cerebral cortex following ischemia in male and female rats. *Neuroscience* 2008;**152**:982–9.
120. Hahn MA, Hahn T, Lee DH, Esworthy RS, Kim BW, Riggs AD, et al. Methylation of polycomb target genes in intestinal cancer is mediated by inflammation. *Cancer Res* 2008;**68**:10280–9.
121. Kim MS, Kondo T, Takada I, Youn MY, Yamamoto Y, Takahashi S, et al. DNA demethylation in hormone-induced transcriptional derepression. *Nature* 2009;**461**:1007–12.
122. Chiang EP, Wang YC, Chen WW, Tang FY. Effects of insulin and glucose on cellular metabolic fluxes in homocysteine transsulfuration, remethylation, S-adenosylmethionine synthesis, and global deoxyribonucleic acid methylation. *J Clin Endocrinol Metab* 2009;**94**:1017–25.
123. Lai JC, Wu JY, Cheng YW, Yeh KT, Wu TC, Chen CY, et al. O6-Methylguanine-DNA methyltransferase hypermethylation modulated by 17beta-estradiol in lung cancer cells. *Anticancer Res* 2009;**29**:2535–40.
124. Singh K, Prasad S. Age- and sex-related analysis of methylation of 5′-upstream sequences of Fmr-1 gene in mouse brain and modulation by sex steroid hormones. *Biogerontology* 2008;**9**:455–65.
125. Hsu PY, Deatherage DE, Rodriguez BA, Liyanarachchi S, Weng YI, Zuo T, et al. Xenoestrogen-induced epigenetic repression of microRNA-9-3 in breast epithelial cells. *Cancer Res* 2009;**69**:5936–45.
126. Sharma P, Kumar J, Garg G, Kumar A, Patowary A, Karthikeyan G, et al. Detection of altered global DNA methylation in coronary artery disease patients. *DNA Cell Biol* 2008;**27**:357–65.
127. Friso S, Pizzolo F, Choi SW, Guarini P, Castagna A, Ravagnani V, et al. Epigenetic control of 11 beta-hydroxysteroid dehydrogenase 2 gene promoter is related to human hypertension. *Atherosclerosis* 2008;**199**:323–7.

128. Ling C, Del Guerra S, Lupi R, Ronn T, Granhall C, Luthman H, et al. Epigenetic regulation of PPARGC1A in human type 2 diabetic islets and effect on insulin secretion. *Diabetologia* 2008;**51**:615–22.
129. Shao WJ, Tao LY, Gao C, Xie JY, Zhao RQ. Alterations in methylation and expression levels of imprinted genes H19 and Igf2 in the fetuses of diabetic mice. *Comp Med* 2008;**58**:341–6.
130. Frieling H, Bleich S, Otten J, Romer KD, Kornhuber J, de Zwaan M, et al. Epigenetic downregulation of atrial natriuretic peptide but not vasopressin mRNA expression in females with eating disorders is related to impulsivity. *Neuropsychopharmacology* 2008;**33**:2605–9.
131. Margueron R, Trojer P, Reinberg D. The key to development: interpreting the histone code? *Curr Opin Genet Dev* 2005;**15**:163–76.
132. Cosgrove MS, Wolberger C. How does the histone code work? *Biochem Cell Biol* 2005;**83**:468–76.
133. Ito T. Role of histone modification in chromatin dynamics. *J Biochem* 2007;**141**:609–14.
134. Trayhurn P, Wood IS. Adipokines: inflammation and the pleiotropic role of white adipose tissue. *Br J Nutr* 2004;**92**:347–55.
135. Watson JA, Watson CJ, McCrohan AM, Woodfine K, Tosetto M, McDaid J, et al. Generation of an epigenetic signature by chronic hypoxia in prostate cells. *Hum Mol Genet* 2009;**18**:3594–604.
136. Rahman I, Marwick J, Kirkham P. Redox modulation of chromatin remodeling: impact on histone acetylation and deacetylation, NF-kappaB and pro-inflammatory gene expression. *Biochem Pharmacol* 2004;**68**:1255–67.
137. Petrelli F, Coderoni S, Moretti P, Paparelli M. Effect of biotin on phosphorylation, acetylation, methylation of rat liver histones. *Mol Biol Rep* 1978;**4**:87–92.
138. Jia J, Chen J. Histone hyperacetylation is involved in the quercetin-induced human leukemia cell death. *Pharmazie* 2008;**63**:379–83.
139. Mukwevho E, Kohn TA, Lang D, Nyatia E, Smith J, Ojuka EO. Caffeine induces hyperacetylation of histones at the MEF2 site on the Glut4 promoter and increases MEF2A binding to the site via a CaMK-dependent mechanism. *Am J Physiol Endocrinol Metab* 2008;**294**: E582–8.
140. Mosley AL, Ozcan S. Glucose regulates insulin gene transcription by hyperacetylation of histone h4. *J Biol Chem* 2003;**278**:19660–6.
141. Mosley AL, Corbett JA, Ozcan S. Glucose regulation of insulin gene expression requires the recruitment of p300 by the beta-cell-specific transcription factor Pdx-1. *Mol Endocrinol* 2004;**18**:2279–90.
142. Waldecker M, Kautenburger T, Daumann H, Busch C, Schrenk D. Inhibition of histone-deacetylase activity by short-chain fatty acids and some polyphenol metabolites formed in the colon. *J Nutr Biochem* 2008;**19**:587–93.
143. Druesne-Pecollo N, Chaumontet C, Pagniez A, Vaugelade P, Bruneau A, Thomas M, et al. In vivo treatment by diallyl disulfide increases histone acetylation in rat colonocytes. *Biochem Biophys Res Commun* 2007;**354**:140–7.
144. Myzak MC, Dashwood WM, Orner GA, Ho E, Dashwood RH. Sulforaphane inhibits histone deacetylase in vivo and suppresses tumorigenesis in Apc-minus mice. *FASEB J* 2006;**20**:506–8.
145. Basak S, Pookot D, Noonan EJ, Dahiya R. Genistein down-regulates androgen receptor by modulating HDAC6–Hsp90 chaperone function. *Mol Cancer Ther* 2008;**7**:3195–202.
146. Murugan RS, Vinothini G, Hara Y, Nagini S. Black tea polyphenols target matrix metalloproteinases, RECK, proangiogenic molecules and histone deacetylase in a rat hepatocarcinogenesis model. *Anticancer Res* 2009;**29**:2301–5.
147. van der Linden AM, Nolan KM, Sengupta P. KIN-29 SIK regulates chemoreceptor gene expression via an MEF2 transcription factor and a class II HDAC. *EMBO J* 2007;**26**:358–70.

148. Tavera-Mendoza LE, Quach TD, Dabbas B, Hudon J, Liao X, Palijan A, et al. Incorporation of histone deacetylase inhibition into the structure of a nuclear receptor agonist. *Proc Natl Acad Sci USA* 2008;**105**:8250–5.
149. Green KN, Steffan JS, Martinez-Coria H, Sun X, Schreiber SS, Thompson LM, et al. Nicotinamide restores cognition in Alzheimer's disease transgenic mice via a mechanism involving sirtuin inhibition and selective reduction of Thr231-phosphotau. *J Neurosci* 2008;**28**:11500–10.
150. Balestrieri ML, Rienzo M, Felice F, Rossiello R, Grimaldi V, Milone L, et al. High glucose downregulates endothelial progenitor cell number via SIRT1. *Biochim Biophys Acta* 2008;**1784**:936–45.
151. Ito K, Lim S, Caramori G, Cosio B, Chung KF, Adcock IM, et al. A molecular mechanism of action of theophylline: induction of histone deacetylase activity to decrease inflammatory gene expression. *Proc Natl Acad Sci USA* 2002;**99**:8921–6.
152. Adams J, Klaidman L. Sirtuins, nicotinamide and aging: a critical review. *Lett Drug Des Discov* 2007;**4**:44–8.
153. Cohen HY, Miller C, Bitterman KJ, Wall NR, Hekking B, Kessler B, et al. Calorie restriction promotes mammalian cell survival by inducing the SIRT1 deacetylase. *Science* 2004;**305**:390–2.
154. Wang F, Nguyen M, Qin FX, Tong Q. SIRT2 deacetylates FOXO3a in response to oxidative stress and caloric restriction. *Aging Cell* 2007;**6**:505–14.
155. Balasubramanyam K, Varier RA, Altaf M, Swaminathan V, Siddappa NB, Ranga U, et al. Curcumin, a novel p300/CREB-binding protein-specific inhibitor of acetyltransferase, represses the acetylation of histone/nonhistone proteins and histone acetyltransferase-dependent chromatin transcription. *J Biol Chem* 2004;**279**:51163–71.
156. Balasubramanyam K, Altaf M, Varier RA, Swaminathan V, Ravindran A, Sadhale PP, et al. Polyisoprenylated benzophenone, garcinol, a natural histone acetyltransferase inhibitor, represses chromatin transcription and alters global gene expression. *J Biol Chem* 2004;**279**:33716–26.
157. Balasubramanyam K, Swaminathan V, Ranganathan A, Kundu TK. Small molecule modulators of histone acetyltransferase p300. *J Biol Chem* 2003;**278**:19134–40.
158. Choi KC, Jung MG, Lee YH, Yoon JC, Kwon SH, Kang HB, et al. Epigallocatechin-3-gallate, a histone acetyltransferase inhibitor, inhibits EBV-induced B lymphocyte transformation via suppression of RelA acetylation. *Cancer Res* 2009;**69**:583–92.
159. Lin C, Kang J, Zheng R. Oxidative stress is involved in inhibition of copper on histone acetylation in cells. *Chem Biol Interact* 2005;**151**:167–76.
160. Friis RM, Wu BP, Reinke SN, Hockman DJ, Sykes BD, Schultz MC. A glycolytic burst drives glucose induction of global histone acetylation by picNuA4 and SAGA. *Nucleic Acids Res* 2009;**37**:3969–80.
161. Shepard BD, Tuma PL. Alcohol-induced protein hyperacetylation: mechanisms and consequences. *World J Gastroenterol* 2009;**15**:1219–30.
162. Honma K, Mochizuki K, Goda T. Inductions of histone H3 acetylation at lysine 9 on SGLT1 gene and its expression by feeding mice a high carbohydrate/fat ratio diet. *Nutrition* 2009;**25**:40–4.
163. Honma K, Mochizuki K, Goda T. Carbohydrate/fat ratio in the diet alters histone acetylation on the sucrase-isomaltase gene and its expression in mouse small intestine. *Biochem Biophys Res Commun* 2007;**357**:1124–9.
164. Adcock IM, Ito K, Barnes PJ. Histone deacetylation: an important mechanism in inflammatory lung diseases. *COPD* 2005;**2**:445–55.

165. Ito K, Barnes PJ, Adcock IM. Glucocorticoid receptor recruitment of histone deacetylase 2 inhibits interleukin-1beta-induced histone H4 acetylation on lysines 8 and 12. *Mol Cell Biol* 2000;**20**:6891–903.
166. Qiao L, Schaack J, Shao J. Suppression of adiponectin gene expression by histone deacetylase inhibitor valproic acid. *Endocrinology* 2006;**147**:865–74.
167. Shi YB, Ritchie JW, Taylor PM. Complex regulation of thyroid hormone action: multiple opportunities for pharmacological intervention. *Pharmacol Ther* 2002;**94**:235–51.
168. Saxena NK, Vertino PM, Anania FA, Sharma D. Leptin-induced growth stimulation of breast cancer cells involves recruitment of histone acetyltransferases and mediator complex to CYCLIN D1 promoter via activation of Stat3. *J Biol Chem* 2007;**282**:13316–25.
169. Weaver IC, Cervoni N, Champagne FA, D'Alessio AC, Sharma S, Seckl JR, et al. Epigenetic programming by maternal behavior. *Nat Neurosci* 2004;**7**:847–54.
170. Berdeaux R, Goebel N, Banaszynski L, Takemori H, Wandless T, Shelton GD, et al. SIK1 is a class II HDAC kinase that promotes survival of skeletal myocytes. *Nat Med* 2007;**13**:597–603.
171. Bedford MT, Clarke SG. Protein arginine methylation in mammals: who, what, and why. *Mol Cell* 2009;**33**:1–13.
172. Bannister AJ, Schneider R, Kouzarides T. Histone methylation: dynamic or static? *Cell* 2002;**109**:801–6.
173. Santos-Rosa H, Caldas C. Chromatin modifier enzymes, the histone code and cancer. *Eur J Cancer* 2005;**41**:2381–402.
174. Dillon SC, Zhang X, Trievel RC, Cheng X. The SET-domain protein superfamily: protein lysine methyltransferases. *Genome Biol* 2005;**6**:227.
175. Shi Y. Histone lysine demethylases: emerging roles in development, physiology and disease. *Nat Rev Genet* 2007;**8**:829–33.
176. Lan F, Nottke AC, Shi Y. Mechanisms involved in the regulation of histone lysine demethylases. *Curr Opin Cell Biol* 2008;**20**:316–25.
177. Johnson AB, Denko N, Barton MC. Hypoxia induces a novel signature of chromatin modifications and global repression of transcription. *Mutat Res* 2008;**640**:174–9.
178. Xia X, Lemieux ME, Li W, Carroll JS, Brown M, Liu XS, et al. Integrative analysis of HIF binding and transactivation reveals its role in maintaining histone methylation homeostasis. *Proc Natl Acad Sci USA* 2009;**106**:4260–5.
179. Yang J, Ledaki I, Turley H, Gatter KC, Montero JC, Li JL, et al. Role of hypoxia-inducible factors in epigenetic regulation via histone demethylases. *Ann NY Acad Sci* 2009;**1177**:185–97.
180. De Santa F, Totaro MG, Prosperini E, Notarbartolo S, Testa G, Natoli G. The histone H3 lysine-27 demethylase Jmjd3 links inflammation to inhibition of polycomb-mediated gene silencing. *Cell* 2007;**130**:1083–94.
181. Ara AI, Xia M, Ramani K, Mato JM, Lu SC. *S*-Adenosylmethionine inhibits lipopolysaccharide-induced gene expression via modulation of histone methylation. *Hepatology* 2008;**47**:1655–66.
182. Li Y, Reddy MA, Miao F, Shanmugam N, Yee JK, Hawkins D, et al. Role of the histone H3 lysine 4 methyltransferase, SET7/9, in the regulation of NF-kappaB-dependent inflammatory genes. Relevance to diabetes and inflammation. *J Biol Chem* 2008;**283**:26771–81.
183. Brasacchio D, Okabe J, Tikellis C, Balcerczyk A, George P, Baker EK, et al. Hyperglycemia induces a dynamic cooperativity of histone methylase and demethylase enzymes associated with gene-activating epigenetic marks that coexist on the lysine tail. *Diabetes* 2009;**58**:1229–36.
184. Agherbi H, Gaussmann-Wenger A, Verthuy C, Chasson L, Serrano M, Djabali M. Polycomb mediated epigenetic silencing and replication timing at the INK4a/ARF locus during senescence. *PLoS ONE* 2009;**4**:e5622.

185. Kikuno N, Shiina H, Urakami S, Kawamoto K, Hirata H, Tanaka Y, et al. Genistein mediated histone acetylation and demethylation activates tumor suppressor genes in prostate cancer cells. *Int J Cancer* 2008;**123**:552–60.
186. Chen H, Ke Q, Kluz T, Yan Y, Costa M. Nickel ions increase histone H3 lysine 9 dimethylation and induce transgene silencing. *Mol Cell Biol* 2006;**26**:3728–37.
187. Sun H, Zhou X, Chen H, Li Q, Costa M. Modulation of histone methylation and MLH1 gene silencing by hexavalent chromium. *Toxicol Appl Pharmacol* 2009;**237**:258–66.
188. Zhou X, Li Q, Arita A, Sun H, Costa M. Effects of nickel, chromate, and arsenite on histone 3 lysine methylation. *Toxicol Appl Pharmacol* 2009;**236**:78–84.
189. Clarke S, Banfield K. *Homocysteine in health and disease.* Cambridge: Cambridge University Press; 2001.
190. Pogribny IP, Tryndyak VP, Muskhelishvili L, Rusyn I, Ross SA. Methyl deficiency, alterations in global histone modifications, and carcinogenesis. *J Nutr* 2007;**137**:216S–22S.
191. Dobosy JR, Fu VX, Desotelle JA, Srinivasan R, Kenowski ML, Almassi N, et al. A methyl-deficient diet modifies histone methylation and alters Igf2 and H19 repression in the prostate. *Prostate* 2008;**68**:1187–95.
192. Davidson JM, Mellott TJ, Kovacheva VP, J.K. B. Gestational choline supply regulates methylation of histone H3, expression of histone methyltransferases G9a (Kmt1c) and Suv39h1 (Kmt1a), and DNA methylation of their genes in rat liver and brain. *J Biol Chem* 2009;**284**:1982–9.
193. Sharif J, Nakamura M, Ito T, Kimura Y, Nagamune T, Mitsuya K, et al. Food restriction in pregnant mice can induce changes in histone modifications and suppress gene expression in fetus. *Nucleic Acids Symp Ser (Oxf)* 2007;125–6.
194. Lillycrop KA, Slater-Jefferies JL, Hanson MA, Godfrey KM, Jackson AA, Burdge GC. Induction of altered epigenetic regulation of the hepatic glucocorticoid receptor in the offspring of rats fed a protein-restricted diet during pregnancy suggests that reduced DNA methyltransferase-1 expression is involved in impaired DNA methylation and changes in histone modifications. *Br J Nutr* 2007;**97**:1064–73.
195. Li J, Bardag-Gorce F, Oliva J, Dedes J, French BA, French SW. Gene expression modifications in the liver caused by binge drinking and *S*-adenosylmethionine feeding. The role of epigenetic changes. *Genes Nutr* 2009.
196. Pal-Bhadra M, Bhadra U, Jackson DE, Mamatha L, Park PH, Shukla SD. Distinct methylation patterns in histone H3 at Lys-4 and Lys-9 correlate with up- & down-regulation of genes by ethanol in hepatocytes. *Life Sci* 2007;**81**:979–87.
197. Bardag-Gorce F, French BA, Joyce M, Baires M, Montgomery RO, Li J, et al. Histone acetyltransferase p300 modulates gene expression in an epigenetic manner at high blood alcohol levels. *Exp Mol Pathol* 2007;**82**:197–202.
198. Burke SJ, Collier JJ, Scott DK. cAMP prevents glucose-mediated modifications of histone H3 and recruitment of the RNA polymerase II holoenzyme to the L-PK gene promoter. *J Mol Biol* 2009;**392**:578–88.
199. El-Osta A, Brasacchio D, Yao D, Pocai A, Jones PL, Roeder RG, et al. Transient high glucose causes persistent epigenetic changes and altered gene expression during subsequent normoglycemia. *J Exp Med* 2008;**205**:2409–17.
200. Hall RK, Wang XL, George L, Koch SR, Granner DK. Insulin represses phosphoenolpyruvate carboxykinase gene transcription by causing the rapid disruption of an active transcription complex: a potential epigenetic effect. *Mol Endocrinol* 2007;**21**:550–63.
201. Villeneuve LM, Reddy MA, Lanting LL, Wang M, Meng L, Natarajan R. Epigenetic histone H3 lysine 9 methylation in metabolic memory and inflammatory phenotype of vascular smooth muscle cells in diabetes. *Proc Natl Acad Sci USA* 2008;**105**:9047–52.

202. Miao F, Smith DD, Zhang L, Min A, Feng W, Natarajan R. Lymphocytes from patients with type 1 diabetes display a distinct profile of chromatin histone H3 lysine 9 dimethylation: an epigenetic study in diabetes. *Diabetes* 2008;**57**:3189–98.
203. Miao F, Wu X, Zhang L, Yuan YC, Riggs AD, Natarajan R. Genome-wide analysis of histone lysine methylation variations caused by diabetic conditions in human monocytes. *J Biol Chem* 2007;**282**:13854–63.
204. Murayama A, Ohmori K, Fujimura A, Minami H, Yasuzawa-Tanaka K, Kuroda T, et al. Epigenetic control of rDNA loci in response to intracellular energy status. *Cell* 2008;**133**:627–39.
205. Kabra DG, Gupta J, Tikoo K. Insulin induced alteration in post-translational modifications of histone H3 under a hyperglycemic condition in L6 skeletal muscle myoblasts. *Biochim Biophys Acta* 2009;**1792**:574–83.
206. Oki M, Aihara H, Ito T. Role of histone phosphorylation in chromatin dynamics and its implications in diseases. *Subcell Biochem* 2007;**41**:319–36.
207. Cappelletti V, Fioravanti L, Miodini P, Di Fronzo G. Genistein blocks breast cancer cells in the G(2)M phase of the cell cycle. *J Cell Biochem* 2000;**79**:594–600.
208. Davie JR, Chadee DN. Regulation and regulatory parameters of histone modifications. *J Cell Biochem Suppl* 1998;**30–31**:203–13.
209. Juan G, Traganos F, James WM, Ray JM, Roberge M, Sauve DM, et al. Histone H3 phosphorylation and expression of cyclins A and B1 measured in individual cells during their progression through G2 and mitosis. *Cytometry* 1998;**32**:71–7.
210. Ahamed S, Foster JS, Bukovsky A, Wimalasena J. Signal transduction through the Ras/Erk pathway is essential for the mycoestrogen zearalenone-induced cell-cycle progression in MCF-7 cells. *Mol Carcinog* 2001;**30**:88–98.
211. Celeste A, Fernandez-Capetillo O, Kruhlak MJ, Pilch DR, Staudt DW, Lee A, et al. Histone H2AX phosphorylation is dispensable for the initial recognition of DNA breaks. *Nat Cell Biol* 2003;**5**:675–9.
212. Hassan YI, Zempleni J. Epigenetic regulation of chromatin structure and gene function by biotin. *J Nutr* 2006;**136**:1763–5.
213. Ke Q, Davidson T, Chen H, Kluz T, Costa M. Alterations of histone modifications and transgene silencing by nickel chloride. *Carcinogenesis* 2006;**27**:1481–8.
214. Vickers MH. Developmental programming and adult obesity: the role of leptin. *Curr Opin Endocrinol Diabetes Obes* 2007;**14**:17–22.
215. Cobiac L. Epigenomics and nutrition. *Forum Nutr* 2007;**60**:31–41.
216. Johnson IT, Belshaw NJ. Environment, diet and CpG island methylation: epigenetic signals in gastrointestinal neoplasia. *Food Chem Toxicol* 2008;**46**:1346–59.
217. Waterland RA, Garza C. Potential mechanisms of metabolic imprinting that lead to chronic disease. *Am J Clin Nutr* 1999;**69**:179–97.
218. Lathe R. The individuality of mice. *Genes Brain Behav* 2004;**3**:317–27.
219. Lesage J, Blondeau B, Grino M, Breant B, Dupouy JP. Maternal undernutrition during late gestation induces fetal overexposure to glucocorticoids and intrauterine growth retardation, and disturbs the hypothalamo-pituitary adrenal axis in the newborn rat. *Endocrinology* 2001;**142**:1692–702.
220. Mairesse J, Lesage J, Breton C, Breant B, Hahn T, Darnaudery M, et al. Maternal stress alters endocrine function of the feto-placental unit in rats. *Am J Physiol Endocrinol Metab* 2007;**292**:E1526–33.
221. Sarwer DB, von Sydow Green A, Vetter ML, Wadden TA. Behavior therapy for obesity: where are we now? *Curr Opin Endocrinol Diabetes Obes* 2009;**16**:347–52.
222. Vige A, Gallou-Kabani C, Junien C. Sexual dimorphism in non-Mendelian inheritance. *Pediatr Res* 2008;**63**:340–7.

223. Yao XH, Gregoire Nyomba BL. Abnormal glucose homeostasis in adult female rat offspring after intrauterine ethanol exposure. *Am J Physiol Regul Integr Comp Physiol* 2007;**292**: R1926–33.
224. Elton CW, Pennington JS, Lynch SA, Carver FM, Pennington SN. Insulin resistance in adult rat offspring associated with maternal dietary fat and alcohol consumption. *J Endocrinol* 2002;**173**:63–71.
225. Joshi S, Rao S, Golwilkar A, Patwardhan M, Bhonde R. Fish oil supplementation of rats during pregnancy reduces adult disease risks in their offspring. *J Nutr* 2003;**133**:3170–4.
226. Joshi S, Rao S, Girigosavi S, Daware M, Kale A, Hegde M. Differential effects of fish oil and folic acid supplementation during pregnancy in rats on cognitive performance and serum glucose in their offspring. *Nutrition* 2004;**20**:465–72.
227. White CL, Braymer HD, York DA, Bray GA. Effect of a high or low ambient perinatal temperature on adult obesity in Osborne-Mendel and S5B/Pl rats. *Am J Physiol Regul Integr Comp Physiol* 2005;**288**:R1376–84.
228. Feinberg AP. Phenotypic plasticity and the epigenetics of human disease. *Nature* 2007;**447**:433–40.
229. Grino M. Prenatal nutritional programming of central obesity and the metabolic syndrome: role of adipose tissue glucocorticoid metabolism. *Am J Physiol Regul Integr Comp Physiol* 2005;**289**:R1233–5.
230. Seckl JR. Prenatal glucocorticoids and long-term programming. *Eur J Endocrinol* 2004;**151** (Suppl. 3):U49–62.
231. Ravelli AC, van Der Meulen JH, Osmond C, Barker DJ, Bleker OP. Obesity at the age of 50 y in men and women exposed to famine prenatally. *Am J Clin Nutr* 1999;**70**:811–6.
232. Eriksson J, Forsen T, Osmond C, Barker D. Obesity from cradle to grave. *Int J Obes Relat Metab Disord* 2003;**27**:722–7.
233. Langley-Evans SC. Fetal programming of cardiovascular function through exposure to maternal undernutrition. *Proc Nutr Soc* 2001;**60**:505–13.
234. Bellinger L, Sculley DV, Langley-Evans SC. Exposure to undernutrition in fetal life determines fat distribution, locomotor activity and food intake in ageing rats. *Int J Obes (Lond)* 2006;**30**:729–38.
235. Woodall SM, Johnston BM, Breier BH, Gluckman PD. Chronic maternal undernutrition in the rat leads to delayed postnatal growth and elevated blood pressure of offspring. *Pediatr Res* 1996;**40**:438–43.
236. de Rooij SR, Painter RC, Phillips DI, Osmond C, Michels RP, Godsland IF, et al. Impaired insulin secretion after prenatal exposure to the Dutch famine. *Diab Care* 2006;**29**:1897–901.
237. Martorell R, Stein AD, Schroeder DG. Early nutrition and later adiposity. *J Nutr* 2001;**131**:874S–80S.
238. Simmons RA. Developmental origins of diabetes: the role of epigenetic mechanisms. *Curr Opin Endocrinol Diabetes Obes* 2007;**14**:13–6.
239. Koletzko B, von Kries R, Monasterolo RC, Subias JE, Scaglioni S, Giovannini M, et al. Infant feeding and later obesity risk. *Adv Exp Med Biol* 2009;**646**:15–29.
240. Young JB. Developmental origins of obesity: a sympathoadrenal perspective. *Int J Obes (Lond)* 2006;**30**(Suppl. 4):S41–9.
241. Langley-Evans SC, Bellinger L, McMullen S. Animal models of programming: early life influences on appetite and feeding behaviour. *Matern Child Nutr* 2005;**1**:142–8.
242. Vickers MH, Breier BH, Cutfield WS, Hofman PL, Gluckman PD. Fetal origins of hyperphagia, obesity, and hypertension and postnatal amplification by hypercaloric nutrition. *Am J Physiol Endocrinol Metab* 2000;**279**:E83–7.
243. Gambling L, Dunford S, Wallace DI, Zuur G, Solanky N, Srai SK, et al. Iron deficiency during pregnancy affects postnatal blood pressure in the rat. *J Physiol* 2003;**552**:603–10.

244. Liang C, Oest ME, Prater MR. Intrauterine exposure to high saturated fat diet elevates risk of adult-onset chronic diseases in C57BL/6 mice. *Birth Defects Res B Dev Reprod Toxicol* 2009;**86**:377–84.
245. Petry CJ, Dorling MW, Pawlak DB, Ozanne SE, Hales CN. Diabetes in old male offspring of rat dams fed a reduced protein diet. *Int J Exp Diabetes Res* 2001;**2**:139–43.
246. Bellinger L, Lilley C, Langley-Evans SC. Prenatal exposure to a maternal low-protein diet programmes a preference for high-fat foods in the young adult rat. *Br J Nutr* 2004;**92**:513–20.
247. Levin BE, Dunn-Meynell AA. Maternal obesity alters adiposity and monoamine function in genetically predisposed offspring. *Am J Physiol Regul Integr Comp Physiol* 2002;**283**: R1087–93.
248. Nehiri T, Duong Van Huyen JP, Viltard M, Fassot C, Heudes D, Freund N, et al. Exposure to maternal diabetes induces salt-sensitive hypertension and impairs renal function in adult rat offspring. *Diabetes* 2008;**57**:2167–75.
249. Dabelea D, Hanson RL, Lindsay RS, Pettitt DJ, Imperatore G, Gabir MM, et al. Intrauterine exposure to diabetes conveys risks for type 2 diabetes and obesity: a study of discordant sibships. *Diabetes* 2000;**49**:2208–11.
250. Al Mamun A, Lawlor DA, Alati R, O'Callaghan MJ, Williams GM, Najman JM. Does maternal smoking during pregnancy have a direct effect on future offspring obesity? Evidence from a prospective birth cohort study. *Am J Epidemiol* 2006;**164**:317–25.
251. Toschke AM, Koletzko B, Slikker Jr. W, Hermann M, von Kries R. Childhood obesity is associated with maternal smoking in pregnancy. *Eur J Pediatr* 2002;**161**:445–8.
252. Oberlander TF, Weinberg J, Papsdorf M, Grunau R, Misri S, Devlin AM. Prenatal exposure to maternal depression, neonatal methylation of human glucocorticoid receptor gene (NR3C1) and infant cortisol stress responses. *Epigenetics* 2008;**3**:97–106.
253. Fresiello A, Grammatikopoulos G, Pignatelli M, Sadile AG. Environmental factors during postnatal period modify activity and non-selective attention in the Naples high-excitability rat. *Behav Brain Res* 2002;**130**:111–5.
254. Matthews SG, Phillips DI. Minireview: transgenerational inheritance of the stress response: a new frontier in stress research. *Endocrinology* 2009;**151**:7–13.
255. Mathers JC. Session 2: personalised nutrition.epigenomics: a basis for understanding individual differences? *Proc Nutr Soc* 2008;**67**:390–4.
256. Weaver IC, Champagne FA, Brown SE, Dymov S, Sharma S, Meaney MJ, et al. Reversal of maternal programming of stress responses in adult offspring through methyl supplementation: altering epigenetic marking later in life. *J Neurosci* 2005;**25**:11045–54.
257. Bjornsson HT, Sigurdsson MI, Fallin MD, Irizarry RA, Aspelund T, Cui H, et al. Intra-individual change over time in DNA methylation with familial clustering. *JAMA* 2008;**299**:2877–83.
258. Burdge GC, Lillycrop KA, Phillips ES, Slater-Jefferies JL, Jackson AA, Hanson MA. Folic acid supplementation during the juvenile-pubertal period in rats modifies the phenotype and epigenotype induced by prenatal nutrition. *J Nutr* 2009;**139**:1054–60.
259. Szeto IM, Das PJ, Aziz A, Anderson GH. Multivitamin supplementation of Wistar rats during pregnancy accelerates the development of obesity in offspring fed an obesogenic diet. *Int J Obes (Lond)* 2009;**33**:364–72.
260. Delage B, Dashwood RH. Dietary manipulation of histone structure and function. *Annu Rev Nutr* 2008;**28**:347–66.
261. Yajnik CS, Deshpande SS, Jackson AA, Refsum H, Rao S, Fisher DJ, et al. Vitamin B12 and folate concentrations during pregnancy and insulin resistance in the offspring: the Pune Maternal Nutrition Study. *Diabetologia* 2008;**51**:29–38.
262. Thorne JL, Campbell MJ, Turner BM. Transcription factors, chromatin and cancer. *Int J Biochem Cell Biol* 2009;**41**:164–75.

263. Suzuki MM, Bird A. DNA methylation landscapes: provocative insights from epigenomics. *Nat Rev Genet* 2008;**9**:465–76.
264. Esteller M. Epigenetics in cancer. *N Engl J Med* 2008;**358**:1148–59.
265. Dolinoy DC, Jirtle RL. Environmental epigenomics in human health and disease. *Environ Mol Mutagen* 2008;**49**:4–8.
266. Walley AJ, Blakemore AI, Froguel P. Genetics of obesity and the prediction of risk for health. *Hum Mol Genet* 2006;**15**(Spec No. 2):R124–30.
267. York D, Bouchard C. How obesity develops: insights from the new biology. *Endocrine* 2000;**13**:143–54.
268. Marti A, Martinez-Gonzalez MA, Martinez JA. Interaction between genes and lifestyle factors on obesity. *Proc Nutr Soc* 2008;**67**:1–8.
269. Terruzzi I, Senesi P, Fermo I, Lattuada G, Luzi L. Are genetic variants of the methyl group metabolism enzymes risk factors predisposing to obesity? *J Endocrinol Invest* 2007;**30**:747–53.
270. Sullivan KE, Reddy AB, Dietzmann K, Suriano AR, Kocieda VP, Stewart M, et al. Epigenetic regulation of tumor necrosis factor alpha. *Mol Cell Biol* 2007;**27**:5147–60.
271. Achard V, Boullu-Ciocca S, Desbriere R, Grino M. Perinatal programming of central obesity and the metabolic syndrome: role of glucocorticoids. *Metab Syndr Relat Disord* 2006;**4**:129–37.
272. El Gazzar M, Yoza BK, Hu JY, Cousart SL, McCall CE. Epigenetic silencing of tumor necrosis factor alpha during endotoxin tolerance. *J Biol Chem* 2007;**282**:26857–64.
273. Arai M, Imazeki F, Sakai Y, Mikata R, Tada M, Seki N, et al. Analysis of the methylation status of genes up-regulated by the demethylating agent, 5-aza-2′-deoxycytidine, in esophageal squamous cell carcinoma. *Oncol Rep* 2008;**20**:405–12.
274. Chang YS, Wang L, Liu D, Mao L, Hong WK, Khuri FR, et al. Correlation between insulin-like growth factor-binding protein-3 promoter methylation and prognosis of patients with stage I non-small cell lung cancer. *Clin Cancer Res* 2002;**8**:3669–75.
275. Hirasawa Y, Arai M, Imazeki F, Tada M, Mikata R, Fukai K, et al. Methylation status of genes upregulated by demethylating agent 5-aza-2′-deoxycytidine in hepatocellular carcinoma. *Oncology* 2006;**71**:77–85.
276. Kim JW, Cheng Y, Liu W, Li T, Yegnasubramanian S, Zheng SL, et al. Genetic and epigenetic inactivation of LPL gene in human prostate cancer. *Int J Cancer* 2009;**124**:734–8.
277. Lai JC, Cheng YW, Chiou HL, Wu MF, Chen CY, Lee H. Gender difference in estrogen receptor alpha promoter hypermethylation and its prognostic value in non-small cell lung cancer. *Int J Cancer* 2005;**117**:974–80.
278. Lazcoz P, Munoz J, Nistal M, Pestana A, Encio I, Castresana JS. Frequent promoter hypermethylation of RASSF1A and CASP8 in neuroblastoma. *BMC Cancer* 2006;**6**:254.
279. Nosho K, Shima K, Irahara N, Kure S, Firestein R, Baba Y, et al. SIRT1 histone deacetylase expression is associated with microsatellite instability and CpG island methylator phenotype in colorectal cancer. *Mod Pathol* 2009;**22**:922–32.
280. Okochi O, Hibi K, Sakai M, Inoue S, Takeda S, Kaneko T, et al. Methylation-mediated silencing of SOCS-1 gene in hepatocellular carcinoma derived from cirrhosis. *Clin Cancer Res* 2003;**9**:5295–8.
281. Tokita T, Maesawa C, Kimura T, Kotani K, Takahashi K, Akasaka T, et al. Methylation status of the SOCS3 gene in human malignant melanomas. *Int J Oncol* 2007;**30**:689–94.
282. Worthley DL, Whitehall VL, Buttenshaw RL, Irahara N, Greco SA, Ramsnes I, et al. DNA methylation within the normal colorectal mucosa is associated with pathway-specific predisposition to cancer. *Oncogene* 2009;**29**:1653–62.

283. Jiang Y, Jiang J, Xiong J, Cao J, Li N, Li G, et al. Homocysteine-induced extracellular superoxide dismutase and its epigenetic mechanisms in monocytes. *J Exp Biol* 2008;**211**:911–20.
284. Melzner I, Scott V, Dorsch K, Fischer P, Wabitsch M, Bruderlein S, et al. Leptin gene expression in human preadipocytes is switched on by maturation-induced demethylation of distinct CpGs in its proximal promoter. *J Biol Chem* 2002;**277**:45420–7.
285. Stoger R. In vivo methylation patterns of the leptin promoter in human and mouse. *Epigenetics* 2006;**1**:155–62.
286. Torrisani J, Hanoun N, Laurell H, Lopez F, Maoret JJ, Souque A, et al. Identification of an upstream promoter of the human somatostatin receptor, hSSTR2, which is controlled by epigenetic modifications. *Endocrinology* 2008;**149**:3137–47.
287. Ronn T, Poulsen P, Hansson O, Holmkvist J, Almgren P, Nilsson P, et al. Age influences DNA methylation and gene expression of COX7A1 in human skeletal muscle. *Diabetologia* 2008;**51**:1159–68.
288. Guan HP, Ishizuka T, Chui PC, Lehrke M, Lazar MA. Corepressors selectively control the transcriptional activity of PPARgamma in adipocytes. *Genes Dev* 2005;**19**:453–61.
289. Boque N, Campion J, Milagro FI, Moreno-Aliaga MJ, Martinez JA. Some cyclin-dependent kinase inhibitors-related genes are regulated by vitamin C in a model of diet-induced obesity. *Biol Pharm Bull* 2009;**32**:1462–8.
290. Mattsson C, Olsson T. Estrogens and glucocorticoid hormones in adipose tissue metabolism. *Curr Med Chem* 2007;**14**:2918–24.
291. Sertic J, Juricic L, Ljubic H, Bozina T, Lovric J, Markeljevic J, et al. Variants of ESR1, APOE, LPL and IL-6 loci in young healthy subjects: association with lipid status and obesity. *BMC Res Notes* 2009;**2**:203.
292. Farmer SR. Transcriptional control of adipocyte formation. *Cell Metab* 2006;**4**:263–73.
293. Musri MM, Gomis R, Parrizas M. Chromatin and chromatin-modifying proteins in adipogenesis. *Biochem Cell Biol* 2007;**85**:397–410.
294. Catalioto RM, Maggi CA, Giuliani S. Chemically distinct HDAC inhibitors prevent adipose conversion of subcutaneous human white preadipocytes at an early stage of the differentiation program. *Exp Cell Res* 2009;**315**:3267–80.
295. Holness MJ, Caton PW, Sugden MC. Acute and long-term nutrient-led modifications of gene expression: potential role of SIRT1 as a central co-ordinator of short and longer-term programming of tissue function. *Nutrition* 2010;**26**(5):491–501.
296. Lomb DJ, Laurent G, Haigis MC. Sirtuins regulate key aspects of lipid metabolism. *Biochim Biophys Acta* 2010;**1804**(8):1652–7.
297. Crujeiras AB, Parra D, Goyenechea E, Martinez JA. Sirtuin gene expression in human mononuclear cells is modulated by caloric restriction. *Eur J Clin Invest* 2008;**38**:672–8.
298. Zulet MA, Puchau B, Navarro C, Marti A, Martinez JA. Inflammatory biomarkers: the link between obesity and associated pathologies. *Nutr Hosp* 2007;**22**:511–27.
299. Stenvinkel P, Karimi M, Johansson S, Axelsson J, Suliman M, Lindholm B, et al. Impact of inflammation on epigenetic DNA methylation—a novel risk factor for cardiovascular disease? *J Intern Med* 2007;**261**:488–99.
300. Kamei Y, Suganami T, Ehara T, Kanai S, Hayashi K, Yamamoto Y, et al. Increased expression of DNA methyltransferase 3a in obese adipose tissue: studies with transgenic mice. *Obesity (Silver Spring)* 2010;**18**:314–21.
301. Geeraert B, Crombe F, Hulsmans M, Benhabiles N, Geuns JM, Holvoet P. Stevioside inhibits atherosclerosis by improving insulin signaling and antioxidant defense in obese insulin-resistant mice. *Int J Obes (Lond)* 2010;**34**:569–77.

302. Wang Z, Zhou YT, Kakuma T, Lee Y, Kalra SP, Kalra PS, et al. Leptin resistance of adipocytes in obesity: role of suppressors of cytokine signaling. *Biochem Biophys Res Commun* 2000;**277**:20–6.
303. Estep JM, Baranova A, Hossain N, Elariny H, Ankrah K, Afendy A, et al. Expression of cytokine signaling genes in morbidly obese patients with non-alcoholic steatohepatitis and hepatic fibrosis. *Obes Surg* 2009;**19**:617–24.
304. Lee SD, Kuo WW, Bau DT, Ko FY, Wu FL, Kuo CH, et al. The coexistence of nocturnal sustained hypoxia and obesity additively increases cardiac apoptosis. *J Appl Physiol* 2008;**104**:1144–53.
305. Felicia Faienza M, Santoro N, Lauciello R, Calabro R, Giordani L, Di Salvo G, et al. Igf2 gene variants and risk of hypertension in obese children and adolescents. *Pediatr Res* 2010;**67** (4):340–4.
306. Singh V, Grotzinger C, Nowak KW, Zacharias S, Goncz E, Pless G, et al. Somatostatin receptor subtype-2-deficient mice with diet-induced obesity have hyperglycemia, nonfasting hyperglucagonemia, and decreased hepatic glycogen deposition. *Endocrinology* 2007;**148**:3887–99.
307. Reinehr T, Panteliadou A, de Sousa G, Andler W. Insulin-like growth factor-I, insulin-like growth factor binding protein-3 and growth in obese children before and after reduction of overweight. *J Pediatr Endocrinol Metab* 2009;**22**:225–33.
308. Gomez-Ruiz A, de Miguel C, Campion J, Martinez JA, Milagro FI. Time-dependent regulation of muscle caveolin activation and insulin signalling in response to high-fat diet. *FEBS Lett* 2009;**583**:3259–64.
309. Furuhashi M, Fucho R, Gorgun CZ, Tuncman G, Cao H, Hotamisligil GS. Adipocyte/macrophage fatty acid-binding proteins contribute to metabolic deterioration through actions in both macrophages and adipocytes in mice. *J Clin Invest* 2008;**118**:2640–50.
310. Moro C, Smith SR. Natriuretic peptides: new players in energy homeostasis. *Diabetes* 2009;**58**:2726–8.
311. Dandona S, Roberts R. Creating a genetic risk score for coronary artery disease. *Curr Atheroscler Rep* 2009;**11**:175–81.
312. Gemma C, Sookoian S, Dieuzeide G, Garcia SI, Gianotti TF, Gonzalez CD, et al. Methylation of TFAM gene promoter in peripheral white blood cells is associated with insulin resistance in adolescents. *Mol Genet Metab* 2010;**100**(1):83–7.
313. He M, Cornelis MC, Franks PW, Zhang C, Hu FB, Qi L. Obesity genotype score and cardiovascular risk in women with type 2 diabetes mellitus. *Arterioscler Thromb Vasc Biol* 2010;**30**:327–32.
314. Kim M, Long TI, Arakawa K, Wang R, Yu MC, Laird PW. DNA methylation as a biomarker for cardiovascular disease risk. *PLoS ONE* 2010;**5**:e9692.

Genetics of the Metabolic Complications of Obesity

Joshua P. Lewis*,† and
Alan R. Shuldiner*,†

*University of Maryland School of Medicine, Baltimore, Maryland, USA
†Geriatric Research and Education Clinical Center, Veterans Administration Medical Center, Baltimore, Maryland, USA

The mechanisms linking obesity to its metabolic complications are extremely complex and remain hotly debated. It is hypothesized that reasons for variability in susceptibility to obesity-mediated metabolic complications involve both environmental and genetic factors. While advances in knowledge of variation in the human genome have led to the identification of susceptibility genes contributing to obesity and related disorders, relatively few studies have focused specifically on interactions between obesity and genetic polymorphisms in the development of metabolic complications. Despite these limited efforts, an increasing amount of evidence, particularly in the study of adipokines and adipose tissue, suggests that the effects of some gene variants on metabolic traits are modified by or present only in the setting of obesity. Furthermore, some of these loci may have larger effects on metabolic phenotypes when provoked by certain dietary or lifestyle factors. In this chapter, we review the small number of genes and their variants for which there is the greatest evidence for a role in obesity-associated metabolic complications. Identification of such variants will provide additional insight into disease

Progress in Molecular Biology
and Translational Science, Vol. 94
DOI: 10.1016/S1877-1173(10)94012-X

1877-1173/10 $35.00

pathophysiology and may provide the scientific rationale for the design of more individualized prevention or treatment strategies for obesity and its metabolic complications.

I. Introduction

It has long been known that obesity, particularly central or visceral obesity, is associated with metabolic complications that include insulin resistance, glucose intolerance/diabetes, dyslipidemia, hypertension, and a prothrombotic milieu.[1] These metabolic conspirators are all risk factors for cardiovascular disease (CVD), a major cause of morbidity and mortality in obese people.[2]

The mechanisms linking obesity to its metabolic complications are extremely complex and remain hotly debated.[3] One leading theory posits that systemic insulin resistance plays a pivotal role. A chronic imbalance between energy intake and energy expenditure results in the systemic overload of triglycerides. This may result in dysfunctional adipose tissue that secretes too much or too little of various adipokines that may have systemic effects on insulin sensitivity and inflammation (e.g., fatty acids, adiponectin, leptin, tumor necrosis factor alpha [TNF], interleukin-6 [IL6]).[4] In many with obesity, excessive deposition of triglycerides occurs in the liver and muscle. This "ectopic" fat deposition may further contribute to insulin resistance and derangements in glucose and lipid metabolism.[5] In a subset of individuals, nonalcoholic fatty liver disease (NAFLD) may progress to chronic inflammation and fibrosis, now the most common cause of end-stage liver disease.

Despite these well-known associations between obesity, insulin resistance, and the metabolic syndrome, not all who are obese develop metabolic complications of obesity. Reasons for variability in susceptibility to obesity-related metabolic complications are likely to involve both environmental and genetic factors. For example, high levels of physical activity may have beneficial effects on metabolic complications of obesity even in the absence of significant weight loss. Furthermore, certain ethnic groups, for example, Asians, appear to develop metabolic complications at significantly lower degrees of obesity, suggesting a lower "tolerance" for excess fat accumulation, perhaps due to genetic factors. African-Americans appear to be less susceptible to NAFLD than Caucasians, due in part to a higher frequency in a variant in the patatin-like phospholipase domain containing 3 (*PNPLA3*) gene.[6]

With advances in knowledge of genetic variation in the human genome, an increasing number of susceptibility genes that contribute to or cause obesity have been identified.[7] Several rare monogenic forms of obesity resulting from mutations in the genes encoding leptin, the leptin receptor, proopiomelanocortin, melanocortin 4-receptor, and peroxisome proliferator-activated receptor gamma (PPARG) have been established.[8] Candidate gene studies and large-scale genome-wide association studies (GWAS) have been instrumental in discovering common genetic variants that contribute to polygenic obesity (e.g., *FTO*).[9] Indeed, similar efforts have resulted in the identification of significant genetic risk factors of several metabolic conditions, including type 2 diabetes (T2D),[10] cardiovascular dyslipidemia,[11] hypertension,[12] and NAFLD.[13]

By contrast, there has been a paucity of *bona fide* genetic variants that have been identified that increase susceptibility to metabolic complications specifically in the setting of obesity, that is, gene variants whose effects on metabolic traits are modified by or present only in the setting of obesity. Yet, another set of gene variants may have pleiotropic effects, increasing susceptibility to obesity *and* one or more metabolic complications. These gene variants might be regarded as those that distinguish between "benign" and more "malignant" forms of obesity and thus would have important diagnostic and prognostic implications. Identification of these gene variants would uncover new insights into disease mechanisms and potential targets for novel interventions. At the time of this writing, there have been no large-scale, comprehensive candidate gene studies or GWAS designed to identify variants associated with metabolic syndrome traits specifically or preferentially in the setting of obesity or studies to identify variants with pleiotropic effects on obesity and related metabolic traits.

In order to demonstrate that a gene variant for a metabolic trait manifests specifically in the setting of obesity, one must show (1) an effect of a genetic variant on a metabolic trait that occurs specifically in an obese subgroup in stratified analysis of high and low BMI, or (2) that in regression analysis between the genetic variant and metabolic syndrome trait, the interaction term (variant × BMI) is statistically significant. Alternatively, prospective or interventional (e.g., weight loss, feeding) studies are powerful approaches to identifying obesity-associated metabolic syndrome traits.

Even studies demonstrating significant interactive effects between a genetic variant and obesity on metabolic syndrome traits, various kinds of biases, and/or type I error due to multiple comparisons could result in false-positive results. Distinguishing a true-positive association from a false-positive finding requires independent replication in other populations and studies. Although replication is the single most important criteria to distinguish a true-positive result from a false-positive result, lack of replication of a true-positive association may be due to differences among studies (e.g., population characteristics, differences in phenotyping methods) or type II error caused by inadequate sample size and power. Indeed, underpowered studies are often more the rule than the exception, since most common gene variants for complex traits like obesity and associated metabolic complications have a very modest effect on the phenotype. This problem is compounded when interactive effects between a modest-effect gene variant and BMI on metabolic traits are being sought. Given the nontrivial requirement for replication of an association in independent study samples, other lines of supporting evidence are often highly desirable such as *in vitro* or *in vivo* (animal or human) functional studies showing a role of the gene (and/or gene variant) on the phenotype.

Although a large number of genes and their variants have been investigated for their effect on metabolic traits that manifest in the presence of obesity, very few (if any) would fulfill the criteria described previously. In this chapter, we review the relatively small number of genes and their variants for which there is the greatest evidence for a role in obesity-associated metabolic complications.

II. Adipokine Genes

Discoveries in the past 15 years have significantly changed the view of adipose tissue in both energy balance and disease pathophysiology. In addition to its function as a major triglyceride depot, adipose tissue, particularly white adipose tissue, is an active endocrine organ that secretes a number of bioactive proteins known as adipokines. Since the discovery of the adipokine leptin in 1994, over 50 adipocyte-derived products have been identified. Adipokines have also been implicated in a broad number of physiological processes, including glucose and lipid homeostasis, appetite regulation, vascular remodeling and reactivity, and immunity.[14] In addition to their paracrine actions in adipose tissue, adipokines mitigate their dysmetabolic actions through endocrine mechanisms in the liver, muscle, brain, and the vascular system, among other organs.[15]

Multiple studies have shown that dysregulation of individual adipokines contribute not only to obesity but also to its complications including metabolic syndrome, diabetes, CVD, asthma, and NAFLD.[15–17] For example, the adipokines TNF and IL6 are associated with insulin resistance and inflammation and are overexpressed with increasing adiposity.[18] Conversely, adiponectin possesses anti-inflammatory and insulin-sensitizing effects and is significantly decreased in obese individuals.[19]

Given the well-documented adipokine level changes associated with obesity and the role that adipokines play in metabolism, many have speculated that genetic variation in adipokine genes might be important determinants of obesity-associated metabolic complications.[20] The following sections review the genetics of the adipokine genes with the greatest evidence for a role in the dysmetabolic complications of obesity.

A. Adiponectin

As already described, it has been well established that (1) adiponectin acts as an insulin sensitizer and protects from metabolic syndrome traits and CVD, and (2) circulating levels of adiponectin are markedly reduced in individuals who are obese, diabetic, and/or have metabolic syndrome. The gene for adiponectin, *ADIPOQ*, is approximately 16 kb, has three exons, and is located on chromosome 3q26, a region that has been previously associated with T2D and metabolic syndrome.[21–23]

Mounting evidence suggests that genetic variation in the adiponectin gene contributes to adiponectin levels and obesity-associated metabolic complications. Four of the most commonly evaluated SNPs in *ADIPOQ* (+276G>T [rs1501299], +45T>G [rs2241766], −11391G>A [rs17300539], and −11377C>G [rs266729]) have all been associated with adiponectin levels, albeit with some inconsistencies.[24–26] In order to better understand some of these discrepancies, a meta-analysis examining over 2000 individuals was conducted.[27] Significant associations with adiponectin levels were still supported for the +276G>T and −11391G>A variants. Like other common expression-modulating SNPs, these polymorphisms explain only a small proportion of the total variation of adiponectin levels (< 10%). Therefore, it is likely that other genes/polymorphisms will be identified that significantly alter circulating levels of adiponectin[27] and/or that interactions of these SNPs with other modulators such as diet and obesity may begin to explain a larger proportion of the variance.

Numerous studies have reported association of the *ADIPOQ* promoter and genic polymorphisms with T2D, insulin resistance, impaired glucose tolerance, and metabolic syndrome. However, there are an equally large number of publications that have not replicated these findings. In an attempt to clarify which SNPs in adiponectin contribute to the development of these metabolic traits, meta-analyses have been conducted using larger sample sizes. In one such meta-analysis, the authors evaluated the +45T>G and +276G>T variants with respect to T2D, insulin resistance, and BMI and the −11377C>G and −11391G>A variants with respect to T2D.[27] From this analysis, the only significant association observed existed between the +276G>T and insulin resistance. In another meta-analysis examining the effects of adiponectin variants in Chinese individuals, no association was found with the +276G>T variant and T2D.[28] Though the same study did report that the +45T>G SNP was associated with of T2D. However, significant heterogeneity was observed among the individual studies, and therefore, these results should be interpreted with caution. In a second meta-analysis performed in Chinese individuals, both the +45T>G and +276G>T polymorphisms were significantly associated with T2D (OR of 1.59 and 1.26, respectively).[29]

Taken together, these meta-analyses suggest that polymorphisms in *ADIPOQ* have significant effects on several metabolic complications such as insulin resistance and T2D. However, it is not quite clear whether any of these variants are clinically relevant. Most likely, the answer to this question will be context-specific depending on variables such as ethnicity, lifestyle choices, and environmental exposures. Indeed, recent investigations evaluating interactions between these variables and *ADIPOQ* polymorphisms support this hypothesis (see below).

Changes in circulating levels of adiponectin, whether caused by genetic variants or the onset of obesity, significantly affect other metabolic complications. However, being obese or genetically predisposed does not guarantee the development of a particular metabolic disorder. Similar to all other complex traits, multiple genetic, environmental, and lifestyle factors contribute to the development of the phenotype. Therefore, it is interesting to ask whether, in the presence of obesity, genetic variants in *ADIPOQ* lead to differences in the risk for metabolic complications. Similarly, in obese individuals with a given risk variant(s), are there environmental exposures or lifestyle changes that modulate metabolic complications susceptibility? Indeed, recent investigations, although limited, have attempted to answer these questions. While additional studies are needed to clarify and confirm such interactions, it is becoming increasingly evident that obesity-linked metabolic complications are influenced by genetic factors and these genetic influences are also modulated by environmental and lifestyle factors.

Because of the independent effects of both *ADIPOQ* polymorphisms and adiposity on circulating adiponectin levels, it has been postulated that the interaction of obesity and *ADIPOQ* variants could further lower adiponectin levels and/or lead to enhanced disease susceptibility. In an Italian population consisting of 220 obese patients and 116 healthy controls, it was found, not surprisingly, that obese individuals had lower serum levels of adiponectin than controls (30.5 ± 28.3 vs. 43.9 ± 35.7 μg/ml, $P < 0.01$).[30] However, obese individuals carrying the G-allele of the −11377C>G SNP had significantly lower circulating adiponectin compared to obese individuals without the G-allele (22.9 ± 20.5 vs. 33.1 ± 29.4 μg/ml, $P < 0.05$), leading to a further decrease in adiponectin concentration in these individuals.[30] In an Iranian population consisting of obese patients with and without T2D, it was found that the G-allele of the SNP at position +45 was associated with an increased risk of developing T2D (OR = 2.57, 95% CI: 1.05–6.30, $P = 0.04$).[31] Similarly, in a Han Chinese population consisting of 338 T2D patients and 460 nondiabetic controls, the −11377C>G polymorphism was associated with T2D in the obese group (OR = 2.45, 95% CI: 1.13–5.31, $P = 0.02$) but not in a nonobese group.[32] Interestingly, these investigators also reported that another SNP in *ADIPOQ*, rs1063539, was associated with T2D only in the nonobese group (OR = 2.34, 95% CI: 1.31–4.21, $P = 0.004$). Together, these results suggest that genetic susceptibility to T2D may differ depending on whether an individual is obese or not. Similar interactions between obesity and *ADIPQ* variants have been reported for CVD risk. An investigation in 449 diabetic women with CVD and 946 with diabetes showed that individuals in the highest quartile of genetic risk score, generated from obesity risk polymorphisms from several genes, had significantly lower adiponectin levels and had a 53% higher CVD risk, suggesting that obesity-related SNPs may jointly affect CVD risk.[33]

There is increasing evidence that metabolic complications in obese individuals with specific *ADIPOQ* polymorphisms can be partially or fully ameliorated with diet-based interventions. For example, 363 Korean individuals with impaired fasting glucose or T2D followed a dietary intervention involving the substitution of refined rice with whole grains for 12 weeks.[34] They found that individuals with the TT genotype at the +45T>G *ADIPOQ* locus had higher adiponectin concentrations and reduced insulin resistance as measured by homeostatic model assessment (HOMA) compared to other genotypes after the dietary intervention. Additionally, haplotype analyses using +45T>G and +276G>T showed that reduction in insulin resistance and increased adiponectin levels associated with the T–G haplotype occurred in overweight–obese patients but not in normal-weight subjects. In other words, obese individuals with the G-allele at position +45 and the T-allele at position +276, both considered risk alleles in most studies, did not show the beneficial changes in insulin resistance or adiponectin levels in response to the dietary intervention seen in individuals with the other haplotypes.[34]

In another investigation, 32 obese Japanese women were treated by meal replacement with a low-calorie diet for 8 weeks, and several *ADIPOQ* polymorphisms were evaluated with respect to components of metabolic syndrome.[35] After treatment, subjects having at least one copy of the G-allele at position +276 had a significant decrease in waist circumference compared to those who had the TT genotype ($P = 0.026$). At position +45, evidence also showed that triglyceride levels were significantly reduced in individuals with the TT genotype compared to those who had the TG genotype after treatment ($P = 0.003$). Finally, these investigators also reported that, after treatment, individuals with the CC genotype at position −11377 of *ADIPOQ* had higher systolic blood pressure and fasting plasma glucose compared to those who had the CG or GG genotypes ($P = 0.04$).[35] Interestingly, the results of this study suggest that adiponectin SNPs lead to different changes in obesity-related metabolic traits during meal replacement with a low-calorie diet.

In another 8-week low-calorie diet intervention study, Goyenechea and colleagues reported interesting findings regarding the *ADIPOQ* promoter variant −11391G>A variant and several traits related to metabolic syndrome in 180 obese Spanish participants.[36] Before treatment, the GG genotype was associated with higher triglyceride and insulin concentrations (both $P < 0.05$) as well as an increased risk of insulin resistance ($OR = 2.44$, $P = 0.03$) and metabolic syndrome ($OR = 3.24$, $P = 0.003$). However, after treatment, all risk associated with the GG genotype disappeared. Interestingly, after 32 and 60 weeks of posttreatment and weight regain, carriers of the risk genotype recovered the risk of metabolic comorbidities ($OR = 2.42$, $P = 0.43$ and $OR = 2.88$, $P = 0.01$, respectively).[36] Together, these results show that enhanced risks associated with the G-allele of adiponectin −11391G>A variant in obese individuals can be nullified through weight loss with a low-calorie diet, though not sustained afterward.

Because of the increasing prevalence of childhood obesity, investigations pertaining to *ADIPOQ* polymorphisms and related metabolic traits have also been conducted in children. One study showed that a significant interaction exists between fiber intake and +276G>T genotype in 991 Greek children (mean age of 11.2 ± 0.6 years).[37] In fact, when dietary fiber intake was low, T-allele carriers had significantly lower adiponectin concentrations compared to GG homozygotes (4.2 ± 2.3 vs. 5.1 ± 2.7 μg/ml, $P = 0.02$). However, as fiber intake increased, no association was found between this locus and levels of adiponectin. In another investigation conducted in children, a relationship between the +276G>T variant and fasting insulin, insulin resistance, and several long-chain polyunsaturated fatty acids (LCPUFAs) were reported.[38] In all instances, T-allele carriers had worse metabolic profiles. Individuals who had the T-allele had higher circulating insulin concentrations (20.6 vs. 12.4 μU/ml, $P = 0.04$) and higher HOMA-IR index values (4.5 vs.2.6, $P = 0.03$). Also, T-allele carriers had significantly different concentrations of several LCPUFAs compared to GG homozygotes. An interaction between the +276G>T variant and $n-6/n-3$ LCPUFAs affecting insulin resistance was also observed. The authors suggest that obese children who carry the T-allele at position +276 of *ADIPOQ* may be at increased risk of metabolic complications because of differences in plasma phospholipid profiles.

B. Tumor Necrosis Factor Alpha

TNF is a pleiotropic regulatory cytokine produced by several cell types, including in adipose tissue, that can exert a variety of effects on cellular and biological processes, including immunity, inflammation, apoptosis, angiogenesis, coagulation, growth promotion and inhibition, and energy homeostasis.[39] Dysregulation of TNF has been implicated in a broad range of pathological conditions. Disturbances in the circulating levels of TNF have not only been shown in individuals with obesity, insulin resistance, diabetes, and metabolic syndrome, but also in subjects with inflammatory diseases, CVD, asthma, AIDS, and some forms of cancer.

Human genetic investigations have contributed to our understanding of the role of TNF in obesity and its related metabolic complications. While multiple genetic studies in ethnically diverse populations have been conducted, there have been inconsistencies regarding the role of *TNF* polymorphisms in obesity and metabolic syndrome traits. The most intensively studied *TNF* polymorphism is a G>A variant in the promoter at position −308 (rs1800629). The variant resides within the consensus sequence of the transcription factor AP-2, and *in vitro* studies have indicated decreased transcription of the G-allele.[40] Some studies have suggested that the −308 G>A variant is associated with serum concentrations of TNF and increased susceptibility of obesity, insulin resistance, diabetes, metabolic syndrome, and several forms of CVD[41–43]; however, replication of these findings has been inconsistent.[44,45]

One explanation for these inconsistencies could be interactions between obesity itself and *TNF* variants on metabolic syndrome traits. For example, Um and coworkers studied *TNF* haplotypes defined by two SNPs, −308G>A and the TNF-β +252A>G. They showed that obese Korean individuals carrying any haplotype other than the A–G haplotype (for the −308 and +252 variants, respectively) had significantly higher waist-to-hip ratios (0.94 vs. 0.92, $P=0.005$), while no difference was reported in nonobese individuals.[46] In addition, haplotype analysis of several *TNF* variants in a French-Canadian population revealed that the *TNF* locus was important in the development of obesity-associated hypertension in nonmorbidly obese individuals but not in morbidly obese patients ($P=0.02$).[47] Interestingly, Pihlajamaki and colleagues showed that the effect of the *TNF* −308G>A SNP on insulin action was dependent on obesity.[48] In this investigation, analyses in all study participants revealed a significant association between this variant and glucose oxidation ($P=0.008$), lipid synthesis ($P=0.037$), and free fatty acid levels during hyperinsulinemia ($P=0.023$). However, when stratifying by BMI (BMI cutoff of 26 kg/m^2), association with *TNF* −308G>A and glucose oxidation was observed only in normal-weight individuals ($P=0.036$). Conversely, in overweight study participants, the *TNF* −308G>A polymorphism was significantly associated with insulin-mediated free fatty acid suppression ($P=0.042$) and rates of lipid synthesis ($P=0.010$) but not glucose oxidation ($P=0.193$). Taken together, these findings suggest complex interactions between variants in *TNF* and obesity on metabolic syndrome traits. In another study, evidence suggesting that variation in *TNF*, both at positions −238 and −308, significantly impacted diabetes-related traits in obese but not in nonobese diabetic patients.[49] For the −308G>A variant, they found that obese A-allele carriers had increased levels of insulin resistance compared to obese patients who were homozygous for the

G-allele (HOMA-IR index 4.36 vs. 2.85, $P = 0.01$). No difference in measures of insulin resistance between genotypes was observed in the entire population or in nonobese diabetic patients. Similarly, in the same study, obese carriers of the −238A allele had reduced ability to suppress postprandial plasma levels of free fatty acids during an 8-h metabolic profile test compared to those with the GG genotype ($P = 0.0004$), while no difference was found in nonobese diabetic patients ($P = 0.60$). Based on these results and the findings of other investigations showing an interaction between polymorphisms in *TNF* and diet on glucose and lipid metabolism,[50,51] the authors suggested that the effects of *TNF* in the progression of T2D are directly related to the additive effects of obesity and/or dietary factors.

Investigations of *TNF* polymorphisms in obese and lean individuals also suggest that this locus interacts with adiposity on T2D susceptibility itself. An interesting study by Kamizono and coworkers evaluated the −857C>T SNP (rs1799724) in the following groups: obese patients with T2D, nonobese patients with T2D, obese individuals with normal glucose tolerance, and nonobese individuals with normal glucose tolerance.[52] They found that obese patients with T2D had a significantly higher frequency of T-allele homozygotes compared to lean individuals with normal glucose tolerance (0.15 vs. 0.04, respectively). Remarkably, none of the obese individuals with normal glucose tolerance had this genotype. This data suggest that the combination of obesity and the TT genotype at position −857 facilitated the progression of T2D in this population. A previous study showed that individuals who were homozygous for the T-allele at position −857 showed higher transcriptional activation of *TNF*.[53] Taken together, obese individuals with this risk genotype may produce more TNF, ultimately leading to insulin resistance and the formation of T2D.

An interaction between obesity and polymorphisms in *TNF* may also contribute to an increased risk of CVD. As mentioned earlier, both obesity and CVD are associated with elevated levels of circulating TNF. This has led some to hypothesize that genetic variants associated with increased TNF levels might also be associated with CVD, particularly in the setting of obesity. The *TNF* −308G>A variant was evaluated in 148 survivors of myocardial infarction and 148 age-, gender-, and race-matched controls.[54] Interactive effects between the −308G>A SNP and smoking, hypertension, dyslipidemia, diabetes, and obesity on myocardial infarction risk were evaluated. No interaction was found between this variant and smoking, hypertension, dyslipidemia, or diabetes. However, evidence of interaction was observed between the A-allele carriers and obesity on CVD risk. The OR for myocardial infarction in obese noncarriers was 2.8 (95% CI: 1.3–6.0), whereas in obese individuals who carried at least one copy of the A-allele, the OR increased to 14.5 (95% CI: 1.8–113). Although this interactive effect was not statistically significant, it was suggestive of an interaction warranting further study in larger populations.

Recently, a larger investigation involving 1213 myocardial infarction survivors and 1561 healthy controls evaluated the effect of obesity, smoking, and several *TNF* promoter variants on myocardial infarction risk.[55] In agreement with previous reports, higher circulating levels of TNF were associated with an increased CVD risk. In fact, men in the highest quartile of TNF levels showed a 70% increase in the risk of myocardial infarction compared to men in the lowest quartile (OR = 1.70, 95% CI: 1.1–2.6). Interestingly, a strong interactive effect in men between TNF levels and obesity was also observed. The risk of myocardial infarction increased significantly in obese men who also had high levels of TNF (OR = 3.4, 95% CI: 2.1–5.6, $P < 0.0001$), with 56% of this increased risk being attributed to an interaction between obesity and TNF levels. In the control population, carriers of the −857T allele had significantly higher TNF levels compared to GG homozygotes ($P = 0.0004$). This association was nominally present in the case population ($P = 0.06$). However, no association was found in men between variants in *TNF* and the risk of myocardial infarction.

There is increasing evidence that obesity is a risk factor for asthma. To investigate whether TNF variants influence this relationship, Castro-Giner and colleagues performed a case–control study with a population consisting of over 9000 individuals.[16] In this study, both obesity and the *TNF* −308G>A SNP were significantly associated with asthma (OR = 2.36, 95% CI: 1.72–3.21 for obesity and OR = 1.34, 95% CI: 1.10–1.62 for −308A allele carriers). The effect of each of these risk factors appeared to be independent of each other. When stratified, the −308A allele was associated with asthma in all BMI strata; however, obese patients were at particularly high risk. Finally, stratified analyses based on atopic status revealed that the increased risk of asthma in obese −308A allele carriers was limited to nonatopic individuals with an OR of 3.15 (95% CI: 1.18–8.54). Even though replication of these findings was confirmed using two independent populations within this study, further investigation is needed to more clearly define the complex interaction of obesity and *TNF* variants on asthma risk.[16]

C. Interleukin-6

IL6 is another pleiotropic adipokine that has been implicated in the pathogenesis of obesity-related metabolic traits. However, in contrast to adiponectin or TNF, which have been generally attributed with beneficial or detrimental effects on obesity and metabolism, respectively, more controversy exists regarding the overall effects of IL6. Historically, IL6 has been grouped with other obesity-predisposing cytokines such as TNF because levels of IL6 are elevated in obese individuals. However, recent work on IL6 in skeletal muscle (as opposed to liver or adipose tissue) has challenged this belief. Despite the conflicting reports regarding the function of IL6 both *in vivo* and *in vitro*, genetic studies have shown that variants in this gene affect several obesity-associated phenotypes, including insulin resistance, diabetes, metabolic syndrome, CVD, sleep apnea, and others.

Several lines of evidence suggest interactions between *IL6* genotype and measures of adiposity on the risk of developing insulin resistance and/or T2D. One such study evaluated 1526 participants of the Framingham Heart Study and showed a significant interaction in men between *IL6* genotype and BMI in predicting insulin resistance ($P < 0.0001$).[56] In this study, obese men (BMI > 27 kg/m^2) with the CC genotype at position − 174 of *IL6* (rs1800795) had higher levels of insulin resistance compared to G-allele carriers, whereas in lean men, individuals with G-allele were more insulin resistant. Results were comparable when using waist circumference instead of BMI as the obesity metric ($P = 0.003$). A significant *IL6* − 174G>C genotype × BMI interaction in men was also observed for both fasting plasma glucose ($p < 0.016$) and fasting insulin levels ($p < 0.001$). Furthermore, the CC genotype was associated with a 30% increase in the prevalence of T2D for each unit increase in BMI, while other genotypes were less affected by changes in BMI.[56] This sex-specific effect of obesity and the *IL6* − 174G>C polymorphism on T2D risk was also shown in a separate cohort of 270 French-Canadians.[57] In 106 obese Caucasian volunteers, those carrying the C-allele showed higher systolic blood pressure, insulin resistance, and plasma insulin concentrations and lower adiponectin levels compared to GG homozygotes.[58] The C-allele was more prevalent in individuals showing HOMA-IR and waist circumference values above the median cutoff ($P = 0.010$ and 0.019, respectively).

Möhlig and coworkers observed an increased risk of developing T2D in obese individuals who were homozygous for the C-allele at position − 174 of *IL6*.[59] In fact, obese patients (BMI ≥ 28 kg/m^2) with the CC genotype had a 17.68-fold increased risk of developing T2D compared to those with BMI values less than 28 kg/m^2 ($P = < 0.01$). This is in contrast to a 2.94-fold and 3.44-fold risk of T2D for obese individuals with the GC or GG genotype, respectively, as compared to nonobese individuals with the corresponding genotype. Furthermore, the − 174 *IL6* variant modified the correlation between BMI and serum IL6 concentrations. These data suggest that obesity is more detrimental (at least in terms of insulin resistance and T2D risk) for individuals carrying the CC genotype due to an increase in IL6 concentration.[59,60]

Interventional studies have also supported a role for *IL6* variants in obesity-related metabolic complications. Razquin and coworkers tested the effects of a Mediterranean diet on the prevention of CVD in 737 subjects.[61] At baseline before one of the three dietary interventions, CC homozygotes of the *IL6* − 174G>C polymorphism had higher body weight ($P = 0.024$), BMI ($P = 0.020$), and a trend toward higher waist circumference ($P = 0.059$). Individuals with the CC genotype lost significantly more weight when placed on a Mediterranean diet supplemented with virgin olive oil (as opposed to a low-fat diet or a Mediterranean diet supplemented with nuts) compared to individuals with the GC or GG genotypes ($P = 0.003$). Furthermore, a significant interaction was observed between the CC genotype and the Mediterranean diet supplemented with virgin olive oil ($P = 0.028$).

In another study, weight loss was associated with a 16.9% increase in plasma IL6 concentrations in individuals who are CC homozygotes, whereas participants with other genotypes showed a 19.9% reduction in IL6 levels ($P = 0.001$).[62] Eklund and colleagues observed that C-reactive protein (CRP) levels are directly affected by a weight loss × *IL6*–174G/C interaction.[63] In this study, no differences in CRP levels were apparent by genotype before weight loss. However, G-allele carriers had significantly lower CRP levels compared to CC homozygotes after weight loss (1.01 vs. 1.93 mg/l, respectively, $P = 0.007$), suggesting that individuals with the CC genotype may derive less benefit *vis-à-vis* reductions in CRP (and possibly CVD risk) with weight loss compared to G-allele carriers.[63] A study of Crohn's disease showed that CC homozygotes have a 6-fold increased risk of developing Crohn's disease and that higher intake of saturated and monosaturated fats exacerbates this effect.[64]

Obstructive sleep apnea syndrome (OSAS) is recognized as a major public health problem and is prevalent in approximately 2–4% of middle-aged Caucasians. In addition to its marked effect on wakefulness and the quality of life, OSAS increases morbidity and mortality in patients with hypertension and CVD and increases the risk of stroke.[65] Obesity and inflammation are both major risk factors for OSAS, and circulating concentrations of IL6 are increased in these patients.[66]

Some recent evidence suggests that variants in *IL6* interact with measures of adiposity in the development of OSAS. Zhang and coworkers reported that variants in *IL6* modified the risk of OSAS in a BMI-dependent manner.[65] In this study, obese participants had a 9.09-fold increased risk of developing OSAS compared to normal-weight individuals,.[65] There was no evidence of association between SNPs in *IL6* and OSAS observed in the entire population. However, after stratifying by BMI, the frequency of the *IL6* − 572G allele was significantly lower in nonobese OSAS patients compared to nonobese controls, suggesting a protective effect of this allele in the development of OSAS in nonobese individuals ($P = 0.014$). Moreover, they observed that the severity of sleep-disordered breathing (as measured by apnea–hypopnea index) increased linearly in nonobese *IL6* − 572C allele carriers (14.3, 22.0, and 34.8 for GG, CG, and CC genotypes, respectively; $P = 0.012$), whereas no difference was shown in obese individuals. Although these data are cross-sectional, it is reasonable to hypothesize that genotyping of *IL6* polymorphisms may be a strategy for detecting the risk of OSAS in nonobese individuals and that obese *IL6* − 572G allele carriers may benefit more from weight reduction compared to CC homozygotes in terms of OSAS risk.

III. Peroxisome Proliferator-Activated Receptor Gamma

PPARG is a member of the nuclear hormone receptor superfamily that transcriptionally regulates multiple important biological functions important in energy, lipid, and glucose homeostasis. Gene activation is determined by numerous factors, including the type of ligand (e.g., thiazolidinediones vs. free fatty acids), the coactivators present, and the sequence of the *PPARG* response element (PPRE) in the promoter of the target gene. Extrinsic factors such as lack of physical exercise and high-calorie diets resulting in increased circulating concentrations of free fatty acids activate *PPARG*-mediated genetic programs that favor the development of obesity.[67] The actions of PPARG not only affect phenotypes important in the development of obesity but may also significantly impact the progression of obesity-linked disorders, including diabetes, metabolic syndrome, CVD, osteoporosis, and cancer.

Although *PPARG* is expressed widely in many cell types and organs, because of the role of PPARG in adipogenesis and lipid homeostasis, this transcriptional pathway has been studied intensively in adipocytes.[68] In addition to regulating differentiation, PPARG is also important in preserving adipocyte function and viability.[69] In skeletal muscle and liver, PPARG plays a central role in glucose and lipid homeostasis, respectively.[70,71] Studies in macrophages have shown that PPARG can affect cholesterol efflux, regulate inflammatory cytokines (e.g., TNF and IL6), and enhance foam cell formation in the presence of oxidized low-density lipoproteins (LDLs), implicating PPARG in the development of CVD.[72–74] This pathway may also play a role in high-fat-diet-induced hypertension, heart hypertrophy, and bone homeostasis.[75]

Given the pleiotropic effects of PPARG action, functional genetic variants may also have pleiotropic effects on obesity and metabolic phenotypes. Several polymorphisms in the human *PPARG* gene have been studied. Of these variants, the Pro12Ala SNP of *PPARG* isoform 2 (PPARG2) has been the most intensively investigated.[76] This missense mutation results in an approximately twofold lower affinity of PPARG2 binding to PPREs, ultimately leading to a decrease in transcriptional activity.[76] This reduction in activity has been observed both in the presence and absence of PPARG agonists.[77,78] For a more comprehensive review of PPARG genetics, readers are directed to reviews by Cecil, He, and higher.[75,79,80]

The Pro12Ala *PPARG2* polymorphism has been extensively evaluated in relation to insulin resistance and T2D. Several groups have shown that individuals with the Pro12 allele (the more common allele in the population) have decreased insulin sensitivity, that is, insulin resistance.[75] A meta-analysis of 57 studies showed that obese individuals homozygous for the Pro12 allele were more insulin resistant, had higher fasting insulin, and had higher fasting glucose compared to Ala12 homozygotes.[81] Consistent with an effect on insulin resistance, compelling replicated studies have indicated that the Pro12 allele is also a susceptibility variant for T2D (OR ~ 1.2)[75,77,81–84]

After initial identification of the *PPARG2* Pro12Ala polymorphism,[76] several studies reported a significant association of the Ala12 allele with lower BMI.[75,77] However, evidence from other studies has suggested that the Ala12 allele is significantly associated with increased BMI.[85,86] More recently, in a meta-analysis of over 32,000 nondiabetic individuals, no association with BMI was observed in the entire population.[81] However, when evaluating only the Caucasian subset (> 15,000 individuals), a significant association was observed between the Ala12 allele and higher BMI.

Significant associations between the Ala12 allele and BMI in opposite directions could potentially be explained by gene × environment interactions. Indeed, there is mounting evidence that extrinsic factors, for example, diet, physical activity, (and perhaps adiposity itself) are important modifiers of the effect of the Pro12Ala *PPARG2* variant on obesity-related metabolic traits and T2D. Since free fatty acids are the endogenous ligands of PPARG, the interaction between dietary fat intake and the *PPARG2* Pro12Ala variant has been heavily studied. For example, Luan and colleagues demonstrated that the effects of the *PPARG2* Pro12Ala variant on metabolic syndrome traits are influenced by the ratio of dietary polyunsaturated fat to saturated fat (*P*:*S* ratio).[87] As the *P*:*S* ratio increases, both BMI and fasting insulin levels decrease in Ala12 carriers but not Pro12 homozygotes. After adjustment for BMI, the interaction between *P*:*S* ratio and genotype on fasting insulin was nonsignificant, suggesting that obesity is driving the insulin effect. Frank and coworkers extended these findings and showed that the association of *P*:*S* ratio with fasting insulin was also modified by physical activity in a *PPARG2* Pro12Ala allele-specific manner.[88] More recently, Nelson and coworkers observed that the Pro12 allele was associated with T2D only among individuals with high polyunsaturated fat intake or who were physically inactive.[89] Significant *PPARG2* Pro12Ala genotype × nutrient interactions regarding T2D risk have also been observed with fish intake, $n-3$ fatty acids, oleic acid, and dietary fiber.[90–92] In a randomized clinical trial testing the effect-modifying role of *PPARG2* Pro12Ala after drug treatment (metformin or troglitazone) or lifestyle modification, a significant genotype × polyunsaturated fatty acid interaction on visceral fat reduction was observed, which was most evident in T2D patients treated with metformin.[93] Results from the Quebec Family Study not only suggest that Pro12 homozygotes have worse lipid profiles with increased fat and saturated fat intake, but also that an interaction between fat intake and genotype explains a significant proportion of the variance of BMI and waist circumference.[94] In fact, increasing total fat or saturated fat intake was associated with increased waist circumference, BMI, and visceral adipose tissue area in Pro12 homozygotes but not Ala12 carriers. Evidence also suggests a modifying effect of polyunsaturated fat intake on the influence of *PPARG2* Pro12Ala genotype on myocardial infarction.[95]

Intervention studies have also supported interactive effects between extrinsic factors and/or BMI and Pro12Ala *PPARG2* genotype on metabolic traits. For example, Nicklas and colleagues observed that after 6 months of a hypocaloric diet and weight loss, women with the Ala12 allele lost equivalent amounts of weight but had a greater increase in insulin sensitivity and carbohydrate oxidation compared to Pro12 homozygotes.[96] In addition, Ala12 carriers had a 19% reduction in fat oxidation, whereas no change was observed in Pro12 homozygotes. Importantly, weight regain after 12 months was greater in women who carried at least one copy of the Ala12 allele. Together, these data suggest that the Ala12 allele may increase the risk of obesity by increasing insulin sensitivity and decreasing fat oxidation, a metabolic milieu well known to promote fat accretion.

In addition to substantial evidence for interactions between adiposity and genetic variants in *PPARG* on the development of insulin resistance and T2D, there are also reports of interactions between *PPARG* variants and adiposity on coronary heart disease, combined hyperlipidemia, bone mass, cancer, and other metabolic disorders.[97–100] Additional investigation is needed to better understand the genetics of this pleiotropic transcription factor, which may lead to the development of new preventive and treatment strategies for several complex diseases.

IV. Other Candidate Genes

A few other genes that have been investigated less intensively for obesity-associated metabolic complications are worth mentioning. Junyent and coworkers recently observed that polymorphisms in the gene PDZ domain containing 1 (*PDZK1*) significantly contribute to measures of dyslipidemia in an obesity-dependent manner.[101] In this investigation, variants in *PDZK1* were evaluated in 1000 participants of the Genetics of Lipid Lowering Drugs and Diet Network Study. They observed that the *PDZK1* i33968C/T variant was associated with metabolic syndrome ($P = 0.034$), an effect that was mainly driven by the association with plasma triglycerides ($P = 0.004$) and very low-density lipoprotein (VLDL) ($P = 0.021$). Importantly, however, a significant gene × BMI × diet interaction was shown in which the association with metabolic syndrome was observed only in obese individuals with diets high in polyunsaturated fatty acids and carbohydrates. On the other hand, the same allele was associated with protection from metabolic syndrome in nonobese participants. Together, these results suggest that variants in *PDZK1* are important in the development of hypertriglyceridemia and that dietary and lifestyle changes may be beneficial for individuals with specific genetic variants.[101]

Other investigations have shown that polymorphisms in several genes can affect the onset of T2D in an obesity-dependent manner. Genetic variants in phosphatidylinositol-3,4,5-triphosphate-dependent Rac exchange factor 1 (*PREX1*) were significantly associated with T2D.[102] Formal mediation testing revealed that approximately 30–40% of the effects these SNPs had on T2D risk were significantly mediated by BMI, showing that variants in this gene may contribute to the disease through their effect on adiposity.[102]

In another study, Muthumala and colleagues observed that variation in the angiotensin I converting enzyme 1 (*ACE*) gene contributed to T2D only in obese individuals.[103] In fact, in lean individuals, no association was observed between an insertion/deletion variant in *ACE* and T2D (HR = 0.75, 95% CI: 0.46–1.22). In obese participants, however, individuals who had at least one copy of the deletion allele had a significantly greater risk of developing T2D (HR = 4.26, 95% CI: 1.30–13.93) with a genotype-obesity interaction of $P = 0.01$.[103] Although limited, these studies suggest great complexity in the genetic architecture of T2D susceptibility.

Plasma levels of apolipoprotein C-III are associated with triglyceride levels and the risk of atherosclerosis and myocardial infarction.[104] Candidate gene and recent GWAS provide compelling evidence that polymorphisms in the promoter of *APOC3* are associated with CVD. Furthermore, the effect of these SNPs on myocardial infarction risk may be altered by abdominal obesity. Haplotype analysis of *APOC3* indicates that the risk of myocardial infarction is strongly influenced by both fasting hyperglycemia and waist circumference. Surprisingly, however, the individuals in the lowest quintile of both of these phenotypes were the ones with the worst outcomes. The ORs for myocardial infarction for the *APOC3**222 haplotype (corresponding to the 3238C/G, −455T>C, and −482C>T variants) were 1.72 (95% CI: 1.16–2.54) and 1.84 (95% CI: 1.31–2.59) in participants in the lowest quintiles of waist circumference and fasting hyperglycemia, respectively. No difference in risk was observed in members in the highest quintiles. The authors suggest that variants in the *APOC3* promoter, in the presence of abdominal obesity and/or hyperglycemia, lead to increased gene expression in response to insulin and decreased cardioprotection.[104]

V. Summary and Conclusions

With GWAS and more comprehensive candidate gene studies, there have been tremendous advances in the identification of common gene variants that have main effects on BMI (and other adiposity measures), lipids, and blood pressure. However, there have been relatively few investigations that have evaluated the interactions between genetic variation and obesity on the development of metabolic syndrome traits. A common theme for common variants for complex diseases and traits, including metabolic syndrome traits, is that individual susceptibility loci have very small main effect sizes. However, it is still possible that some of these loci have larger effects on metabolic phenotypes in the presence of obesity or when provoked by a certain dietary or other lifestyle factor. Identification of these functional variants will provide additional insight into disease pathophysiology and may provide the scientific rationale for the design of more individualized prevention and treatment strategies for obesity and its metabolic complications. In addition to replicated cross-sectional studies, prospective intervention studies will be necessary to provide the evidence base to translate this new knowledge into useful clinical paradigms.

References

1. Bruce KD, Hanson MA. The developmental origins, mechanisms, and implications of metabolic syndrome. *J Nutr* 2010;**140**(3):648–52.
2. Mathieu P, Lemieux I, Despres JP. Obesity, inflammation, and cardiovascular risk. *Clin Pharmacol Ther* 2010;**87**(4):407–16.
3. O'Rourke RW. Molecular mechanisms of obesity and diabetes: at the intersection of weight regulation, inflammation, and glucose homeostasis. *World J Surg* 2009;**33**(10):2007–13.
4. Sethi JK, Vidal-Puig AJ. Thematic review series: adipocyte biology. Adipose tissue function and plasticity orchestrate nutritional adaptation. *J Lipid Res* 2007;**48**(6):1253–62.
5. Lettner A, Roden M. Ectopic fat and insulin resistance. *Curr Diab Rep* 2008;**8**(3):185–91.
6. Romeo S, Kozlitina J, Xing C, Pertsemlidis A, Cox D, Pennacchio LA, et al. Genetic variation in PNPLA3 confers susceptibility to nonalcoholic fatty liver disease. *Nat Genet* 2008;**40**(12):1461–5.
7. Vimaleswaran KS, Loos RJ. Progress in the genetics of common obesity and type 2 diabetes. *Expert Rev Mol Med* 2010;**12**:e7.
8. Farooqi IS. Monogenic human obesity syndromes. *Prog Brain Res* 2006;**153**:119–25.
9. Frayling TM, Timpson NJ, Weedon MN, Zeggini E, Freathy RM, Lindgren CM, et al. A common variant in the FTO gene is associated with body mass index and predisposes to childhood and adult obesity. *Science* 2007;**316**(5826):889–94.
10. Rampersaud E, Damcott CM, Fu M, Shen H, McArdle P, Shi X, et al. Identification of novel candidate genes for type 2 diabetes from a genome-wide association scan in the Old Order Amish: evidence for replication from diabetes-related quantitative traits and from independent populations. *Diabetes* 2007;**56**(12):3053–62.
11. Kathiresan S, Manning AK, Demissie S, D'Agostino RB, Surti A, Guiducci C, et al. A genome-wide association study for blood lipid phenotypes in the framingham heart study. *BMC Med Genet* 2007;**8**(Suppl 1):S17.
12. Org E, Eyheramendy S, Juhanson P, Gieger C, Lichtner P, Klopp N, et al. Genome-wide scan identifies CDH13 as a novel susceptibility locus contributing to blood pressure determination in two european populations. *Hum Mol Genet* 2009;**18**(12):2288–96.
13. Chalasani N, Guo X, Loomba R, Goodarzi MO, Haritunians T, Kwon S, et al. Genome-wide association study identifies variants associated with histologic features of nonalcoholic fatty liver disease. *Gastroenterology* 2010 [Epub ahead of print].

14. Maury E, Brichard SM. Adipokine dysregulation, adipose tissue inflammation and metabolic syndrome. *Mol Cell Endocrinol* 2010;**314**(1):1–16.
15. Lago F, Gomez R, Gomez-Reino JJ, Dieguez C, Gualillo O. Adipokines as novel modulators of lipid metabolism. *Trends Biochem Sci* 2009;**34**(10):500–10.
16. Castro-Giner F, Kogevinas M, Imboden M, de Cid R, Jarvis D, Machler M, et al. Joint effect of obesity and TNFA variability on asthma: two international cohort studies. *Eur Respir J* 2009;**33**(5):1003–9.
17. Tsochatzis EA, Papatheodoridis GV, Archimandritis AJ. Adipokines in nonalcoholic steatohepatitis: from pathogenesis to implications in diagnosis and therapy. *Mediators Inflamm* 2009;**2009**:831670.
18. Lehrke M, Broedl UC, Biller-Friedmann IM, Vogeser M, Henschel V, Nassau K, et al. Serum concentrations of cortisol, interleukin 6, leptin and adiponectin predict stress induced insulin resistance in acute inflammatory reactions. *Crit Care* 2008;**12**(6):R157.
19. Han SH, Sakuma I, Shin EK, Koh KK. Sep-Oct) Antiatherosclerotic and anti-insulin resistance effects of adiponectin: basic and clinical studies. *Prog Cardiovasc Dis* 2009;**52**(2):126–40.
20. Walley AJ, Blakemore AI, Froguel P. Genetics of obesity and the prediction of risk for health. *Hum Mol Genet* 2006;**15**(Spec No. 2):R124–30.
21. Vionnet N, Hani EH, Dupont S, Gallina S, Francke S, Dotte S, et al. Genomewide search for type 2 diabetes-susceptibility genes in french whites: evidence for a novel susceptibility locus for early-onset diabetes on chromosome 3q27-qter and independent replication of a type 2-diabetes locus on chromosome 1q21-q24. *Am J Hum Genet* 2000;**67**(6):1470–80.
22. Kissebah AH, Sonnenberg GE, Myklebust J, Goldstein M, Broman K, James RG, et al. Quantitative trait loci on chromosomes 3 and 17 influence phenotypes of the metabolic syndrome. *Proc Natl Acad Sci USA* 2000;**97**(26):14478–83.
23. Mori Y, Otabe S, Dina C, Yasuda K, Populaire C, Lecoeur C, et al. Genome-wide search for type 2 diabetes in Japanese affected sib-pairs confirms susceptibility genes on 3q, 15q, and 20q and identifies two new candidate loci on 7p and 11p. *Diabetes* 2002;**51**(4):1247–55.
24. Pollin TI, Tanner K, O'connell JR, Ott SH, Damcott CM, Shuldiner AR, et al. Linkage of plasma adiponectin levels to 3q27 explained by association with variation in the APM1 gene. *Diabetes* 2005;**54**(1):268–74.
25. Vasseur F, Helbecque N, Lobbens S, Vasseur-Delannoy V, Dina C, Clement K, et al. Hypoadiponectinaemia and high risk of type 2 diabetes are associated with adiponectin-encoding (ACDC) gene promoter variants in morbid obesity: evidence for a role of ACDC in diabesity. *Diabetologia* 2005;**48**(5):892–9.
26. Buzzetti R, Petrone A, Zavarella S, Zampetti S, Spoletini M, Potenziani S, et al. The glucose clamp reveals an association between adiponectin gene polymorphisms and insulin sensitivity in obese subjects. *Int J Obes (Lond)* 2007;**31**(3):424–8.
27. Menzaghi C, Trischitta V, Doria A. Genetic influences of adiponectin on insulin resistance, type 2 diabetes, and cardiovascular disease. *Diabetes* 2007;**56**(5):1198–209.
28. Li S, Shin HJ, Ding EL, van Dam RM. Adiponectin levels and risk of type 2 diabetes: a systematic review and meta-analysis. *JAMA* 2009;**302**(2):179–88.
29. Li S, Li L, Li K, Qi X, Hoekema D, Liu H, et al. Association of adipose most abundant transcript 1 gene (apM1) with type 2 diabetes mellitus in a chinese population: a meta-analysis of case–control studies. *Clin Endocrinol (Oxf)* 2008;**68**(6):885–9.
30. Daniele A, Cammarata R, Pasanisi F, Finelli C, Salvatori G, Calcagno G, et al. Molecular analysis of the adiponectin gene in severely obese patients from southern Italy. *Ann Nutr Metab* 2008;**53**(3–4):155–61.

31. Mohammadzadeh G, Zarghami N. Associations between single-nucleotide polymorphisms of the adiponectin gene, serum adiponectin levels and increased risk of type 2 diabetes mellitus in Iranian obese individuals. *Scand J Clin Lab Invest* 2009;**69**(7):764–71.
32. Wang X, Zhang S, Chen Y, Liu H, Lan C, Chen X, et al. APM1 gene variants -11377C/G and 4545G/C are associated respectively with obesity and with non-obesity in chinese type 2 diabetes. *Diabetes Res Clin Pract* 2009;**84**(3):205–10.
33. He M, Cornelis MC, Franks PW, Zhang C, Hu FB, Qi L. Obesity genotype score and cardiovascular risk in women with type 2 diabetes mellitus. *Arterioscler Thromb Vasc Biol* 2010;**30**(2):327–32.
34. Chung HK, Chae JS, Hyun YJ, Paik JK, Kim JY, Jang Y, et al. Influence of adiponectin gene polymorphisms on adiponectin level and insulin resistance index in response to dietary intervention in overweight–obese patients with impaired fasting glucose or newly diagnosed type 2 diabetes. *Diab Care* 2009;**32**(4):552–8.
35. Tsuzaki K, Kotani K, Nagai N, Saiga K, Sano Y, Hamada T, et al. Adiponectin gene single-nucleotide polymorphisms and treatment response to obesity. *J Endocrinol Invest* 2009;**32**(5):395–400.
36. Goyenechea E, Collins LJ, Parra D, Abete I, Crujeiras AB, O'Dell SD, et al. The − 11391 G/A polymorphism of the adiponectin gene promoter is associated with metabolic syndrome traits and the outcome of an energy-restricted diet in obese subjects. *Horm Metab Res* 2009;**41**(1):55–61.
37. Ntalla I, Dedoussis G, Yannakoulia M, Smart MC, Louizou E, Sakka SD, et al. ADIPOQ gene polymorphism rs1501299 interacts with fibre intake to affect adiponectin concentration in children: the GENe-diet attica investigation on childhood obesity. *Eur J Nutr* 2009;**48**(8):493–7.
38. Verduci E, Scaglioni S, Agostoni C, Radaelli G, Biondi M, Manso AS, et al. The relationship of insulin resistance with SNP 276G>T at adiponectin gene and plasma long-chain polyunsaturated fatty acids in obese children. *Pediatr Res* 2009;**66**(3):346–9.
39. Cawthorn WP, Sethi JK. TNF-alpha and adipocyte biology. *FEBS Lett* 2008;**582**(1):117–31.
40. Wilson AG, Symons JA, McDowell TL, McDevitt HO, Duff GW. Effects of a polymorphism in the human tumor necrosis factor alpha promoter on transcriptional activation. *Proc Natl Acad Sci USA* 1997;**94**(7):3195–9.
41. Elahi MM, Asotra K, Matata BM, Mastana SS. Tumor necrosis factor alpha − 308 gene locus promoter polymorphism: an analysis of association with health and disease. *Biochim Biophys Acta* 2009;**1792**(3):163–72.
42. Heijmans BT, Westendorp RG, Droog S, Kluft C, Knook DL, Slagboom PE. Association of the tumour necrosis factor alpha − 308G/A polymorphism with the risk of diabetes in an elderly population-based cohort. *Genes Immun* 2002;**3**(4):225–8.
43. Lee BC, Ahn SY, Doo HK, Yim SV, Lee HJ, Jin SY, et al. Susceptibility for ischemic stroke in Korean population is associated with polymorphisms of the interleukin-1 receptor antagonist and tumor necrosis factor-alpha genes, but not the interleukin-1beta gene. *Neurosci Lett* 2004;**357**(1):33–6.
44. Ko GT, Lee SC, Pu YB, Ng MC, So WY, Thomas N, et al. Tumour necrosis factor-alpha promoter gene polymorphism at − 308 (genotype AA) in Chinese subjects with type 2 diabetes. *Diabet Med* 2003;**20**(2):167–8.
45. Pereira TV, Rudnicki M, Franco RF, Pereira AC, Krieger JE. Effect of the G-308A polymorphism of the tumor necrosis factor alpha gene on the risk of ischemic heart disease and ischemic stroke: a meta-analysis. *Am Heart J* 2007;**153**(5):821–30.
46. Um JY, Park JH, Kim HM. Gene polymorphisms in tumor necrosis factor locus and waist–hip ratio in obese Koreans. *Clin Chim Acta* 2003;**338**(1–2):117–22.

47. Pausova Z, Deslauriers B, Gaudet D, Tremblay J, Kotchen TA, Larochelle P, et al. Role of tumor necrosis factor-alpha gene locus in obesity and obesity-associated hypertension in French Canadians. *Hypertension* 2000;**36**(1):14–9.
48. Pihlajamaki J, Ylinen M, Karhapaa P, Vauhkonen I, Laakso M. The effect of the – 308A allele of the TNF-alpha gene on insulin action is dependent on obesity. *Obes Res* 2003;**11**(7):912–7.
49. Fontaine-Bisson B, Wolever TM, Chiasson JL, Rabasa-Lhoret R, Maheux P, Josse RG, et al. Tumor necrosis factor alpha – 238G>A genotype alters postprandial plasma levels of free fatty acids in obese individuals with type 2 diabetes mellitus. *Metabolism* 2007;**56**(5):649–55.
50. Grimble RF, Howell WM, O'Reilly G, Turner SJ, Markovic O, Hirrell S, et al. The ability of fish oil to suppress tumor necrosis factor alpha production by peripheral blood mononuclear cells in healthy men is associated with polymorphisms in genes that influence tumor necrosis factor alpha production. *Am J Clin Nutr* 2002;**76**(2):454–9.
51. Kubaszek A, Pihlajamaki J, Komarovski V, Lindi V, Lindstrom J, Eriksson J, et al. Promoter polymorphisms of the TNF-alpha (G-308A) and IL-6 (C-174G) genes predict the conversion from impaired glucose tolerance to type 2 diabetes: the Finnish diabetes prevention study. *Diabetes* 2003;**52**(7):1872–6.
52. Kamizono S, Yamada K, Seki N, Higuchi T, Kimura A, Nonaka K, et al. Susceptible locus for obese type 2 diabetes mellitus in the 5′-flanking region of the tumor necrosis factor-alpha gene. *Tissue Antigens* 2000;**55**(5):449–52.
53. Higuchi T, Seki N, Kamizono S, Yamada A, Kimura A, Kato H, et al. Polymorphism of the 5′-flanking region of the human tumor necrosis factor (TNF)-alpha gene in Japanese. *Tissue Antigens* 1998;**51**(6):605–12.
54. Padovani JC, Pazin-Filho A, Simoes MV, Marin-Neto JA, Zago MA, Franco RF. Gene polymorphisms in the TNF locus and the risk of myocardial infarction. *Thromb Res* 2000;**100**(4):263–9.
55. Bennet AM, van Maarle MC, Hallqvist J, Morgenstern R, Frostegard J, Wiman B, et al. Association of TNF-alpha serum levels and TNFA promoter polymorphisms with risk of myocardial infarction. *Atherosclerosis* 2006;**187**(2):408–14.
56. Herbert A, Liu C, Karamohamed S, Liu J, Manning A, Fox CS, et al. BMI modifies associations of IL-6 genotypes with insulin resistance: the Framingham study. *Obesity (Silver Spring)* 2006;**14**(8):1454–61.
57. Berthier MT, Paradis AM, Tchernof A, Bergeron J, Prud'homme D, Despres JP, et al. The interleukin 6-174G/C polymorphism is associated with indices of obesity in men. *J Hum Genet* 2003;**48**(1):14–9.
58. Goyenechea E, Parra D, Martinez JA. Impact of interleukin 6-174G>C polymorphism on obesity-related metabolic disorders in people with excess in body weight. *Metabolism* 2007;**56**(12):1643–8.
59. Mohlig M, Boeing H, Spranger J, Osterhoff M, Kroke A, Fisher E, et al. Body mass index and C-174G interleukin-6 promoter polymorphism interact in predicting type 2 diabetes. *J Clin Endocrinol Metab* 2004;**89**(4):1885–90.
60. Stephens JW, Hurel SJ, Lowe GD, Rumley A, Humphries SE. Association between plasma IL-6, the IL6-174G>C gene variant and the metabolic syndrome in type 2 diabetes mellitus. *Mol Genet Metab* 2007;**90**(4):422–8.
61. Razquin C, Martinez JA, Martinez-Gonzalez MA, Fernandez-Crehuet J, Santos JM, Marti A. A Mediterranean diet rich in virgin olive oil may reverse the effects of the association – 174G/C IL6 gene variant on 3-year body weight change. *Mol Nutr Food Res* 2010;**54**(Suppl 1):S75–82.
62. Poitou C, Lacorte JM, Coupaye M, Bertrais S, Bedel JF, Lafon N, et al. Relationship between single nucleotide polymorphisms in leptin, IL6 and adiponectin genes and their circulating product in morbidly obese subjects before and after gastric banding surgery. *Obes Surg* 2005;**15**(1):11–23.

63. Eklund C, Nenonen A, Kukkonen-Harjula K, Borg P, Fogelholm M, Laine S, et al. Association of the IL6-174(G/C) polymorphism with C-reactive protein concentration after weight loss in obese men. *Eur Cytokine Netw* 2006;**17**(2):131–5.
64. Guerreiro CS, Ferreira P, Tavares L, Santos PM, Neves M, Brito M, et al. Fatty acids, IL6, and TNFalpha polymorphisms: an example of nutrigenetics in Crohn's disease. *Am J Gastroenterol* 2009;**104**(9):2241–9.
65. Zhang X, Liu RY, Lei Z, Zhu Y, Huang JA, Jiang X, et al. Genetic variants in interleukin-6 modified risk of obstructive sleep apnea syndrome. *Int J Mol Med* 2009;**23**(4):485–93.
66. Imagawa S, Yamaguchi Y, Ogawa K, Obara N, Suzuki N, Yamamoto M, et al. Interleukin-6 and tumor necrosis factor-alpha in patients with obstructive sleep apnea–hypopnea syndrome. *Respiration* 2004;**71**(1):24–9.
67. Knouff C, Auwerx J. Peroxisome proliferator-activated receptor-gamma calls for activation in moderation: lessons from genetics and pharmacology. *Endocr Rev* 2004;**25**(6):899–918.
68. Rosen ED, Sarraf P, Troy AE, Bradwin G, Moore K, Milstone DS, et al. PPAR gamma is required for the differentiation of adipose tissue *in vivo* and *in vitro*. *Mol Cell* 1999;**4**(4):611–7.
69. Imai T, Takakuwa R, Marchand S, Dentz E, Bornert JM, Messaddeq N, et al. Peroxisome proliferator-activated receptor gamma is required in mature white and brown adipocytes for their survival in the mouse. *Proc Natl Acad Sci USA* 2004;**101**(13):4543–7.
70. Norris AW, Chen L, Fisher SJ, Szanto I, Ristow M, Jozsi AC, et al. Muscle-specific PPAR-gamma-deficient mice develop increased adiposity and insulin resistance but respond to thiazolidinediones. *J Clin Invest* 2003;**112**(4):608–18.
71. Matsusue K, Haluzik M, Lambert G, Yim SH, Gavrilova O, Ward JM, et al. Liver-specific disruption of PPARgamma in leptin-deficient mice improves fatty liver but aggravates diabetic phenotypes. *J Clin Invest* 2003;**111**(5):737–47.
72. Moore KJ, Rosen ED, Fitzgerald ML, Randow F, Andersson LP, Altshuler D, et al. The role of PPAR-gamma in macrophage differentiation and cholesterol uptake. *Nat Med* 2001;**7**(1):41–7.
73. Vats D, Mukundan L, Odegaard JI, Zhang L, Smith KL, Morel CR, et al. Oxidative metabolism and PGC-1beta attenuate macrophage-mediated inflammation. *Cell Metab* 2006;**4**(1):13–24.
74. Tontonoz P, Nagy L, Alvarez JG, Thomazy VA, Evans RM. PPARgamma promotes monocyte/macrophage differentiation and uptake of oxidized LDL. *Cell* 1998;**93**(2):241–52.
75. He W. PPARgamma2 polymorphism and human health. *PPAR Res* 2009;**2009**:849538.
76. Yen CJ, Beamer BA, Negri C, Silver K, Brown KA, Yarnall DP, et al. Molecular scanning of the human peroxisome proliferator activated receptor gamma (hPPAR gamma) gene in diabetic Caucasians: identification of a Pro12Ala PPAR gamma 2 missense mutation. *Biochem Biophys Res Commun* 1997;**241**(2):270–4.
77. Deeb SS, Fajas L, Nemoto M, Pihlajamaki J, Mykkanen L, Kuusisto J, et al. A Pro12Ala substitution in PPARgamma2 associated with decreased receptor activity, lower body mass index and improved insulin sensitivity. *Nat Genet* 1998;**20**(3):284–7.
78. Masugi J, Tamori Y, Mori H, Koike T, Kasuga M. Inhibitory effect of a proline-to-alanine substitution at codon 12 of peroxisome proliferator-activated receptor-gamma 2 on thiazolidinedione-induced adipogenesis. *Biochem Biophys Res Commun* 2000;**268**(1):178–82.
79. Cecil JE, Watt P, Palmer CN, Hetherington M. Energy balance and food intake: the role of PPARgamma gene polymorphisms. *Physiol Behav* 2006;**88**(3):227–33.
80. Celi FS, Shuldiner AR. The role of peroxisome proliferator-activated receptor gamma in diabetes and obesity. *Curr Diab Rep* 2002;**2**(2):179–85.
81. Tonjes A, Scholz M, Loeffler M, Stumvoll M. Association of Pro12Ala polymorphism in peroxisome proliferator-activated receptor gamma with pre-diabetic phenotypes: meta-analysis of 57 studies on nondiabetic individuals. *Diab Care* 2006;**29**(11):2489–97.

82. Altshuler D, Hirschhorn JN, Klannemark M, Lindgren CM, Vohl MC, Nemesh J, et al. The common PPARgamma Pro12Ala polymorphism is associated with decreased risk of type 2 diabetes. *Nat Genet* 2000;**26**(1):76–80.
83. Gouda HN, Sagoo GS, Harding AH, Yates J, Sandhu MS, Higgins JP. The association between the peroxisome proliferator-activated receptor-gamma2 (PPARG2) Pro12Ala gene variant and type 2 diabetes mellitus: a HuGE review and meta-analysis. *Am J Epidemiol* 2010;**171**(6):645–55.
84. Huguenin GV, Rosa G. The ala allele in the PPAR-gamma2 gene is associated with reduced risk of type 2 diabetes mellitus in Caucasians and improved insulin sensitivity in overweight subjects. *Br J Nutr* 2010;**104**(4):488–97.
85. Beamer BA, Negri C, Yen CJ, Gavrilova O, Rumberger JM, Durcan MJ, et al. Chromosomal localization and partial genomic structure of the human peroxisome proliferator activated receptor-gamma (hPPAR gamma) gene. *Biochem Biophys Res Commun* 1997;**233**(3):756–9.
86. Wei Q, Jacobs Jr. DR, Schreiner PJ, Siscovick DS, Steffes MW, Fornage M. Patterns of association between PPARgamma genetic variation and indices of adiposity and insulin action in African-Americans and whites: the CARDIA study. *J Mol Med* 2006;**84**(11):955–65.
87. Luan J, Browne PO, Harding AH, Halsall DJ, O'Rahilly S, Chatterjee VK, et al. Evidence for gene-nutrient interaction at the PPARgamma locus. *Diabetes* 2001;**50**(3):686–9.
88. Franks PW, Luan J, Browne PO, Harding AH, O'Rahilly S, Chatterjee VK, et al. Does peroxisome proliferator-activated receptor gamma genotype (Pro12ala) modify the association of physical activity and dietary fat with fasting insulin level? *Metabolism* 2004;**53**(1):11–6.
89. Nelson TL, Fingerlin TE, Moss LK, Barmada MM, Ferrell RE, Norris JM. Association of the peroxisome proliferator-activated receptor gamma gene with type 2 diabetes mellitus varies by physical activity among non-Hispanic whites from Colorado. *Metabolism* 2007;**56**(3):388–93.
90. Ylonen SK, Salminen I, Lyssenko V, Virtanen SM, Groop L, Aro A, et al. The Pro12Ala polymorphism of the PPAR-gamma2 gene affects associations of fish intake and marine $n-3$ fatty acids with glucose metabolism. *Eur J Clin Nutr* 2008;**62**(12):1432–9.
91. Soriguer F, Morcillo S, Cardona F, Rojo-Martinez G, de la Cruz Almaraz M, de Adana Rui, et al. Pro12Ala polymorphism of the PPARG2 gene is associated with type 2 diabetes mellitus and peripheral insulin sensitivity in a population with a high intake of oleic acid. *J Nutr* 2006;**136**(9):2325–30.
92. Lindi VI, Uusitupa MI, Lindstrom J, Louheranta A, Eriksson JG, Valle TT, et al. Association of the Pro12Ala polymorphism in the PPAR-gamma2 gene with 3-year incidence of type 2 diabetes and body weight change in the Finnish diabetes prevention study. *Diabetes* 2002;**51**(8):2581–6.
93. Franks PW, Jablonski KA, Delahanty L, Hanson RL, Kahn SE, Altshuler D, et al. The Pro12Ala variant at the peroxisome proliferator-activated receptor gamma gene and change in obesity-related traits in the diabetes prevention program. *Diabetologia* 2007;**50**(12):2451–60.
94. Robitaille J, Despres JP, Perusse L, Vohl MC. The PPAR-gamma P12A polymorphism modulates the relationship between dietary fat intake and components of the metabolic syndrome: results from the Quebec family study. *Clin Genet* 2003;**63**(2):109–16.
95. Ruiz-Narvaez EA, Kraft P, Campos H. Ala12 variant of the peroxisome proliferator-activated receptor-gamma gene (PPARG) is associated with higher polyunsaturated fat in adipose tissue and attenuates the protective effect of polyunsaturated fat intake on the risk of myocardial infarction. *Am J Clin Nutr* 2007;**86**(4):1238–42.
96. Nicklas BJ, van Rossum EF, Berman DM, Ryan AS, Dennis KE, Shuldiner AR. Genetic variation in the peroxisome proliferator-activated receptor-gamma2 gene (Pro12Ala) affects metabolic responses to weight loss and subsequent weight regain. *Diabetes* 2001;**50**(9):2172–6.

97. Sookoian S, Garcia SI, Porto PI, Dieuzeide G, Gonzalez CD, Pirola CJ. Peroxisome proliferator-activated receptor gamma and its coactivator-1 alpha may be associated with features of the metabolic syndrome in adolescents. *J Mol Endocrinol* 2005;**35**(2):373–80.
98. Pischon T, Pai JK, Manson JE, Hu FB, Rexrode KM, Hunter D, et al. Peroxisome proliferator-activated receptor-gamma2 P12A polymorphism and risk of coronary heart disease in US men and women. *Arterioscler Thromb Vasc Biol* 2005;**25**(8):1654–8.
99. Swarbrick MM, Chapman CM, McQuillan BM, Hung J, Thompson PL, Beilby JP. A Pro12Ala polymorphism in the human peroxisome proliferator-activated receptor-gamma 2 is associated with combined hyperlipidaemia in obesity. *Eur J Endocrinol* 2001;**144**(3):277–82.
100. Ackert-Bicknell CL, Demissie S, Marin de Evsikova C, Hsu YH, DeMambro VE, Karasik D, et al. PPARG by dietary fat interaction influences bone mass in mice and humans. *J Bone Miner Res* 2008;**23**(9):1398–408.
101. Junyent M, Arnett DK, Tsai MY, Kabagambe EK, Straka RJ, Province M, et al. Genetic variants at the PDZ-interacting domain of the scavenger receptor class B type I interact with diet to influence the risk of metabolic syndrome in obese men and women. *J Nutr* 2009;**139**(5):842–8.
102. Lewis JP, Palmer ND, Ellington JB, Divers J, Ng MC, Lu L, et al. Analysis of candidate genes on chromosome 20q12–13.1 reveals evidence for BMI mediated association of PREX1 with type 2 diabetes in European Americans. *Genomics* 2010;**96**(4):211–9.
103. Muthumala A, Gable DR, Palmen J, Cooper JA, Stephens JW, Miller GJ, et al. Is the influence of variation in the ACE gene on the prospective risk of type 2 diabetes in middle-aged men modified by obesity? *Clin Sci (Lond)* 2007;**113**(12):467–72.
104. Ruiz-Narvaez EA, Sacks FM, Campos H. Abdominal obesity and hyperglycemia mask the effect of a common APOC3 haplotype on the risk of myocardial infarction. *Am J Clin Nutr* 2008;**87**(6):1932–8.

Obesity: From Animal Models to Human Genetics to Practical Applications

Craig H. Warden[*,†,‡] and
Janis S. Fisler[§]

[*]Rowe Program in Genetics, University of California, Davis, California, USA
[†]Department of Neurobiology, Physiology and Behavior, University of California, Davis, California, USA
[‡]Department of Pediatrics, University of California, Davis, California, USA
[§]Department of Nutrition, University of California, Davis, California, USA

Although many animal models are used in genetic studies, the mouse is most common. Analysis of single-gene mutations, linkage analysis in crossbred strains, and gene targeting are the primary techniques used to associate obesity phenotypes with specific genes or alleles. The orthologous human gene can then be tested, either in linkage studies in families or in genome-wide association studies (GWAS), for effect on the phenotype. Frequent lack of concordance between mouse and human obesity genes may be due to the difference in phenotypes

Progress in Molecular Biology
and Translational Science, Vol. 94
DOI: 10.1016/S1877-1173(10)94013-1

measured in humans (body mass index) versus mouse (fat mass or % body fat), lack of intermediate phenotypes, and the fact that identified genes account for only a small percentage of the heritability of common obesity, suggesting that many genes remain unknown. New technology allows analysis of individual genomes at a reasonable cost, making large-scale obesity genome projects in humans feasible. Such projects could identify common allelic variants that contribute to obesity and to variable individual response to obesity therapy. Currently, family history may be more predictive than genetics for risk of obesity, but individual testing could ultimately guide therapy and, in the aggregate, guide public health policy. The primary limitation to development of genotype-based diets is that successful randomized diet trials of widely ranging macronutrient content, adequately powered for finding rare Mendelian mutations, have not been performed.

I. Introduction

All members of the animal kingdom can be used as obesity models, although different phyla may provide different types of information. Animal models have served two main purposes: discovery of novel mechanisms for obesity and discovery of specific genes that regulate obesity.

All organisms from yeast to humans can store triglycerides as lipid droplets.[1] Since triglyceride storage is evolutionarily ancient, it is likely that many mechanisms underlying obesity are conserved across species and that any eukaryote could be used as a model to study lipid storage. Since many of the genes for storage and composition remain unknown for mammals,[1] studies in models where genetics can be manipulated easily will likely lead to identification of the genes needed to regulate many aspects of fat synthesis, storage, use, and variations of composition.

Any multicellular animal can be used to study interactions between tissues that promote obesity. *Caenorhabditis elegans* store fat in gut granules. Neurons and defined adipose bodies are present in popular model organisms such as zebra fish and *Drosophila*. Defined fat pads appear to be present in all vertebrates, so any vertebrate can potentially be used to model the complex brain–body interactions that influence fat accumulation in humans.

Continued sequencing of genomes from an ever-increasing collection of species means that many animal models can be used for discovery of genes that cause obesity. This provides novel opportunities for gene discovery because many animals, such as hibernating mammals and migrating birds, have natural cycles of weight gain and loss that can be used to understand the process in that specific animal and perhaps generalized to other species.

II. Methods of Genetic Analysis in Animal Models

Because of more readily available genetic tools, most genetic studies in animals are done in rodents, generally mice, although agricultural animals have also been used. In animals, genetic effects are studied by a variety of methods, including examination of spontaneous and induced mutations, linkage analysis, and insertion or deletion of a candidate gene.

A. Single-Gene Mutations

Evidence of the biological basis of the regulation of body fat stores comes from the identification of more than 50 single-gene mutations and Mendelian syndromes that result in spontaneous massive obesity or in adipose tissue atrophy[2] (see also http://www.informatics.jax.org/). The first such gene $Lep^{ob/ob}$ identified in the ob/ob mouse,[3] led to the discovery of the hormone leptin and the dissection of a major pathway of appetite regulation in the brain.

Most human obesity is not due to mutations in single genes but is inherited as a complex, polygenic, and quantitative trait. Many of the genes causing single-gene obesity in humans, however, are implicated in polygenic obesity as well (e.g., *LEP, LEPR, POMC*, and *MC4R*). In common obesity, there are also likely to be complex interactions among genes and between genes and environmental factors.[4]

B. Linkage Analysis

An extensively used method for analysis of complex traits is quantitative trait locus (QTL) screening for natural allelic variation using crosses of strains varying in the trait of interest. QTL mapping is a hypothesis-free method for mapping Mendelian factors that underlie complex traits, in virtually any animal model, by using genetic linkage maps in crossbred animals.[5,6] QTL mapping also allows dissection of the genetic architecture of complex diseases, a difficult task in humans. QTL studies of obesity in mice have been published since the late 1990s, with a large number of QTLs linked variously to body weight, body fat, energy expenditure, food intake, leptin levels, or weight gain[2] (see also http://www.informatics.jax.org/).

Meta-analyses of existing data increase the effective sample size and provide more power to find loci. A meta-analysis of QTLs associated with body weight and adiposity from a total of 34 published crossbreeding experiments included 42 different parental stains and more than 14,500 mice and identified highly significant linkage of body weight and/or adiposity with segments of the mouse genome that include at least 34 candidate genes for body weight and/or adiposity.[7]

However, there has been little effort to dissect the complex architecture of obesity using QTL models. Bayesian analysis of the data from the backcross BSB model ((C57BL/6J F X *Mus spretus* M) F X C57BL/6J M) identified additional QTLs when allowing for epistatic interactions among the six chromosomal regions previously shown to be associated with obesity.[8] More recently, use of advanced intercross generations (i.e., crossing through F_{10}) in the LG × SM mouse model allowed mapping of obesity phenotypes with five times the resolution of the earlier F_2 generation-based studies and identified additional QTLs, inheritance patterns, and pleiotropic and epistatic QTLs in that model.[9] Much of the potential of linkage analysis in mice to unravel the complex genetic architecture of common obesity has so far not been tapped, suggesting that this is an area of future productive research.

C. Gene Targeting

QTLs are valuable for identifying candidate genes to be further evaluated by gene targeting experiments. Expression of genes can be altered by targeted insertion or deletion in the whole animal or by strategies that cause gene deficiencies in only those tissues of interest.[10,11]

A recent example of using gene targeting to confirm obesity genes identified by linkage analysis is from the work of Yang *et al.*[12] From QTLs found in linkage analysis, these authors had predicted a large number of genes causal for abdominal obesity from an F2 BXD intercross (C57BL/6J X DBA/2J). Knockout or transgenic models were then made of nine candidate genes from loci with the highest significance in the F2 cross and obesity phenotypes and gene expression in liver were measured. The nine models demonstrated shared metabolic pathways involved in the TCA cycle, fatty acid metabolism, and obesity. Development of obesity models with under- or overexpression of a gene not only confirmed that the gene contributes to obesity in the BXD cross but also provided models to investigate the resulting related phenotypes including energy balance, feeding and activity behaviors, and fat and carbohydrate metabolism.

Examination of a dataset of almost 2000 knockout mouse strains found that one-third of the viable knockout strains differed in body size or weight from controls with most weighing less and only 3% weighing more than control.[13] This study did not examine obesity or fat mass, but only body weight and size. Reed and colleagues concluded that if their survey was a representative set of knockout genes, then as many as 6000 genes could influence body size in the mouse.

There are several possible confounding factors involved in gene insertion/deletion studies such as the redundancy of gene function,[14] gene–gene interactions requiring mutations in two or even three independent genes to have the phenotype expressed,[4] and gene–environment interactions necessary for obesity to develop.[15,16] Also, the source of DNA surrounding the altered gene generally began as a 129-strain embryonic stem cell, and it is possible that these gene-targeted models also include 129-strain alleles of obesity genes within the retained donor region. Nevertheless, insertion/deletion models continue to be a valuable tool in the dissection of the genetic causes of obesity.

III. From Animal Models to Human Genetics

Because of evolutionary relationships between animal models and humans, many ancestral chromosomal segments have been retained, although some regions of homology are made more complex by chromosomal rearrangements within the region of homology. Thus, identification of a QTL or gene in another species gives the chromosomal location of the same gene in the human.

This evolutionary relationship among species was used to test whether an obesity QTL on mouse chromosome 2, identified in an intercross of NZB/BINJ and SM/J strains, contains a gene that contributes to human obesity. Linkage analysis between markers located within the homologous region on human chromosome 20 and measures of obesity was performed in more than 150 French Canadian families: a locus on 20q13 that contributes to body fat and fasting insulin was found.[17] This locus was later confirmed in a second population group.[18]

This same locus was identified by QTLs for natural variation in triglyceride storage in *Drosophila*,[19] then fine-mapped to identify candidate genes, which included an uncoupling protein (*Ucp4c*) and laminin A (*LanA*).[20] Female *Drosophila* flies heterozygous for mutant *LanA* had significantly lower triglyceride storage, body weight, and protein content than control flies. The orthologous human gene, *LAMA5*, maps the obesity linkage region on chromosome 20q13.2–q13.3[2] and significant association was found between SNPs in *LAMA5* and measures of body composition in women.[20]

A. Linkage Studies in Humans

Initial whole genome scans in humans were linkage studies in large extended families or nuclear families. The last Human Obesity Gene Map containing data up through October 2005[2] listed more than 300 QTLs for various measures of adiposity, respiratory quotient, metabolic rate, and plasma leptin levels in humans. However, a meta-analysis of 37 published genome-wide linkage studies was unable to find strong evidence for specific loci for body mass index (BMI) or obesity despite including data from more than 10,000 families and 31,000 individuals.[21] Saunders and colleagues speculate that locus heterogeneity, epistasis, and age-dependent effects may explain why no strong loci were found. Although linkage studies are an excellent tool for finding obesity genes anywhere in the genome, they are not well suited to discovering loci with small effects.[22,23] In contrast, association studies have proven powerful for detection of alleles with small effects.

Linkage studies, although they do not allow sufficient fine mapping to identify any specific gene, are useful in identifying candidate genes for further study. Family studies may also be useful in detecting rare alleles, parent-of-origin effects, and gene–gene interactions.[24]

B. Association Studies in Humans

Linkage and limited association studies in humans have been superseded by genome-wide association studies (GWAS) that more accurately identify the genomic locations of obesity genes. The GWAS approach has the same advantage as linkage studies in that no prior assumptions about the identity of the underlying genes need to be made. Also, the overall data can be evaluated to identify the most significant results, including gene–gene interactions. However, GWAS also have disadvantages in that they restrict the number of markers per gene, although improved technologies will overcome this. For recent reviews of GWAS in humans, see Refs. 25,26.

GWAS involving large cohorts of European ancestry confirmed the fat mass- and obesity-associated gene (*FTO*) and the melanocortin-4 receptor gene (*MC4R*) as genes associated with BMI[27–29] [reviewed in Ref. 30]. Of the 15 other significant genes/loci identified, most were nonoverlapping: 11 loci were significantly associated with BMI in only one of the three studies. The two larger studies, which were not designed to examine obesity, both found four other loci significantly associated with BMI (*NEGR1*, *TMEM18*, *SH2B1*, and *KCTD15*).[28,29] The third smaller study, where ascertainment was for extreme obesity, only overlapped with the studies of Willer and Thorleifsson in confirming *FTO* and *MC4R* as obesity genes.

IV. Concordance Between Animal Models and Humans

Animal models have clearly aided in the discovery of some of the genes associated with human obesity and identified underlying mechanisms. The primary example is the cloning of the *Lep* gene in ob/ob mice, then humans,[3] and the subsequent identification of the hormone leptin. Additional studies confirmed that severe leptin deficiency causes human obesity and led directly to treatment for the rare affected individuals. Whether the *LEP* gene contributes to common human obesity is as yet uncertain.

Another gene of considerable interest that is conserved across species and was recently implicated in obesity is the clock homolog gene (*CLOCK*). In mammals, the transcription factor CLOCK pairs with the protein BMAL1 to turn on genes controlling circadian rhythms.[31] Homozygous *Clock* mutant mice have altered diurnal feeding rhythm, are hyperphagic and obese, and develop characteristics of the metabolic syndrome.[32] Subsequent human studies showed that specific *CLOCK* polymorphisms are associated with obesity[33,34] and the metabolic syndrome.[35] Night eating in humans is associated with, and may contribute to, obesity.[36] This was confirmed in mice when combining a high-fat diet with feeding of nocturnal mice only during the light phase of the diurnal cycle contributed to weight gain.[37] Although *CLOCK* is not yet confirmed as an obesity gene in humans, variants in *CLOCK* appear to contribute to obesity risk.

Other examples of genes identified in another species and found to affect energy homeostasis in humans are *SIRT1* and the *UCPs*. Sirtuins including SIRT1, which act as cellular energy sensors and target UCP2 among other proteins,[38] were initially identified in yeast and then found to be conserved in *C. elegans*, *Drosophila*, rodents, and humans. The function of *UCP1* has been extensively studied since the 1960s in animal models of energy efficiency and thermogenesis. The *UCP* homologues, *UCP2* and *UCP3*, were identified in 1997 in mice based on their sequence similarity to *UCP1*.[39–41] Despite the extensive studies of these genes in mouse models, the *UCP* genes are not conclusively associated with obesity in mice but are commonly accepted as obesity genes in humans.

However, concordance between obesity genes identified in mice and humans is often not found. In fact, the Mouse Genome Informatics (MGI) database (http://www.informatics.jax.org/) lists just six genes where mouse and human orthologous genes are associated with obesity (*ADRB2*, *ADRB3*, *MC3R*, *MC4R*, *POMC*, and *SIM1*), although this certainly is an underestimate. The MGI database also lists 21 genes that are associated with obesity in mice but not in humans and 16 human specific obesity genes or loci that are not associated with obesity in mice. In the study by Yang *et al.* discussed above,[12] none of the nine abdominal obesity causal genes in the mouse insertion/deletion models were identified in GWAS in humans.

A major problem in using mouse genetics to direct the search for human obesity genes is a difference in phenotypes measured. In mice, obesity is usually measured as fat pad weights or as % of body fat from chemical analysis. Subjects in most human obesity studies to date were ascertained for some disease or phenotype other than obesity. Since the subjects had height and weight measured, BMI could be easily calculated and used as a surrogate for obesity. However, BMI reflects a mix of lean and adipose tissues, is not an appropriate measure of obesity in certain groups, and does not reflect body fat distribution. QTLs influencing BMI often do not overlap with QTLs influencing body fat, even within the same study, in either humans or mice.[2,42] Thus, some of the lack of concordance between mouse and human obesity genes could be due to differences in phenotypes measured.

Another factor is that BMI and % body fat are endpoint phenotypes and are influenced by many genes. Studies in both animals and humans that look at intermediate phenotypes, such as eating behavior[43] or fat distribution[44] might prove more successful. Investigators have started to integrate gene expression or proteomic or metabolomic phenotypes into QTL studies in mice,[45] but this is usually not available in large human studies.

A potentially useful approach for integrating human and animal model studies is that used by English and Butte.[46] These authors used publicly available datasets of different types (i.e., microarray, proteomics, genetics, and gene-targeted models from humans, rodents, and worm) to prioritize genes most likely to be associated with common obesity. Their model is based on counting the number of experiments in which a gene is implicated in the trait and they predict that more complex models[47,48] could be even stronger in identifying candidate genes for further validation.

V. The Problem of "Missing" Heritability of Obesity

Statistical genetic studies in humans suggest that as much as 70% of obesity may be genetic. However, all existing methods have identified genes that cause at most a few percent of all genetic variation that exists.[49] Even the gene most strongly associated with human obesity, *FTO*, has only about a 1% effect on variance in BMI.[50] Low heritability of all known genes and loci is true in both human and mouse studies of common, complex diseases, including obesity.

A possible explanation for the missing heritability is that existing studies have missed rare variants that have larger effects than common variants.[24,49] One such rare allele was associated with protection from obesity by examining severely obese adults and children.[51] Rare variants with larger effect are more likely to be found by studying large cohorts. However, rare variants are likely to be missed using the GWAS approach.

Some of the missing heritability may also be due to genes that are unknown because they influence obesity by mechanisms that have not been tested in humans, such as maternal genetic effects, or not often been accounted for in studies such as imprinting or other unusual patterns of inheritance.[52–54] Common copy number polymorphisms should be detected by the GWAS approach, but rare ones may not be.[24,55]

Imprinting occurs when there is differential expression of a gene depending on whether it was inherited from the mother or the father. Maternal genetic effects occur when maternal genotype influences phenotypes of progeny independent of progeny genotype. Studies in animal models have shown that both imprinted loci[56] and genes[57,58] and maternal effects genes[59–61] influence obesity. In one study of a cross of LG and SM mice, maternal genotype may have a threefold greater impact on obesity of adult progeny than direct genetic effects.[62] Studies of imprinting revealed that there are many different "flavors" of imprinting in addition to the classic differential expression of the gene dependent on the parent-of-origin. Maternal genetic effects may produce the same patterns of phenotype variation as seen in imprinting and, thus, be confounded with imprinting.[60]

There is also the possibility, as recently proposed, that the high heritability estimates from earlier studies are wrong.[63] Although observations of nuclear families and twin pairs suggest heritability levels of 50–90% for obesity, other studies including adoptions indicate considerably lower heritability. Whether the heritability estimates for obesity and other common diseases are inflated may become clear as additional large genome studies are done.

VI. Practical Applications of Genetics to the Diagnosis and Treatment of Obesity

A. Individual Variation in Response to Treatment

Interindividual variation in response to change in energy balance was demonstrated in males 50 years ago by Sims and colleagues.[64] Twenty years ago, Bouchard and colleagues showed variable, individual changes of fat mass and distribution by submitting male identical twin pairs to overfeeding or to exercise training.[65,66] Individual factors, genetic and behavioral, that lead to variable weight loss response to exercise may be gender dependent.[67]

B. Genetic Influence on Response to Treatment

Attention has now turned to identifying polymorphisms that predict response to lifestyle interventions. As reviewed by Deram and Villares, many candidate genes have been examined with different weight loss strategies with both positive and negative results.[68] Of interest, the two genes consistently associated with human obesity, *MC4R* and *FTO*, do not appear to be associated with response to lifestyle intervention,[69–72] although an allele of *FTO* may interact with insulin-induced gene 2 (*INSIG2*) to inhibit weight loss.[73] Among several genes shown to affect response to the weight loss drug sibutramine, homozygous C allele carriers of guanine nucleotide binding protein (G-protein), beta polypeptide 3 (*GNB3*) showed reduced weight loss in two studies.[74,75] And both individual genes (*IL6* and *UCP2*) and gene networks are correlated with weight loss following bariatric surgery.[76–78]

Although numerous intervention studies have examined the association between risk alleles for various obesity genes and response to lifestyle intervention, unfortunately, the studies are too dissimilar in populations studied, therapy, length of intervention, etc., to make comparisons across studies feasible.[68] In addition, because of small sample sizes, statistical power of these studies is weak. Future studies will need to be large and will need to consider both diet and exercise and measure multiple risk alleles.

C. Using Genotype to Guide Therapy

If human obesity, like other common diseases, is due to multiple common variants of small effect, will genome analysis ever be a practical guide to diagnosis or therapy? In this event, each person is a unique blend of lifestyle-responsive genes causing obesity, genes promoting thinness, and genes promoting obesity that are not lifestyle sensitive. Effects of all will need to be included in an individualized model that then recommends a range of healthy choices for that person. Individualized treatment is likely to be possible for some people and not for others: some will have weight determined primarily by diet-responsive genes, whereas others will have weight determined by diet-independent obesity genes.

The core assumption in gene × treatment approaches is that direct effects are the predominant cause of obesity. However, treatment based on individual genotype is much more complex if indirect effects such as imprinting or maternal genetic effects have significant impact on obesity, in which case obesity treatment based on genotype is a family affair best conducted by genotyping mothers and children.

As more variants are identified and prediction of genetic risk becomes more practical, large-scale, randomized trials will be needed to link specific variants to the outcome of drug and/or lifestyle interventions in the treatment of obesity.[79,80] At present, the Women's Health Initiative (WHI) is the only randomized diet trial with a large enough sample size to detect genetic effects on treatment response. However, all the WHI diet trials failed to find effects, perhaps because the diets actually consumed were not sufficiently different. New diet trials need large enough sample sizes to find rare alleles, should they exist. They also need to test a wider range of diets than previously tested, since smaller scale, randomized trials indicate individual differences in response to low-carbohydrate or low-fat diets.

For common diseases, even in combination, these small relative risk alleles have low ability to discriminate among those who will or will not get the disease.[81] Lango and associates[82] found that a panel of 18 type 2 diabetes genes was no better at predicting risk of diabetes than an individual's age, BMI, or sex. It appears that a personalized approach to diagnosis and therapy based on the individual's genetic makeup is still in the future.

VII. Using Genotype to Guide Drug Development

Novel genes and hormones that regulate obesity seem to be discovered every month. Indeed, every animal model paper claims practical applications. Discovery of additional causes always opens up the possibility of finding a novel drug target, yet, to date, no genes found in animal models have led to practical therapies for common human obesity.

The rare individuals with monogenic disorders of leptin deficiency are effectively treated with recombinant leptin.[83–85] Although there was much initial hope that leptin therapy would be efficacious in treating common obesity, such has not been the case. Since leptin is produced in adipose tissue, most obese individuals have high levels of leptin and appear to be resistant to its effects.[86] Recombinant leptin has been effective in restoring reproductive function in women with relative leptin deficiency due to undernutrition or strenuous exercise.[87] Combination treatment with recombinant leptin and the amylin analog pramlintide produces greater weight loss than either drug alone in both rats and humans,[88] suggesting that combination therapies including leptin may soon be developed for common obesity.

Sirt1 overexpression reduces body weight and fat mass and improves metabolic phenotypes in mice.[89,90] Resveratrol, a phytoalexin, is a *Sirt1* activator that protects against insulin resistance and obesity in mice on a high-calorie diet.[91] Since resveratrol is a component of red wine, it has gotten much popular attention and is readily available as a dietary supplement, although dosage requirements in humans make such supplements impractical. Other small-molecule *Sirt1* activators with greater potency than resveratrol have been described and may prove therapeutically useful.[92]

VIII. Can Genome Analysis of Obesity Guide Public Policy?

Under what conditions would information on genotype be useful for preventing obesity in a public health context? Would genetic information change school lunches, the USDA pyramid, and other national-level recommendations? Clearly, this would not be targeted at people with specific alleles, but rather to increase the flexibility of the current national recommendations in order to accommodate larger ranges of potentially healthy diets and to include more people with different gene–environment responses. Although all recommendations include some flexibility, current ranges of recommended diets remain relatively limited in regard to macronutrients like fat and carbohydrate, as well as composition issues such as simple sugars. The core problem with changing public health recommendations is that the present science base does not exist to determine the range of obesity-prevention diets. Public health recommendations will be hard to change without large-scale, randomized trials testing widely variable diet composition effects. A model similar to the International Cancer Genome Consortium[93] may be needed to coordinate large-scale obesity genome studies in order to allow classification of obesity for most efficacious management.

IX. Cost and Ethical Concerns of Individual Genome Analysis

Since the common variant alleles (genes) found by GWAS explain only a small percentage of variance in obesity, it is possible that rare variants in many genes contribute to obesity. Sequencing of only the protein coding exons (exomic sequencing) covers only about 1% of the genome yet would catch approximately 90% of large effect variants.[94,95] Whole-exome sequencing costs are currently estimated at $4000 per subject[95] and are expected to drop as new technologies are developed, thus making genome sequencing practical for all patients.

This will raise the question of who should have genome testing. Since obesity itself is not a hidden disease, at least one ethical problem is addressed—genetic tests for obesity are less likely than other tests to reveal the existence of a hidden future problem, although they might reveal a potential for future adult weight gain or post-pregnancy weight gain. If gene panels can be developed that predict response to therapy, then genome analysis could benefit the individual obese patient.

A potentially more serious ethical problem relates to prenatal testing. If obesity gene tests are available, then genetic testing for obesity could be conducted as part of prenatal screens, with the consequent potential that couples might selectively abort children that are predisposed to obesity.

Ethical issues related to the large-scale sequencing projects, including informed consent, patient care and confidentiality, access to samples and data, and intellectual property rights, were discussed by the International Cancer Genome Consortium.[93]

References

1. Kadereit B, Kumar P, Wang WJ, Miranda D, Snapp EL, Severina N, et al. Evolutionarily conserved gene family important for fat storage. *Proc Natl Acad Sci USA* 2008;**105**(1):94–9.
2. Rankinen T, Zuberi A, Chagnon YC, Weisnagel SJ, Argyropoulos G, Walts B, et al. The human obesity gene map: the 2005 update. *Obesity (Silver Spring)* 2006;**14**(4):529–644.
3. Zhang Y, Proenca R, Maffei M, Barone M, Leopold L, Friedman JM. Positional cloning of the mouse obese gene and its human homologue. *Nature* 1994;**372**(6505):425–32.

4. Warden CH, Yi N, Fisler J. Epistasis among genes is a universal phenomenon in obesity: evidence from rodent models. *Nutrition* 2004;**20**(1):74–7.
5. Silver LM. *Mouse genetics.* New York: Oxford University Press; 1995 p. 362.
6. Warden CH, Fisler JS. Molecular genetics of obesity. In: Bray GA, Bouchard C, James WPT, editors. *Handbook of obesity.* New York: Marcel Dekker, Inc.; 1998. p. 223–42.
7. Wuschke S, Dahm S, Schmidt C, Joost HG, Al-Hasani H. A meta-analysis of quantitative trait loci associated with body weight and adiposity in mice. *Int J Obes (Lond)* 2007;**31**(5):829–41.
8. Yi N, Diament A, Chiu S, Kim K, Allison DB, Fisler JS, et al. Characterization of epistasis influencing complex spontaneous obesity in the BSB model. *Genetics* 2004;**167**(1):399–409.
9. Fawcett GL, Jarvis JP, Roseman CC, Wang B, Wolf JB, Cheverud JM. Fine-mapping of obesity-related quantitative trait loci in an f(9/10) advanced intercross line. *Obesity (Silver Spring)* 2009;**18**(7):1383–92.
10. Glaser S, Anastassiadis K, Stewart AF. Current issues in mouse genome engineering. *Nat Genet* 2005;**37**(11):1187–93.
11. McMinn JE, Liu SM, Liu H, Dragatsis I, Dietrich P, Ludwig T, et al. Neuronal deletion of Lepr elicits diabesity in mice without affecting cold tolerance or fertility. *Am J Physiol Endocrinol Metab* 2005;**289**(3):E403–11.
12. Yang X, Deignan JL, Qi H, Zhu J, Qian S, Zhong J, et al. Validation of candidate causal genes for obesity that affect shared metabolic pathways and networks. *Nat Genet* 2009;**41**(4):415–23.
13. Reed DR, Lawler MP, Tordoff MG. Reduced body weight is a common effect of gene knockout in mice. *BMC Genet* 2008;**9**:4.
14. Bachman ES, Dhillon H, Zhang CY, Cinti S, Bianco AC, Kobilka BK, et al. BetaAR signaling required for diet-induced thermogenesis and obesity resistance. *Science* 2002;**297**(5582):843–5.
15. Andreasen CH, Andersen G. Gene–environment interactions and obesity—further aspects of genomewide association studies. *Nutrition* 2009;**25**(10):998–1003.
16. Hetherington MM, Cecil JE. Gene–environment interactions in obesity. *Forum Nutr* 2010;**63**:195–203.
17. Lembertas AV, Perusse L, Chagnon YC, Fisler JS, Warden CH, Purcell-Huynh DA, et al. Identification of an obesity quantitative trait locus on mouse chromosome 2 and evidence of linkage to body fat and insulin on the human homologous region 20q. *J Clin Invest* 1997;**100**(5):1240–7.
18. Lee JH, Reed DR, Li WD, Xu W, Joo EJ, Kilker RL, et al. Genome scan for human obesity and linkage to markers in 20q13. *Am J Hum Genet* 1999;**64**(1):196–209.
19. De Luca M, Yi N, Allison DB, Leips J, Ruden DM. Mapping quantitative trait loci affecting variation in *Drosophila* triacylglycerol storage. *Obes Res* 2005;**13**(9):1596–605.
20. De Luca M, Chambers MM, Casazza K, Lok KH, Hunter GR, Gower BA, et al. Genetic variation in a member of the laminin gene family affects variation in body composition in Drosophila and humans. *BMC Genet* 2008;**9**:52.
21. Saunders CL, Chiodini BD, Sham P, Lewis CM, Abkevich V, Adeyemo AA, et al. Meta-analysis of genome-wide linkage studies in BMI and obesity. *Obesity (Silver Spring)* 2007;**15**(9):2263–75.
22. Risch N, Merikangas K. The future of genetic studies of complex human diseases. *Science* 1996;**273**(5281):1516–7.
23. Sham PC, Cherny SS, Purcell S, Hewitt JK. Power of linkage versus association analysis of quantitative traits, by use of variance-components models, for sibship data. *Am J Hum Genet* 2000;**66**(5):1616–30.
24. Manolio TA, Collins FS, Cox NJ, Goldstein DB, Hindorff LA, Hunter DJ, et al. Finding the missing heritability of complex diseases. *Nature* 2009;**461**(7265):747–53.
25. Vimaleswaran KS, Loos RJ. Progress in the genetics of common obesity and type 2 diabetes. *Expert Rev Mol Med* 2010;**12**:e7.

26. Walley AJ, Asher JE, Froguel P. The genetic contribution to non-syndromic human obesity. *Nat Rev Genet* 2009;**10**(7):431–42.
27. Meyre D, Delplanque J, Chevre JC, Lecoeur C, Lobbens S, Gallina S, et al. Genome-wide association study for early-onset and morbid adult obesity identifies three new risk loci in European populations. *Nat Genet* 2009;**41**(2):157–9.
28. Thorleifsson G, Walters GB, Gudbjartsson DF, Steinthorsdottir V, Sulem P, Helgadottir A, et al. Genome-wide association yields new sequence variants at seven loci that associate with measures of obesity. *Nat Genet* 2009;**41**(1):18–24.
29. Willer CJ, Speliotes EK, Loos RJ, Li S, Lindgren CM, Heid IM, et al. Six new loci associated with body mass index highlight a neuronal influence on body weight regulation. *Nat Genet* 2009;**41**(1):25–34.
30. Hofker M, Wijmenga C. A supersized list of obesity genes. *Nat Genet* 2009;**41**(2):139–40.
31. Silver R, Lesauter J. Circadian and homeostatic factors in arousal. *Ann NY Acad Sci* 2008;**1129**:263–74.
32. Turek FW, Joshu C, Kohsaka A, Lin E, Ivanova G, McDearmon E, et al. Obesity and metabolic syndrome in circadian Clock mutant mice. *Science* 2005;**308**(5724):1043–5.
33. Garaulet M, Corbalan MD, Madrid JA, Morales E, Baraza JC, Lee YC, et al. CLOCK gene is implicated in weight reduction in obese patients participating in a dietary programme based on the Mediterranean diet. *Int J Obes (Lond)* 2010;**34**(3):516–23.
34. Sookoian S, Gemma C, Gianotti TF, Burgueno A, Castano G, Pirola CJ. Genetic variants of Clock transcription factor are associated with individual susceptibility to obesity. *Am J Clin Nutr* 2008;**87**(6):1606–15.
35. Scott EM, Carter AM, Grant PJ. Association between polymorphisms in the Clock gene, obesity and the metabolic syndrome in man. *Int J Obes (Lond)* 2008;**32**(4):658–62.
36. Colles SL, Dixon JB, O'Brien PE. Night eating syndrome and nocturnal snacking: association with obesity, binge eating and psychological distress. *Int J Obes (Lond)* 2007;**31**(11):1722–30.
37. Arble DM, Bass J, Laposky AD, Vitaterna MH, Turek FW. Circadian timing of food intake contributes to weight gain. *Obesity (Silver Spring)* 2009;**17**(11):2100–2.
38. Chaudhary N, Pfluger PT. Metabolic benefits from Sirt1 and Sirt1 activators. *Curr Opin Clin Nutr Metab Care* 2009;**12**(4):431–7.
39. Boss O, Samec S, Paoloni-Giacobino A, Rossier C, Dulloo A, Seydoux J, et al. Uncoupling protein-3: a new member of the mitochondrial carrier family with tissue-specific expression. *FEBS Lett* 1997;**408**(1):39–42.
40. Fleury C, Neverova M, Collins S, Raimbault S, Champigny O, Levi-Meyrueis C, et al. Uncoupling protein-2: a novel gene linked to obesity and hyperinsulinemia. *Nat Genet* 1997;**15**(3):269–72.
41. Vidal-Puig A, Solanes G, Grujic D, Flier JS, Lowell BB. UCP3: an uncoupling protein homologue expressed preferentially and abundantly in skeletal muscle and brown adipose tissue. *Biochem Biophys Res Commun* 1997;**235**(1):79–82.
42. Warden CH, Fisler JS, Shoemaker SM, Wen PZ, Svenson KL, Pace MJ, et al. Identification of four chromosomal loci determining obesity in a multifactorial mouse model. *J Clin Invest* 1995;**95**(4):1545–52.
43. de Krom M, Bauer F, Collier D, Adan RA, la Fleur SE. Genetic variation and effects on human eating behavior. *Annu Rev Nutr* 2009;**29**:283–304.
44. Liem ET, Vonk JM, Sauer PJ, van der Steege G, Oosterom E, Stolk RP, et al. Influence of common variants near INSIG2, in FTO, and near MC4R genes on overweight and the metabolic profile in adolescence: the TRAILS (TRacking Adolescents' Individual Lives Survey) Study. *Am J Clin Nutr* 2010;**91**(2):321–8.
45. Schadt EE, Lamb J, Yang X, Zhu J, Edwards S, Guhathakurta D, et al. An integrative genomics approach to infer causal associations between gene expression and disease. *Nat Genet* 2005;**37**(7):710–7.

46. English SB, Butte AJ. Evaluation and integration of 49 genome-wide experiments and the prediction of previously unknown obesity-related genes. *Bioinformatics* 2007;**23**(21):2910–7.
47. Aerts S, Lambrechts D, Maity S, Van Loo P, Coessens B, De Smet F, et al. Gene prioritization through genomic data fusion. *Nat Biotechnol* 2006;**24**(5):537–44.
48. Calvo S, Jain M, Xie X, Sheth SA, Chang B, Goldberger OA, et al. Systematic identification of human mitochondrial disease genes through integrative genomics. *Nat Genet* 2006;**38**(5):576–82.
49. Goldstein DB. Common genetic variation and human traits. *N Engl J Med* 2009;**360**(17):1696–8.
50. Frayling TM, Timpson NJ, Weedom MN, Zeggini E, Freathy RM, Lindgren CM, et al. A common variant in the FTO gene is associated with body mass index and predisposes to childhood and adult obesity. *Sciencexpress* 2007;1–5.
51. Blakemore AI, Meyre D, Delplanque J, Vatin V, Lecoeur C, Marre M, et al. A rare variant in the visfatin gene (NAMPT/PBEF1) is associated with protection from obesity. *Obesity (Silver Spring)* 2009;**17**(8):1549–53.
52. Dong C, Li WD, Geller F, Lei L, Li D, Gorlova OY, et al. Possible genomic imprinting of three human obesity-related genetic loci. *Am J Hum Genet* 2005;**76**(3):427–37.
53. Guo YF, Shen H, Liu YJ, Wang W, Xiong DH, Xiao P, et al. Assessment of genetic linkage and parent-of-origin effects on obesity. *J Clin Endocrinol Metab* 2006;**91**(10):4001–5.
54. Wermter AK, Scherag A, Meyre D, Reichwald K, Durand E, Nguyen TT, et al. Preferential reciprocal transfer of paternal/maternal DLK1 alleles to obese children: first evidence of polar overdominance in humans. *Eur J Hum Genet* 2008;**16**(9):1126–34.
55. McCarroll SA, Kuruvilla FG, Korn JM, Cawley S, Nemesh J, Wysoker A, et al. Integrated detection and population-genetic analysis of snps and copy number variation. *Nat Genet* 2008;**40**(10):1166–74.
56. Cheverud JM, Hager R, Roseman C, Fawcett G, Wang B, Wolf JB. Genomic imprinting effects on adult body composition in mice. *Proc Natl Acad Sci USA* 2008;**105**(11):4253–8.
57. Dolinoy DC, Weidman JR, Waterland RA, Jirtle RL. Maternal genistein alters coat color and protects Avy mouse offspring from obesity by modifying the fetal epigenome. *Environ Health Perspect* 2006;**114**(4):567–72.
58. Xie T, Chen M, Gavrilova O, Lai EW, Liu J, Weinstein LS. Severe obesity and insulin resistance due to deletion of the maternal Gsalpha allele is reversed by paternal deletion of the Gsalpha imprint control region. *Endocrinology* 2008;**149**(5):2443–50.
59. Casellas J, Farber CR, Gularte RJ, Haus KA, Warden CH, Medrano JF. Evidence of maternal QTL affecting growth and obesity in adult mice. *Mamm Genome* 2009;**20**(5):269–80.
60. Hager R, Cheverud JM, Wolf JB. Maternal effects as the cause of parent-of-origin effects that mimic genomic imprinting. *Genetics* 2008;**178**(3):1755–62.
61. Warden CH, Fisler JS, Espinal G, Graham J, Havel PJ, Perroud B. Maternal influence of prolyl endopeptidase on fat mass of adult progeny. *Int J Obes (Lond)* 2009;**33**(9):1013–22.
62. Wolf JB, Vaughn TT, Pletscher LS, Cheverud JM. Contribution of maternal effect QTL to genetic architecture of early growth in mice. *Heredity* 2002;**89**(4):300–10.
63. Bouchard C. Defining the genetic architecture of the predisposition to obesity: a challenging but not insurmountable task. *Am J Clin Nutr* 2010;**91**(1):5–6.
64. Sims EA, Danforth Jr. E, Horton ES, Bray GA, Glennon JA, Salans LB. Endocrine and metabolic effects of experimental obesity in man. *Recent Prog Horm Res* 1973;**29**:457–96.
65. Bouchard C, Tremblay A, Despres JP, Nadeau A, Lupien PJ, Theriault G, et al. The response to long-term overfeeding in identical twins. *N Engl J Med* 1990;**322**(21):1477–82.
66. Bouchard C, Tremblay A, Despres JP, Theriault G, Nadeau A, Lupien PJ, et al. The response to exercise with constant energy intake in identical twins. *Obes Res* 1994;**2**(5):400–10.

67. Boutcher SH, Dunn SL. Factors that may impede the weight loss response to exercise-based interventions. *Obes Rev* 2009;**10**(6):671–80.
68. Deram S, Villares SM. Genetic variants influencing effectiveness of weight loss strategies. *Arq Bras Endocrinol Metabol* 2009;**53**(2):129–38.
69. Grau K, Hansen T, Holst C, Astrup A, Saris WH, Arner P, et al. Macronutrient-specific effect of FTO rs9939609 in response to a 10-week randomized hypo-energetic diet among obese Europeans. *Int J Obes (Lond)* 2009;**33**(11):1227–34.
70. Hainerova I, Larsen LH, Holst B, Finkova M, Hainer V, Lebl J, et al. Melanocortin 4 receptor mutations in obese Czech children: studies of prevalence, phenotype development, weight reduction response, and functional analysis. *J Clin Endocrinol Metab* 2007;**92**(9):3689–96.
71. Haupt A, Thamer C, Heni M, Tschritter O, Machann J, Schick F, et al. Impact of variation near MC4R on whole-body fat distribution, liver fat, and weight loss. *Obesity (Silver Spring)* 2009;**17**(10):1942–5.
72. Muller TD, Hinney A, Scherag A, Nguyen TT, Schreiner F, Schafer H, et al. 'Fat mass and obesity associated' gene (FTO): no significant association of variant rs9939609 with weight loss in a lifestyle intervention and lipid metabolism markers in German obese children and adolescents. *BMC Med Genet* 2008;**9**:85.
73. Reinehr T, Hinney A, Toschke AM, Hebebrand J. Aggravating effect of INSIG2 and FTO on overweight reduction in a one-year lifestyle intervention. *Arch Dis Child* 2009;**94**(12):965–7.
74. Hauner H, Meier M, Jockel KH, Frey UH, Siffert W. Prediction of successful weight reduction under sibutramine therapy through genotyping of the G-protein beta3 subunit gene (GNB3) C825T polymorphism. *Pharmacogenetics* 2003;**13**(8):453–9.
75. Hsiao DJ, Wu LS, Huang SY, Lin E. Weight loss and body fat reduction under sibutramine therapy in obesity with the C825T polymorphism in the GNB3 gene. *Pharmacogenet Genomics* 2009;**19**(9):730–3.
76. Chen HH, Lee WJ, Wang W, Huang MT, Lee YC, Pan WH. Ala55Val polymorphism on UCP2 gene predicts greater weight loss in morbidly obese patients undergoing gastric banding. *Obes Surg* 2007;**17**(7):926–33.
77. Kim K, Perroud B, Espinal G, Kachinskas D, Austrheim-Smith I, Wolfe BM, et al. Genes and networks expressed in perioperative omental adipose tissue are correlated with weight loss from Roux-en-Y gastric bypass. *Int J Obes (Lond)* 2008;**32**(9):1395–406.
78. Sesti G, Perego L, Cardellini M, Andreozzi F, Ricasoli C, Vedani P, et al. Impact of common polymorphisms in candidate genes for insulin resistance and obesity on weight loss of morbidly obese subjects after laparoscopic adjustable gastric banding and hypocaloric diet. *J Clin Endocrinol Metab* 2005;**90**(9):5064–9.
79. Bray MS. Implications of gene–behavior interactions: prevention and intervention for obesity. *Obesity (Silver Spring)* 2008;**16**(Suppl. 3):S72–8.
80. Shuldiner AR. Obesity genes and gene–environment–behavior interactions: recommendations for a way forward. *Obesity (Silver Spring)* 2008;**16**(Suppl. 3):S79–81.
81. Janssens AC, van Duijn CM. Genome-based prediction of common diseases: advances and prospects. *Hum Mol Genet* 2008;**17**(R2):R166–73.
82. Lango H, Palmer CN, Morris AD, Zeggini E, Hattersley AT, McCarthy MI, et al. Assessing the combined impact of 18 common genetic variants of modest effect sizes on type 2 diabetes risk. *Diabetes* 2008;**57**(11):3129–35.
83. Farooqi IS, Jebb SA, Langmack G, Lawrence E, Cheetham CH, Prentice AM, et al. Effects of recombinant leptin therapy in a child with congenital leptin deficiency. *N Engl J Med* 1999;**341**(12):879–84.
84. Farooqi IS, Matarese G, Lord GM, Keogh JM, Lawrence E, Agwu C, et al. Beneficial effects of leptin on obesity, T cell hyporesponsiveness, and neuroendocrine/metabolic dysfunction of human congenital leptin deficiency. *J Clin Invest* 2002;**110**(8):1093–103.

85. Licinio J, Caglayan S, Ozata M, Yildiz BO, de Miranda PB, O'Kirwan F, et al. Phenotypic effects of leptin replacement on morbid obesity, diabetes mellitus, hypogonadism, and behavior in leptin-deficient adults. *Proc Natl Acad Sci USA* 2004;**101**(13):4531–6.
86. Heymsfield SB, Greenberg AS, Fujioka K, Dixon RM, Kushner R, Hunt T, et al. Recombinant leptin for weight loss in obese and lean adults: a randomized, controlled, dose-escalation trial. *JAMA* 1999;**282**(16):1568–75.
87. Welt CK, Chan JL, Bullen J, Murphy R, Smith P, DePaoli AM, et al. Recombinant human leptin in women with hypothalamic amenorrhea. *N Engl J Med* 2004;**351**(10):987–97.
88. Roth JD, Roland BL, Cole RL, Trevaskis JL, Weyer C, Koda JE, et al. Leptin responsiveness restored by amylin agonism in diet-induced obesity: evidence from nonclinical and clinical studies. *Proc Natl Acad Sci USA* 2008;**105**(20):7257–62.
89. Bordone L, Cohen D, Robinson A, Motta MC, van Veen E, Czopik A, et al. SIRT1 transgenic mice show phenotypes resembling calorie restriction. *Aging Cell* 2007;**6**(6):759–67.
90. Pfluger PT, Herranz D, Velasco-Miguel S, Serrano M, Tschop MH. Sirt1 protects against high-fat diet-induced metabolic damage. *Proc Natl Acad Sci USA* 2008;**105**(28):9793–8.
91. Lagouge M, Argmann C, Gerhart-Hines Z, Meziane H, Lerin C, Daussin F, et al. Resveratrol improves mitochondrial function and protects against metabolic disease by activating SIRT1 and PGC-1alpha. *Cell* 2006;**127**(6):1109–22.
92. Milne JC, Lambert PD, Schenk S, Carney DP, Smith JJ, Gagne DJ, et al. Small molecule activators of SIRT1 as therapeutics for the treatment of type 2 diabetes. *Nature* 2007;**450**(7170):712–6.
93. Consortium, T.I.C.G. International network of cancer genome projects. *Nature* 2010;**464**(7291):993–8.
94. Jones S, Hruban RH, Kamiyama M, Borges M, Zhang X, Parsons DW, et al. Exomic sequencing identifies PALB2 as a pancreatic cancer susceptibility. *Science* 2009;**324**(5924):217.
95. Lifton RP. Individual genomes on the horizon. *N Engl J Med* 2010;**362**(13):1235–6.

Index

A

B

C

M

N

O

P

Q

R

S

U

V

W

FSC
www.fsc.org
MIX
Papier | Fördert
gute Waldnutzung
FSC® C083411

Kolibri 360 GmbH
Ferdinand-Jühlke-Straße 7
99095 Erfurt, Deutschland
produktsicherheit@kolibri360.de